高等学校“十三五”规划教材

配位化学

Coordination Chemistry

（双语版）

第三版

李 晖 主编

化学工业出版社

·北京·

《配位化学》（第三版）在前两版的基础上进行了修订，编入了超分子化学的基本概念与X射线单晶与粉末衍射技术，同时增加了纳米配合物的制备与纳米表征技术。全书共分5章，第1、2章简单介绍了配位化学的发展、基本概念和基本理论，第3章为配合物结构的谱学研究方法，第4、5章为配合物的物理化学性质和配位反应平衡。

《配位化学》（第三版）可作为高等院校化学及相关专业高年级本科生和研究生的教材，也可供化学教师及科研工作者参考。

图书在版编目（CIP）数据

配位化学：双语版：英汉对照/李晖主编. —3版.
—北京：化学工业出版社，2020.5（2024.7 重印）
高等学校“十三五”规划教材
ISBN 978-7-122-36346-6

Ⅰ.①配… Ⅱ.①李… Ⅲ.①络合物化学-英、汉
Ⅳ.①O641.4

中国版本图书馆CIP数据核字（2020）第034379号

责任编辑：宋林青　　文字编辑：刘志茹
责任校对：王　静　　装帧设计：刘丽华

出版发行：化学工业出版社(北京市东城区青年湖南街13号 邮政编码100011)
印　　装：北京机工印刷厂有限公司
787mm×1092mm 1/16 印张19¼ 彩插1 字数470千字 2024年7月北京第3版第3次印刷

购书咨询：010-64518888　　售后服务：010-64518899
网　　址：http://www.cip.com.cn
凡购买本书，如有缺损质量问题，本社销售中心负责调换。

定　　价：49.80元

前　言

配位化学 100 多年的发展历程鲜活呈现了多学科交叉与融合过程，这种交叉与融合使得配位化学成为一个非常活跃的研究领域，创新成果层出不穷。

此次修订注重将超分子化学的基本概念与 X 射线单晶与粉末衍射技术编入其中。这是因为，超分子化学自 1987 年诞生以来，大量的研究表明分子间作用力（或非共价相互作用力）在自组装与功能调控方面都具有重要的作用。同时，这一发展的重要性也体现在配合物或配位聚合物的结构与性质的调控方面。对分子间作用力的深刻认识是基于 X 射线衍射技术的推广与普及。《配位化学》（第二版）出版至今又过去了 8 年多时间，期间，配位化学与纳米科学和技术发生了深度交融，制备了各类纳米尺寸的配合物和配位聚合物，各种新型的纳米表征技术应用到了配位化学研究中。

因此，《配位化学》（第三版）的推出也就势在必行。《配位化学》（第三版）是在第二版的基础上将纳米表征技术在配位化学中的应用编入第 3 章 3.8 节。这一部分内容主要是由北京理工大学化学与化工学院赵扬长聘副教授完成。她在纳米材料领域的研究与教学中都做出了非常出色的工作。因此，由她来撰写这一部分将为本教材增色。北京理工大学分析测试中心的马宏伟博士基于他在 X 射线衍射方面扎实的功底，对 X 射线单晶与粉末衍射部分进行了修订。施如菲在第二版贡献的基础上持续对第三版做出了她的贡献，特别是第 1 章和第 4 章的文字加工与仔细校读，为本书增色不少。在此，作者谨向他们表示衷心的感谢。另外，第三版的修改部分还包括对第 5 章内容的优化。

当然，即使一版再版，本教材的内容尚难免无法及时跟上科学技术创新探索的步伐。唯有不断地学习与掌握配位化学的国际前沿状况以及与其他学科的交叉融合，才能跟上科学技术的发展步伐，也才能更好地传承和完善配位化学的知识谱系。

书中的疏漏之处，敬请专家和读者指正。

李晖

于北京理工大学化学与化工学院

2020 年 1 月 8 日

第一版前言

无机化学、有机化学、分析化学和物理化学等都是经典的化学学科分支。配位化学作为无机化学和有机化学的交叉领域，迄今已有一百多年的历史。配位化学的兴起和迅速发展不仅给古老的无机化学带来了生机，也为化学领域中其他分支学科的发展开辟了更广阔的天地。20 世纪 80 年代后期发展起来的超分子化学与配位化学之间更有着紧密的联系，可以看成是广义的配位化学（游效曾院士在全国配位化学会议上的报告）。因此，配位化学已经成为化学领域中的重要分支，是与材料科学、生命科学、物理学等众多学科相互渗透、高度融汇的重要学科领域。

目前，许多高等院校都为化学及相关专业的高年级本科生和研究生开设了配位化学课。也有一些配位化学的专著出版，如游效曾院士的《配位化合物的结构和性质》等。但尚未有合适的教材适应当前相关层次的教学。作者在多年配位化学的教学实践过程中，收集了大量的资料，结合作者多年的国内外的科研经历，编撰了这本双语教材。

本教材有以下几个特点。

（1）英、中双语是本教材的首要特点。作为自然科学的学习，英语是一种非常重要的工具，是我们了解国际最新发展动态的重要窗口，作为 21 世纪的大学生和研究生，对英语熟练地听、说、读、写已是一项基本要求。教育部也大力提倡在高等教育中运用双语教学。本教材正是为适应这一新的需求而编撰的。而且，本教材在北京理工大学的高年级本科生和研究生教学中，深受学生欢迎，已取得了良好的教学效果。

（2）深入浅出，重点突出，是本教材的另一特点。配位化学是一门相当成熟的学科，有着一整套的新理论、新概念、新方法和新反应等。本教材以配位化学中最为重要的概念、理论、方法和性质为主体，层次分明地展开叙述。对于涉及量子化学的内容，如分子的对称性——群论、分子轨道理论和配位场理论等部分，只给出结论并注重这些结论在解决化学问题中的应用，而不涉及量化计算。

（3）选材新颖，具有时代性，是本教材的第三个特点。将配位化学一些新近发展的成果融入教材和整个教学实践中一直是作者努力的方向，但由于配位化学的发展非常迅速，所以，也不太可能囊括所有的最新研究成果，只能适当地编入一些材料，以便读者进一步跟踪有关发展动态。

本书在编撰过程中得到了多方的支持，尤其是研究生——郭明、田红、何飞跃，本科生白萌等为本书的中文输入和制图付出了辛勤的劳动，化学工业出版社的编辑在本书的出版过程中提供了有益的建议和大量的帮助。在此谨向他们表示衷心的感谢。

由于时间仓促及作者水平有限，书中不妥之处，敬请读者批评指正。使用过程中如有问题可与作者联系，E-mail:lihui@bit.edu.cn。

李晖

北京理工大学理学院化学系

2005 年 9 月

第二版前言

本书自2006年出版以来，受到了高校师生和广大读者的欢迎，至2010年底已经发行了上万册。这也反映出配位化学在化学、生命科学、材料学等相关专业的教学和科研中的重要性和强烈需求。

近五年来，作者在配位化学的教学与科研过程中，深深体会到配位化学的迅速发展及其与其他学科领域的交叉与融合，尤其是与超分子化学的相互渗透。因此，在第二版中，作者觉得不得不将超分子化学的一些基本概念编入到教材中（第1、2章）。同时，X射线衍射技术（包括粉末衍射和单晶衍射）在物质结构分析中的重要性也被每年大量报道的新颖配合物的晶体结构所证实，说明X射线衍射技术已成为研究配合物结构的最重要的方法。所以，将X射线衍射的基本原理编入第3章。同时，鉴于分子轨道理论在复杂配位化合物的成键与结构应用中的局限性，将原第2章中的这部分内容进行了适当的删减。第4、5章没有做明显的改动，并不是这部分没有新的发展，而是因为一方面时间有限，没有充足的精力做太多的改动；另一方面，第二版应该保持第一版的主要特色。

第二版的修订与编撰同样得到了很多人的帮助与支持，我的女儿施如菲作为加州大学尔湾分校（UC Irvine）化学生物学专业的学生，对超分子化学和配位化学具有浓厚的兴趣，为第二版的修订提出了很多中肯的建议，我的博士研究生汤贝贝为第二版的中英文校对付出了辛勤的劳动，出版社的编辑也为本版的出版付出了大量的心血。在此作者谨向他们表示衷心的感谢。

由于时间仓促及作者水平有限，书中的不妥和疏漏之处在所难免，敬请专家和读者指正。

李晖
于北京理工大学化学学院
2011年6月

Contents

Chapter 1 An Introduction to Coordination Chemistry

1.1 The History of Coordination Chemistry

1.1.1 The Origin of Coordination Chemistry

One of the most productive areas of research in the twentieth century was Alfred Werner's development of coordination chemistry. It is a measure of Werner's impact on the realm of inorganic chemistry that the number, variety, and complexity of coordination compounds continues to grow even as we pass the centennial anniversary of his original work.

The first coordination compound was most likely prepared in the late 1700s by Tassaert, a French chemist. He observed that ammonia combined with a cobalt ore to yield a reddish brown product. Over the next century, many compounds were synthesized and characterized, but little progress was made in formulating and accounting for their molecular structures(Fig.1.1). The discovery and explanation of coordination compounds should be viewed against the larger picture of progress in understanding atomic structure, the periodic table, and molecular bonding.

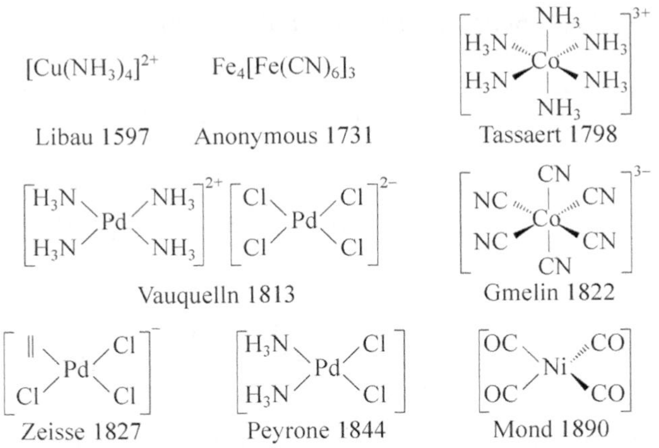

$[Cu(NH_3)_4]^{2+}$ Libau 1597

$Fe_4[Fe(CN)_6]_3$ Anonymous 1731

$[Co(NH_3)_6]^{3+}$ Tassaert 1798

$[Pd(NH_3)_4]^{2+}$ $[PdCl_4]^{2-}$ Vauquelln 1813

$[Co(CN)_6]^{3-}$ Gmelin 1822

Zeisse 1827

Peyrone 1844

Mond 1890

Fig.1.1 Some important compounds as landmarks in inorganic chemistry

The contributions of Proust and Lavoisier led Dalton to formulate the first concrete atomic theory in 1808. Mendeleev published his first periodic table in 1869. With the discoveries of X-rays, radioactivity, electrons, and the nucleus at the beginning of the twentieth century, the modern quantum-mechanical picture of the atom started to emerge in the 1920s. This model gives a theoretical explanation for atomic line spectra and the modern periodic table. However, no theoretical basis was developed to satisfactorily account for these wondrous compounds at that time.

Given the success of organic chemists in describing the structural units and fixing atomic valences found in carbon-based compounds, it was natural that these ideas be applied to the ammonates. The results, however, were disappointing; for example, considering the typical data for the cobalt ammonate chlorides listed in the Table 1.1. The formulas used in the last few decades of the nineteenth century indicated the ammonia-to-cobalt mole ratio but left the nature of the bonding between them to the imagination. This uncertainty was reflected in the dot used in the formula to connect, for example, $CoCl_3$ to the appropriate number of ammonias. Conductivities measured when these compounds were dissolved in water are given qualitatively, which was just then starting to be taken as a measure of the number of ions produced in solution. The "number of chloride ions precipitated" was determined by the addition of aqueous silver nitrate, as represented in [equation (1.1)]:

$$AgNO_3(aq) + Cl^-(aq) \longrightarrow AgCl(s) + NO_3^-(aq) \tag{1.1}$$

Table 1.1 The cobalt ammonate chlorides

Formula	Conductivity	Number of Cl^- precipitated
$CoCl_3 \cdot 6NH_3$	High	3
$CoCl_3 \cdot 5NH_3$	Medium	2
$CoCl_3 \cdot 4NH_3$	Low	1
$IrCl_3 \cdot 3NH_3$	Zero	0

Now how might you explain such data? In 1869, Christian Wilhelm Blomstrand firstly formulated his theory to account for the cobalt ammonate chlorides and other series of ammonates. He produced a picture of $CoCl_3 \cdot 6NH_3$ have shown in Fig.1.2(a).

$CoCl_3 \cdot 6NH_3$ Co(—NH_3—NH_3—Cl)(—NH_3—NH_3—Cl)(—NH_3—NH_3—Cl)

(a) Blomstrand's representation of $CoCl_3 \cdot 6NH_3$

(1) $CoCl_3 \cdot 6NH_3$ Co(—NH_3—Cl)(—NH_3—NH_3—NH_3—NH_3—Cl)(—NH_3—Cl)

(2) $CoCl_3 \cdot 5NH_3$ Co(—Cl)(—NH_3—NH_3—NH_3—NH_3—Cl)(—NH_3—Cl)

(3) $CoCl_3 \cdot 4NH_3$ Co(—Cl)(—NH_3—NH_3—NH_3—NH_3—Cl)(—Cl)

(4) $IrCl_3 \cdot 3NH_3$ Ir(—Cl)(—NH_3—NH_3—NH_3—Cl)(—Cl)

(b) Jorgensen's representations of four members of the series with the iridium subsitituted for the intended cobalt in compound (4)

Fig.1.2 Representations of the cobalt ammonate chlorides by Blomstrand and Jorgensen

Based on the prevailing ideas of that time, this was a perfectly reasonable structure. The divalent ammonia he proposed was consistent with a view of ammonium chloride written as H—NH_3—Cl. The valence of 3 for cobalt was satisfied and nitrogen atoms were chained together much like carbon in organic compounds. The three monovalent chlorides were far enough removed from the cobalt atom to be available to be precipitated by aqueous silver chloride.

In 1884, S. M. Jørgensen proposed some amendments to his mentor's picture(Table 1.2). First, he had new evidence that correctly indicated that these compounds were monomeric. Second, he adjusted the distance of the chloride groups from the cobalt to account for the rates at which various chlorides were precipitated. The first chloride is precipitated much more rapidly than the others and so was put farther away and therefore less under the influence of the cobalt atom. His diagrams for the first three cobalt ammonate chlorides are shown in Fig.1.2(b). Note that, in the second compound, one chloride is now directly attached to the cobalt, therefore, unavailable to be precipitated by silver nitrate. In the third compound, two chlorides are similarly pictured. These changes are significant. It appeared that the Blomstrand-Jørgensen theory was on the right track.

Table 1.2 The historical setting of coordination compounds

Atomic structure and the periodic table	Molecular structure and bonding	Coordination chemistry
1750		
1774: Law of conservation of matter: Lavoisier 1799: Law of definite composition: Proust		1798: First cobalt ammonates observed: Tassaert
1800		
1808: Dalton's atomic theory published in *New System of Chemical Philosophy*	1830: The radical theory of structure: Liebig, Wöhler, Berzelius, Dumas (organic compounds composed of methyl, ethyl, etc, radicals)	1822: Cobalt ammonate oxalates prepared: Gmelin
	1852: Concept of valence: Frankland(all atoms have a fixed valence) 1854: Tetravalent carbon atom: Kekulé	1851: $CoCl_3 \cdot 6NH_3$, $CoCl_3 \cdot 5NH_3$, and other cobalt ammonates prepared: Genth, Claudet, Fremy
1859: Spectroscope developed: Bunsen and Kirchhoff 1869: Mendeleev's first periodic table organizes 63 known elements 1885: Balmer formula for visible H spectrum 1894: First "inert gas"discovered 1895: X-rays discovered: Roentgen 1896: Radioactivity discovered: Becquerel	1874: Tetrahedral carbon atom: Le Bel and Van't Hoff 1884: Dissociation theory of electrolytes: Arrhenius	1869: Chain theory of ammonates: Blomstrand 1884: Amendments to chain theory: Jørgensen 1892: Werner's dream about coordination compounds
1900		
1902: Discovery of the electron: Thomson 1905: Wave-particle duality of light: Einstein 1911: α-particle/gold foil experiment; nuclear model of the atom: Rutherford 1913: Bohr model of the atom (quantization of electron energy) 1923: Wave-particle duality of electrons: De Broglie 1926: Schrödinger quantum-mechanical atom (electrons in orbitals about nucleus; electron spectroscopy explained as transitions among orbitals)	1923: Electron-dot diagrams: Lewis 1931: Valence-bond theory: Pauling, Heitler, London, Slater Early 1930s: Molecular orbital theory: Hund, Bloch, Mulliken, Hückel 1940: Valence-shell electron-pair repulsion (VSEPR) theory: Sidgwick	1902: Three postulates of coordination theory proposed: Werner 1911: Optical isomers of *cis*-$[CoCl(NH_3)(en)_2]X_2$ resolved: Werner 1914: Non-carbon-containing optical isomers resolved: Werner 1927: Lewis ideas applied to coordination compounds: Sidgwick 1933: Crystal field theory: Bethe and Van Vleck
Modern periodic table including trends in periodic properties	Modern concepts of chemical bonding	Modern coordination theory

But was there a compound with only three ammonias? As shown in Fig.1.2(b)(4), the theory predicted that it should exist and, furthermore, should have one ionizable chloride. But this critical compound was not available. After considerable time and effort, the analogous iridium ammonate chloride was found to be a neutral compound with no ionizable chlorides. The theory was in trouble.

1.1.2 The Modern Coordination Chemistry—Werner Coordination Chemistry

Alfred Werner(1866—1919), as a young unsalaried lecturer in organic chemistry, was torn between organic and inorganic chemistry. His first contributions (the *stereochemistry*, or spatial arrangements, of atoms in nitrogen compounds) were in the organic field, but so many intriguing inorganic questions were being raised in those days. He observed the difficulties that inorganic chemists were having in explaining coordination compounds, and he was aware that the established ideas of organic chemistry seemed to lead only into blind alleys and dead ends. In 1892, his coordination theory came to him. But his new theory broke with the earlier traditions, and he had essentially no experimental proof to support his ideas. Werner's theory was considered to be audacious fiction. Werner spent the rest of his life directing a systematic and thorough research program to prove that his intuition was correct.

Werner decided that the idea of a single fixed valence could not apply to cobalt and other similar metals. Working with the cobalt ammonates and other related series involving chromium and platinum, he proposed instead that these metals have two types of valence, a primary valence and a secondary valence. The primary, or ionizable, valence corresponded to what we call today the *oxidation state*; for cobalt, it is the 3+ state. The secondary valence is more commonly called the *coordination number*; for cobalt, it is 6. Werner maintained that this secondary valence was directed toward fixed geometric positions in space.

Fig.1.3 shows Werner's early proposals for the bonding in the cobalt ammonates. He said that the cobalt must simultaneously satisfy both its primary and secondary valences. The solid lines show the groups that satisfy the primary valence. The dashed lines, always directed toward the same fixed positions in space, showing how the secondary valence was satisfied. In compound (1), all three chlorides satisfy only the primary valence, and the six ammonias satisfy only the secondary. In compound (2), one chloride must do double duty and help satisfy both valences. The chloride that satisfies the secondary valence (and is directly bound to the Co^{3+} ion) was concluded to be unavailable for precipitation by silver nitrate. Compound (3) has two chlorides doing double duty and only one available for precipitation. Compound (4), according to Werner, should be a neutral compound with no ionizable chlorides. This was exactly what Jørgensen had found with the iridium compound.

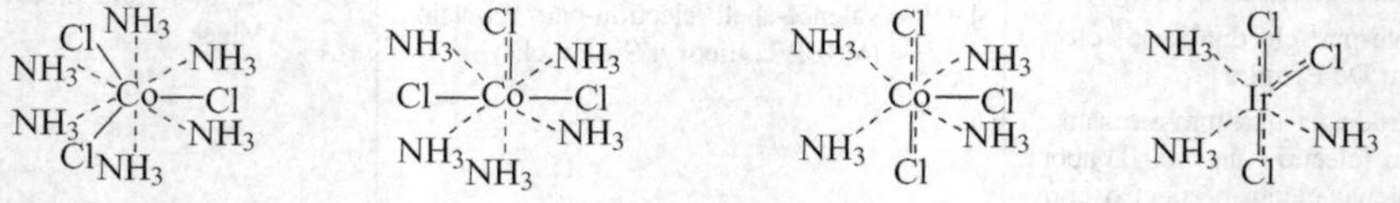

(1) $CoCl_3 \cdot 6NH_3$ (2) $CoCl_3 \cdot 5NH_3$ (3) $CoCl_3 \cdot 4NH_3$ (4) $IrCl_3 \cdot 3NH_3$

Fig.1.3 Werner's representations of the cobalt ammonate chlorides. The solid lines represent groups that satisfy the primary valence or oxidation state (3+) of cobalt, and the dashed lines represent those that satisfy the secondary valence, or coordiantion number (6). The secondary valence occupies fixed positions in space

Werner next turned to the geometry of the secondary valence (or coordination number). As shown in Table 1.3, six ammonias about a central metal atom or ion might assume one of several different common geometries, including hexagonal planar, trigonal prismatic, and octahedral. The table compares some information about the predicted and actual number of isomers for a variety of substituted coordination compounds.

Table 1.3 The number of actual versus predicted isomers for three different geometries

	Geometries of Coordination Number 6			
	Hexagonal planar 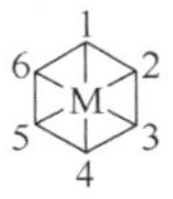	Trigonal prism 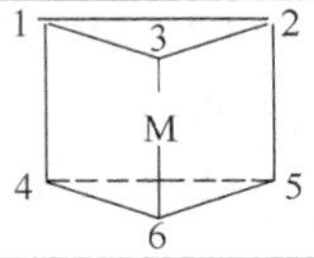	Octahedral 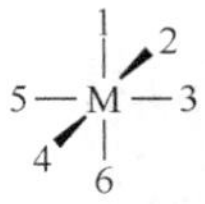	
Formula	No. of predicted isomers (numbers in parentheses indicate position of the B ligands)			No. of actual isomers
MA_5B	One	One	One	One
MA_4B_2	Three	Three	Two	Two
	(1, 2)	(1, 2)	(1, 2)	
	(1, 3)	(1, 4)	(1, 6)	
	(1, 4)	(1, 6)		
MA_3B_3	Three	Three	Two	Two
	(1, 2, 3)	(1, 2, 3)	(1, 2, 3)	
	(1, 2, 4)	(1, 2, 4)	(1, 2, 6)	
	(1, 3, 5)	(1, 2, 6)		

A few comments about the information in this table needed to be given before discussing. (1) The symbols for the compounds use M for the central metal and A's and B's for the various ligands. (2) The numbers in parentheses for each isomer refer to the relative positions of the B ligands.

Isomers are defined here as compounds that have the same numbers and types of chemical bonds but differ in the spatial arrangements of those bonds (A more detailed discussion of isomers is presented in the following sections).

For the MA_5B case in Table 1.3, only one isomer could actually be prepared experimentally, a result consistent with all three of the proposed geometries. For the MA_4B_2 case, however, Werner could prepare only two isomers (Fig.1.4). For the octahedral case, this actual number matched the possible number, but for the hexagonal planar and trigonal prism cases, there were three possible isomers. Assuming that Werner had not missed the preparation of an isomer someplace along the line, the data indicated that the "fixed positions in space" for six ligands is octahedral. The same type of analysis for the MA_3B_3 case gives a similar result. Only the octahedral configuration gives the same number of isomers as were actually prepared.

Given these results, Werner could predict that two isomers would be found for the $CoCl_3 \cdot 4NH_3$ case. These proved somewhat difficult to prepare, but in 1907 Werner was finally successful. He found two isomers, one a bright green and the other a violet color. By comparing the actual number

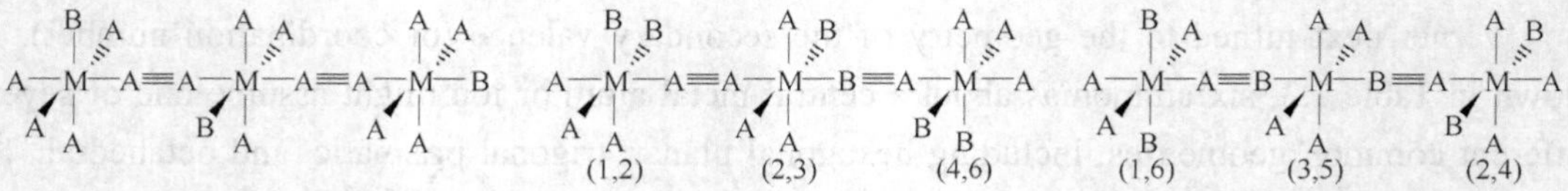

(1,2) (2,3) (4,6) (1,6) (3,5) (2,4)

(a) MA_5B (b) MA_4B_2(Isomer 1) (c) MA_4B_2(Isomer 2)

Fig.1.4 Equivalent configurations for some octahedral isomers

Fig.1.5 A copy of the cover of book 《Alfred Werner—Founder of Coordination Chemistry》

of known isomers with the number that should exist for various geometries, Werner concluded that the six ligands in the cobalt ammonates were in an octahedral arrangement. So, the coordination theory was growing stronger.

All of this goes to demonstrate, as so often is the case in science. Sometimes, we need to take risks. We must occasionally follow our intuitions and advocate a new and sometimes poorly supported way of thinking about a phenomenon in order to make a truly revolutionary advance. Blomstrand and Jørgensen tried to extend the established ideas of organic chemistry to account for the newer coordination compounds. In doing so, one could argue, they actually impaired progress in the understanding of this branch of chemistry. The trick, of course, is to know when to stick to the established ideas and when to break away from them. Werner chose the latter course. 20 years later in 1913, he received the Nobel Prize in chemistry.

At the start of the 20th century, inorganic chemistry was not a prominent field until Werner studied the metal-amine complexes such as $[Co(NH_3)_6Cl_3]$—Werner compound. He is a founder of coordination chemistry. Fig.1.5 is a copy of the cover of book 《Alfred Werner—Founder of Coordination Chemistry》.

1.1.3 Extending Coordination Chemistry—Supramolecular Chemistry

Supramolecular chemistry refers to the area of chemistry that focuses on the the weaker and reversible non-covalent interactions between molecules. These non-covalent interactions include hydrogen bonding, π-π stacking, electrostatic effects, hydrophobic forces and Van der Waals forces. Recently, the coordination bonding in coordination complexes of metal is accepted as an important force in supramolecular chemistry. At this point, the supramolecular chemistry can be considered as extended coordination chemistry. The concepts such as molecular self-assembly, molecular recognition, preorganization, and macrocyclic effects, have been demonstrated by supramolecular chemistry. Biological systems are often the inspiration for supramolecular research. The study of non-covalent interactions is crucial to understanding many biological processes from cell structure to vision that rely on these forces for structure and function.

The importance of supramolecular chemistry was established by the 1987 Nobel Prize for Chemistry which was awarded to Donald J. Cram, Jean-Marie Lehn, and Charles J. Pedersen in recognition of their work in this area. The development of selective "host-guest" complexes in particular, in which a host molecule recognizes and selectively binds a certain guest, was cited as an important contribution.

Supramolecular chemistry is a new rapidly progressing field on the crossroads among chemistry, biochemistry, physics and technology. Exciting phenomena unthinkable within the realm of classical organic chemistry (for example, alkali metal anions) not only provide the basis for revolutionizing numerous branches of industry but also improve our understanding of the functioning of living organisms and of the origin of life.

In the 1990s, supramolecular chemistry became even more sophisticated, with researchers such as James Fraser Stoddart developing molecular machinery and highly complex self-assembly structures, and Itamar Willner developing sensors and methods of electronic and biological interfacing. The emerging science of nanotechnology also had a strong influence on the subject, with building blocks such as fullerenes, nanoparticles and dendrimers becoming involved in synthetic systems.

(1) Some Concepts in Supramolecular Chemistry

- Molecular self-assembly Molecular self-assembly is the construction of systems without guidance or management from an outside source (other than to provide a suitable environment). The molecules are directed to assemble through noncovalent interactions. Self-assembly may be subdivided into intermolecular self-assembly and intramolecular self-assembly.
- Molecular recognition Molecular recognition is the specific binding of a guest molecule to a complementary host molecule to form a host-guest complex. Often, the definition of which species is the "host" and which is the "guest" is arbitrary. The molecules are able to identify each other using noncovalent interactions.
- Building blocks of supramolecular chemistry Supramolecular systems are rarely designed from first principles. Rather, chemists have a range of well-studied structural and functional building blocks that they are able to use to build up larger functional architectures. Many of these exist as whole families of similar units, from which the analog with the exact desired properties can be chosen.
- Synthetic recognition motifs Synthetic recognition motifs are included, the formation of carboxylic acid dimers and other simple hydrogen bonding interactions; the use of crown ether binding with metal or ammonium cations, the π-π charge-transfer interactions of bipyridinium with dioxyarenes or diaminoarenes, et al.
- Macrocycles Macrocycles are very useful in supramolecular chemistry, as they provide whole cavities that can completely surround guest molecules and may be chemically modified to fine-tune their properties.
- Structural units Many supramolecular systems require their components to have suitable spacing and conformations relative to each other, and therefore easily-employed structural units are required.

● Dynamic covalent chemistry In dynamic covalent chemistry, covalent bonds are broken and formed in a reversible reaction under thermodynamic control. While covalent bonds are keys to the process, the system is directed by noncovalent forces to form the lowest energy structures.

● Biomimetics Many synthetic supramolecular systems are designed to copy functions of biological systems. These biomimetic architectures can be used to learn about both the biological model and the synthetic implementation.

● Imprinting Molecular imprinting describes a process by which a host is constructed from small molecules using a suitable molecular species as a template. After construction, the template is removed leaving only the host. The template for host construction may be subtly different from the guest that the finished host bind. In its simplest form, imprinting utilizes only steric interactions, but more complex systems also incorporate hydrogen bonding and other interactions to improve binding strength and specificity.

● Molecular machinery Molecular machinery is molecules or molecular assemblies that can perform functions such as linear or rotational movement, switching, and entrapment. These devices exist at the boundary between supramolecular chemistry and nanotechnology, and prototypes have been demonstrated using supramolecular concepts.

(2) The Application of Supramolecular Chemistry Designing supramolecular systems with desired properties makes chemical industry cleaner and safer, electronics smaller by developing devices composed of single molecule or molecular aggregate. It will also entirely change the way we use energy resources. It will also transform pharmaceutical industry and medicine by developing new ways of drugs administration and new composite biocompatible materials which will serve as implants of new generation changing dentistry, surgery, and other branches of medicine.

● Materials Molecular self-assembly process of supramolecular chemistry in particular has been applied to the development of new materials. Large structures can be readily accessed using bottom-up synthesis as they are composed of small molecules requiring fewer steps to synthesize. Thus most of the bottom-up approaches to nanotechnology are based on supramolecular chemistry.

● Catalysis A major application of supramolecular chemistry is the design and understanding of catalysis and catalysts. Noncovalent interactions are extremely important in catalysis, binding reactants into conformations suitable for reaction and lowering the transition state energy of reaction. Template-directed synthesis is a special case of supramolecular catalysis.

● Medicine Supramolecular chemistry is also important to the development of new pharmaceutical therapies by understanding the interactions at a drug binding site. The area of drug delivery has also made critical advances as a result of supramolecular chemistry providing encapsulation and targeted release mechanisms.

● Green chemistry Research in supramolecular chemistry also has application in green chemistry where reactions have been developed which proceed in the solid state directed by non-covalent bonding. Such procedures are highly desirable since they reduce the need for solvents during the production of chemicals.

- Data storage and processing Supramolecular chemistry has been used to demonstrate computation functions on a molecular scale. In many cases, photonic or chemical signals have been used in these components, but electrical interfacing of these units has also been shown by supramolecular single transduction devices. Data storage has been accomplished by the use of molecular switches with photochromic and photoisomerizable units, by electrochromic and redox-switchable units, and even by molecular motion.
- Other Functional Devices Supramolecular chemistry is often pursued to develop new functions that cannot appear from a single molecule. These functions also include magnetic properties, light responsiveness, self-healing polymers, molecular sensors, etc. Supramolecular research has been applied to develop high-tech sensors.

1.2 The Key Features of Coordination Complex

1.2.1 The Concepts of Coordination Complex

Today, the molecular formulas of coordination compounds are represented in a manner that makes it clearer which groups are part of the coordination sphere and which are not. The metal atom or ion and the ligands coordinated to it are enclosed in brackets. It follows that the cobalt ammonate chlorides can be represented as

(1) $CoCl_3 \cdot 6NH_3$ $[Co(NH_3)_6]Cl_3$

(2) $CoCl_3 \cdot 5NH_3$ $[Co(NH_3)_5Cl]Cl_2$

(3) $CoCl_3 \cdot 4NH_3$ $[Co(NH_3)_4Cl_2]Cl$

(4) $CoCl_3 \cdot 3NH_3$ $[Co(NH_3)_3Cl_3]$

The ammonia molecules and chloride ions inside the brackets satisfy the coordination number of cobalt. The chlorides in the coordination sphere also help to satisfy the +3 oxidation state of the cobalt. The chlorides outside the brackets, sometimes called counterions, help satisfy only the oxidation state. They are the only ionic chlorides available to be precipitated by silver nitrate. For example, if compound (2) is placed in water and treated with aqueous silver ions, the resulting reaction would be that represented by equation (1.2).

$$[Co(NH_3)_5Cl]Cl_2(s) + 2Ag^+(aq) \longrightarrow 2AgCl(s) + [Co(NH_3)_5Cl]^{2+}(aq) \tag{1.2}$$

Consider the following series of platinum ammonates presented in their modern format, the series is extended to include anionic complex ions. The counterions in these latter cases, compounds (10) and (11), are potassium cations.

(5) $[Pt(NH_3)_6]Cl_4$

(6) $[Pt(NH_3)_5Cl]Cl_3$

(7) $[Pt(NH_3)_4Cl_2]Cl_2$

(8) $[Pt(NH_3)_3Cl_3]Cl$

(9) $[Pt(NH_3)_2Cl_4]$

(10) $K[Pt(NH_3)Cl_5]$

(11) $K_2[PtCl_6]$

Based on these facts, the definitions of the coordination chemistry and coordination compounds could be concluded.

Coordination chemistry is the study of compounds formed between metal ions and other neutral or negatively charged molecules.

A coordination complex is the product of a Lewis acid-base reaction in which neutral molecules or anions (called ligands) bond to a central metal atom (or ion) by coordinate bonds.

Ammonia was certainly one of the most famous ligands to be investigated. It is referred to as a monodentate ligand, defined as one that shares only a single pair of electrons with a metal atom or ion. The word monodentate comes from the Greek monos and the Latin dentis and, not unexpectedly, literally means “one tooth”. A monodentate ligand, then, has only one pair of electrons with which to “bite” the metal. Some other common ligands are shown in Table 1.4 (The nomenclature for these ligands is discussed in the next section).

Not surprisingly, there are bidentate, tridentate, and, in general, multidentate ligands as well. In general, the denticity of a ligand is defined as the number of pairs of electrons it shares with a metal atom or ion. A few of the other common multidentate ligands are also given in Table 1.4. The denticities of these ligands are given in parentheses. For example, the denticity of ethylenediamine (en) is 2 (Fig.1.6).

Table 1.4 Common monodentate, multidentate, bridging, and ambidentate ligands

Usually monodentate ligands			
F^-	fluoro		
Br^-	bromo		
I^-	iodo		
CO_3^{2-}	carbonato		
NO_3^-	nitrato		
SO_3^{2-}	sulfito		
$S_2O_3^{2-}$	thiosulfato		
SO_4^{2-}	sulfato		
CO	carbonyl		
Cl^-	chloro		
O^{2-}	oxo		
O_2^{2-}	peroxo	Common bridging ligands (sulfato to nitro)	
OH^-	hydroxo		
NH_2^-	amino		
CN^-	cyano		
SCN^-	thiocyanato		Ambidentate ligands (thiocyanato, nitro)
NO_2^-	nitro		
H_2O	aqua		
NH_3	ammine		
CH_3NH_2	methylamine		
$P(C_6H_5)_3$	triphenylphosphine		
$As(C_6H_5)_3$	triphenylarsine		
N_2	dinitrogen		

countinued

Usually monodentate ligands		
O_2	dioxygen	
NO	nitrosyl	
C_2H_4	ethylene	
C_5H_5N	pyridine	
Multidentate ligands		
$NH_2CH_2CH_2NH_2$	ethylenediamine(en)	(2)
$CH_3C(=O)\bar{C}HC(=O)CH_3$	acetylacetonato(acac)	(2)
$C_2O_4^{2-}$	oxalato(ox)	(2)
$NH_2CH_2COO^-$	glycinato(gly)	(2)
$NH_2CH_2CH_2NHCH_2CH_2NH_2$	diethylenetriamine(dien)	(3)
$N(CH_2COO)_3^{3-}$	nitrilotriacetato(NTA)	(4)
$(OOCCH_2)_2NCH_2CH_2N(CH_2COO)_2^{4-}$	ethylenediamine tetraacetato(EDTA)	(6)

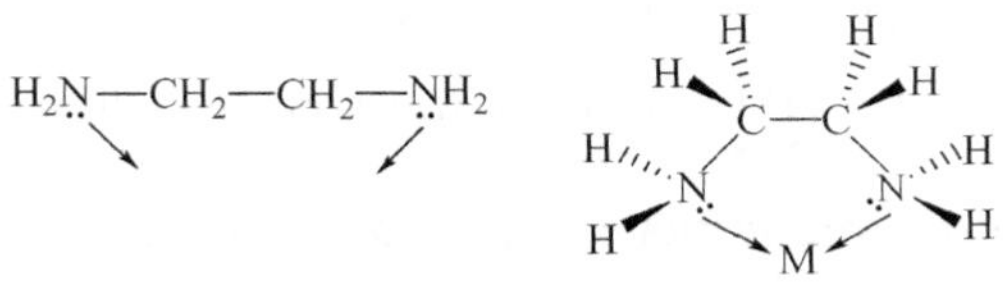

$H_2\ddot{N}—CH_2—CH_2—\ddot{N}H_2$

(a) (b)

Fig.1.6 The bidentate ethylenediamine ligand

1.2.2 Classification of Ligand

(1) Monodentate ligands Monodentate ligands have only one point of attachment to the metal center and occupy only one coordination site. Examples of monodentate ligands include NH_3, H_2O, PMe_3, CO and other neutral two-electron donors.

(2) Bidentate ligand As shown in Fig.1.6, ethylenediamine was a bidentate ligand of particular importance in the work of both Werner and Jørgensen. Both of the nitrogen atoms in this compound have a lone pair of electrons that can be shared with a metal. A bidentate ligand may act in a monodentate fashion if one end is not attached to the metal center, in which case it is called ambidentate(see the next part). Examples of bidentate ligands are 2,2'-bipyridine (bipy), acetylacetonate (acac), diphenylphosphinoethane (dppe) and oxalate (ox)(Fig.1.7).

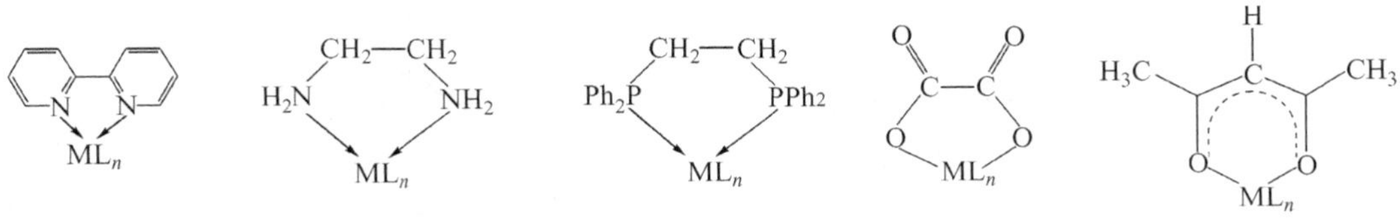

bipy en dppe ox acac

Fig.1.7 Some examples of common bidentate ligands

(3) Ambidentate ligands Two other general types of ligands are represented in Table 1.4 and should be brief mentioned here. ①The common bridging ligands, defined as those containing two pairs of electrons shared with two metal atoms simultaneously. The interaction of such ligands with metal atoms can be represented as M←:L:→M. Ligands of the bridging type include amide (NH_2^-), carbonyl (CO), chloro (Cl^-), cyano (CN^-), hydroxo (OH^-), nitro (NO_2^-), oxo (O^{2-}), peroxo (O_2^{2-}), sulphato (SO_4^{2-}) and thiocyanato (SCN^-). ②The ligand to include at this point is the ambidentate ligand. Depending on the experimental conditions and the metals involved, these ligands can use one of two different atoms to share a pair of electrons with a metal. If we represent this type of ligand as :AB:, then it can form one of two possible coordinate-covalent bonds, either M←:AB: or :AB:→M, with a metal atom. Common ambidentate ligands include cyano, thiocyanato, and nitro. The classic example is the NO_2^- ligand as shown (Fig.1.8). The SCN^- ligand can bind through the S (thiocyanato) or through the N (iso thiocyanato).

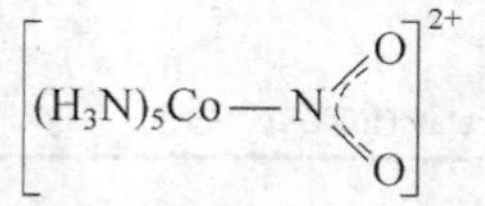

$[(H_3N)_5Co{-}NO_2]^{2+}$ Nitro pentamminenitrocobalte(Ⅲ)

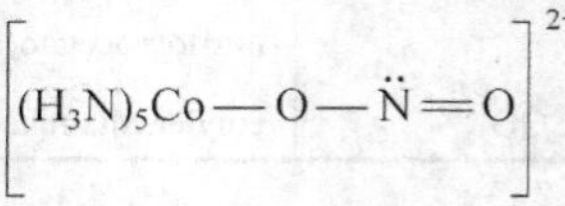

$[(H_3N)_5Co{-}O{-}\ddot{N}{=}O]^{2+}$ Nitrito pentamminenitritocobalte(Ⅲ)

Fig.1.8 The coordination modes of NO_2^- ligand

(4) Multidentate ligands Multidentate ligands(also called polydentate ligands) have more than one point of attachment to the metal center and occupy more than one coordination site. By definition, bidentate ligands as well as ambidentate ligands are subsets of polydentate ligands. A 6-coordinate tetraanionic ligand is ethylenediamineteraacetato (EDTA) shown below in its uncomplexed and complexed states (with hydrogens omitted for clarity). The salen ligand is versatile dianionic, tetradendate ligand. The tridentate tris(pyrazolyl)borate ligand is an example of a monoanionic tridentate ligand which is often used as a Cp analog (Fig.1.9).

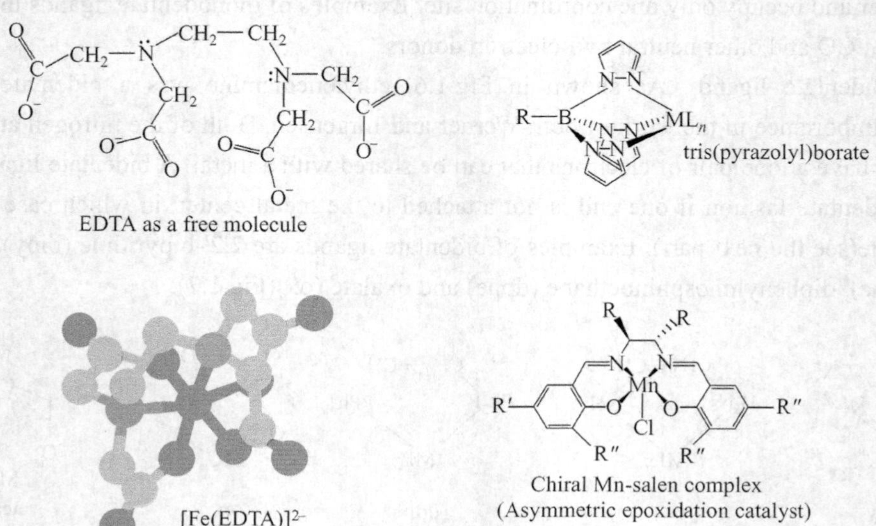

Fig.1.9 Some examples of multidentate ligands

(5) Chelating Multidentate ligands that form one or more rings with a metal atom in the manner that the resulting configuration rather resembles a crab clutching at its prey are called chelates or chelating agents, derived from the Greek chele, meaning "claw". A chelating ligand also called a chelate. The examples shown above for bidentate ligands are additional examples of chelating ligands.

(6) Bite angle A bite angle is the ligand-metal-ligand angle formed when a bidentate or polydentate ligand coordinates to a metal center. This number can be used to characterize the distortion from an idealized geometry or to gauge the degree of ring strain. In the example in the following picture, we can see that the tetrahedral metal center is distorted from the ideal 109.5° (Fig.1.10).

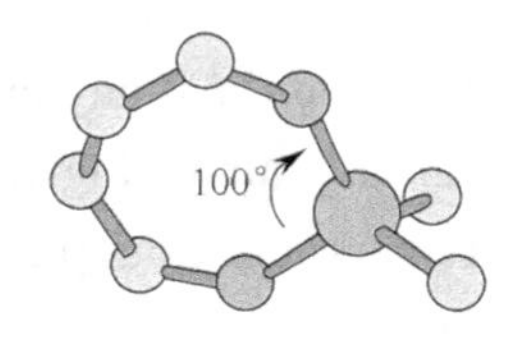

Fig.1.10 The distorted bite angle of ideal

1.2.3 Coordination Number and Coordination Geometry

Coordination number is termed as the total number of points of attachment to the central element. This can vary from 2 to as many as 16, but is usually 6.

(1) Coordination number 2 This arrangement is not very common for first row transition metal ion complexes and some of the best known examples are for silver(Ⅰ). For instance, a method often employed for the detection of chloride ions involves the formation of the linear diamminesilver(Ⅰ) complex. The first step is:

$$Ag^+ + Cl^- \longrightarrow AgCl$$

and to ensure that the precipitate is really the chloride salt, two further tests must be done:

$$AgCl + 2NH_3 \cdot H_2O \longrightarrow [Ag(NH_3)_2]^+ + Cl^- + 2H_2O$$

$$\Big\| HNO_3$$

$$AgCl(\text{re-ppts}) + NH_4^+ + NO_3^- + 2H_2O$$

The reaction of a bidentate ligand such as 1,2-diaminoethane with Ag^+ does not lead to chelated ring systems, but instead to linear two coordinate complexes. One reason for this is that bidentate ligands cannot exist in *trans* arrangements, which they cannot span 180°.

(2) Coordination number 3 Once again, this is not very common for first row transition metal ions. Examples with three different geometries have been identified(Fig.1.11).

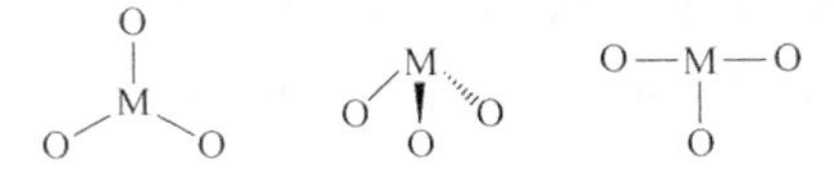

Fig.1.11 The three different geometries of coordination number 3

- Trigonal planar: Well known for main group species like CO_3^{2-} etc, this geometry has the four atoms in a plane with the bond angles between the donor atoms at 120 degrees.
- Trigonal pyramid: More common with main group ions.
- T-shaped: The first example of a T-shaped molecule was found in 1977.

(3) Coordination number 4 Two different geometries are possible. The tetrahedron is the more common while the square planar is found almost exclusively with metal ions having a d^8

electronic configuration (Fig.1.12).

- Tetrahedral: The chemistry of molecules centred around a tetrahedral C atom is covered in organic courses. To be politically correct, please change all occurrences of C to Co. There are large numbers of tetrahedral cobalt(Ⅱ) complexes known.

- Square planar: This is fairly rare and is included only because some extremely important molecules exist with this shape.

(4) Coordination number 5 Two possible geometries are available: square pyramid and trigonal bipyramid (Fig.1.13). The structure of $[Cr(en)_3][Ni(CN)_5]\cdot 1.5H_2O$ was reported in 1968 to be a remarkable example of a complex exhibiting both types of geometry in the same crystal. The reaction of cyanide ion with Ni^{2+} proceeds via several steps:

$$Ni^{2+} + 2CN^- \longrightarrow Ni(CN)_2$$

$$Ni(CN)_2 + 2CN^- \longrightarrow [Ni(CN)_4]^{2-}\text{(orange-red)},\ \lg\beta_4 = 30.1$$

$$[Ni(CN)_4]^{2-} + CN^- \longrightarrow [Ni(CN)_5]^{3-}\text{(deep red)}$$

Oxovanadium salts (vanadyl, VO^{2+}) often show square pyramidal geometry, for example, $VO(acac)_2$. Note that the vanadium(Ⅳ) can be considered coordinatively unsaturated and addition of pyridine leads to the formation of an octahedral complex.

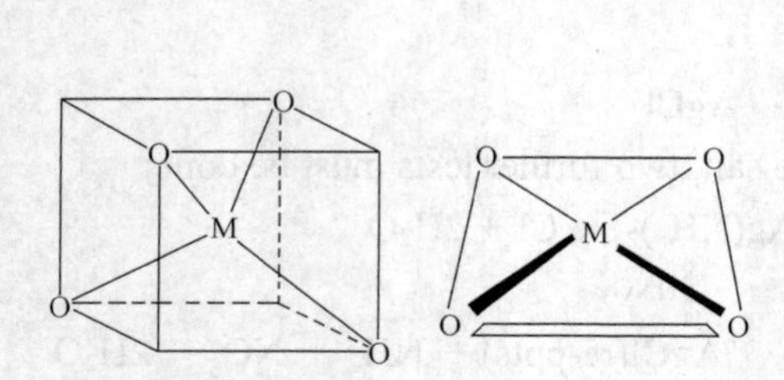

Fig.1.12 The two different geometries of coordination number 4

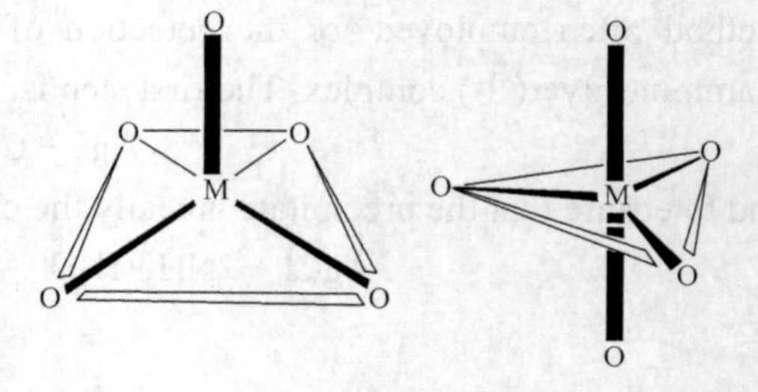

Fig.1.13 The two different geometries of coordination number 5

(5) Coordination number 6 (Fig.1.14)

- Hexagonal planar: Unknown for first row transition metal ions, although the arrangement of six groups in a plane is found in some higher coordination number geometries.

- Trigonal prism: Most trigonal prismatic compounds have three bidentate ligands such as oxalates and few are known for first row transition metal ions.

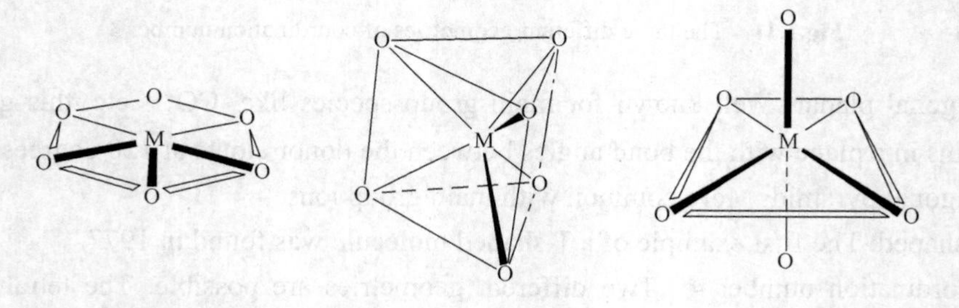

Fig.1.14 The three possible geometries of coordination number 6

- Octahedral: The most common geometry found for first row transition metal ions, including all aqua ions. In some cases distortions are observed and these can sometimes be explained in terms of the John-Teller theorem.

(6) Coordination number 7 Three geometries are possible: capped octahedron, capped trigonal prism, pentagonal bipyramid (Fig.1.15).

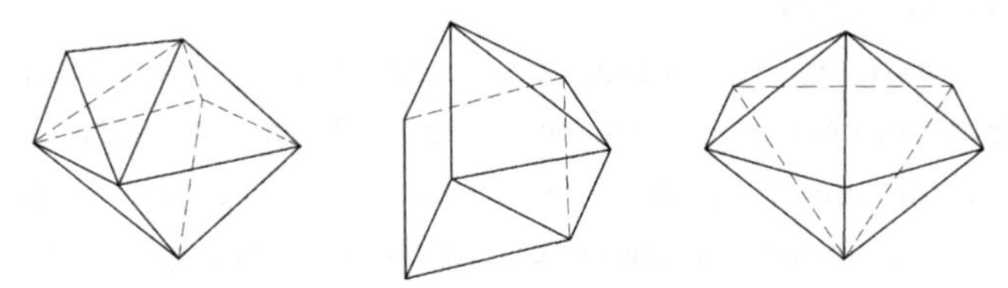

Fig.1.15 The three possible geometries of coordination number 7

(7) Coordination number 8 Dodecahedron (D_{2d}), cube (O_h), square antiprism (D_{4d})(Fig.1.16).

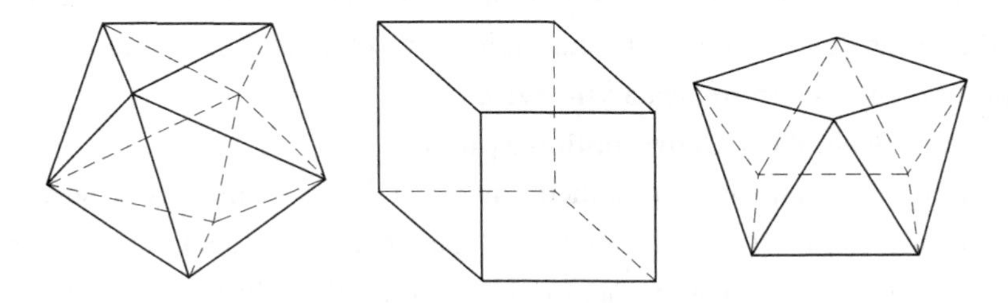

Fig.1.16 The three possible geometries of coordination number 8

(8) Coordination Number 9 Three-face centred trigonal prism (D_{3h}).

(9) Coordination Number 10 Bicapped square antiprism (D_{4d}).

(10) Coordination Number 11 All-faced capped trigonal prism (D_{3h}).

(11) Coordination Number 12 Cuboctahedron(O_h).

There are many definitions of the term coordination number, but there is no one simple unambiguous definition that works in all cases. For simple monodentate and chelating ligands, the coordination number can be defined as the number of atoms or ligands directly bonded to the metal atom. For example, $[Fe(NH_3)_6]^{3+}$ and $[Fe(en)_3]^{3+}$ are both 6-coordinate complexes. But complications quickly arise when one considers the bonding of alkene or alkenyl fragments such as ethylene and cyclopentadienyl.

For example, in ferrocene all ten cyclopentadienyl carbons are equally bonded to the iron center, but few would call this a 10-coordinate complex. Given the reactivity and bonding of ferrocene, calling it a 2-coordinate complex doesn't seem quite right either. The general consensus is to consider ferrocene a six coordinate species because there are six electron pairs on the ligands that participate in bonding.

While this method seems to alleviate the problem, it adds further complications. Here are some examples:

- Alkoxide, imido and oxo ligands can donate one, two or three pairs of electrons to a metal center, but in all cases each only occupies one coordination site.

- Cp_2TiCl_2 would not be considered an 8-coordinate complex by most chemists.
- Alkene complexes can be drawn as neutral two electron donors but are sometimes considered dianionic bidentate ligands. Either way, alkenes are usually considered to occupy only one coordination site.

1.2.4 Coordinative Unsaturation

We use the term coordinative unsaturation to describe a complex that has one or more open coordination sites where another ligand can be accommodated. Typically, most complexes with a CN of less than 6 fall into this category, although this can vary. Coordinative unsaturation is an important concept in organotransition metal chemistry. For example, for an olefin to undergo insertion, metathesis or polymerization, there must be room for the olefin to approach or bond to the metal. A common means of creating the requisite vacant site is by loss of a coordinated ligand or a change in the bonding mode of a ligand (such as conversion of a trihapto allyl ligand to the monohapto form). Such changes are often accompanied by a reduction in the electron count as well. Hence, while the terms "coordinative unsaturation" and "electronic saturation" are different, they often work hand-in-hand as part of a catalytic cycle.

1.2.5 Primary and Secondary Coordination Sphere

The primary coordination sphere of a metal involves the set of ligands closest to the metal that are directly attached [Fig.1.17(a)]. The nature of the interaction between the centre metal and organic ligand is covalent bond. The primary coordination sphere has fixed coordination geometry (see 1.2.3 section). The secondary (or outer) coordination sphere is much fuzzier and often constructed by mobile cations, counterions or solvents [Fig.1.17(b)]. The interactions between the secondary sphere and centre metal ions are noncovalent interactions, such as H-bonding, ion-dipole interaction, dipole-dipole interactions, and so on.

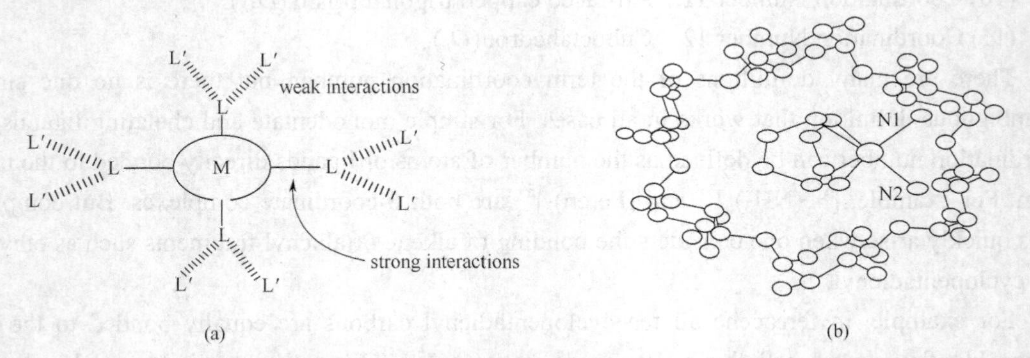

Fig.1.17 (a) Primary coordination sphere by coordination bonding and (b) secondary coordination sphere by noncovalent interactions

The primary and secondary coordination sphere is very important in the metalloprotein that perform many of life's most important processes. Metalloproteins, proteins that contain one or more metal cofactors at their active-sites, can be thought of as the ultimate transition metal complex. The ligand environment about the metal-center in a metalloprotein is often characterized by low symmetry, an unusual coordination geometry, and unique metal-ligand bonding.

The cytochrome P450 (P450) is a good example of the issue. The most notable feature of the P450 active site is the heme-thiolate bond formed by proximal cysteine ligation to the heme iron [Fig.1.18(a)]. The subtle roles that certain amino acids play in the formation of a secondary coordination sphere around the heme-thiolate on the proximal side and around the substrate-binging site in the distal pocket are just as important; they allow the enzymes to function efficiently as monooxygenases and help fine tune their selectivity. For the detail, secondary sphere interactions on the distal side of P450s are primarily involved in forming a structured H-bonding network and providing a proton source for the activation of dioxygen as well as substrate bonding. An additional mutation of the distal noncoordinating histidine to either a valine or isoleucine resulted in a P450-like protein with an iron(Ⅲ) resting state [Fig.1.18(b)]. This effect, termed the *trans* effect by the authors, mimics the more hydrophobic environment of the Cyt P450 and contributes significantly to the binding of the axial ligand and to the stability. Similarly, mutation of the axial His ligand to Cys in CcP resulted in a very unstable ligand that was rapidly oxidized to cysteic acid. It was then recognized that a nonpolar residue next to the Cys unit is conserved in P450 proteins, while the analogous amino acid in CcP is an aspartic acid [Fig.1.18(c)]. Stable coordination of cysteine thiolate in the ferrous carbonyl derivative of an engineered protein was also obtained by mutations of the axial His and Met ligands in cytochrome b562 to Cys and Gly, respectively [Fig.1.18(d)].

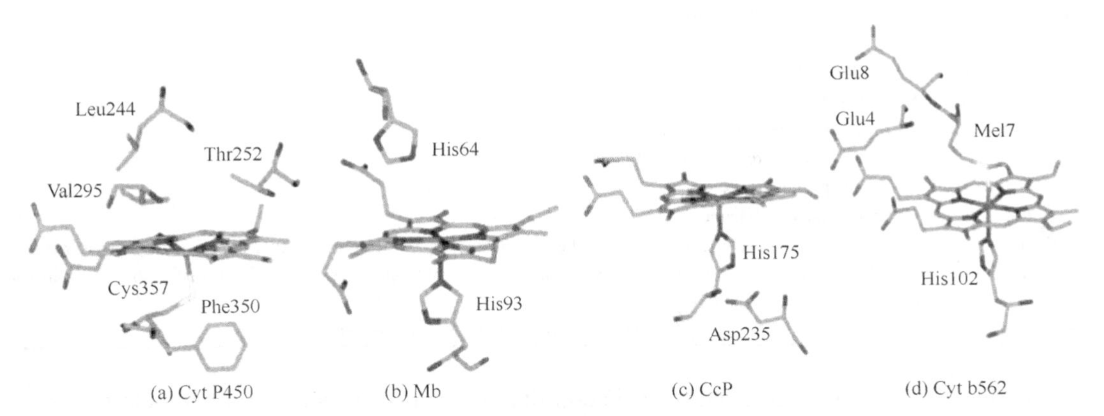

Fig.1.18 Key structural differences of Cyt P450(a), Mb(b), CcP(c) and Cyt b562(d)

1.3 Nomenclature of Coordination Complex

Structures of coordination compounds can be very complicated, and their names is long because the ligands may already have long names. Knowing the rules of nomenclature not only enable you to understand what the complex is, but also let you give appropriate names to them.

Here, the nomenclature of coordination compounds is introduced based on the basics of naming ligands (including multidentate, ambidentate, and bridging) and simple coordination compounds (Table 1.5).

Table 1.5 Nomenclature Rules for Simple Coordination Compounds

Ligands

1. Anionic ligands end in –o

F^-	Fluoro	NO_2^-	nitro	SO_3^{2-}	sulfito	OH^-	hydroxo
Cl^-	chloro	ONO^-	nitrito	SO_4^{2-}	sulfato	CN^-	cyano
Br^-	bromo	NO_3^-	nitrato	$S_2O_3^{2-}$	thiosulfato	NC^-	isocyano
I^-	iodo	CO_3^{2-}	carbonato	ClO_3^-	chlorato	SCN^-	thiocyanato
O^{2-}	oxo	$C_2O_4^{2-}$	oxalato	CH_3COO^-	acetato	NCS^-	isothiocyanato

2. Neutral ligands are named as the neutral molecule

C_2H_4	ethylene	$(C_6H_5)_3P$	triphenylphosphine
$NH_2CH_2CH_2NH_2$	ethylenediamine	CH_3NH_2	methylamine

3. There are special names for four neutral ligands

H_2O aqua NH_3 ammine CO carbonyl NO nitrosyl

4. Cationic ligands end in -ium

$NH_2NH_3^+$ hydrazinium

5. Ambidentate ligands are indicated by

(a) Using special names for the two forms, for example, nitro and nitrito for NO_2^- and ONO^-

(b) Placing the symbol of the coordinating atom in front of the name of the ligand, for example, *S*-thiocyanato and *N*-thiocyanato for SCN^- and NCS^-

6. Bridging ligands are indicated by placing a μ- before the name of the ligand

Simple coordination compounds

1. Name the cation first, then the anion
2. List the ligands alphabetically
3. Indicate the number(2, 3, 4, 5, 6)of each type of ligand by:

(a) The prefixes di-, tri-, tetra-, penta-, hexa- for:

① All monoatomic ligands;

② Polyatomic ligands with short names;

③ Neutral ligands with special names.

(b) The prefixes bis-, tris-, tetrakis-, pentakis-, hexakis- for:

① Ligands whose names contain a prefix of the first type (di-, tri-, etc.);

② Neutral ligands without special names;

③ Ionic ligands with particularly long names.

4. If the anion is complex, add the suffix -ate to the name of the metal(Sometimes the -ium or other suffix of the normal name is removed before adding the -ate suffix. Some metals, such as copper, iron, gold, and silver, use the Latin stem for the metal and become cuprate, ferrate, aurate, and argentate, respectively)
5. Put the oxidation state in Roman numerals in parentheses after the name of the central metal

The rules for naming ligands and simple coordination compounds are outlined below.

- In naming the entire complex, the name of the cation is given first and the anion second (just as for sodium chloride), no matter whether the cation or the anion is the complex species.
- In the complex ion, the name of the ligand or ligands precedes that of the central metal atom (This procedure is reversed for writing formulae).
- Ligand names generally end with "o" if the ligand is negative (chloro for Cl^-, cyano for CN^-, hydrido for H^-). For neutral ligands, their names are not changed (methylamine for $MeNH_2$).

Special ligand names are as follow:

H_2O, aqua

NH_3, ammine (two m's, amine with one m is for organic compounds)

CO, carbonyl

NO, nitrosyl

The last "e" in names of negative ions is changed to "o" in names of complexes. Sometimes "ide" is changed to "o". Example for anionic ligands:

bromide (Br^-) →bromo
chloride (Cl^-)→chloro
hydroxide (OH^-)→hydroxo
oxide (O^{2-})→oxo
peroxide (O_2^{2-})→peroxo
cyanide (CN^-)→cyano
azide (N_3^-)→axido
amide (NH_2^-)→amido
carbonate (CO_3^{2-})→carbonato
nitrate (NO_2^-)→nitrito (M—O bond)or nitro (M—N bond)
nitrate (NO_3^-)→nitro (when bonded through N)
sulfide (S^{2-})→sulfido
thiocyanate (SCN^-)→*S*-thiocyanato
thiocyanate (NCS^-)→*N*-thiocyanato
oxalate ($C_2O_4^{2-}$)→oxalato
sulphate (SO_4^{2-})→sulfato
thiosulfate ($S_2O_3^{2-}$)→thiosulfato
ethylenediaminetetraacetato[CH_2—$N(CH_2COO^-)_2]_2$ (EDTA)

- A Greek prefix (mono, di, tri, tetra, penta, hexa, etc.) indicates the number of each ligand (mono is usually omitted for a single ligand of a given type). If the name of the ligand itself contains the terms mono, di, tri, for example, triphenylphosphine, then the ligand name is enclosed in parentheses and its number is given with the alternate prefixes bis, tris, tetrakis instead.

For example, $Ni(PPh_3)_2Cl_2$ is named dichlorobis(triphenylphosphine)nickel(Ⅱ).

Greek prefix mono-, di- (or bis), tri-, tetra-, penta-, hexa, hepta-, octa-, nona-, (ennea-), deca- etc for 1, 2, 3, ⋯, 10.

- A Roman numeral or a zero in parentheses is used to indicate the oxidation state of the central metal atom.
- If the complex ion is negative, the name of the metal ends in "ate". For example:

scandium, Sc=scandate
titanium, Ti=titanate
vanadium, V=vanadate
chromium, Cr=chromate
manganese, Mn=manganate
iron, Fe=ferrate
cobalt, Co=cobalatate
nickel, Ni=nickelate
copper, Cu=cuprate

zinc, Zn=zincate

- If more than one ligand is present in the species, then the ligands are named in alphabetical order regardless of the number of each. For example, NH_3 (ammine) would be considered an "a" ligand and come before Cl^- (chloro) (This is where the 1971 rules differ from the 1957 rules. Some texts still say that ligands are named in the order: neutral then anionic).
- Some additional notes:

(a) Some metals in anions have special names. For example:

B→Borate, Au→Aurate, Ag→Argentate, Fe→Ferrate, Pb→Plumbate, Sn→Stannate, Cu→Cuprate

(b) Use of brackets or enclosing marks.

Square brackets are used to enclose a complex ion or neutral coordination species. Examples:

$[Co(en)_3]Cl_3$; $[Co(NH_3)_3(NO_2)_3]$; $K_2[CoCl_4]$

Note that it is not necessary to enclose the halogens in brackets.

Normal names that will not change are as follow:

C_5H_5N, pyridine

$H_2NCH_2CH_2NH_2$, ethylenediamine

C_5H_4N- C_5H_4N, dipyridyl

$P(C_6H_5)_3$, triphenylphosphine

$NH_2CH_2CH_2NHCH_2CH_2NH_2$, diethylenetriamine

(c) A bridging ligand is indicated by placing a "μ" before its name. So a bridging hydroxide (OH^-), amide (NH_2^-) or peroxide (O^{2-}) ligand becomes μ-hydroxo, μ-amido, or μ-peroxo, respectively. If there is more than one of a given bridging ligand, the prefix indicating the number of ligands is placed after the μ. For example, if there are two bridging chloride ligands, they are indicated as μ-dichloro. For example, $(H_3N)_3Co(OH)_3Co(NH_3)_3$, Triamminecobalt(Ⅲ)-$\mu$-trihydroxo triamminecobalt(Ⅲ)。

(d) There are two ways to handle ambidentate ligands. One is to use a slightly different form of the name, depending on the atom that is donating the electron pair to the metal. The second is to actually put the symbol of the donating atom before the name of the ligand. So —SCN might be called thiocyanato or *S*-thiocyanato, whereas —NCS would be isothiocyanato or *N*-thiocyanato. —NO_2^- and —ONO, however, are most always referred to as nitro and nitrito, respectively.

Incidentally, the rules for writing these formulas, as for all chemical compounds, are determined by the International Union of Pure and Applied Chemistry (IUPAC). The IUPAC rules concerning the order in which the formulas of the ligands in a coordination compound should be written are unexpectedly complicated and generally not treated in a textbook at this level.

1.4 Isomerism of Coordination Complex

1.4.1 Definition of Isomers

Compounds with the same molecular formula but different arrangements of atoms in space are called isomers.

Werner has been predicted that the complex $[Co(en)_2ClNH_3]^{2+}$ should exist in two forms, which are mirror images of each other. Werner isolated solids of the two forms, and structural studies confirmed his interpretations.

Because of the prevailing view that optical activity was integrally related to C atoms, Werner decided that it was necessary to prepare a completely inorganic complex (i.e., no C atoms) to prove that "carbon free inorganic compounds can also exist as mirror-image isomers"(Fig.1.19).

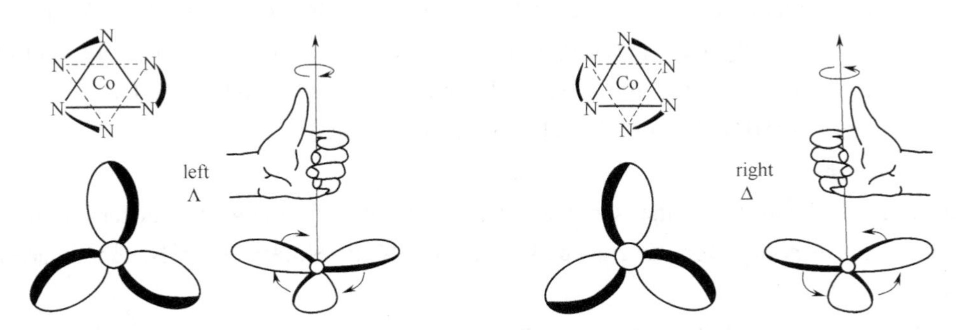

Fig.1.19 The absolute configuration of $[Co(en)_3]^{3+}$

Two principal types of isomerism are known among coordination compounds. Each of which can be further subdivided.

(1) Structural isomerism a. Coordination isomerism; b. Ionisation isomerism; c. Hydrate isomerism; d. Linkage isomerism.

(2) Stereoisomerism a. Geometrical isomerism; b. Optical isomerism.

1.4.2 Structural (or Constitutional) Isomers

This kind of isomer differs in how the atoms are joined together (have different bonds). There are several types of this isomerism frequently encountered in coordination chemistry and the following represents some of them.

(1) Coordination isomerism Where compounds containing complex anionic and cationic parts can be thought of as occurring by interchange of some ligands from the cationic part to the anionic part.

one isomer $[Co(NH_3)_6][Cr(C_2O_4)_3]$

another isomer $[Co(C_2O_4)_3][Cr(NH_3)_6]$

(2) Ionization isomers Different in an anion bonded to the metal, the isomers can be thought of as occurring because of the formation of different ions in solution.

one isomer $[Co(NH_3)_5Br]Cl$, Cl^- anions in solution

another isomer $[Co(NH_3)_5Cl]Br$, Br^- anions in solution

Notice that both anions are necessary to balance the charge of the complex, and that they differ in that one ion is directly attached to the central metal but the other is not. A very similar type of isomerism results from replacement of a coordinated group by a solvent molecule (Solvate isomerism). In the case of water, this is called hydrate isomerism.

(3) Hydrate isomerism The best known example of this occurs for chromium chloride

"$CrCl_3 \cdot 6H_2O$" which may contain 4, 5, or 6 coordinated water molecules.

$[Cr(H_2O)_4Cl_2]Cl \cdot 2H_2O$	bright-green
$[Cr(H_2O)_5Cl]Cl_2 \cdot H_2O$	grey-green
$[Cr(H_2O)_6]Cl_3$	violet

These isomers have very different chemical properties and on reaction with $AgNO_3$ to test for Cl^- ions, would find 1, 2, and 3 Cl^- ions in solution respectively.

(4) Linkage isomers   Differ in an atoms of a ligand bonded to the metal. Linkage isomerism occurs with ambidentate ligands. These ligands are capable of coordinating in more than one way. The best known cases involve the monodentate ligands SCN^-/NCS^- and NO_2^-/ONO^-.

For example: $[Co(NH_3)_5(ONO)]Cl$ and $[Co(NH_3)_5(NO_2)]Cl$.

1.4.3 Stereoisomers

This kind of isomer has the same atoms, same sets of bonds, but differ in the relative orientation of these bonds. Ignoring special cases involving esoteric ligands, then there are two isomerisms.

(1) Diastereoisomers(Geometric isomers)   Diastereoisomers are possible for both square planar and octahedral complexes, but not tetrahedral. For example: *cis*- or *trans*- $[Pt(NH_3)_2Cl_2]$. *cis*- and *trans*- refer to the position of two groups relative to each other. In the *cis*- isomer they are "next to each other", i.e. at 90° in relation to the central metal ion, whereas in the *trans*- isomer they are "opposite each other", i.e. at 180° relative to the central metal ion.

a
|
M······b
cis-

a······M······b
trans-

The first report of the three geometric isomers being isolated and characterised for complexes of the type [Mabcd] was by Il'ya Chernyaev in 1928. The examples reported by Anna Gel'man in 1948, were shown in the Fig.1.20.

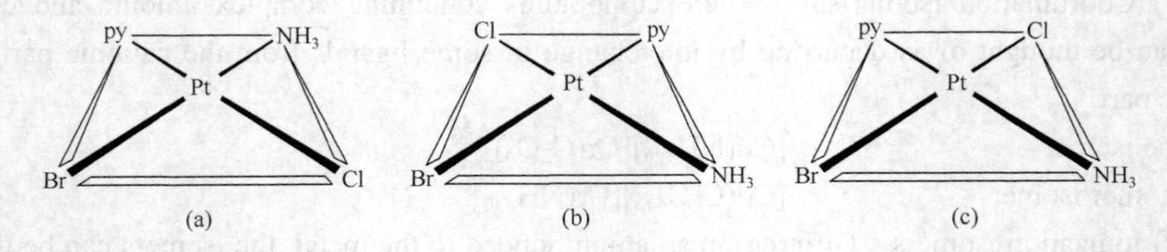

Fig.1.20 The three geometric isomers of a square planar complex $[PtNH_3BrClpy]$

The number of geometric isomers expected for square planar complexes is as follows:

Compound type	No. of isomers
Ma_2b_2	2(*cis*- and *trans*-)
Mabcd	3(use *cis*- and *trans*- relations)

(a, b, c, and d refer to monodentate ligands)

A number of examples of these types have been isolated and characterised and they show very different chemical and biological properties. Thus for example, *cis*-$Pt(NH_3)_2Cl_2$ is an anti-cancer agent whereas the *trans*- isomer is inactive against cancer (it is toxic), and so not useful in chemotherapy(Fig.1.21).

Fig.1.21 Structural scheme of geometric isomers of *cis*- or *trans*- $[Pt(NH_3)_2Cl_2]$

The number of geometric isomers expected for octahedral complexes is as follows:

Compound type	No. of isomers
Ma_4b_2	2(*cis*- and *trans*-)
Ma_3b_3	2(*fac*- and *mer*-)
$M(AA)_2b_2$	3(2 *cis*- and 1 *trans*-)

(a and b are monodentate ligands and AA is a bidentate ligand)

In the second example, new labels are introduced to reflect the relative positions of the ligands around the octahedral structure. Thus, placing the 3 groups on one face of the ocathedral gives rise to the facial isomer and placing the 3 groups around the centre gives rise to the meridinal isomer.

[Mabcdef] is expected to give 15 geometric isomers. In the case of $[Pt(NH_3)BrClIpyNO_2]$, several of these were isolated and characterised by Anna Gel'man and reported in 1956. Optical isomers are possible for each of these 15 forms, making a total of 30 isomers.

(2) Enantiomers (Optical isomers) Enantiomers are related as non-superimposable mirror images and differ in the direction with which they rotate plane-polarised light. These isomers are referred to as enantiomers or enantiomorphs of each other and their non-superimposable structures are described as being asymmetric.

Enantiomers are possible for both tetrahedral and octahedral complexes, but not square planar. They have nonsuperimposable mirror images, The *cis*- isomer of $M(AA)_2b_2$ may also exhibit optical isomerism although we will concentrate largely on optical isomers of the type $M(AA)_3$, for example, *cis*-$[Co(en)_2Cl_2]^+$(Fig.1.22). In 1911, the first resolution of optical isomers was reported by Werner and King just for the complexes *cis*-$[CoNH_3(en)_2X]^{2+}$, where X=Cl^- or Br^-.

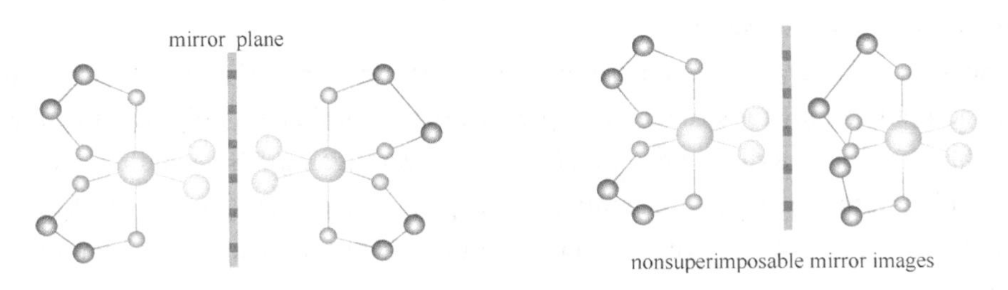

Fig.1.22 Optical isomers of *cis*-$[Co(en)_2Cl_2]^+$

Various methods have been used to denote the absolute configuration of optical isomers such as D or L, *R* or *S*. The 1970 IUPAC rules suggest that helixes are defined and then designated Lambda L (left- D) and Delta D (right-L) in much the same way as we would look at three-bladed propellors and determine whether they would trace out a left or right-hand helix.

The two isomers have identical chemical properties and just denoting their absolute configuration doesn't give any information regarding the direction in which they rotate plane-polarised light. This can only be determined from measurement and then the isomers are further distinguished by using the prefixes [(−) or L] and [(+) or D] depending on whether they rotate left or right. The use of *l*- and *d*- is not recommended since it may appear to conflict with L and D. The L(D)-$[Co(en)_3]^{3+}$ isomer gives a rotation to the right and therefore corresponds to the (+) isomer.

Note that, although it is predicted that tetrahedral complexes with 4 different ligands should be able to give rise to optical isomers (compare carbon chemistry), in general they are too labile and cannot be isolated.

Diastereoisomers have different chemical and physical properties: different colours, melting points, polarities, solubilities, chemical reactivities, etc. However, Most of the properties of two optical isomers (enantiomers) are the same, for example: solubility, melting point, boiling point, chemical reactivity (with nonchiral reagents). Two optical isomers differ in their reactivity with other "chiral" reagents and interactions with plane polarized light (Fig.1.23).

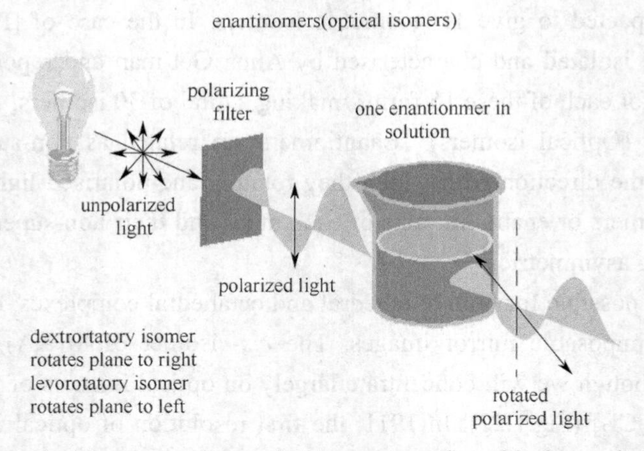

Fig.1.23 The illustration of the optical isomers reacting with plane polarized light

1.4.4 Supramolecular Isomerism

(1) The structural supramolecular isomerism The structural supramolecular isomerism can have profound implications for structure and properties of coordination polymers that are generated by one of the simplest building blocks and linear spacer ligand. These building blocks can be regarded as being based upon self-assembly of T-shaped nodes. There already exists a surprisingly diverse range of structures that have been observed in this context. The following points should be noted about such structures.

- These compounds are not true polymorphs since guest or solvent molecules are present in the lattice. However, neither are they solvates in the conventional sense.

- The diversity of network structures and hence bulk properties is remarkable.
- None of these architectures occurs naturally in minerals.
- The network structures themselves are entirely predictable based upon simple structural considerations.
- Some of these structures can occur from the same building blocks under almost identical crystallization conditions.

In summary, it seems likely that use of appropriate templates or guest molecules facilitates recipes that can be used to reliably generate all supramolecular isomers that are possible for a given node. Therefore, one might assert that there are a finite number of superstructures possible for a given molecular moiety and that it will eventually be possible to determine the crystallization conditions under which each one will occur.

(2) The conformational supramolecular isomerism Conformational changes in flexible ligands such as bis(4-pyridyl)ethane generate a different but often related network architecture since flexibility in ligands can lead to subtle or dramatic changes in architecture, for example, 1, 2-bis(pyridyl)ethane, bipy-eta, can readily adapt gauche- or anti-conformations. Supramolecular isomers are reported for T-shaped nodes linked by linear bifunctional exodentate ligands as 1D ladder, 3D lincoln logs, 2D herringbone, 2D bilayer, 2D brick wall and 3D frame.

A dramatic illustration of how conformational variability can influence crystal packing in organic compounds is illustrated by the compound 5-methyl-2-[(2-nitrophenyl)amino]-3-thiophenecarbonitile. This compound exists in at least six polymorphic phases. The primary difference between the six phases lies with the torsion angle between the thiophene moiety and the *o*-nitroaniline fragment, which varies from 21.7° to 104.7°.

(3) Catenane The different manner and degrees in which networks interpenetrate or interweave can afford significant variations in overall structure and properties depending upon the molecular building blocks that are utilized. Interpenetrated and noninterpenetrated structures are effectively different compounds because their bulk properties will be so different.

The existence of independent interpenetrating networks is surprisingly common if relatively large cavities are generated within a network. The existence of interpenetration has been regarded as a factor that mitigates strongly against the generation of stable open framework structures. However, it is becoming clear that appropriate use of templates can afford either open framework or interpenetrated structures for the same network. This is exemplified by the prototypal diamondoid and square grid networks $Cd(CN)_2$ and $M(bipy)_2X_2$. Both of these compounds have been prepared as interpenetrated and noninterpenetrated forms. Furthermore, some interpenetrated structures can also be regarded as open framework since if interpenetration will not necessarily afford close-packing.

(4) Optical Networks can be inherently chiral and can therefore crystallize in chiral (enantiomorphic) space groups. Therefore, an analogy can be drawn with homochiral compounds. This type of supramolecular isomerism lies at the heart of an important issue: spontaneous resolution of chiral solids.

Chapter 2 The Symmetry and Bonding of Coordination Complex

2.1 Symmetry in Chemistry—Group Theory

Group theory is one of the most powerful mathematical tools used in quantum chemistry and spectroscopy. It allows the user to predict, interpret, rationalize, and often simplify complex theory and data. At its heart is the fact that the set of operations associated with the symmetry elements of a molecule constitutes a mathematical set called a group. This allows the application of the mathematical theorems associated with such groups to the symmetry operations.

2.1.1 Symmetry Elements

The symmetry elements associated with a molecule are:

(1) Proper axis of rotation (C_n) where n=1, 2, ···. This implies n-fold rotational symmetry about the axis.

(2) Plane of reflection (σ) It implies bilateral symmetry about the plane. These planes are further classified as: σ_h—horizontal plane that is perpendicular to the principal axis of rotation (i.e. axis with highest value of n). If no principal axis exists, σ_h is defined as the molecular plane. σ_v or σ_d—vertical plane that contains the principal axis of rotation and is perpendicular to a σ_h plane, if it exists. When both σ_v or σ_d planes are present, the σ_v planes contain the greater number of atoms, the σ_d planes contain bond angle bisectors. If only one type of vertical plane is present, σ_v or σ_d may be used depending on the total symmetry of the molecule.

(3) Center of inversion (I) This is a central point through which all C_n and σ elements must pass. If no such common point exists there is no center of symmetry.

(4) Improper axis (S_n) This is made up of two parts: C_n and σ_h, both of which may or may not be true symmetry elements of the molecule. If both the C_n and the σ_h are present, then S_n must also exist. The following relations are helpful in this regard.

- If n is even, $S_n^m = E$ (m=n), including $C_n/2$.
- If n is odd, $S_n^m = s$ and $S_n^{2m} = E$ (m=n), including both C_n and σ perpendicular to C_n.

(5) The identity operation (E) Every molecule possesses at least one symmetry element: the identity. The identity operation amounts to doing nothing to a molecule and so leaves any molecule completely unchanged.

2.1.2 Symmetry Operation

All symmetry operations associated with isolated molecules can be characterized as rotations:

(1) Proper rotations C_n^k; k=1, ···, n. When k=n, C_n^k=E, the identity operation n indicates a rotation of 360°/n where n=1, ···

(2) Improper rotations S_n^k; k=1, ···, n. When k=1, n=1, S_n^k =s, reflection operation. When

k=1, n=2, S_n^k =i, inversion operation.

In general practice, we distinguish five types of operation:

- E, identity operation.
- C_n^k , proper rotation about an axis.
- σ, reflection through a plane.
- i, inversion through a center.
- S_n^k , rotation about an axis followed by reflection through a plane perpendicular to that axis.

Each of these symmetry operations is associated with a symmetry element that is a point, a line, or a plane about which the operation is performed such that the molecule's orientation and position before and after the operation are indistinguishable.

2.1.3 Molecular Point Group

Molecules can be categorized as linear, planar or non-planar. Knowing the symmetry elements of the molecule, we can now use the following flow chart to determine the molecular point group (Fig.2.1).

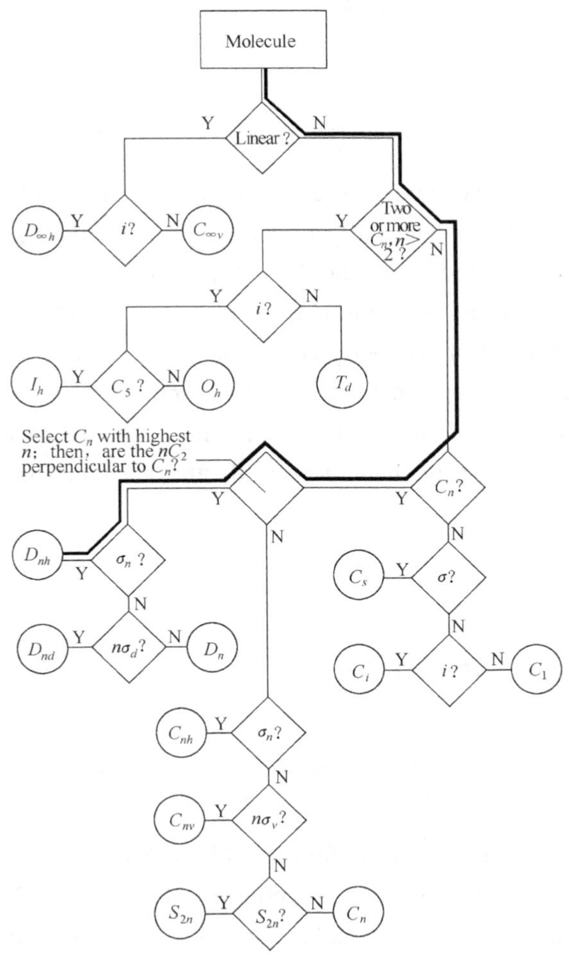

Fig.2.1 Programmed scheme of determination of molecular point group

(1) Linear molecules We will examine two linear molecules: chloromethane and ethyne.

- All linear molecules possess an axis lying along the line-of-centers of the atoms that is labeled C_∞ because a rotation about this axis by any angle, which is some multiple of 360°/inf, will produce an equivalent structure (Fig.2.2).

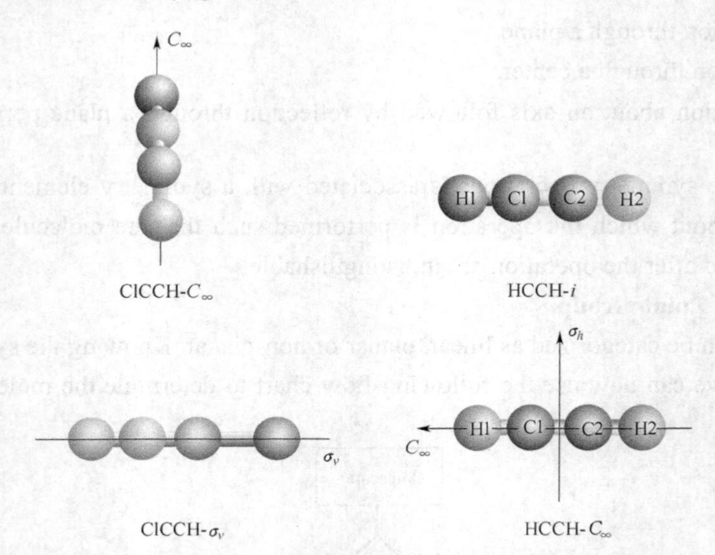

Fig.2.2 Proper and improper rotation axis of the linear molecules

- In addition to the above symmetry elements, a linear molecule may possess a center of symmetry, *i*. If the molecule has a center of symmetry, it will also possess an infinite number of C_2 axes passing through the center and perpendicular to the principal axis (C_∞);

An improper axis, S_∞, corresponding to rotation about the C_∞ axis followed by reflection in a plane passing through the center of symmetry and perpendicular to the C_∞ axis.

(2) Planar molecules These constitute an important class of molecules characterized by having all of the atoms lying in a common plane that will be a symmetry element. We will consider the following four molecules: formaldehyde, *t*-glyoxal, ethane and benzene (Fig.2.3).

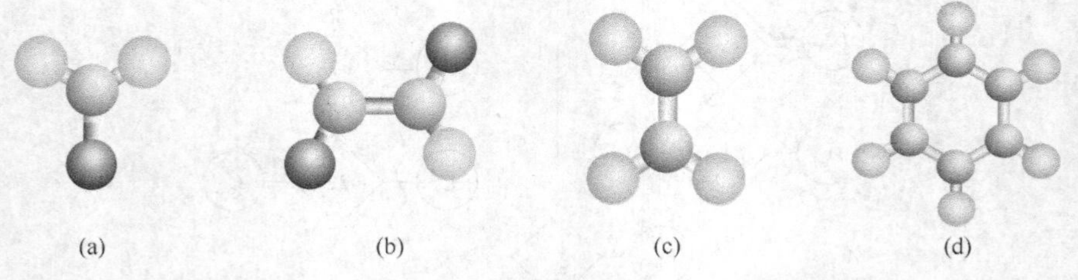

Fig.2.3 Molecular structure modes of the (a)formaldehyde; (b)*t*-glyoxal; (c) ethene; (d)benzene

- Formaldehyde It should be clear that formaldehyde has a 2-fold rotation axis, C_2 passing through the carbon and oxygen atoms and bisecting the HCH angle. This molecule also has a plane of symmetry corresponding to the molecular plane, and containing the C_2 axis, i.e. a σ_v plane. There is, in addition, another plane of symmetry, perpendicular to the molecular plane, bisecting the HCH angle, and containing the C_2 axis, i.e. also a σ_v plane. Formaldehyde has no center of symmetry and no improper axes. So, formaldehyde's symmetry elements are one C_2 axis and two σ_v planes.

- *Trans*-glyoxal This molecule possesses a center of symmetry, i. It also has a C_2 axis which must pass through the center of symmetry, and which is perpendicular to the molecular plane. Since the molecular plane is perpendicular to the C_2 axis(the principal axis)it constitutes a σ_h(horizontal)plane. *Trans*-glyoxal has no improper axes. *Trans*-glyoxal's symmetry elements are one C_2 axis;one σ_h plane and one i, center of inversion.
- Ethene This molecule also has a center of symmetry, i. It has three mutually perpendicular C_2 that must pass through the center. Associated with the three C_2 axes are three mutually perpendicular σ_h planes(each perpendicular to one of the C_2 axes). Ethene has no improper axes. Ethene's symmetry elements are three C_2 axes, three σ_h planes and one i, center of inversion.
- Benzene Benzene has a center of symmetry, i. It also has a rotational axis passing through the center and perpendicular to the molecular plane. This axis in fact combines a C_6, a C_3, and a C_2 axis. The molecule has two sets of C_2 axes in the molecular plane:

(a) $3C_2'$ axes passing through the center and containing two C—H bonds.

(b) $3C_2''$ axes passing through the center and through the center of pairs of bonds on opposite sides of the hexagon. Benzene has a plane corresponding to the molecular plane that is perpendicular to the C_6 principal axis of rotation, i.e. a σ_h plane. It also has three σ_v planes associated with the three C_2' axes and containing the principal axis (C_6) and the center (i). In addition there are three σ_d planes associated with the three C_2'' and containing the principal axis (C_6) and the center (i). Finally, because there is a C_6, C_3 and a C_2 axis perpendicular to the molecular plane, there also exists an S_6, S_3, and an S_2 axis but the latter is redundant since it is equivalent to the already defined center of inversion, i. Benzene's symmetry elements are: one i, one C_6, one C_3, one C_2 , three C_2', three C_2'', one σ_h, three σ_v, three σ_d, one S_6 and one S_3.

(3) Non-planar molecules This simply means that the atoms do not all lie in one plane. The molecule is "three-dimensional". We will consider the examples: chloromethane, methane and allene (Fig.2.4).

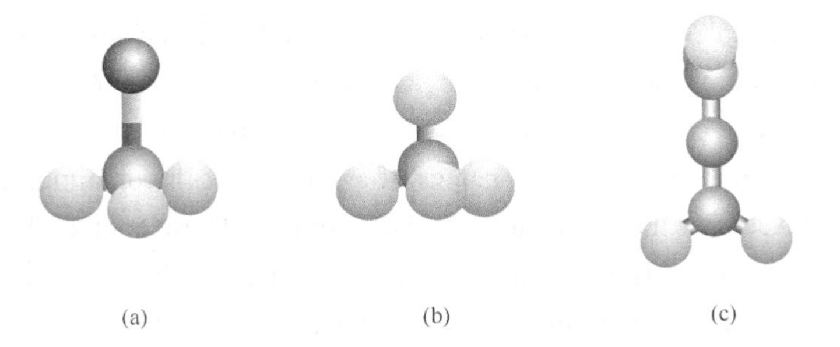

Fig.2.4 Molecular structure modes of (a) Chloromethane; (b) Methane and (c) Allene

- Chloromethane In this molecule the three hydrogen atoms are symmetrically arranged about a line drawn though the line-of-centers of the carbon and chlorine atoms, a C_3 axis. There are also three (σ_v) planes containing the chlorine, the carbon, and one of the hydrogen. These planes also contain the C_3 axis. Chloromethane's symmetry elements are one C_3 axis of rotation and three σ_v planes of reflection containing the C_3 axis.

- Methane In methane, we can draw a line through the line-of-centers of the carbon and any hydrogen atom and the remaining three hydrogen atoms will be symmetrically arranged about that line—a C_3 axis. There will therefore be four C_3 axes. The plane containing the carbon and any two hydrogen atoms will bisect the HCH angle involving the other two hydrogen atoms. This gives us the following planes: CH1H2, CH1H3, CH1H4, CH2H3, CH2H4, CH3H4. Each of these planes contains the C_3 axis and is a σ_d plane [Fig.2.5(a)]. If the methane is drawn within a cube with the hydrogen atoms at the vertices as shown in [Fig.2.5(b)]. We can identify three mutually perpendicular C_2 axes.

Using this same model, we see that rotation perpendicular to any face by 90° followed by reflection in a plane perpendicular to the rotation axis results in equivalent structures i.e. there are three S_4 improper axes. Note that neither the rotation axis nor the reflection plane is, in themselves, symmetry elements of the molecule. Methane's symmetry elements are four C_3 axes of rotation, three C_2 axes of rotation, three S_4 improper axes of rotation and six σ_d reflection plane.

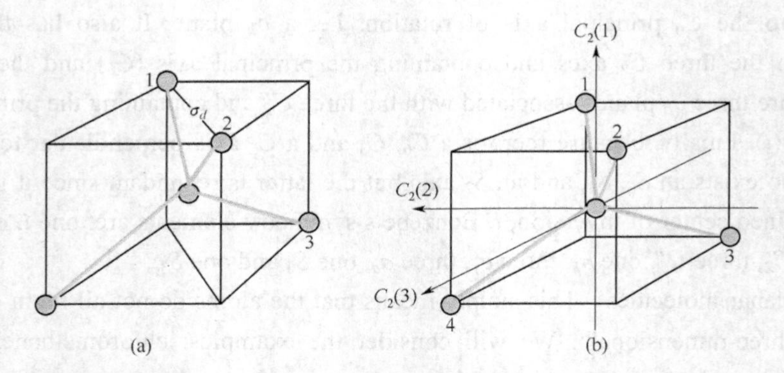

Fig.2.5 The illustrations of the symmetry elements in the methane molecule

- Allene This molecule can be most readily studied by inscribing it within a rectangle in which the hydrogen atoms are placed at the vertices. There is a C_2 axis lying along the line-of-centers of the three carbons. There are two C_2' axes perpendicular to the C_2 axis and passing through the central carbon. There are two planes of symmetry each containing one of the pairs of hydrogen atoms and bisecting the angle between the other pair. These contain the principal axis, C_2, and are σ_d planes. Finally there is an S_4 improper axis coincident with the C_2 axis with its reflection plane perpendicular to the C_2 axis and passing through the central carbon. Allene's symmetry elements are one C_2 axis of rotation, two C_2' axes of rotation perpendicular to the C_2, two σ_d planes of symmetry and one S_4 improper axis coincident with the C_2. Some important molecular point groups are shown in Table 2.1.

Table 2.1 Some important molecular point groups

Nonaxial groups	C_1	C_s	C_i	—	—	—	—
C_n groups	C_2	C_3	C_4	C_5	C_6	C_7	C_8
D_n groups	D_2	D_3	D_4	D_5	D_6	D_7	D_8
C_{nv} groups	C_{2v}	C_{3v}	C_{4v}	C_{5v}	C_{6v}	C_{7v}	C_{8v}
C_{nh} groups	C_{2h}	C_{3h}	C_{4h}	C_{5h}	C_{6h}	—	—
D_{nh} groups	D_{2h}	D_{3h}	D_{4h}	D_{5h}	D_{6h}	D_{7h}	D_{8h}
D_{nd} groups	D_{2d}	D_{3d}	D_{4d}	D_{5d}	D_{6d}	D_{7d}	D_{8d}
S_n groups	S_2	S_4	S_6	S_8	S_{10}	S_{12}	—
Cubic groups	T	T_h	T_d	O	O_h	I	I_h
Linear groups	$C_{\infty v}$	$D_{\infty h}$	—	—	—	—	—

2.1.4 Character Tables

All the character tables are laid out in the same way.

- The top row and first column consist of the symmetry operations and irreducible representations, respectively.
- The table elements are the characters.
- The final two columns show the first and second order combinations of Cartesian coordinates.

It often occurs for a point group that there are non-equivalent operations of the same type. For example there are three C_2 operations in the D_{2d} point group, two of which are not equivalent to the third. In such cases the different operations may be distinguished with a "prime" or by indicating some Cartesian reference (such as the x, y, and z related C_2 operations in D_2) (Table 2.2, Table 2.3).

Table 2.2 Character table for point group C_2

C_2	E	C_2	linear functions, rotations	Quadratic functions	Cubic functions
A	+1	+1	z, R_z	x^2, y^2, z^2, xy	z^3, xyz, y^2z, x^2z
B	+1	−1	x, y, R_x, R_y	yz, xz	$xz^2, yz^2, x^2y, xy^2, x^3, y^3$

Table 2.3 Character table for point group D_2

D_2	E	$C_2(z)$	$C_2(y)$	$C_2(x)$	linear functions, rotations	Quadratic functions	Cubic functions
A	+1	+1	+1	+1	—	x^2, y^2, z^2	xyz
B_1	+1	+1	−1	−1	z, R_z	xy	z^3, y^2z, x^2z
B_2	+1	−1	+1	−1	y, R_y	xz	yz^2, x^2y, y^3
B_3	+1	−1	−1	+1	x, R_x	yz	xz^2, xy^2, x^3

2.2 Valence Bond Theory

Mixing a number of atomic orbitals to form the same number of hybrid orbitals to explain chemical bonding and shapes and structures of molecules is a rather recent myth. The most significant development in the first half of the 20th century is the human's ability to understand the structure of atoms and molecules. Computation has made mathematical concepts visible to the extent that we now can see the atomic and molecular orbitals. On the other hand, Using everyday encountered materials can also generate beautiful illustrations of hybrid atomic orbitals. The several parts are included in this section as follow:

- Describe the valence bond (VB) approach to chemical bonding.
- Demonstrate hybridization of atomic orbitals for VB.
- Correlate the molecular shape to the hybrid atomic orbitals of some central atoms.
- Combine the concepts of hybrid orbitals, valence bond theory, VSEPR, resonance structures, and octet rule to describe the shapes and structures of some common molecules.

The valence-bond approach considers the overlap of the atomic orbitals (AO) of the participation atoms to form a chemical bond. Due to the overlapping, electrons are localized in the bond region. The overlapping AOs can be of different types, for example, a sigma bond may be formed by the overlapping the following AOs (Table 2.4).

Table 2.4 Chemical bonds formed due to overlap of atomic orbitals

s-s	s-p	s-d	p-p	p-d	d-d
H—H	H—C	H—Pd in	C—C	F—S	Fe—Fe
Li—H	H—N	palladium	P—P	in SF_6	
	H—F	hydride	S—S		

However, the atomic orbitals for bonding may not be"pure"atomic orbitals directly from the solution of the Schrödinger Equation. Often, the bonding atomic orbitals have a character of several possible types of orbitals. The method to get an AO with the proper character for the bonding is called hybridization. The resulting atomic orbitals are called hybridized atomic orbitals or simply hybrid orbitals. We shall look at the shapes of some hybrid orbitals first, because these shapes determine the shapes of the molecules.

2.2.1 Hybridization of Atomic Orbitals

The solution to the Schrodinger Equation provides the wavefunctions for the following atomic orbitals:

1s, 2s, 2p, 3s, 3p, 3d, 4s, 4p, 4d, 4f, etc.

For atoms containing two or more electrons, the energy levels are shifted with respect to those of the H atom. An atomic orbital is really the energy state of an electron bound to an atomic nucleus. The energy state changes when one atom is bonded to another atom.

Quantum mechanical approaches by combining the wave functions to give new wavefunctions are called hybridization of atomic orbitals. Hybridization has a sound mathematical fundation, but it is a little too complicated to show the details here. Therefore, a set of atomic orbitals to a new set of hybrid atomic orbitals or hybrid orbitals are shown only in the text. At this level, we consider the following hybrid orbitals:

$$sp;\ sp^2;\ sp^3;\ sp^3d;\ sp^3d^2.$$

(1) sp hybrid orbitals The sp hybrid atomic orbitals are possible states of electron in an atom, especially when it is bonded to others. These electron states have half 2s and half 2p characters. From a mathematical viewpoint, there are two ways to combine the 2s and 2p atomic orbitals:

$$sp_1 = 2s + 2p;\ sp_2 = 2s - 2p$$

These energy states (sp_1 and sp_2) have a region of high electron probability each, and the two atomic orbitals are located opposite to each other, centered on the atom. The sp hybrid orbitals are represented by Fig.2.6.

For example, the molecule H—Be—H is formed due to the overlapping of two 1s orbitals of two H atoms and the two sp hybridized orbitals of Be. Thus, the H—Be—H molecule is linear.

The ground state electronic configuration of Be is $1s^2 2s^2$, and one may think of the electronic configuration "before" bonding as $1s^2[sp]^2$. The two electrons in the sp hybrid orbitals have the same energy. Probably, the concept of hybridizing AOs for the bonding is just a story made up to explain the molecular shape of Cl—Be—Cl. In general, when two and only two atoms bond to a third atom and the third atom makes use of the sp hybridized orbitals, the three atoms are on a straight line.

(2) sp^2 hybrid orbitals The energy states of the valence electrons in atoms of the second period are in the 2s and 2p orbitals. If we mix two of the 2p orbitals with a 2s orbital, we end up with three sp^2 hybridized orbitals. These three orbitals lie on a plane, as shown in Fig.2.7, they point to the vertices of a equilateral triangle as shown here.

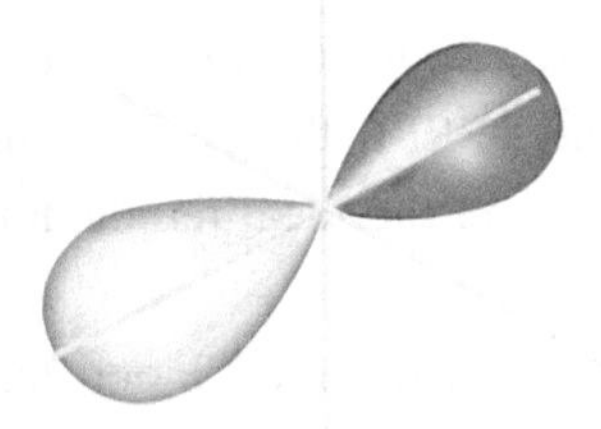

Fig.2.6 The representation of the sp hybrid orbitals

Fig.2.7 The representation of the sp^2 hybrid orbitals

When the central atom makes use of sp^2 hybridized orbitals, the compound so formed has a trigonal shape. BF_3 is such a molecule (Table 2.5).

Table 2.5 Molecules with sp^2 hybrid orbitals

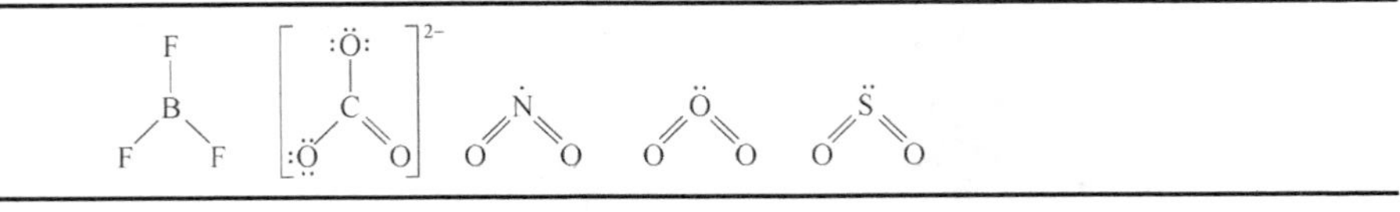

Not all three sp^2 hybridized orbitals have to be used in bonding. One of the orbitals may be occupied by a pair or a single electron. If we do not count the unshared electrons, these molecules are bent, rather than linear. The three molecules shown together with the BF_3 molecule are such molecules. Carbon atoms also makes use of the sp^2 hybrid orbitals in the compound $H_2C{=}CH_2$. In this molecule, the remaining p orbital from each of the carbon overlap to form the additional π-π bond. Other ions such as CO_3^{2-}, and NO_3^-, can also be explained in the same way (Table 2.6).

(3) sp^3 hybrid orbitals Mixing one s and all three p atomic orbitals produces a set of four equivalent sp^3 hybrid atomic orbitals. The four sp^3 hybrid orbitals points towards the vertices of a tetrahedron, as shown in Fig.2.8.

Table 2.6 Planar molecules with sp^2 hybrid orbitals

H H \ / C=C / \ H H	$[O_2C{=}O]^{2-}$ (O, O, C=O)	[O, O, N=O]

When sp^3 hybrid orbitals are used for the central atom in the formation of molecule, the molecule is said to have the shape of a tetrahedron. The typical molecule is CH_4 , in which the 1s orbital of a H atom overlap with one of the sp^3 hybrid orbitals to form a C—H bond. Four H atoms form four such bonds, and they are all equivalent. The CH_4 molecule is the most cited molecule to have a tetrahedral shape. Other molecules and ions having tetrahedral shapes are SiO_4^{4-} , SO_4^{2-} .

As in the cases with sp^2 hybrid orbitals, one or two of the sp^3 hybrid orbitals may be occupied by non-bonding electrons. Water and ammonia are such molecules. The C, N and O atoms in CH_4, NH_3, H_2O molecules use the sp^3 hybrid orbtals, however, a lone pair occupy one of the orbitals in NH_3, and two lone pairs occupy two of the sp^3 hybrid orbitals in H_2O. The lone pairs must be considered in the VSEPR model, and we can represent a lone pair by E, and two lone pairs by E_2. Thus, we have NH_3E and OH_2E_2 respectively.

The VSEPR number is equal to the number of bonds plus the number of lone pair electrons. It does not matter what is the order of the bond, any bonded pair is considered on bond. Thus, the VSEPR number is 4 for all of CH_4, $:NH_3$, $:\ddot{O}H_2$. According the VSEPR theory, the lone electron pairs require more space, and the H—O—H angle is 105°, less than the ideal tetrahedral angle of 109.5°.

(4) dsp^3 hybrid orbitals The five dsp^3 hybrid orbitals resulted when one 3d, one 4s, and three 4p atomic orbitals are mixed. When an atom makes use of five dsp^3 hybrid orbitals to bond to five other atoms, the geometry of the molecule is often a trigonal-bipyramidal. For example, the molecule $PClF_4$ displayed (Fig.2.9) forms such a structure. In this diagram, the Cl atom takes up an axial position of the trigonal-bipyramid. There are structures in which the Cl atom may take up the equatorial position. The change in arrangement is accomplished by simply change the bond angles.

Fig.2.8 The representation of the sp^3 hybrid orbitals

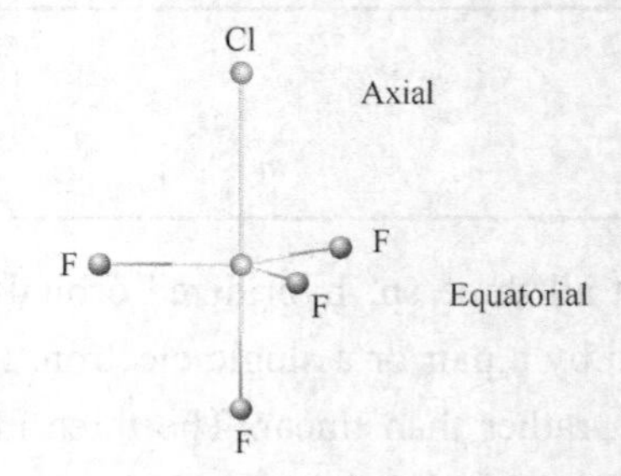

Fig.2.9 The representation of the dsp^3 hybrid orbitals based on the $PClF_4$ molecule

Some of the dsp^3 hybrid orbitals may be occupied by electron pairs. The shapes of these molecules are interesting. In $TeCl_4$, only one of the hybrid dsp^3 orbitals is occupied by a lone pair of electrons. This kind of structure may be represented by $TeCl_4E$, where E represents a lone pair of electrons. Two lone pairs occupy two such orbitals in the molecule BrF_3, or BrF_3E_2 . The compound SF_4 is another AX_4E type, and many interhalogen compounds ClF_3 and IF_3 are AX_3E_2 type. The ion I_3^- is of the type AX_2E_3 (Fig.2.10).

(5) d^2sp^3 hybrid orbitals The six d^2sp^3 hybrid orbitals resulted when two 3d, one 4s, and three 4p atomic orbitals are mixed. When an atom makes use of six d^2sp^3 hybrid orbitals to bond to six other atoms, the molecule takes the shape of an octahedron, in terms of molecular geometry (Fig.2.11). The gas compound SF_6 is a typical such structure.

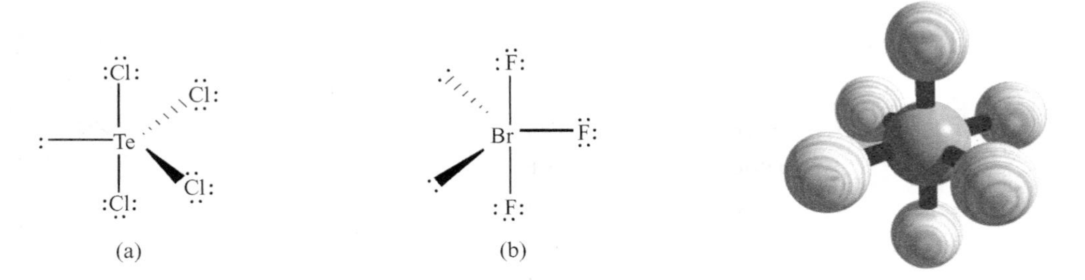

Fig.2.10 The molecular structure mode of (a) $TeCl_4E$ molecule and (b) BrF_3E_2 molecule (E represents a lone pair of electrons)

Fig.2.11 The representation of the d^2sp^3 hybrid orbitals

There are also cases that some of the d^2sp^3 hybrid orbitals are occupied by lone pair electrons leading to the structures of the following types:

AX_6, AX_5E, AX_4E_2, AX_3E_3, AX_2E_4, IOF_5, IF_5E, XeF_4E_2.

2.2.2 Molecular Shapes

While the hybridized orbitals were introduced, in the foregoing discussion, Valence-shell electron-pair repulsion (VSEPR) model was included to suggest the shapes of various molecules. Specifically, the VSEPR model counts unshared electron pairs and the bonded atoms as the VSEPR number. A single-, double- and triple-bond are considered as 1. After having considered the hybridized orbitals and the VSEPR model, we cannot take a systematic approach to rationalize the shapes of many molecules based on the number of valence electrons.

A summary in the form of a table is given in Table 2.7 to account for the concepts of hybrid orbitals, valence bond theory, VSEPR, resonance structures, and octet rule. In this table, the geometric shapes of the molecules are described by linear, trigonal planar, tetrahedral, trigonal bypyramidal, and octahedral. The hybrid orbitals used are sp, sp^2, sp^3, dsp^3, and d^2sp^3. The VSEPR number is the same for all molecules of each group. Instead of using NH_3E, and OH_2E_2, we use $:NH_3$, $:\ddot{O}H_2$ to emphasize the unshared (or lone) electron pairs.

Only Be and C atoms are involved in linear molecules. In gas phase, BeH_2 and BeF_2 are stable, and these molecules do not satisfy the octet rule. The element C makes use of sp hybridized orbitals and it has the ability to form double and triple bonds in these linear molecules. Carbon compounds are present in trigonal planar and tetrahedral molecules, using different hybrid orbitals. The extra electron in nitrogen for its compounds in these groups appears as lone unpaired electron or lone electron pairs. More electrons in O and S lead to compounds with lone electron pairs. The five-atom anions are tetrahedral, and many resonance structures can be written for them.

Table 2.7 A summary of hybrid orbitals, valence bond theory, VSEPR, resonance structures, and octet rule

Linear	Trigonal planar	Tetrahedral	Trigonal bipyramidal	Octahedral
sp	sp^2	sp^3	dsp^3	d^2sp^3
BeH_2	BH_3	CH_4	PF_5	SF_6
BeF_2	BF_3	CF_4	PCl_5	IOF_5
CO_2	CH_2O	CCl_4	$PFCl_4$	PF_6^-
HCN	(>C=O)	CH_3Cl	:SF_4	SiF_6^{2-}
HC≡CH	>C=C<	NH_4^+	:TeF_4	:BrF_5
	CO_3^{2-}	:NH_3	::ClF_3	:IF_5
	benzene	:PF_3	::BrF_3	::XeF_4
	graphite	:SOF_2	:::XeF_2	
	fullerenes	::OH_2	::: I_3^-	
	·NO_2	::SF_2	(:::I I_2^-)	
	N_3^-		::: ICl_2^-	
	:$OO_2(O_3)$	SiO_4^{4-}		
	:SO_2	PO_4^{3-}		
	SO_3	SO_4^{2-}		
		ClO_4^-		

· a lone odd electron   : a lone electron pair

Trigonal bipyramidal and octahedral molecules have 5 and 6 VSEPR pairs. When the central atoms contain more than 5 or 6 electrons, the extra electrons form lone pairs. The number of lone pairs can easily be derived using Lewis dot structures for the valence electrons.

In describing the shapes of these molecules, we often ignore the lone pairs. Thus, ·NO_2, N_3^-, :OO_2 (O_3), and :SO_2 are bent molecules whereas :NH_3, :PF_3, and :SOF_2 are pyramidal. The lone electron pair takes up the equatorial location in : SF_4, which has the same structure as :TeF_4 described earlier. If you lay a model of this molecule on the side, it looks like a butterfly. By the same reason, :$\ddot{\mathrm{Cl}}\mathrm{F}_3$ and :$\ddot{\mathrm{B}}\mathrm{rF}_3$ have a T shape, and :$\underset{..}{\ddot{\mathrm{X}}}\mathrm{eF}_2$, :$\underset{..}{\ddot{\mathrm{I}}}\,\mathrm{I}_2^-$, and :$\underset{..}{\ddot{\mathrm{I}}}\,\mathrm{Cl}_2^-$ are linear. Similarly, :BrF_5 and :IF_5 are square pyramidal whereas :$\ddot{\mathrm{X}}\mathrm{eF}_4$ is square planar.

A brilliant question is" Which atom in the formula is usually the center atom? "

Usually, the atom in the center is more electropositive than the terminal atoms. However, the H and halogen atoms are usually at the terminal positions because they form only one bond. However, the application of VSEPR theory can be expanded to complicated molecules. By applying the VSEPR theory, one deduces the following results:

- H—C—C bond angle=109°.
- H—C═C bond angle=120°, geometry around C trigonal planar.
- C═C═C bond angle=180°, in other words linear.
- H—N—C bond angle=109°, tetrahedral around N.
- C—O—H bond angle=105° or 109°, 2 lone electron pairs around O.

2.3 Crystal Field Theory

This theory largely replaced VB theory for interpreting the chemistry of coordination compounds. It was proposed by the physicist Hans Bethe in 1929. Subsequent modifications were proposed by J. H. Van Vleck in 1935 to allow for some covalency in the interactions. These modifications are often referred to as ligand field theory (LFT). Pure crystal field theory (CFT) assumes that the interactions between the metal ion and the ligands are purely electrostatic (ionic). The ligands are regarded as point charges. Although somewhat unrealistic, it uses symmetry considerations that are valid for ligand field theory as well as MO theory. It is of outmost importance to draw the d orbitals correctly in order to get an idea of which of these orbitals will interact with the ligands (point charges).

The five d orbitals in an isolated gaseous metal are degenerate. If a spherically symmetric field of negative charges is placed around the metal, these orbitals remain degenerate, but all of them are raised in energy as a result of the repulsion between the negative charges on the ligands and in the d orbitals. If rather than a spherical field, discrete point charges are allowed to interact with the metal ion, the degeneracy of the d orbitals is removed.

The splitting of d orbital energies and its consequences are at the heart of crystal field theory.

2.3.1 CFT for Octahedral Geometry

On going from a spherical to an octahedral symmetry, all d orbitals are raised in energy, relative to the free ion. However, not all d orbitals will interact to the same extent with the six point charges located on the $+x$, $-x$, $+y$, $-y$, $+z$ and $-z$ axes respectively. The orbitals which lie along these axes (i.e. x^2-y^2, z^2) will be destabilized more than the orbitals that lie in-between the axes (i.e. xy, xz, yz). Referring to the chara- cter table for the O_h point group reveals that the x^2-y^2, z^2 orbitals belong to the E_g irreducible representation and xy, xz, yz belong to the T_{2g} irreducible representation. The extent to which these two sets of orbitals are split (the e_g and the t_{2g} use lower case for orbitals) is denoted by Δ_o or alternatively $10D_q$ (Fig.2.12). As the barycenter must be conserved on going from a spherical field to an octahedral field, the t_{2g} set must be stabilized as much as the e_g set is destabilized.

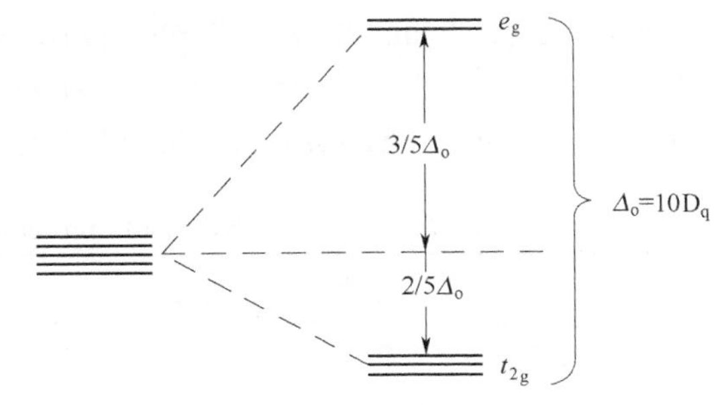

Fig.2.12 The representation of splitting of two sets of orbitals—t_{2g} and e_g, in octahedral geometry

For example, $[Ti(H_2O)_6]^{3+}$. This is a d^1 complex and the electron occupies the lowest energy orbital available, i.e. one of the three degenerate t_{2g} orbitals. The purple colour is the result of the absorption of light that results in the promotion of this t_{2g} electron into the e_g level: $t_{2g}^1e_g^0 \rightarrow t_{2g}^0e_g^1$. The UV-Vis absorption spectrum reveals that this transition occurs with a maximum at 20300 cm^{-1} that corresponds to Δ_o=243kJ/mol (Fig.2.13). Typical Δ_o values are of the same order of magnitude as the energy of a chemical bond (1000cm^{-1} =11. 96kJ/mol or 2.86kcal/mol or 0. 124eV).

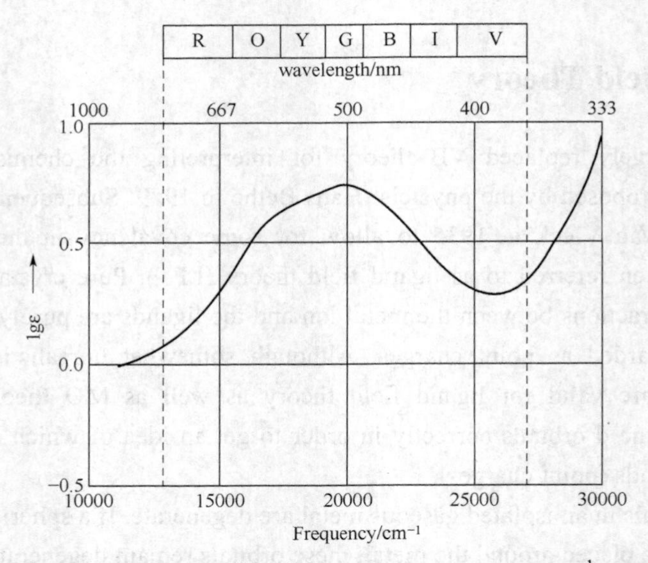

Fig.2.13 The UV-Vis absorption spectrum of transition metal complex with d^1 electronic configuration

What happens when more than one electron in d orbitals?

- **For d^2-d^9 systems the electron-electron interactions must be taken into account.**
- **For d^1-d^3 systems Hund's rule predicts that the electrons will not pair and occupy the t_{2g} set.**
- **For d^4-d^7 systems There are two possibilities: either put the electrons in the t_{2g} set and therefore pair the electrons. This is the so-called low spin case or strong field situation. Or put the electrons in the e_g set, which lies higher in energy, but the electrons do not pair. This is the so-called high spin case or weak field situation. Therefore, there are two important parameters to consider: the pairing energy (P) and the e_g -t_{2g} splitting (referred to as Δ_o, $10D_q$).**

For both the high spin and low spin situations, it is possible to compute the crystal field stabilization energy (CFSE) as a function of electron count and spin state (Table 2.8).

Table 2.8 CFSE as a function of electron count and spin state

	Weak field			Strong field		
d^n	Configuration	Unparied electrons	CFSE	Configuration	Unpaired electrons	CFSE
d^1	t_{2g}^1	1	$0.4\Delta_o$	t_{2g}^1	1	$0.4\Delta_o$
d^2	t_{2g}^2	2	$0.8\Delta_o$	t_{2g}^2	2	$0.8\Delta_o$
d^3	t_{2g}^3	3	$1.2\Delta_o$	t_{2g}^3	3	$1.2\Delta_o$
d^4	$t_{2g}^3e_g^1$	4	$0.6\Delta_o$	t_{2g}^4	2	$1.6\Delta_o$
d^5	$t_{2g}^3e_g^2$	5	$0.0\Delta_o$	t_{2g}^5	1	$2.0\Delta_o$
d^6	$t_{2g}^4e_g^2$	4	$0.4\Delta_o$	t_{2g}^6	0	$2.4\Delta_o$
d^7	$t_{2g}^5e_g^2$	3	$0.8\Delta_o$	$t_{2g}^6e_g^1$	1	$1.8\Delta_o$
d^8	$t_{2g}^6e_g^2$	2	$1.2\Delta_o$	$t_{2g}^6e_g^2$	2	$1.2\Delta_o$
d^9	$t_{2g}^6e_g^3$	1	$0.6\Delta_o$	$t_{2g}^6e_g^3$	1	$0.6\Delta_o$
d^{10}	$t_{2g}^6e_g^4$	0	$0.0\Delta_o$	$t_{2g}^6e_g^4$	0	$0.0\Delta_o$

Note: This table is somewhat simplified because pairing energies and electron-electron effects have been neglected.

The electron pairing energy (P) is composed of two terms (Table 2.9):

Table 2.9 Pairing Energy as a Function of Metal and Electron Count

d^n	Ion	P_{coul}	P_{ex}	P_T
d^4	Cr^{2+}	71.2(5950)	171.3(14475)	244.3(20425)
	Mn^{3+}	87.9(7350)	213.7(17865)	301.6(25215)
d^5	Cr^+	67.3(5625)	144.3(12062)	211.6(17687)
	Mn^{2+}	91.0(7610)	194.0(16215)	285.0(23825)
	Fe^{3+}	120.2(10050)	237.1(19825)	357.4(29875)
d^6	Mn^+	73.5(6145)	100.6(8418)	174.2(14563)
	Fe^{2+}	89.2(7460)	139.8(11690)	229.1(19150)
	Co^{3+}	113.0(9450)	169.6(14175)	282.6(23625)
d^7	Fe^+	87.9(7350)	123.6(10330)	211.5(17680)
	Co^{2+}	100(8400)	150(12400)	250(20800)

- The Coulombic repulsion This repulsion must be overcome when forcing electrons to occupy the same orbital. As 5d orbitals are more diffuse than 4d orbitals that are more diffuse than 3d orbitals, the pairing energy becomes smaller as one goes down a period. As a rule, 4d and 5d transition metals are always low spin.
- The loss of exchange energy The exchange energy that is known as Hund's rule is proportional to the number of electrons having parallel spins. The greater this number, the more difficult it is to pair electrons. Therefore, the d^5 (Fe^{3+}, Mn^{2+}) configuration is the most enclined to form high spin complexes.

2.3.2 CFT for Tetrahedral Geometry

The tetrahedral symmetry can be derived from a cubic symmetry where only four of the eight corners are occupied by point charges. In such a situation, it is the xy, yz, xz orbitals that are destabilized as they point towards the incoming point charges. The x^2-y^2 and z^2 are stabilized so that the barycenter is preserved. Please note that the irreducible representations are t_2 and e in the T_d point group. The crystal field splitting in a tetrahedral symmetry is intrinsically smaller than in the octahedral symmetry as there are only four ligands (instead of six ligands in the octahedral symmetry)interacting with the transition metal ion (Fig.2.14).

The point charge model predicts that Δ_t=4/9Δ_o. As a result, low spin configurations are rarely observed. Usually, if a strong field ligand is present, the square planar geometry will be favoured.

2.3.3 CFT for Square Planar Geometry

In order to derive the square planar geometry, let us consider a tetragonal distortion of the octahedral field. Removing both point charges from the z axis, stabilizes the z^2 orbital of the e_g set. As the bonding of the remaining four point charges increases, the x^2-y^2 orbitals rises in energy. Again here, as the symmetry is lowered, the degeneracy is lifted and the orbitals have different irreducible representations.

As shown in the Fig.2.15, an octahedral complex (a) undergoing z axis elongation such that it becomes tetragonally distorted (b) and finally reaches the square planar limit(c). The d_z^2 orbital may lie below the d_{xz} and d_{zy} orbitals in the square planar complex.

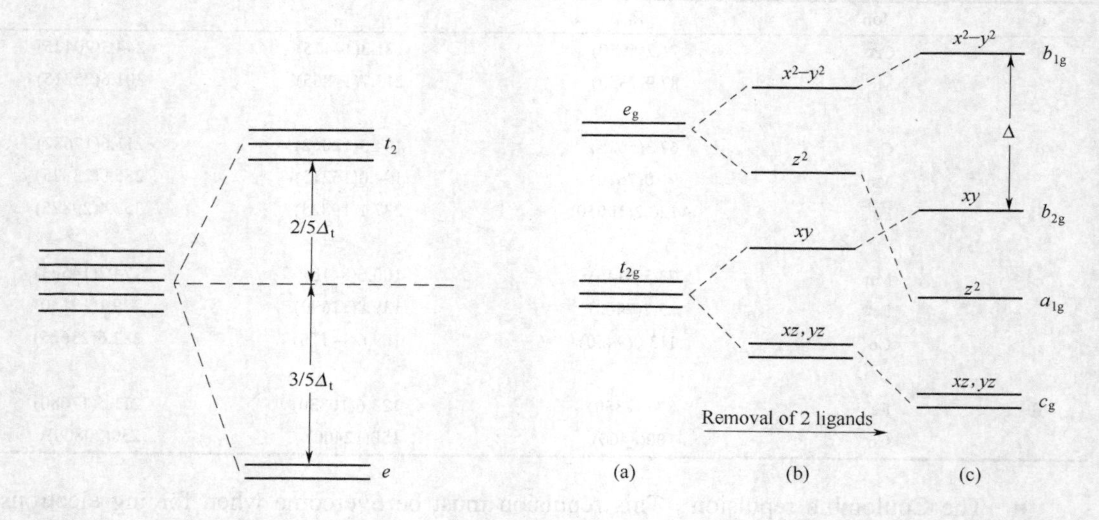

Fig.2.14 The representation of splitting of two sets of orbitals—*e* and t_2, in tetrahedral geometry

Fig.2.15 The representation of splitting of orbitals in square planar geometry

2.3.4 Factors Influencing the Magnitude of Δ

- Oxidation state of the metal: The higher the charge of the metal, the greater the Δ. As a rule of thumb, Δ increases by ca. 50% when the oxidation state increase by one unit.
- Nature of the metal ion: Δ increases on going from 3d<4d<5d. (ca. 50% going from Co to Rh. ca. 25% going from Rh to Ir.)
- Number and geometry of the ligands: Δ_o is ca. 50% larger than Δ_t.
- Nature of the ligands: Based on similar data for a wide variety of complexes, it is possible to list ligands in order of increasing field strength in the so-called spectrochemical series. It is

$$I^- < Br^- < S^{2-} < SCN^- < Cl^- < N_3^-, F^- < \text{urea}, OH^- < \text{ox}, O^{2-} < H_2O < NCS^- < \text{py}, NH_3 < \text{en} < \text{bpy}, \text{phen} < NO_2^- < CH_3^-, C_6H_5^- < CN^- < CO.$$

It is possible to arrange the metals according to a spectrochemical series as well. The approximate order is

$$Mn^{2+} < Ni^{2+} < Co^{2+} < Fe^{2+} < V^{2+} < Fe^{3+} < Co^{3+} < Mn^{3+} < Mo^{3+} < Rh^{3+} < Ru^{3+} < Pd^{4+} < Ir^{3+} < Pt^{4+}$$

Jørgensen has proposed a formula that allows to estimate the Δ

$$\Delta = fg$$

Where, f is a function of the ligand and g is a function of the metal (Table 2.10).

2.3.5 Applications of CFT

(1) Ionic radii For a given oxidation state, the ionic radius decreases steadily on going from left to right in a transition series. Populating antiboding orbitals (i.e. filling the e_g levels in an octahedron) leads to an increase in ionic radius. Therefore, the ionic radius depends on the spin state of the metal (i.e. high spin or low spin) (Fig.2.16).

Table 2.10 Estimation of the Δ based on the function of the ligand and metal

Complex	Oxidotion state of metal	Symmetry	Δ/cm^{-1}	Complex	Oxidotion state of metal	Symmetry	Δ/cm^{-1}
$[VCl_6]^{2-}$	4	O_h	15400	$[Ru(ox)_3]^{3-}$	3	O_h	28700
VCl_4	4	T_d	7900	$[Ru(H_2O)_6]^{2+}$	2	O_h	19800
$[CrF_6]^{2-}$	4	O_h	22000	$[Ru(CN)_6]^{4-}$	2	O_h	33800
$[CrF_6]^{3-}$	3	O_h	15060	$[CoF_6]^{2-}$	4	O_h	20300
$[Cr(H_2O)_6]^{3+}$	3	O_h	17400	$[CoF_6]^{3-}$	3	O_h	13100
$[Cr(en)_3]^{3+}$	3	O_h	22300	$[Co(H_2O)_6]^{3+}$	3	O_h	20760
$[Cr(CN)_6]^{3-}$	3	O_h	26600	$[Co(NH_3)_6]^{3+}$	3	O_h	22870
$[Mo(H_2O)_6]^{3+}$	3	O_h	26000	$[Co(en)_3]^{3+}$	3	O_h	23160
$[MnF_6]^{2-}$	4	O_h	21800	$[Co(H_2O)_6]^{2+}$	2	O_h	9200
$[TcF_6]^{2-}$	4	O_h	28400	$[Co(NH_3)_6]^{3+}$	2	O_h	10200
$[ReF_6]^{2-}$	4	O_h	32800	$[Co(NH_3)_4]^{2+}$	2	T_d	5900
$[Fe(H_2O)_6]^{3+}$	3	O_h	14000	$[RhF_6]^{+}$	4	O_h	20500
$[Fe(ox)_3]^{3+}$	3	O_h	14140	$[Rh(H_2O)_6]^{3+}$	3	O_h	27200
$[Fe(CN)_6]^{3-}$	3	O_h	35000	$[Rh(NH_3)_6]^{3+}$	3	O_h	34100
$[Fe(CN)_6]^{4-}$	2	O_h	32200	$[IrF_6]^{2+}$	4	O_h	27000
$[Ru(H_2O)_6]^{3+}$	3	O_h	28600	$[Ir(NH_3)_6]^{3+}$	3	O_h	41200

Ligand	f factor	Metal ion	g factor	Ligand	f factor	Metal ion	g factor
Br^-	0.72	Mn(Ⅱ)	8.0	NCS^-	1.02	Ru(Ⅱ)	20
SCN^-	0.73	Ni(Ⅱ)	8.7	$gly^-=NH_3CH_2CO_2^-$	1.18	Mn(Ⅳ)	23
Cl^-	0.78	Co(Ⅱ)	9	$py=C_5H_5N$	1.23	Mo(Ⅲ)	24.6
N_3^-	0.83	V(Ⅱ)	12.0	NH_3	1.25	Rh(Ⅲ)	27.0
F^-	0.9	Fe(Ⅲ)	14.0	$en=NH_2CH_2CH_2NH_2$	1.28	Te(Ⅳ)	30
$ox=C_2O_4^{2-}$	0.99	Cr(Ⅲ)	17.4	bpy=2, 2′-bipyridine	1.33	Ir(Ⅲ)	32
H_2O	1.00	Co(Ⅲ)	18.2	CN^-	1.7	Pt(Ⅳ)	36

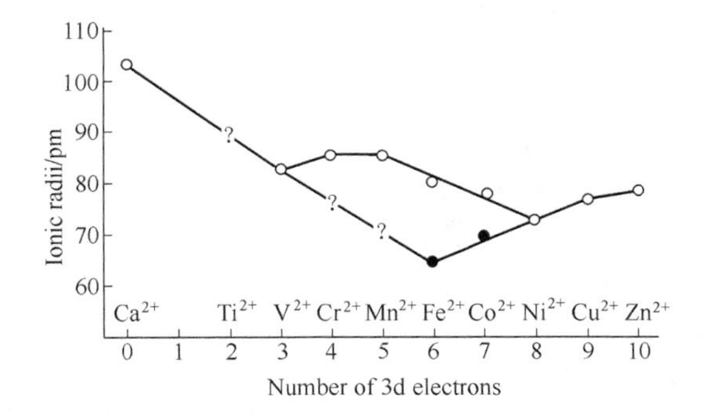

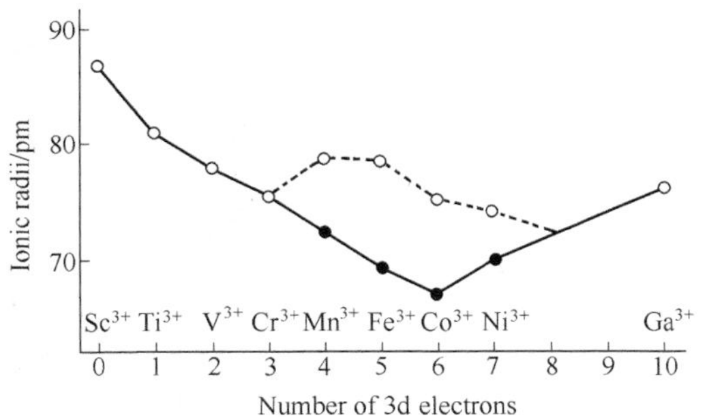

Fig.2.16 The relationships between the ionic radius and the spin state of the metal ions

(2) Hydration enthalpy Let us look at the variation of enthalpy of M^{2+} ions. Since water is a weak field ligand, the complexes are high spin.

$$M^{2+}(g) + 6H_2O(l) \longrightarrow [M(OH_2)_6]^{2+}(aq)$$

Plotting the enthalpy across the first transition series (Fig.2.17).

The straight lines show the trend when the ligand field stabilization energy has been subtracted from the observed values. Note the general trend to greater hydration enthalpy (more exothermic hydration) on crossing the period from left to right.

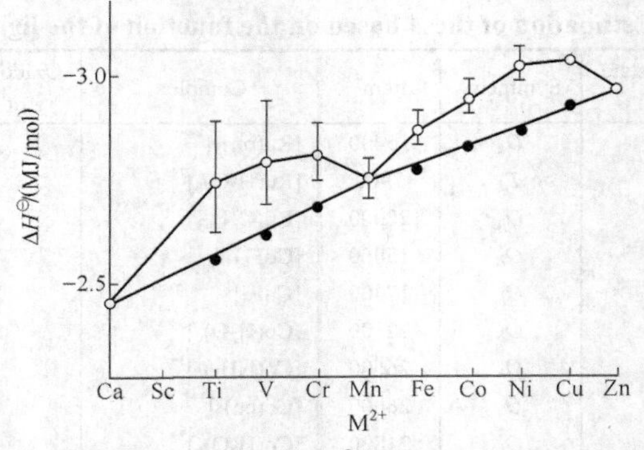

Fig.2.17 The hydration enthalpy of M^{2+} ions of the first row of the d block

This double-humped pattern is very frequent when properties are plotted for transition metals across a transition series. The straight line (full black circles) is obtained by substracting the CFSE for the given electron count. The linear increase in stability on going through a transition series is caused primarily by the increasing acidity (largely due to a decreasing size of the metal cation and electrostatic effects). This forms the basis of the Irving-Williams series.

The Irving-Williams series states that for a given ligand, the stability of the complexes increases from Ba to Cu (and then drops for Zn) (Fig.2.18). It should be noted however that as one moves towards the right, the late transition metals prefer softer ligands.

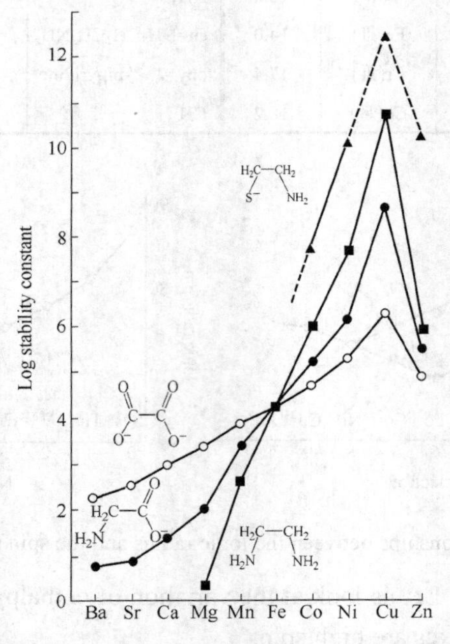

Fig.2.18 The Irving-Williams effect: the stability increase in the series Ba-Cu, decrease with Zn

2.4 Molecular Orbital Theory

A good theory should predict physical and chemical properties of the molecule such as shape, bond energy, bond length, and bond angles. One model does not describe all the properties of molecular bonds. Each model describes a set of properties better than the others. The final test for any theory is experimental data. Because arguments based on atomic orbitals focus on the bonds formed between valence electrons on an atom, they are often said to involve a valence-bond theory. The valence-bond model can't adequately explain the fact that some molecules contains two equivalent bonds with a bond order between that of a single bond and a double bond. The best it can do is suggestion that these molecules are mixtures, or hybrids, of the two Lewis structures that can be written for these molecules.

This problem, and many others, can be overcome by using a more sophisticated model of bonding based on molecular orbitals. Molecular orbital theory (MO) is more powerful than valence-bond theory because the orbitals reflect the geometry of the molecule to which they are applied.

The molecular orbital theory does a good job of predicting electronic spectra and paramagnetism, when the VB theories don't. The MO theory does not need resonance structures to describe molecules, as well as being able to predict bond length and energy. The major draw back is that we are limited to talking about diatomic molecules(molecules that have only two atoms bonded together), or the theory gets very complex. The MO theory treats molecular bonds as a sharing of electrons between nuclei. Unlike the VB theory, which treats the electrons as localized balls of electron density, the MO theory says that the electrons are delocalized. That means that they are spread out over the entire molecule.

Qualitative MO theory is a way of building up a simple picture of the molecular orbitals of a molecule using a few basic rules. The ideas discussed in this section will allow us to make "back-of-the-envelope"predictions and allow us to interpret the results from sophisticated calculations.

2.4.1 Molecular Orbital

We start with the simple principle that atomic orbitals can be combined to give molecular orbitals. In building up our combinations, the overlap between atomic orbitals is determined by their energy, orientation and size. In the simple example of the hydrogen molecule, the two 1s atomic orbitals (one on each H atom) are combined to give two molecular orbitals.

A hydrogen atom consists of a nucleus (a proton) and an electron. It is not possible to accurately determine the position of the electron, but it is possible to calculate the probability of finding the electron at any point around the nucleus. With one hydrogen atom the probability distribution is spherical around the nucleus and it is possible to draw a spherical boundary surface, inside which there is a 95% possibility of finding the electron. The electron has a fixed energy and a fixed spatial distribution called an orbital. In the helium atom there are two electrons associated with the helium nucleus. The electrons have the same spatial distribution and energy (i.e. they occupy the same orbital), but they differ in their spin (Pauli exlusion principle). In general, electrons in atomic nuclei occupy orbitals of fixed energy and spatial distribution, and each orbital only contains a maximum of two electrons with anti-parallel spins. In physics, periodic phenomena are associated with a"wave equation", and in atomic theory the relevant equation is called the"Schrödinger Equation". The wave equation predicts discrete solutions in one dimension for a particle confined to a box within finite walls. The solutions can be shown as in the Fig.2.19.

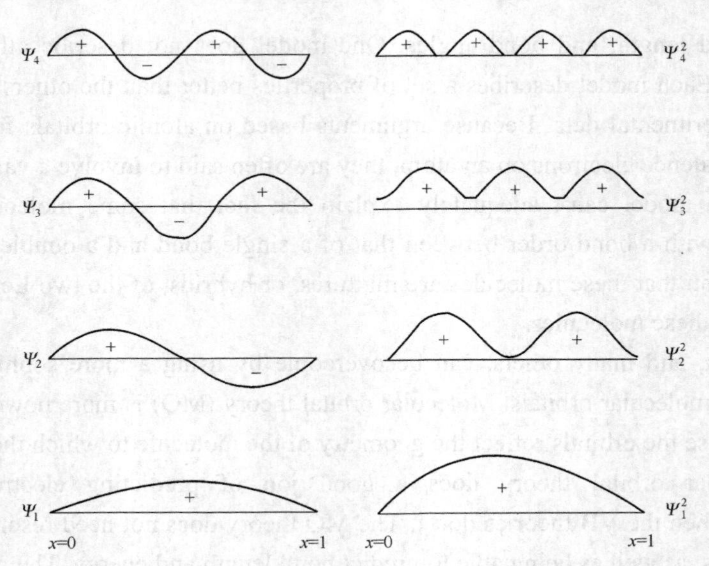

Fig.2.19 The illustration of the solutions of the wave equation

$\Psi_1 \sim \Psi_4$ represent solutions of increasing energy. In three dimensions, the equation determines the energy and defines the spatial distribution of each electron. Solutions of the wave equations in three-dimensions allows calculation of the"shape"of each orbital. The first five solutions of the wave equation for an electron associated with a proton can be shown in the Fig.2.20.

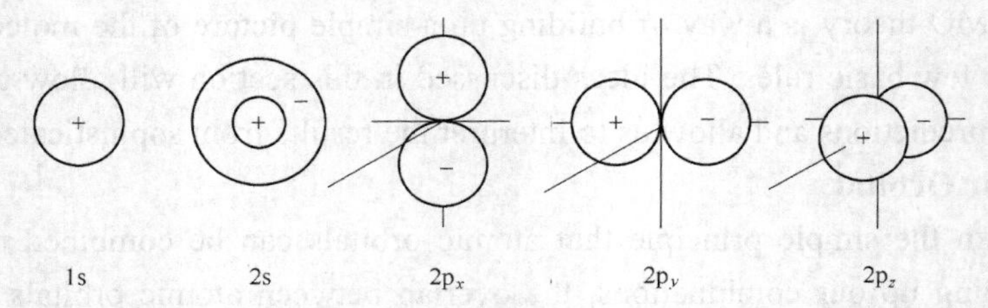

Fig.2.20 The illustration of the first five solutions of the wave equation for an electron associated with a proton

In the hydrogen atom, the 1s atomic orbital has the lowest energy, while the remainder (2s, $2p_x$, $2p_y$ and $2p_z$) are of equal energy (i.e. degenerate), but for all other atoms, the 2s atomic orbital is of lower energy than the $2p_x$, $2p_y$ and $2p_z$ orbitals, which are degenerate. In atoms, electrons occupy atomic orbitals, but in molecules they occupy similar molecular orbitals that surround the molecule.

The simplest molecule is hydrogen, which can be made up of two separate protons and electrons. There are two molecular orbitals for each hydrogen molecule, the lower energy orbital has greater electron density between the two nuclei. This is the bonding molecular orbital—and is of lower energy than the two 1s atomic orbitals of hydrogen atoms and making this orbital more stable than two separated atomic hydrogen orbitals. The upper molecular orbital has a node in the electronic wave function and the electron density is low between the two positively charged nuclei. The energy of the upper orbital is greater than that of the 1s atomic orbital, and such an orbital is called an antibonding molecular orbital.

Normally, the two electrons in hydrogen occupy the bonding molecular orbital, with anti-parallel spins. If molecular hydrogen is irradiated by ultra-violet (UV) light, the molecule may absorb the energy, and promote one electron into its antibonding orbital (σ^*), and the atoms will separate. The energy levels in a hydrogen molecule can be represented in a diagram—showing how the two 1s atomic orbitals combine to form two molecular orbitals, one bonding (σ) and one antibonding (σ^*). Below is a picture (Fig.2.21) of the molecular orbitals that two hydrogen atoms come together to form a hydrogen molecule:

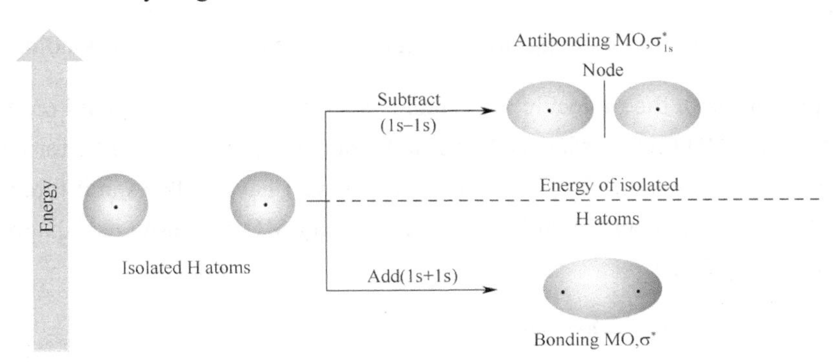

Fig.2.21 The molecular orbitals of two hydrogen atoms coming together to form a hydrogen molecule

In this case, the energy, orientation and size of the 1s atomic orbitals on each H atom are the same. In the ground state of the hydrogen molecule, the two electrons present occupy the bonding MO, and so hydrogen is a nice stable molecule.

2.4.2 Basic Rules of MO Theory

Now, when two atoms come together, their two atomic orbitals react to form two possible molecular orbitals. One of the molecular orbitals is lower in energy. It is called the bonding orbital and stabilizes the molecule because electrons in this orbital spend most of their time in the region directly between the two nuclei. It is called a sigma (σ) molecular orbital because it looks like an s orbital when viewed along the H—H bond. The other orbital is called an anti-bonding orbital because electrons spend most of their time away from the region between the two nuclei. It is higher in energy than the original atomic orbitals and destabilizes the molecule. This orbital is therefore an antibonding, or sigma star (σ^*), molecular orbital.

The MO theory has five basic rules:

- The number of molecular orbitals=the number of atomic orbitals combined;
- Of the two MO's, one is a bonding orbital (lower energy)and one is an anti-bonding orbital (higher energy);
- Electrons enter the lowest orbital available;
- The maximum number of electrons in an orbital is 2 (Pauli Exclusion Principle);
- Electrons spread out before pairing up (Hund's Rule).

Below is a molecular orbital energy diagram for the hydrogen molecule (Fig.2.22). Notice that the two AO's or atomic orbitals combine to form 2

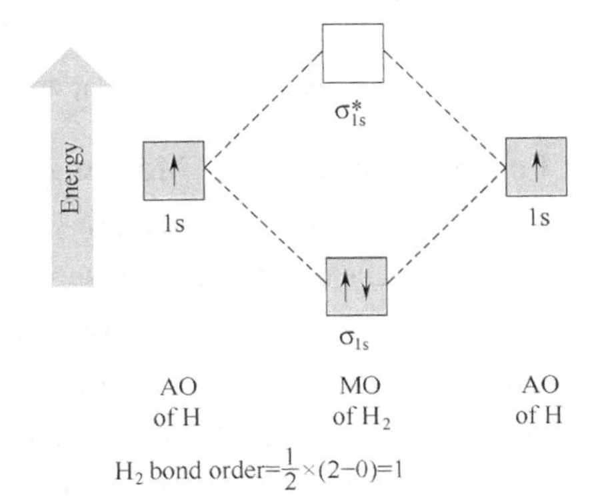

Fig.2.22 A molecular orbital energy diagram for the hydrogen molecule

MO's—the bonding and the anti-bonding molecular orbitals. Also, notice that the five rules have been followed, the electrons having been placed in the lowest energy orbital (rule 3) and have paired up (rule 4) and there are only two electrons in the orbitals (rule 5).

If you notice at the very bottom of the above picture, "bond order"is mentioned. If a molecule is to be stable, it must have a bond order greater than 0. Bond order is calculated as:

$$\frac{1}{2}\text{(numbers of electrons in bonding orbitals−numbers of electrons in anti-bonding orbitals)}$$

If the bond order is 0, the molecule is unstable and won't form. If the bond order is 1, a single bond is formed. If the BO (bond order) is 2 or 3, a double or triple bond will be formed respectively. When the 2nd period atoms are bonded to one another, you have both 2s and three 2p orbitals to contend with. When this happens, you have twice as many MO's! Below is a diagram of the new molecular orbitals (Fig.2.23).

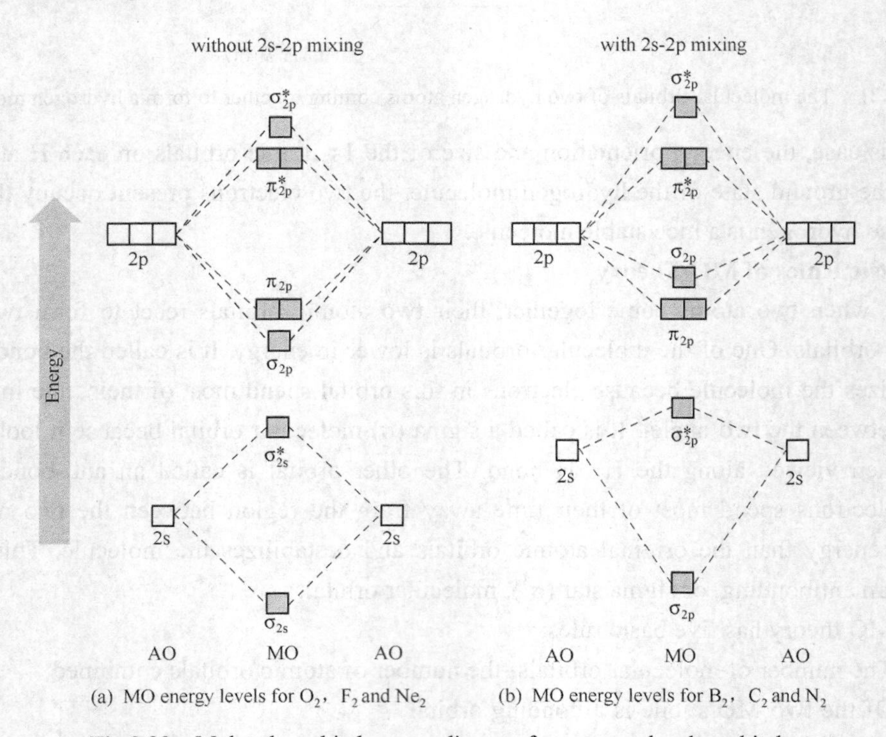

Fig.2.23 Molecular orbital energy diagram for some molecular orbitals

Finally, we can put the Molecular Orbital Theory to use! Would you predict that dilithium or diberylium is more likely to form, based on the Fig.2.24?

The answer is dilithium because it has a bond order of 1 that is stable and diberylium has a BO of 0 that is unstable and therefore will not form. Combining a pair of helium atoms with $1s^2$ electron configurations would produce a molecule with a pair of electrons in both the σ bonding and the σ^* antibonding molecular orbitals. The total energy of a He_2 molecule would be essentially the same as the energy of a pair of isolated helium atoms, and there would be nothing to hold the helium atoms together to form a molecule.

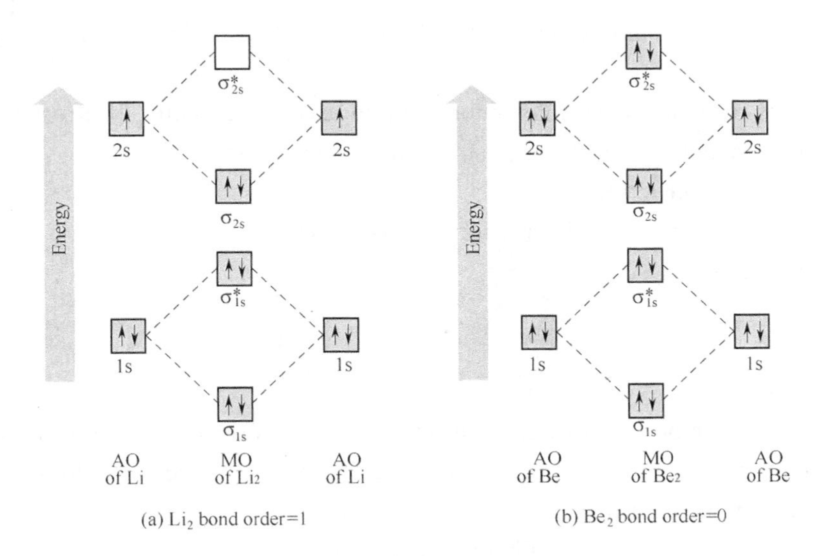

Fig.2.24 Molecular orbital energy diagram for Li_2 and Be_2 molecules

The fact that a He_2 molecule is neither more nor less stable than a pair of isolated helium atoms illustrates an important principle: The core orbitals on an atom make no contribution to the stability of the molecules that contain this atom. The only orbitals that are important in our discussion of molecular orbitals are those formed when valence-shell orbitals are combined. The molecular orbital diagram for an O_2 molecule would therefore ignore the 1s electrons on both oxygen atoms and concentrate on the interactions between the 2s and 2p valence orbitals.

2.5 Intermolecular Interaction

The interactions between the molecules are considerably weaker (by 1 or 2 orders of magnitude) than covalent bond and termed as non-covalent interactions or intermolecular interactions that were first recognized by J. D. Van der Waals in the last century.

Covalent and non-covalent interactions have completely different origins. A covalent bond is formed when partially occupied orbitals of interacting atoms overlap and consists of a pair of electrons shared by these atoms. Covalent interactions are of short range and covalent bonds are generally shorter than 2Å. Noncovalent interactions are known to act at distances of several angstroms or even tens of angstroms and overlap is thus unnecessary (in fact overlap between occupied orbitals leads only to repulsion), which originate from interaction between permanent multipoles, between a permanent multipole and an induced multipole, and finally, between an instantaneous time variable multipole and an induced multipole.

The role of noncovalent interactions in nature was fully recognized only in the last two decades; they play an important role in chemistry and physics and, moreover, are of key importance in the bio-disciplines. The structures of liquids, solvation phenomena, molecular crystals, physisorption, bio-macromolecules such as DNA and proteins, and molecular recognition are only a few phenomena determined by noncovalent interactions. Noncovalent interactions play a special role in supramolecular chemistry, which has been defined by Lehn as "chemistry beyond the molecule".

Various types of noncovalent interactions can be used, specifically, electrostatic interactions, hydrogen bonding, stacking interactions, van de Waals interactions.

2.5.1 Electrostatic Interactions

The examples of electrostatic interactions are as follow.

(1) Ion-ion interaction

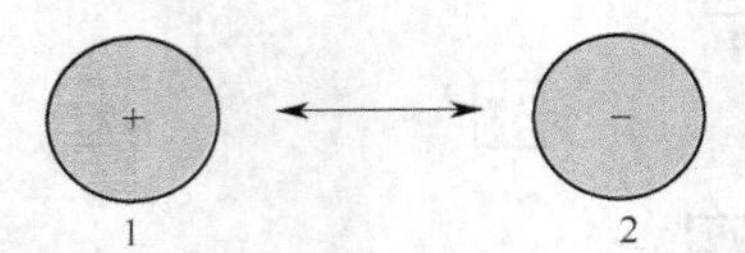

This interaction can be very strong, even stronger than covalent bonds in some cases. It can be an attractive or a repulsive force in long range ($1/r$) and non-directional force. Ion-ion interaction is highly dependants on the dielectric constant of the medium.

$$\text{Energy}=(kz_1z_2e^2)/(\varepsilon r_{12})$$

$$k=1/4\pi\varepsilon_0=\text{Coulomb constant}=9\times10^9(\text{N}\cdot\text{m}^2/\text{C}^2)$$

e=elementary charge, ε=dielectric constant, r_{12}=the distance between the charges.

The energy of an ion-ion interaction only decreases at a rate proportional to $1/r$. Therefore these are very long range forces.

When designing a host/guest complex, the energy to bring two oppositely charged species to distance of 3nm of one another in water is −0.59kJ/mol and −1.77kJ/mol for 1nm distance. The energy will be −28.8kJ/mol if two oppositely charged species are separated from 1nm in chloroform.

(2) Ion-dipole interaction The features of ion-dipole interactions as follow: non-directional force, can be attractive or repulsive, medium range interactions ($1/r^2$), significantly weaker then ion-ion interactions.

$$\text{Energy}=-kQu\cos q/(er^2)$$

Maximum when q=0° or 180°; Zero when q=90°.

$u=ql$, u=dipole moment, l=length of the dipole, q=partial charge on dipole, r=distance from charge to center of dipole, Q=charge on ion (Fig.2.25).

(3) Dipole-dipole interaction The features of dipole-dipole interactions as follow: non-directional forces, can be attractive or repulsive, short range ($1/r^3$), significantly weaker than ion-dipole interactions, occur between molecules that have permanent net dipoles (polar molecules). For example, dipole-dipole interactions occur between SCl_2 molecules, PCl_3 molecules and $(CH_3)_2CO$ molecules.

$$\text{Energy}=-(ku_1u_2/er^3)^2\cos\theta_1\cos\theta_2-\sin\theta_1\sin\theta_2\cos\Phi$$

Maximum in inline configuration where θ=0.

Simplifies to: $-2(ku_1u_2/er^3)$.

Example: Two acetone molecules in chloroform in head to tail arrangement separated by 0.5nm. The energy is −1.68kJ/mol (Fig. 2.26).

Fig.2.25 The illustration of crown ether complex with alkai metal ions

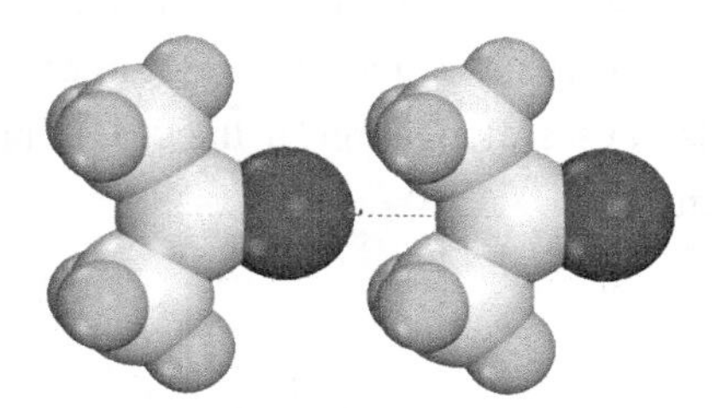

Fig.2.26 The illustration of dipole-dipole interaction between two acetone molecules

2.5.2 Hydrogen Bonding

H-bonding is a special case of dipole-dipole interactions, which have a certain amount of covalent character and directionality. It can be significantly stronger (short range, 2.5~3.5Å) than typical dipole-dipole, probably the most important of all intermolecular interactions. Forms when a hydrogen atom is positioned between two electronegative atoms, i.e. O, N, F: D—H···A (Fig.2.27).

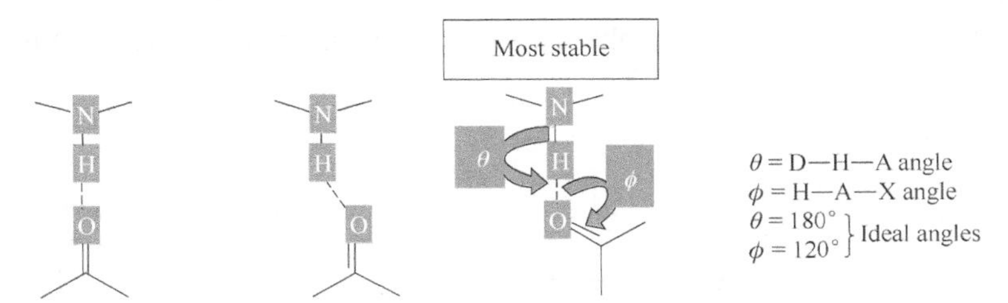

Fig.2.27 Geometry of hydrogen bond

Van der Waals radius of H and O are 1.1Å and 1.5Å respectively. Therefore, closest approach should be 2.6Å. Actual separation (1.76Å) is about 1Å less, which is intermediate between vdw distance and typical O—H covalent bond of 0.96Å (Table 2.11).

Table 2.11 Typical distances and angles (Energy 20~40kJ/mol)

NH···O 1.80 to 2.00Å
OH···O 1.60 to 1.80Å
θ (D—H—A) 150°~160°
ϕ (H—A—X) 120°~130°

The types of H-bonding are conventional H-Bond and non-conventional H-Bond.

Conventional H-Bond: H-bonded complexes are by far the most important and numerous non-covalent complexes. Most H-bonds have electronegative atoms as X, with Y either an electronegative atom having one or two lone electron pairs or a group with a region of excess of electron density (e. g. , π-electrons of aromatic systems). H-bonds with X, Y = F, O, N are best known, for example, O—H···O and N—H···O (N) bond (Fig.2.28).

What is the driving force for geometrical and spectral manifestations of H-bonding? By natural bond orbital analysis it was shown that it is charge transfer (CT) from the lone pairs or π-molecular orbitals of the electron donor (proton acceptor) to the antibonding orbitals of the X—H bond of the electron acceptor (proton donor). An increase of electron density in these antibonding orbitals causes elongation of the X—H bonds, which causes the red-shift of the X—H stretching frequency.

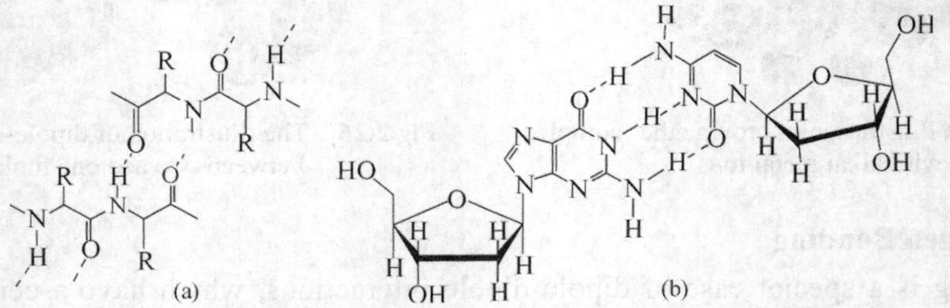

Fig.2.28 The conventional H-bond in (a) proteins and (b) DNA

2.5.3 π-π Stacking

π-π stacking (0~50kJ/mol) is weak electrostatic interaction between aromatic rings. There are two general types: face-to-face and edge-to-face (Fig.2.29). Face-to-face π-stacking interactions are responsible for the slippery feel of graphite. Similar π-stacking interactions help stabilize DNA double helix.

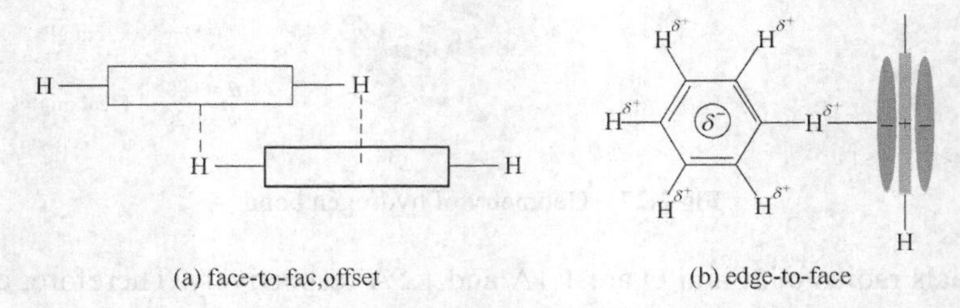

Fig.2.29 The two general types of π-π stacking

2.5.4 Van de Waals Interactions

Van de Waals interactions exist between almost all atoms and molecules and arise from atomic or molecular dipoles. This interactions between the fluctuating induced dipoles (due to instantaneous and short-lived vibrational distortions). Bond energy is very weak (0.1~3kJ/mol).

This transient polarization will continue to fluctuate because of the electron movement, but the induced polarities are synchronized so that the attraction from the polarity is maintained as long as the molecules are close together (Fig.2.30).

Strength of interaction is essentially a function of the surface area of contact and the polarizability of electron shells. The larger the surface area, the stronger the interaction will be. Regardless of other interactions found within a complex there will always be a contribution from VDW. This is what drives molecules to eliminate spaces or vacuums and makes it difficult to engineer porous or hollow structures and gives rise to the phrase "nature abhors a vacuum".

Normally, there are three types of VDW interactions, ion-induced dipole, dipole-induced dipole and dispersion (Fig.2.31).

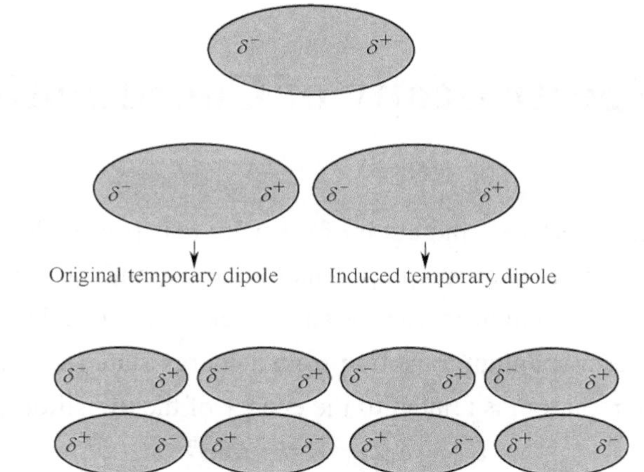

Fig.2.30 The illustration of the orginal temporary dipole, induced temporary dipole and array of the molecules with temporary dipole

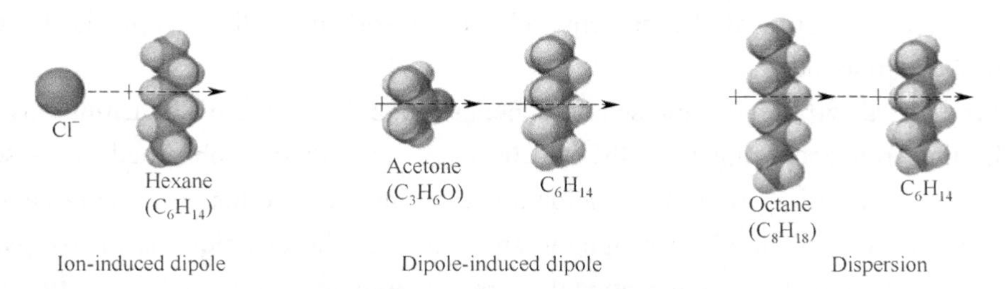

Fig.2.31 Three main types of VDW interactions

Chapter 3 Spectroscopy of Coordination Complex

When a beam of electromagnetic radiation of intensity I_0 is passed through a substance, it can either be absorbed or transmitted, depending upon its frequency, and the structure of the molecule it encounters. Electromagnetic radiation is energy and hence when a molecule absorbs radiation it gains energy as it undergoes a quantum transition from a energy state ($E_{initial}$) to another (E_{final}). The frequency of the absorbed radiation is related to the energy of the transition by Planck's law:

$$E_{final}-E_{initial}=E=h\nu=hc/\lambda$$

Thus, if a transition exists that is related to the frequency of the incident radiation by Planck' s constant, then the radiation can be absorbed. Conversely, if the frequency does not satisfy the Planck expression, then the radiation will be transmitted. A plot of the frequency of the incident radiation vs. some measure of the percent radiation absorbed by the sample is the absorption spectrum of the compound.

The type of absorption spectroscopy depends upon the type of transition involved and accordingly the frequency range of the electromagnetic radiation absorbed. If absorption is accompanied by a transition from one rotational energy level to another then the radiation is from the microwave portion of the electromagnetic spectrum and the technique is microwave spectroscopy. If the transition is from one vibrational energy level to another, then the radiation is from the infrared portion of the electromagnetic spectrum and the technique is known as infrared spectroscopy. If the transition alters the configuration of the valence electrons in the molecule, then the radiation is from the ultraviolet-visible portion of the spectrum and the technique is ultraviolet-visible, or electronic absorption spectroscopy.

3.1 Ultraviolet and Visible Absorption Spectroscopy (UV-Vis)

Many molecules absorb ultraviolet or visible light. The absorbance of a solution increases as attenuation of the beam increases. Absorbance is directly proportional to the path length, b, and the concentration, c, of the absorbing species. Beer's Law states that

$$A=\varepsilon bc$$

where ε is a constant of proportionality, called the absorbtivity.

Different molecules absorb radiation of different wavelengths. An absorption spectrum will show a number of absorption bands corresponding to structural groups within the molecule. For example, the absorption that is observed in the UV region for the carbonyl group in acetone is of the same wavelength as the absorption from the carbonyl group in diethyl ketone.

3.1.1 Electronic Transitions

The absorption of UV or visible radiation corresponds to the excitation of outer electrons.

There are three types of electronic transition that can be considered:

- Transitions involving π, σ, and n electrons.
- Transitions involving charge-transfer electrons.
- Transitions involving d and f electrons (not covered in this section).

When an atom or molecule absorbs energy, electrons are promoted from their ground state to an excited state. In a molecule, the atoms can rotate and vibrate with respect to each other. These vibrations and rotations also have discrete energy levels, which can be considered as being packed on top of each electronic level (Fig.3.1).

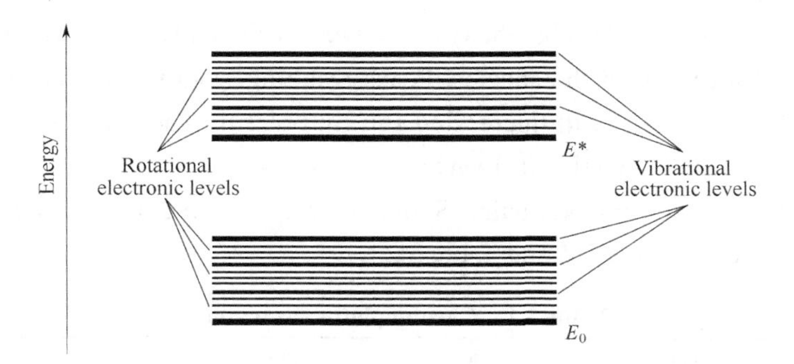

Fig.3.1 The illustration of the discrete energy level of vibrations and rotations in molecule

3.1.2 Absorbing Species Containing π, σ and n Electrons

Absorption of ultraviolet and visible radiation in organic molecules is restricted to certain functional groups (chromophores) that contain valence electrons of low excitation energy. The spectrum of a molecule containing these chromophores is complex. This is because the superposition of rotational and vibrational transitions on the electronic transitions gives a combination of overlapping lines. This appears as a continuous absorption band (Fig.3.2).

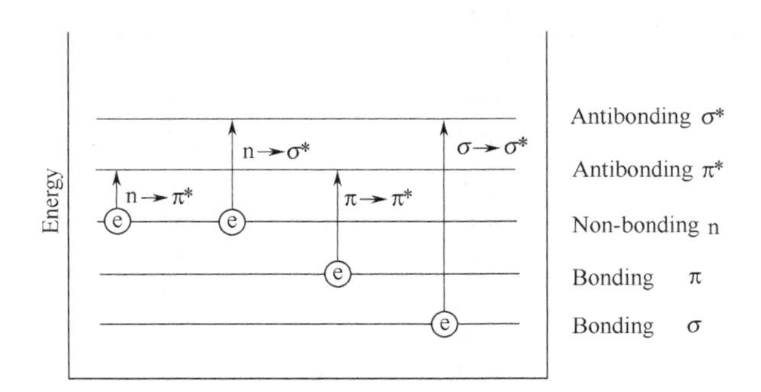

Fig.3.2 Possible electronic transitions of π, σ and n electrons

(1) $\sigma\to\sigma^*$ transitions An electron in a bonding σ orbital is excited to the corresponding antibonding orbital. The energy required is large. For example, methane (which has only C—H bonds, and can only undergo $\sigma\to\sigma^*$ transitions) shows an absorbance maximum at 125nm. Absorption maxima due to $\sigma\to\sigma^*$ transitions are not seen in typical UV-Vis spectra (200-700nm).

(2) n→σ* transitions Saturated compounds containing atoms with lone pairs (non-bonding electrons) are capable of n→σ* transitions. These transitions usually need less energy than σ→σ* transitions. They can be initiated by light whose wavelength is in the range 150-250nm. The number of organic functional groups with n→σ* peaks in the UV region is small.

(3) n→π* and π→π* transitions Most absorption spectroscopy of organic compounds is based on transitions of n or π electrons to the π* excited state. This is because the absorption peaks for these transitions fall in an experimentally convenient region of the spectrum (200-700nm). These transitions need an unsaturated group in the molecule to provide the π electrons. Molar absorbtivities from n→π* transitions are relatively low, and range from 10 to 100L/(mol·cm). π→π* transitions normally give molar absorbtivities between 1000 and 10000L/(mol·cm). The solvent has an effect on the spectrum of the species. Peaks resulting from n→π* transitions are shifted to shorter wavelengths (blue shift) with increasing solvent polarity. The reverse (i.e. red shift) is seen for π→π* transitions. This effect also influences n→π* transitions but is overshadowed by the blue shift resulting from solvation of lone pairs. Some common chromophore absorptions and organic chromophores are listed in Table 3.1 and Table 3.2.

Table 3.1 Chromophore absorptions

Chromophore	Example	Excitation	λ_{max}/nm	ε	Solvent
C=C	ethene	π→π*	171	15000	hexane
C≡C	1-hexyne	π→π*	180	10000	hexane
C=O	ethanal	n→π*	290	15	hexane
		π→π*	180	10000	hexane
N=O	nitromethane	n→π*	275	17	ethanol
		π→π*	200	5000	ethanol
C—X X=Br	methyl bromide	n→σ*	205	200	hexane
X=I	methyl iodide	n→σ*	255	360	hexane

Table 3.2 Organic chromophores

Chromophore	Transition	λ_{max}/nm	lgε
Nitrile(—C≡N)	n to π*	160	<1. 0
Alkyne(—C≡C—)	π to π*	170	3. 0
Alkene(—C=C—)	π to π*	175	3. 0
Alcohol(ROH)	n to σ*	180	2. 5
Ether(ROR)	n to σ*	180	3. 5
Ketone[—C(R)=O]	π to π*	180	3. 0
	n to π*	280	1. 5
Aldehyde[—C(H)=O]	π to π*	190	2. 0
	n to π*	290	1. 0
Amine(—NR_2)	n to σ*	190	3. 5
Acid(—COOH)	n to π*	205	1. 5
Ester(—COOR)	n to π*	205	1. 5
Amide[—C(=O)NH_2]	n to π*	210	1. 5
Thiol(—SH)	n to σ*	210	3. 0
Nitro(—NO_2)	n to π*	271	<1. 0
Azo(—N=N—)	n to π*	340	<1. 0

(4) Charge-transfer absorption Many inorganic species show charge-transfer absorption and are called charge-transfer complexes. For a complex to demonstrate charge-transfer behaviour, one of its components must have electron donating properties and another component must be able to accept electrons. Absorption of radiation then involves the transfer of an electron from the donor to an orbital associated with the acceptor. Molar absorbtivities from charge-transfer absorption are large[greater that 10000L/(mol·cm)].

3.1.3 Electronic Absorption Spectrum of Coordination Complex

There are three kinds of the electronic transitions involved in coordination complexes both in solution and in solid state. They are d→d transition based on the metal ion, metal-to-ligand charge transfer [MLCT, more often ligand-to-metal charge transfer (LMCT)] and charge transfer based on the ligands (LC, ligand centered transitions).

(1) d→d transition based on the metal ion As we have discussed in chapter 2, application of ligand field to both octahedral and tetrahedral complexes predicts a splitting of the d-orbital degeneracy. It seems reasonable, therefore, to expect that all transition metal complexes will demonstrate absorption lines corresponding to d→d transitions. However, in practice we find that this is not the case. Some complexes show strong absorptions, others only show very weak ones. For example, while T_d complexes show intense d→d absorptions, O_h complexes do not. We can explain this in terms of selection rules.

In order to demonstrate the idea of a selection rule, let us consider a somewhat simpler system—the hydrogen atom. The energy levels of a hydrogen atom may be shown on a Grotrian diagram in Fig.3.3.

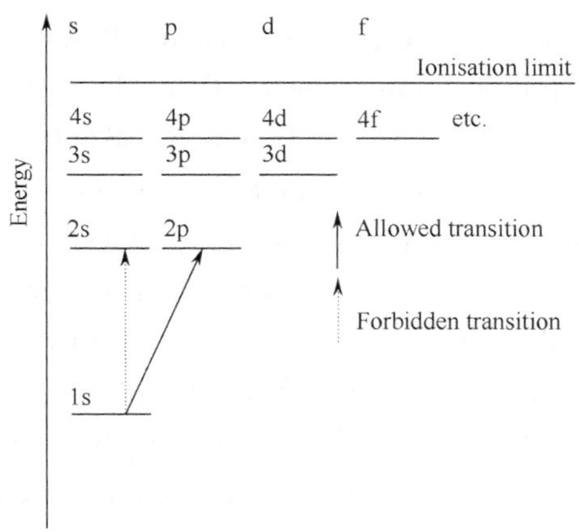

Fig.3.3 A Grotrian diagram for the energy levels of a hydrogen atom

Obviously, the 1s→2p transition is allowed, whereas the 1s→2s transition is "symmetry forbidden". The reason for this is that the hydrogen atom possesses a centre of inversion. This carries through into the SALC (Symmetry Adapted Linear Combinations) combinations of d-orbitals in the O_h and T_d symmetry complexes. The same rule applies in that if there is a centre of inversion, then we require the initial and final states to have different parity. This means that d→d transitions are symmetry forbidden—i.e. the transition dipole moment is zero. However, looking at the symmetry operations of the point group T_d we see that it does not have a centre of inversion. This means that d→d transitions are allowed in T_d complexes. In O_h complexes, there is a centre of inversion, so d→d transitions are symmetry forbidden. This explains why the observed intensity of T_d complexes is much higher than in O_h complexes.

However, this does not explain why we still observe some absorption due to d-d transitions in O_h complexes. There must be some process that is relaxing this symmetry selection rule. The most obvious way to do this is to break the O_h symmetry. This can happen when an essentially O_h

complex vibrates. Some of the modes of vibration will move the ligands in such a way that there is no longer a centre of inversion. Imagine the modes of vibration along the ligand-metal axis for each ligand. These will result in the ligand-metal distances changing—somewhat at random. This vibration is considerably slower than an electronic transition—so to all intents and purposes the molecule may be considered to be stationary during the transition. Such an uncoordinated motion will mean that in a large number of molecules there will be some with asymmetric differences in their ligand-metal distances. These differences will be spread in a distribution. When this asymmetry is present, the above algrebra does not apply—meaning that in the molecules with temporary asymmetric arrangements there is the possibility of absorption. This results in a weak absorption. It also means that the absorption line is broadened somewhat because there is a distribution of different environments and hence dipoles and potential fields in the different asymmetric molecules. This weak relaxation of the Laporte selection rule is known as vibronic coupling because it arises from the interaction of vibrational modes with the electronic transition modes.

Interesting, this weak absorptions fall in the visible region, which can be used in the explanation of coordination complexes' colors.

(2) Metal-to-ligand charge transfer (MLCT or LMCT transition) These transitions of lowest excited energy between the metal centered dπ ground state and ligand π^* states are observed in the visible. The choice of the ligands allows to tune the MLCT absorption. Indeed π-acceptors ligands will present a low-lying π^* orbital and in the same time stabilize the dπ orbital centered on the metal by retro-coordination. This allows also to tune the redox properties as the lowest lying π^* orbital of the ligands will be involved in the reduction processes and the dπ orbital centered on the metal will be involved in the oxidation processes.

In order to understand this, it is necessary to understand the nature of the orbitals involved in the absorption of light. For an octahedral d^6 transition metal complexes (Fig.3.4), the highest occupied molecular orbital (HOMO) is predominantly metal dπ orbital based and the lowest unoccupied molecular orbital (LUMO) is often predominantly ligand π^* orbital based. The nature and energy of the HOMO and LUMO are especially important in understanding the excited state properties of a light absorber molecule since they are the orbitals involved in the lowest energy excited state.

The major electronic transitions that occur in d^6 metal complexes with unsaturated ligands are ligand based $n \to \pi^*$ and $\pi \to \pi^*$ transitions, metal to ligand charge transfer (MLCT) transitions and ligand field transitions (Fig.3.5). The intensity of a transition is determined by selection rules. In order for a transition to be fully allowed, the transition must be both Laporte and spin allowed. Ligand based $\pi \to \pi^*$ transition and MLCT transitions are both spin and Laporte allowed, usually have molar extinction coefficients between 10^3 to 10^5L/(mol·cm). Ligand field transition is spin allowed but Laporte forbidden, typically have 100 and 1000L/(mol·cm).

The photophysical properties of $[Ru(bpy)_3]^{2+}$ has been discovered and there are a vast amount of research focused on the use of this chromophore for photoinitiated energy and electron transfer. This complex is photostable, has a long emission lifetime at room temperature (τ=640ns in acetonitrile) and has a relatively high emission quantum yield (ϕ=0. 062 in acetonitrile). Electronic absorption spectrum of the $[Ru(bpy)_3](PF_6)_2$ in acetonirtile is shown in the Fig.3.6. The electronic absorption spectrum of $[Ru(bpy)_3]^{2+}$ shows $n\rightarrow\pi^*$ and $\pi\rightarrow\pi^*$ transitions in the UV region and a manifold of MLCT transition in the visible region. The MLCT transition centered at 450 nm is actually a manifold of transitions arising from several electronic absorption transitions differing slightly in energy.

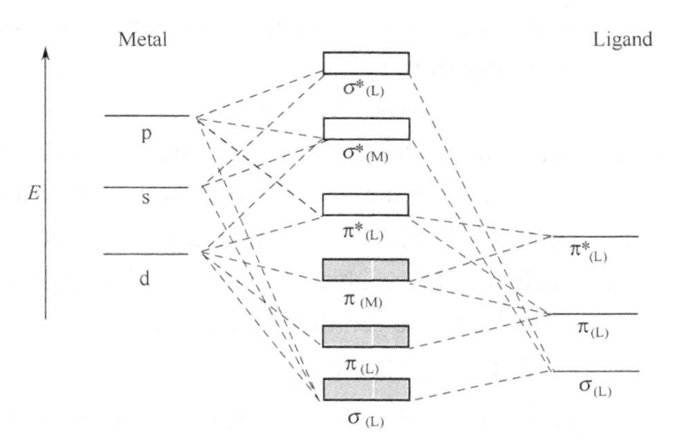

Fig.3.4 Block molecular orbital diagram for a d^6 octahedral metal complex

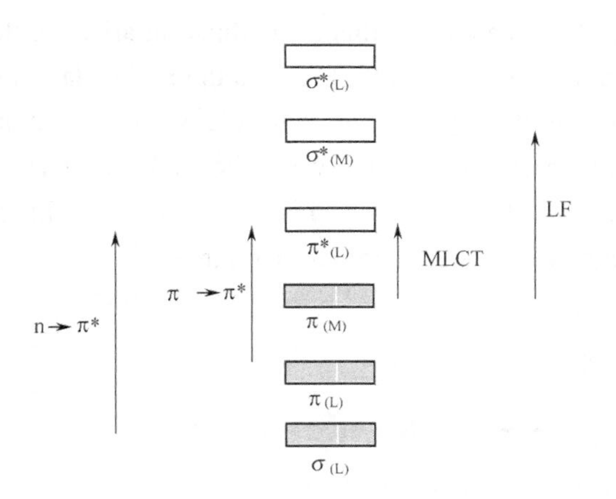

Fig.3.5 Some possible electronic transitions in an octahedral d^6 complex

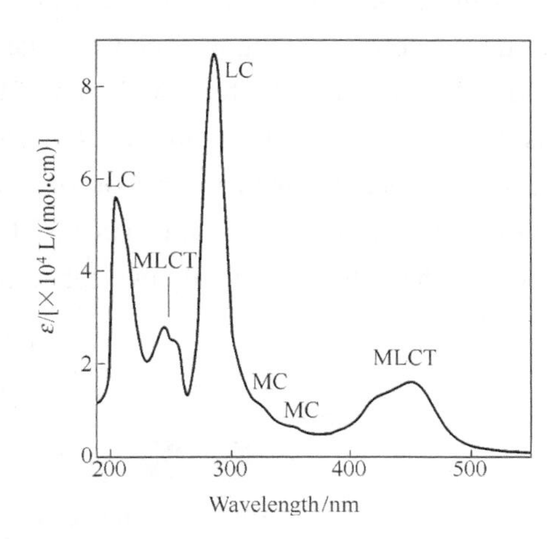

Fig.3.6 Electronic absorption spectrum of the$[Ru(bpy)_3](PF_6)_2$ in acetonirtile

3.2 Infrared Spectroscopy and Raman Spectroscopy

3.2.1 Motion of Molecule

(1) Translational motion The molecule as a whole may move through space in some arbitrary direction and with a particular velocity. This type of motion is called translational motion and with it we associate the translational kinetic energy of the molecule, given by $mv^2/2$ (v=velocity of the center of mass of the molecule). The velocity with which a molecule translates may be resolved into

components along each of the three axes of a Cartesian coordinate system, so that we may write

$$mv^2/2 = mv_x^2/2 + mv_y^2/2 + mv_z^2/2$$

where v_x is the x-component of velocity, etc. , and m is the mass of the molecule.

This equation tells us that we may consider the total translational KE of the molecule to be made up of three parts, each of which represents the kinetic energy of the molecule along one of the reference directions. Since any translation of the molecule may be considered to arise from the vector sum of its motions along the three axes, the kinetic energy may always be broken up into the sum of three contributions, one arising from motion along the x axis, one from motion along the y axis, and one from motion along the z axis. We characterize this situation by saying that the molecule has 3 translational degrees of freedom, one corresponding to each Cartesian axis.

(2) Rotational motion The molecule may rotate about some internal axis. Again, any such axis may be resolved into components along the x, y, and z axes of a Cartesian coordinate system, so that any rotation of the molecule may be resolved into three mutually perpendicular components. The rotational kinetic energy of the molecule can then be written

$$KE_{\text{rot}} = I_x\omega_x^2/2 + I_y\omega_y^2/2 + I_z\omega_z^2/2$$

where I_x, I_y and I_z are the moments of inertia about the x, y and z axes, and ω_x, ω_y and ω_z are the angular velocities about these axes.

Again, then, we see that there are in general 3 degrees of freedom associated with rotational motion, one corresponding to each Cartesian axis. An exception to this generalization arises in the case of a linear molecule, for which one of the three axes is normally taken as the molecular axis. Consider, for example, the diatomic molecule shown in the Fig.3.7. The molecule is shown with its bond axis coincident with the z axis of a coordinate system. In this case, the molecule has no rotational energy about the z axis, because the moment of inertia about this axis is zero. For linear molecules, then, there exist only two rotational degrees of freedom, rather than three.

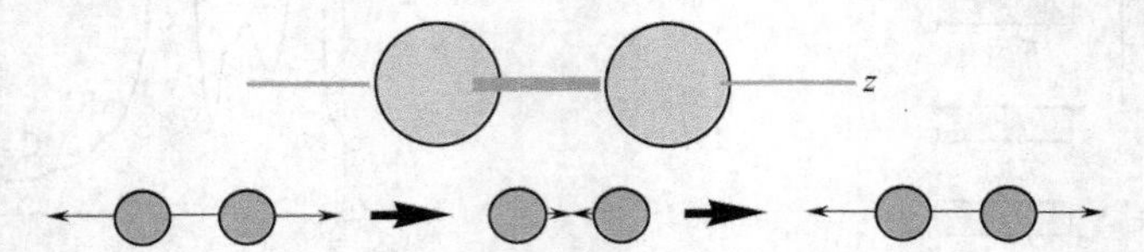

Fig.3.7 A diatomic molecule with its bond axis as the z axis of a coordinate system, only two rotational degrees of freedom

(3) Vibrational motion The molecule may vibrate. A polyatomic molecule, for example, vibrates by repeated stretching and contraction of the bond joining the two atoms, as shown in Fig 3.8.

Such a molecule has, in addition to 3 translational and 3 rotational degrees of freedom, one vibrational degree of freedom. In general, for a polyatomic molecule, we may deduce the number of vibrational degrees of freedom (sometimes called vibrational modes) by subtracting the number of translational and rotational degrees of freedom from the total number of degrees of freedom possessed by the molecule. The latter number is simply $3N$, where N=the number of atoms in the molecule (Realize that each atom may independently move in any of three directions. Each atom has, therefore, 3 degrees of freedom available to it, for a total of $3N$ for the molecule). This leads us to the general rule that the number of vibrational degrees of freedom is given by $3N-6$ for a non-linear polyatomic molecule;and by $3N-5$ for a linear polyatomic molecule. According to this rule, the water molecule, which is a non-linear molecule with $N=3$, should have 3 vibrational degrees of freedom—in other words, should undergo three independent types of vibration. These are shown in Fig.3.9.

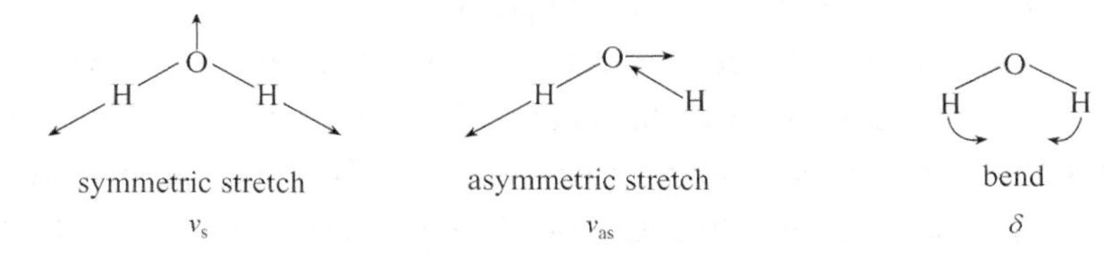

Fig.3.8 A non-linear polyatomic molecule has three independent types of vibration

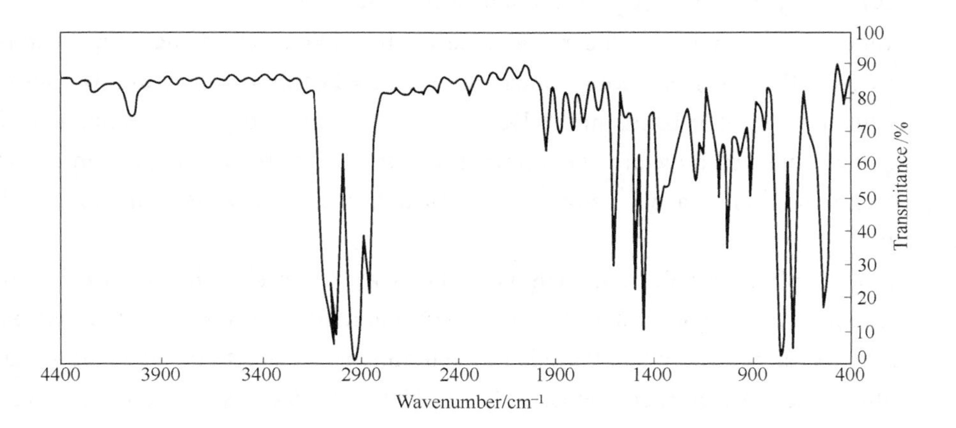

Fig.3.9 A typical IR spectrum

Each of the vibrational motions of a molecule occurs with a certain frequency, which is characteristic of the molecule and of the particular vibration. The energy involved in a particular vibration is characterized by the amplitude of the vibration, so that the higher the vibrational energy, the larger the amplitude of the motion.

Associated with each of the vibrational motions of the molecule, there is a series of vibrational energy states. The molecule may be made to go from one energy level to a higher one by absorption of a quantum of electromagnetic radiation, such that $E_{final}-E_{initial}=h\nu$. In undergoing such a transition, the molecule gains vibrational energy, and this is manifested in an increase in the amplitude of the vibration.

Since most vibrational motions in molecules occur at frequencies of about 10^{14}Hz, then light of wavelength $\lambda=c/\nu=3\times10^{10}\text{cm}\cdot\text{s}^{-1}/10^{14}\text{s}^{-1}=3\times10^{-4}\text{cm}=3\mu\text{m}$ will be required to cause transitions. Light of this wavelength lies in the so-called infrared region of the spectrum. IR spectroscopy, then, deals with transitions between vibrational energy levels in molecules, and is therefore also called vibrational spectroscopy.

An IR spectrum is generally displayed as a plot of the energy of the infrared radiation (expressed either in microns or wave numbers) versus the percent of light transmitted by the compound. This is indicated schematically in Fig.3.9.

3.2.2 IR Spectroscopy of Coordination compound

Organic functional groups differ from one another both in the strength of the bond(s) involved, and in the masses of the atoms involved. The presence of a strong, broad band between 3400 and 3200cm^{-1} indicates the presence of an O—H group in the molecule. The presence of a strong band around 1700cm^{-1} confirms the presence of a C=O group.

For organic molecules, the infrared spectrum can be divided into three regions. Absorptions between 4000 and 1300cm^{-1} are primarily due to specific functional groups and bond types. Those between 1300 and 909cm^{-1}, the fingerprint region, are primarily due to more complex interactions in the molecules; and those between 909 and 650cm^{-1} are usually associated with the presence of benzene rings in the molecule.

Some particularly important regions are indicated in Table 3.3.

At this time we wish to discuss the applications of IR spectroscopy in inorganic chemistry. It is true that many so-called inorganic compounds are in reality largely organic, and for these we look for the same functional group bands in the IR as we do for purely organic compounds. However, what we wish to consider now are the infrared spectra of relatively simple, purely inorganic compounds containing only a few atoms—specifically, inorganic salts containing polyatomic (complex) ions.

Consider a simple inorganic salt, such as KNO_2. On the basis of the empirical formula, we might naively expect there to be a total of six (3×4−6=6) normal modes of vibration associated with this material. However, we know that there is no such thing as a discrete molecule of KNO_2, so our naive calculation leads to incorrect results. In fact, KNO_2 consists of an ionic lattice with K^+ ions and NO_2^- ions arranged in an infinite and very regular array. The crystal consists of essentially isolated K^+ ions and NO_2^- ions. Thus we might expect to be able to consider the vibrational modes of the cation and anion independently of one another, and it turns out that, at least to a first approximation, this can indeed be done. In this case, since the potassium ions are monatomic, they will give rise to no vibrations (3×1−3=0), so we need only consider the nitrite anions. The VSEPR theory leads to the following structure for the nitrite ion (Fig.3.10).

It is a bent species, very much like the water molecule in shape. We thus anticipate three normal vibrational modes for NO_2^-, corresponding to the diagrams drawn earlier for the H_2O molecule, and they should all be infrared active.

In agreement with this expectation, three bands are observed in the IR spectrum of KNO_2: the symmetric stretch at 1335cm^{-1}, the asymmetric stretch at 1250cm^{-1}, and the bending vibration at 830cm^{-1} (bending vibrations occur in general at lower frequencies than do stretching vibrations).

Moreover, the frequencies of these vibrations are about the same regardless of counter ion, substantiating the essential independence of the anion and cation in the crystal. This independence is only an approximation, but we will not worry about the complicating factors now. We can, therefore, diagnose the presence of nitrite ion in a salt from the infrared spectrum of the material. This diagnostic application can be readily implemented only when the spectrum of the material is relatively uncomplicated, but despite this restriction it is an enormously useful application.

Table 3.3 Vibrational Frequencies for Organic Functional Groups

Bond Type	Specific Context	Stretching Frequencies (ν)/cm^{-1}
C—H	$C(sp^3)$—H	2800-3000
	$C(sp^2)$—H	3000-3100
	C(sp)—H	3300
C—C	C—C	1150-1250
	C═C	1600-1670
	C≡C	2100-2260
C—N	C—N	1030-1230
	C═N	1640-1690
	C≡N	2210-2260
C—O	C—O	1020-1275
	C═O	1650-1800
C—X	C—F	1000-1350
	C—Cl	800-850
	C—Br	500-680
	C—I	200-500
N—H	RNH_2,R_2NH	3400-3500(two)
	RNH_3^+,$R_2NH_2^+$,R_3NH^+	2250-3000
	$RCONH_2$,RCONHR′	3400-3500
O—H	ROH	3610-3640(free)
		3200-3400(H-bonded)
	RCO_2H	2500-3000
N—O	RNO_2	1350,1560
	$RONO_2$	1620-1640, 1270-1285
	RN═O	1500-1600
	RO—N═O	1610-1680(two), 750-815
	C═N—OH	930-960
	R_3N—O^+	950-970
S—O	R_2SO	1040-1060
	R_2S(═O)O	1310-1350
		1120-1160
	R—S(═O)$_2$—OR′	1330-1420, 1145-1200
Cumulated systems	C═C═C	1950
	C═C═O	2150
	R_2C═N═N	2090-3100
	RN═C═O	2250-2275
	RN═N═N	2120-2160
Out-of-plane bending vibrations		
Alkynes	C≡C—H	600-700
Alkenes	RCH═CH_2	910,990
	R_2C═CH_2	890
	trans-RCH═CHR	970
	cis-RCH═CHR	725,675
	R_2C═CHR	790-840

Bond Type	Specific Context	Stretching Frequencies (ν)/cm^{-1}
Aromatic	mono-	730-770,690-710(two)
	o-	735-770
	m-	750-810, 690-710(two)
	p-	810-840
	1,2,3-	760-780,705-745(two)
	1,3,5-	810-865,675-730(two)
	1,2,4-	805-825,870,885(two)
	1,2,3,4-	800-810
	1,2,4,5-	855-870
	1,2,3,5	840-850
	penta-	870
Carbonyl Stretching Frequencies		
Aldehydes	RCHO	1725
	C═CCHO	1685
	ArCHO	1700
Ketones	R_2C═O	1715
	C═C—C═O	1675
	Ar—C═O	1690
	four-membered cyclic	1780
	five-membered cyclic	1745
	six-membered cyclic	1715
Carboxylic Acids	RCOOH	1760(monomer)
		1710(dimer)
	C═C—COOH	1720(monomer)
		1690(dimer)
	RCO_2^-	1550-1610,1400(two)
Esters	RCOOR	1735
	C═C—COOR	1720
	ArCOOR	720
	γ-lactone	1770
	δ-lactone	1735
Amides	$RCONH_2$	1690(free)
		1650(associated)
	RCONHR′	1680(free)
		1655(associated)
	$RCONR_2'$	1650
	β-lactam	1745
	γ-lactam	1700
	δ-lactam	1640
Acid anhydrides	RCOOCOR′	1820,1760(two)
Acyl halides	RCOX	1800

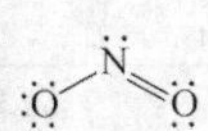

Fig.3.10 The structure for the nitrite ion

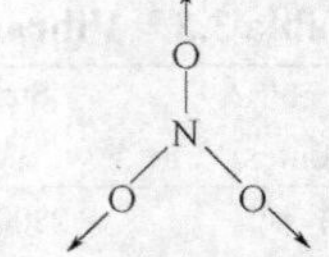

Fig.3.11 The symmetric stretch mode of NO_3^- anion

Let us turn now to a somewhat more complex case, that of $NaNO_3$. Here we anticipate 3×4−6=6 normal vibrational modes. However, the IR spectrum, exhibits only three fundamental bands, at 831cm^{-1}, 1405cm^{-1}, and 692cm^{-1}. There is no doubt that there are 6 normal vibrational modes. The formula is always valid. In the case of NO_3^-, however (or for that matter, any similar planar, triangular ion or molecule), the symmetric stretch (Fig.3.11) is not IR active.

This is because this particular motion does not cause a change in the dipole moment of the ion, and hence cannot give rise to absorption of IR radiation. This eliminates, then, one of the anticipated bands from the IR spectrum.

Among the remaining 5, there are two sets of doubly degenerate vibrations, i.e. , two instances in which 2 vibrations occur with exactly the same frequency. Thus although 5 vibrations absorb IR radiation, they are manifested in only three spectral bands. The symmetric stretch can be observed in the Raman spectrum (see the next part). However, these absorptions can be used in a diagnostic sense just as for nitrite.

In similar fashion, other relatively simple anionic (and cationic) species can be identified via their IR spectra. Spectral data for some of the more common polyatomic ions are given in the Table 3.4.

Table 3.4 IR frequencies of inorganic salts (more common polyatomic ions)

Salt	Band Position/cm^{-1}	Relative Intensity	Salt	Band Position/cm^{-1}	Relative Intensity
NaSCN	758	w	K_2SO_4	1110	vs
	940	vw	$NaClO_3$	935	s
	1620	m		965	vs
	2020	s		990	
	3330	m	$KClO_3$	938	w
KSCN	746	m		962	vs
	945	vw	$NaClO_4$	1100	vs
	1630	m		1630(H_2O)	s
	2020	s		2030	vw
	3400	m	$KClO_4$	627	w
$NaNO_2$	831	m		940	vw
	1358	vs		1075	s
	1790	vw		1140	s
	2428	vw		1990	vw
$NaNO_3$	836	m	$NaBrO_3$	807	vs
	1358	vs	$KBrO_3$	790	vs
	1790	vw	$NaIO_3$	767	vs
	2428	vw		775	
Na_2SO_4	645	w		800	m
	1110	vs			

Note: vs, very strong; s, strong; m, middle; w, weak; vw, very weak.

We have seen that the IR spectrum of the unperturbed (or "ionic") nitrite anion, NO_2^-, consists of three bands at $1335cm^{-1}$, $1250cm^{-1}$ and $830cm^{-1}$. This is what we will observe in the IR spectrum of any inorganic nitrite salt, such as KNO_2 or $NaNO_2$.

However, that the nitrite ion can function as a ligand, the coordination of NO_2^- to a metal ion could occur in either of two ways. These are shown in Fig.3.12.

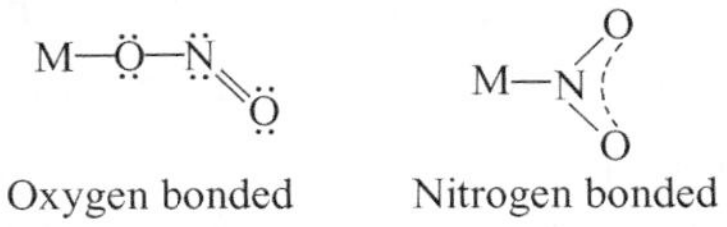

Oxygen bonded　　Nitrogen bonded

Fig.3.12 The two coordination modes of the NO_2^- anion

Can we diagnose the mode of coordination by IR spectrum? The answer is yes. When NO_2^- bonds through an oxygen atom, one of the NO bonds is very nearly a double bond. On the other hand, both NO bonds are intermediate between single and double bonds. The vibrational frequency of a bond increases as its strength increases, so we would expect the frequencies of the NO bonds in NO_2^- to increase in the order:

N-single bond-O (in O-bonded) < NO (in N-bonded) < N-double bond-O (in O-bonded)

In agreement with this, it has been found that in complexes in which NO_2^- is bonded through oxygen, the two N—O stretching frequencies lie in the ranges $1500\text{-}1400cm^{-1}$ for N═O and $1100\text{-}1000cm^{-1}$ for N—O.

In complexes in which NO_2^- is bonded through nitrogen, the bands occur at similar frequencies which are intermediate between the ranges above;namely, $1340\text{-}1300cm^{-1}$ and $1430\text{-}1360cm^{-1}$. Thus it is relatively easy to tell whether a nitrite ion is coordinated through N or O on the basis of IR whether it is coordinated.

As a second example, consider the nitrate ion, which functions as a ligand in the following ways (Fig.3.13).

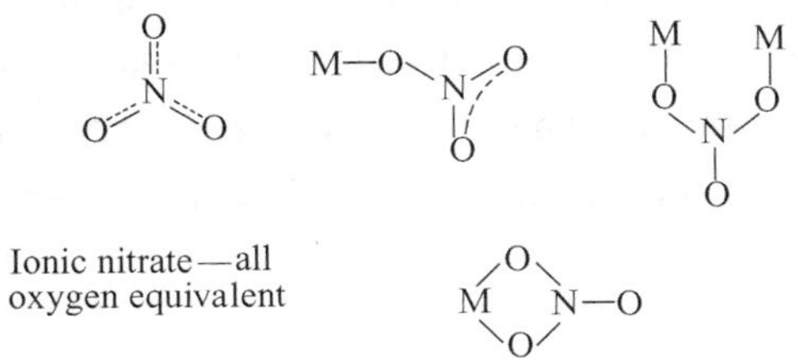

Ionic nitrate—all oxygen equivalent

Fig.3.13 The three coordination modes of the NO_3^- anion

The three types of coordination occurs, the three oxygen atoms, which are identical in the free nitrate ion, are no longer identical in the coordinated NO_3^- ion. In all three cases, two of the oxygen atoms are equivalent and the third is unique. We say that coordination has "reduced the symmetry"of the nitrate ion. The net effect is to change what was originally an AB_3 type species to an AB_2C type. Since C is different from B, chances are that the molecule now has a dipole moment. More importantly, the symmetric stretch will now most likely cause a change in dipole moment, and will therefore become IR active. Similarly, whereas in AB_3 the asymmetric stretch is doubly degenerate, in AB_2C there will be two asymmetric stretches of different energy (Fig.3.14).

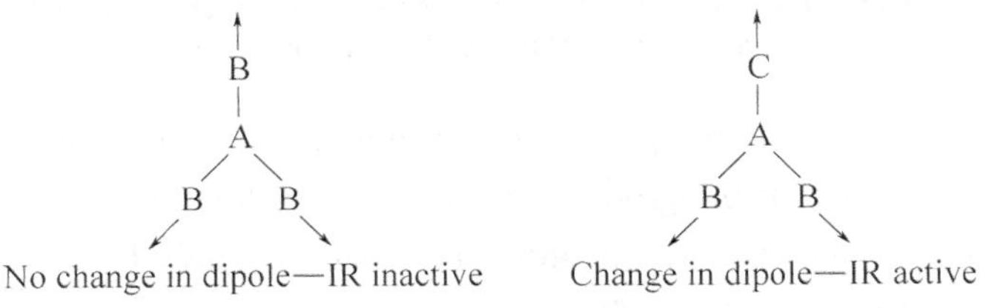

Fig.3.14 The illustration of the symmetric stretch modes of the coordinated NO_3^- ion

free NO_3^-: the single band

coordinated NO_3^- (any of the modes): two bands

It is very easy indeed to distinguish coordinated from free (that is, ionic) nitrate. The more symmetrical a molecule or ion is, the fewer the number of bands that will appear in the IR spectrum.

3.2.3 The Raman Effect and Raman Scattering

When light is scattered from a molecule most photons are elastically scattered. The scattered photons have the same energy (frequency) and, therefore, wavelength, as the incident photons. However, a small fraction of light (approximately 1 in 10^7 photons) is scattered at optical frequencies different from, and usually lower than, the frequency of the incident photons. The process leading to this inelastic scatter is termed the Raman effect. Raman scattering can occur with a change in vibrational, rotational or electronic energy of a molecule. Chemists are concerned primarily with the vibrational Raman effect, which will be meant by the term Raman effect only in here.

The difference in energy between the incident photon and the Raman scattered photon is equal to the energy of a vibration of the scattering molecule. A plot of intensity of scattered light versus energy difference is a Raman spectrum.

(1) Scattering process The Raman effect arises when a photon is incident on a molecule and interacts with the electric dipole of the molecule. It is a form of electronic (more accurately, vibronic) spectroscopy, although the spectrum contains vibrational frequencies. In classical terms, the interaction can be viewed as a perturbation of the molecule's electric field. In quantum mechanics the scattering is described as an excitation to a virtual state lower in energy than a real electronic transition with nearly coincident de-excitation and a change in vibrational energy. The scattering event occurs in 10^{-14} seconds or less. The virtual state description of scattering is shown in Fig.3.15.

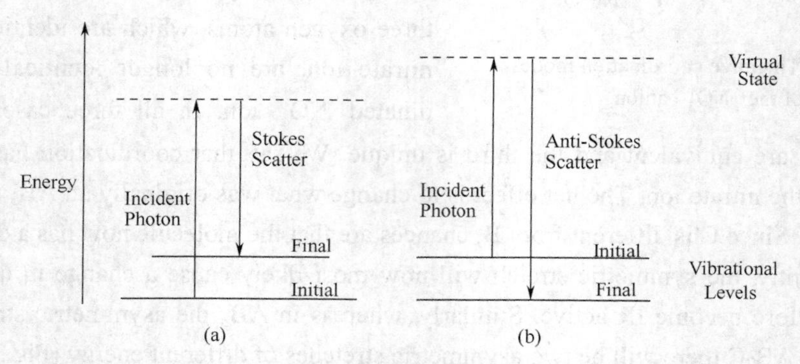

Fig.3.15 The energy level diagram for Raman scattering: (a)Stokes Raman scattering(b)anti-Stokes Raman scattering

At room temperature the thermal population of vibrational excited states is low, although not zero. Therefore, the initial state is the ground state, and the scattered photon will have lower energy (longer wavelength) than the exciting photon. This Stokes shifted scatter is what is usually observed in Raman spectroscopy. Fig.3.15(a) depicts Raman Stokes scattering.

A small fraction of the molecules are in vibrationally excited states. Raman scattering from vibrationally excited molecules leaves the molecule in the ground state. The scattered photon appears at higher energy, as shown in Fig.3.15(b). This anti-Stokes-shifted Raman spectrum is always weaker than the Stokes-shifted spectrum, but at room temperature it is strong enough to be useful for vibrational frequencies less than about $1500cm^{-1}$.

The Stokes and anti-Stokes spectra contain the same frequency information. The ratio of anti-Stokes to Stokes intensity at any vibrational frequency is a measure of temperature. Anti-Stokes Raman scattering is used for contactless thermometry. The anti-Stokes spectrum is also used when the Stokes spectrum is not directly observable, because of poor detector response.

(2) Vibration energy The energy of a vibrational mode depends on molecular structure and environment. Atomic mass, bond order, molecular substituents, molecular geometry and hydrogen bonding all effect the vibrational force constant which, in turn dictates the vibrational energy. For example, the stretching frequency of a phosphorus-phosphorus bond ranges from $460cm^{-1}$ to $610cm^{-1}$ to $775cm^{-1}$ for the single, double and triple bonded moieties, respectively.

Much effort has been devoted to estimation or measurement of force constants. For small molecules, and even for some extended structures such as peptides, reasonably accurate calculations of vibrational frequencies are possible with commercially available software.

Vibrational Raman spectroscopy is not limited to intramolecular vibrations. Crystal lattice vibrations and other motions of extended solids are Raman-active. Their spectra are important in such fields as polymers and semiconductors. In the gas phase, rotational structure is resolvable on vibrational transitions. The resulting vibration/rotation spectra are widely used to study combustion and gas phase reactions generally. Vibrational Raman spectroscopy in this broad sense is an extraordinarily versatile probe into a wide range of phenomena ranging across disciplines from physical biochemistry to materials science.

3.2.4 Raman Selection Rules and Intensities

The Raman selection rule is analogous to the more familiar selection rule for an infrared-active vibration, which states that there must be a net change in permanent dipole moment during the vibration. From group theory it is straightforward to show that if a molecule has a center of symmetry, vibrations that are Raman-active will be silent in the infrared, and vice versa.

Raman scattering occurs because a molecular vibration can change the polarizability. If a vibration does not greatly change the polarizability, the intensity of the Raman band will be low. The vibrations of a highly polar moiety, such as the O—H bond, are usually weak. An external electric field can not induce a large change in the dipole moment and stretching or bending the bond does not change this.

Typical strong Raman scatterers are moieties with distributed electron clouds, such as carbon-carbon double bonds. The pi-electron cloud of the double bond is easily distorted in an external electric field. Bending or stretching the bond changes the distribution of electron density substantially, and causes a large change in induced dipole moment.

Chemists generally prefer a quantum-mechanical approach to Raman scattering theory, which relates scattering frequencies and intensities to vibrational and electronic energy states of the molecule. The standard perturbation theory treatment assumes that the frequency of the incident light is low compared to the frequency of the first electronic excited state. The small changes in the ground state wave function are described in terms of the sum of all possible excited vibronic states of the molecule.

3.2.5 Polarization Effects

Raman scatter is partially polarized, even for molecules in a gas or liquid, where the individual molecules are randomly oriented. The effect is most easily seen with an exciting source which is plane polarized. In isotropic media polarization arises because the induced electric dipole has components that vary spatially with respect to the coordinates of the molecule. Raman scatter from totally symmetric vibrations will be strongly polarized parallel to the plane of polarization of the incident light. The scattered intensity from non-totally symmetric vibrations is 3/4 as strong in the plane perpendicular to the plane of polarization of the incident light as in the plane parallel to it.

The situation is more complicated in a crystalline material. In that case the orientation of the crystal is fixed in the optical system. The polarization components depend on the orientation of the crystal axes with respect to the plane of polarization of the input light, as well as on the relative polarization of the input and the observing polarizer.

3.3 X-ray Diffraction Analysis

X-rays are short wave electromagnetic radiation with wavelengths in the range of 0.001-10nm, which are commensurate with both the atomic size and the shortest inter-atomic distances in crystals. In 1912, Max von Laue showed that the periodicity of crystal lattice allows atoms in a crystal to be located with exceptionally high resolution and precision by means of X-ray diffraction.

3.3.1 Symmetry in Crystals

In section 2.1, we know that molecules are symmetrical objects. As defined in dictionaries, symmetry is the "beauty of form arising from balanced proportions" and to be symmetrical is to have the "correspondence in size, shape and relative position or parts on opposite sides of a dividing line or median plane or about a center or axis." Although symmetry is an intuitive perception to everyone, it has multiple applications in chemistry and other sciences. We observe symmetry at every scale in nature, from the orbits of celestial bodies to the behavior of electrons in atoms. The concept of symmetry is also broad in its application, providing a fundamental similarity between such diverse and apparently unrelated phenomena as the formation of snowflakes, the rhythms of music, and the behavior patterns of bees. More important to us is the relationship of physical properties to the arrangement of atoms on an atomic scale. Once the symmetry of a molecule or crystal is known, we can use the knowledge of symmetry to predict many of the molecule or crystal's physical properties before these are measured or even observed. In crystal structure analysis, symmetry of crystals are used in various ways to derive the lattice constants and the position of atoms.

Crystals have regular external form and can be viewed as symmetrical bodies, suggesting the possible application of group theory to crystal classification and physical property assessment of crystals. Furthermore, since crystals are composed of ions, atoms or molecules stacked together in a manner that is responsible for the external symmetry of crystals, it is reasonable to expect that group theory can provide a way of discussing the local symmetry of molecules in crystal.

Due to the restrains of the internal structure of crystal, the number of finite crystallographic symmetry elements is limited to a total of eight: 1, 2, 3, 4, 6, −1, −2, and −4. Symmetry elements and operations interact with one another only at certain angles, generating new symmetry elements and operations, respectively. The limited number of symmetry elements and the ways in which they may interact with each other lead to a limited number of symmetry groups. And actually, the number is 32 for three-dimensional crystallographic point groups. Based on the main symmetry elements or operations of each point group, the 32 crystallographic point groups are divided into a total of 7 crystal systems. The combination of crystallographic symmetry elements and their orientations with respect to one another in a group defines the crystallographic axis, and the lengths of the axes and angles between them have unique features for each crystal system (Table 3.5).

Table 3.5 Crystallographic Point Groups and Crystal Systems

Crystal System	Unit Cell	Point Groups	Group Order
Triclinic	$a \neq b \neq c$ $\alpha \neq \beta \neq \gamma$	C_1 C_i	1 2
Monoclinic	$a \neq b \neq c$ $\alpha = \beta = \pi/2 \neq \gamma$	C_s (C_{1h}) C_2 C_{2h}	2 2 4
Orthorhombic	$a \neq b \neq c$ $\alpha = \beta = \gamma = \pi/2$	C_{2v} D_2 D_{2h}	4 4 8
Tetragonal	$a = b \neq c$ $\alpha = \beta = \gamma = \pi/2$	C_4 S_4 C_{4h} D_{2d} C_{4v} D_4 D_{4h}	4 4 8 8 8 8 16
Trigonal	$a = b = c$ $\alpha = \beta = \gamma < 2\pi/3 \neq \pi/2$	C_3 S_6 C_{3v} D_3 D_{3d}	3 6 6 6 12
Hexagonal	$a = b \neq c$ $\alpha = \beta = \pi/2$, $\gamma = 2\pi/3$	C_{3h} C_6 C_{6h} D_{3h} C_{6v} D_6 D_{6h}	6 6 12 12 12 12 24
Cubic	$a = b = c$ $\alpha = \beta = \gamma = \pi/2$	T T_h T_d O O_h	12 24 24 24 48

A single crystal is a solid object in which a basic unit of ions, atoms or molecules is repeated over and over periodically in all three dimensions. Such a infinite periodic structure can be conveniently and completely described in terms of a space lattice, which consists of a set of geometric points having identical environments. The smallest repeating unit of the lattice is called the unit cell. A 3D space lattice is characterized by a set of 3 non-coplanar vectors ***a***,***b*** and ***c***, which are usually coincidence with the crystal axes. The lengths of the axes and inter-axial angles are a, b, c, α, β and γ as shown in Fig. 3.16.

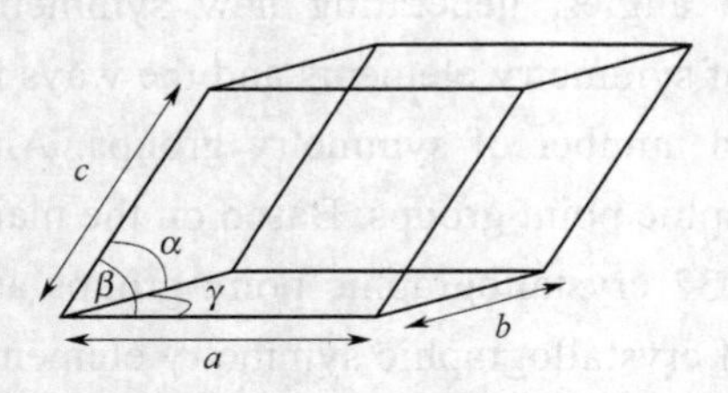

Fig.3.16 The illustration of unit cell lengths and angles

The systematic work of describing and enumerating the 3D space lattice was done by Bravais in 1850, showing that all 3D lattice can be categorized into 14 distinct types, and this is the 14 Bravais Lattice shown in Fig.3.17.

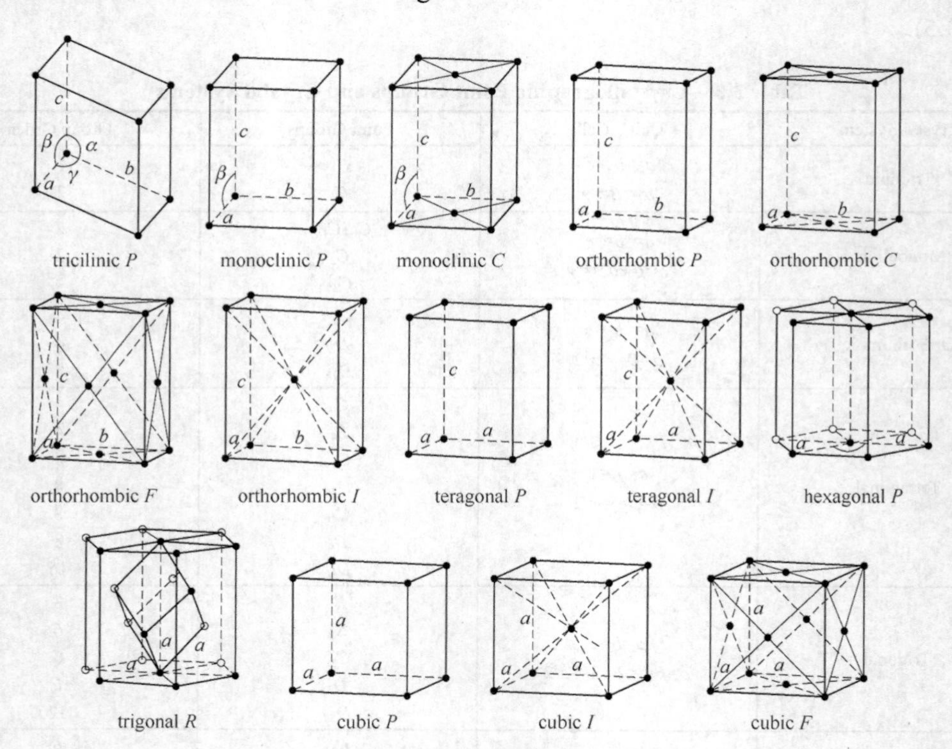

Fig.3.17 The 14 Bravais Lattice

We recall that a point group describes the symmetry of a finite body, and that a lattice constitutes a mechanism for repetition, to an infinite extent, by translation parallel to 3 non-coplanar vectors. So the question is: what is the result of repeating a point group pattern by the translations of the 14 Bravais lattice? And this is actually the real situation of the arrangement of atoms in a crystal. An infinite periodic structure can be brought into self-coincidence by point group symmetry, translation along specific directions, or the combination of point group symmetry and translation. The last one could be operations called screw rotation or glide reflection, respectively. And the corresponding symmetry elements are screw axis and glide planes. Infinite symmetry elements interact with one another and produce new symmetry elements too, and the symmetry elements in a lattice don't have to cross a common point. Various combination of finite and infinite symmetry elements result in crystallographic space groups. Symmetry elements in a space group are spread over the space of an infinite body, therefore each point of a continuous object can be moved in a periodic fashion through space by the operation of symmetry elements of the space group. The number of 3D crystallographic space groups is limited to 230 due to the limited number of allowed rotations and translations. Details for every crystallographic space groups can be found in the International Tables for Crystallography, Volume A.

3.3.2 Single Crystal X-ray Diffraction

Single crystal X-ray diffraction was first observed by Max von Laue in 1912 and used to analyze the structures of crystalline materials by the Braggs farther and son immediately. It is the most powerful method to determine crystal structures at atomic resolution. Structure knowledge gives chemists and physicists access to a large number of information, including connectivity, conformation, accurate bond lengths and angles, symmetry, and 3D packing of atoms. In addition, structures imply the stoichiometry and density of the crystalline materials.

3.3.2.1 X-ray Diffraction by Crystal

X-rays are scattered by electrons because electrons oscillates in the electric field of incoming X-ray beam and an oscillating electric charge radiates electromagnetic waves. As a result, X-rays are radiated from the electrons at the same frequency as the incident beam. X-rays scattered from electrons in different part of an atom reduces with increasing angles, and this phenomena is described by the atomic scattering factor.

Each atom in the crystal contributes to the overall scattering pattern by acting as a source of scattered X-rays. The waves will interference constructively or destructively in varying amount depending upon the direction of the diffracted beam and atomic positions. At least, two quantities are needed to describe a diffracted beam, ie, the direction and amplitude of the beam. The former is given by the well-known Laue Equation and the later by the structure factor.

The direction of the diffracted beam, or in other words the relationship between the directions of the incident and diffracted beam, was first given by Max von Laue in the form of 3 simultaneous equations, which describing the condition under which diffraction can take place are well known as Laue Equations [equation (3.1)]:

$$\begin{aligned} \boldsymbol{a} \cdot s &= h \\ \boldsymbol{b} \cdot s &= k \\ \boldsymbol{c} \cdot s &= l \end{aligned} \tag{3.1}$$

where $\boldsymbol{a},\boldsymbol{b},\boldsymbol{c}$ are the lattice parameters, $\boldsymbol{s}$ is the scattering vector and h,k,l are 3 integers. These equations are of central importance to the subject of X-ray crystallography.

The intensities are affected by many factors, and the one that depends only upon the crystal structure is called the structure factor, which can be expressed in terms of the contents of a single unit cell as [equation (3.2)]:

$$F(hkl) = \sum_{j=1}^{N} f_j \exp[2\pi i(hx_j + ky_j + lz_j)] \tag{3.2}$$

where f_j, x_j, y_j, z_j is the atomic scattering factor and fractional coordinates of atom j, h, k, l are the 3 integers in the Laue equation, and N is the number of atoms in the unit cell. It can be proved mathematically that $F(hkl)$ is the Fourier transform of electron density, and the electron densities at any point (xyz) in the unit cell can be calculated as [equation (3.3)]:

$$\rho(xyz) = \frac{1}{V} \sum_{hkl} F(hkl) \exp[-2\pi i(hx + ky + lz)] \tag{3.3}$$

where the summation is over all the $F(hkl)$s and V is the volume of the unit cell. Once we know electron density of each point in the unit cell, we can locate the peak positions which are assumed to be the position of atoms in the crystal (Fig.3.18).

The general steps involved in crystal structure determination are shown in Fig.3.19.

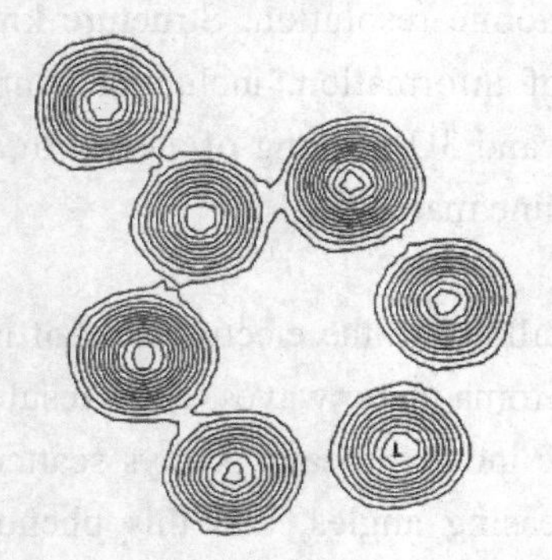

Fig.3.18 Electron Density Map

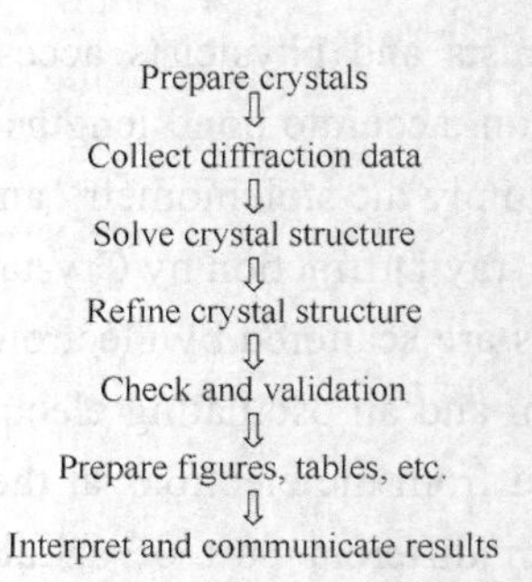

Fig.3.19 The steps involved in a crystal structure determination

3.3.2.2 Crystal Growth

The quality of the crystal from which the diffraction data are collected determines the final quality of the structure. The object of crystal growth for diffraction work is to obtain several relatively large, usually 0.1-0.4mm, high quality single crystals. It sounds not hard, but crystal growing is not a routine work, it is kind of like art. The chosen techniques largely depend on the chemical properties of the compound, and some techniques are listed below.

- Slow Evaporation This is the simplest way to grow crystals and works best for compounds which are not sensitive to ambient conditions in the laboratory.

- Slow Cooling This is good for solute-solvent systems which are less than moderately soluble and the solvent's boiling point is less than 100℃.

- Variations on Slow Evaporation and Slow Cooling If the above two techniques doesn't yield suitable crystals from single solvent systems, one may expand these techniques to binary or tertiary solvent systems. (to record the solvent composition you use!)

- Vapor Diffusion A solution of the substance is prepared using solvent S_1 and placed in test tube T. A second solvent, S_2, is placed in a beaker, B. The test tube containing S_1 is then placed in the beaker and the beaker is sealed. Slow diffusion of S_2 into T and S_1 out of T will cause crystals to form.

- Solvent Diffusion (Layering Technique) This method also is good for milligram amounts of materials which are sensitive to ambient laboratory conditions (air, moisture). Dissolve the solute in S_1 and place in a test tube. Slowly dribble S_2 into the tube so that S_1 and S_2 form discreet layers (CH_2Cl_2/Et_2O).
- Reactant Diffusion This similar to the other diffusion methods except that solutions of the reactants are allowed to diffuse into one another. If the product of the reaction is insoluble, crystals of the product will form where the reactants mix (in silica gels).
- Sublimation The way to grow crystals of somewhat volatile air sensitive crystals is to simply seal a sample under vacuum into a glass tube and placing the tube into an oven for a few days or weeks.
- Convection. The idea behind this method is that the solution becomes more saturated in the warm part of the vessel and is transferred to the cooler region where nucleation and crystal growth occur. The velocity of the convection current is proportional to the thermal gradient across the vessel.
- Co-crystallants This method of crystallization is not for all compounds. If you are not getting suitable crystals from other crystallization methods, you can try to cocrystallize your compound with a "crystallization aid". For example, triphenylphosphine oxide (TPPO) has been used as a cocrystallant for organic molecules which are proton donors. Essentially this method of crystal growing is the same as slow evaporation except that an equimolar amount of TPPO is added to the solution.
- Counterions Very often the crystal quality will improve when a different counterion is used. Ions of similar size tend to pack together better and subsequently give better crystals. In addition, you might also get more desirable physical properties for your crystal. For example, the sodium salt of your compound might be hygroscopic, but the tetramethyl ammonium salt might not be or at least, less so. A word of warning: *Make sure that whatever ion you use for your metathesis does not react with your ion of interest.*
- Ionization of Neutral Compounds If compound is neutral and contains proton donor or acceptor groups, better crystals may grown by first protonating or deprotonating the compound. The ionic form of your compound could then take advantage of hydrogen bonding to yield better crystals. This will alter the electronic properties of your compound, but if a general confirmation of the success of a synthesis is what is needed from the structure determination, then this is not a problem.

3.3.2.3 Single Crystal X-ray Diffraction Experiments

You may start your diffraction experiments once crystals have appeared. The first step is usually to pick up several crystals with suitable size under microscope, and then check the quality of these crystals. Pick the best one and then glue it on top of a thin glass fiber, mount it on the diffractometer, take a diffraction image to confirm the quality of the crystal. If the crystal is good, set up the experimental conditions such as wavelength, temperature and data collection strategies, and then collect the diffraction images, which will be integrated to extract the diffraction intensities. Diffraction intensity for a given *hkl* is directly proportional to the square of the modulus of the structure factor. To calculate $|F|$ for structure solution, several factor such as temperature factor, LP factor and absorption factor must be corrected. After all these corrections, a set of diffraction data is ready for structure analysis.

An outstanding instrument allowing one to measure accurately the diffraction intensities is the four circle serial diffractometer. Until the mid-1990s, nearly all single crystal diffraction data for chemical crystallography was collected using four circle diffractometer. The four circle diffractometer has four circles, $2\theta,\omega,\chi$ and ϕ as shown in Fig.3.20. ω,χ and ϕ bring the crystal into an orientation in which the diffraction occurs, and 2θ brings the detector to the right position to receive the diffraction intensities.

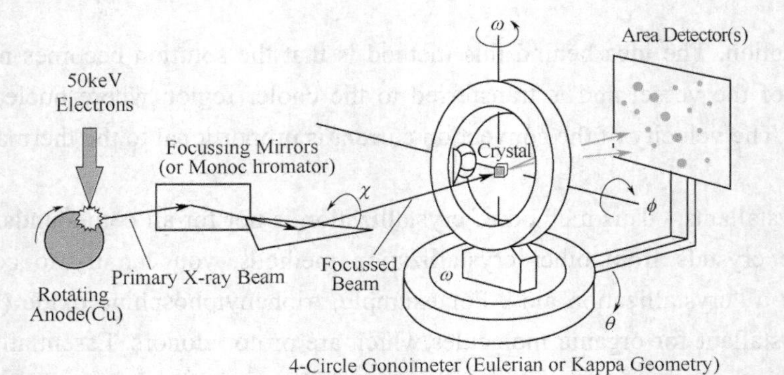

Fig.3.20 The Scheme of a Four-Circle Diffractometer

Fig 3.21 Picture of typical diffractometer.

Apparatus currently on the market is usually equipped with area detectors such CCD, CMOS or IP. Some keeps the four circles, and others have 3 circles 2θ, ω and ϕ, while the fourth χ is fixed at specified angles. One of our single crystal diffractometer is shown in Fig.3.21.

Below is an outline for a typical diffraction experiments on a CCD based diffractometer.

① Mount and center the crystal, Record the crystal colour, shape and dimensions.

② Check crystal quality using still or limited-oscillation frames.

③ Collect some frames, index, refine orientation matrix and determine Bravais lattice. Check whether the unit cell is known or not sensible.

④ Determine data collection strategies and collect frames.

⑤ Integrate the frames, scale the raw data and apply corrections as required.

3.3.2.4 Crystal Structure Analysis

Note that in equation (3.2) electron density is a Fourier transform of structures factors $F(hkl)$. But the measured X-ray intensities yield only the structure factor amplitude $|F(hkl)|$ and not their phases. The Fourier transformation cannot therefore be performed directly from experimental measurements, and the phases must be obtained by other means such as Patterson function, direct methods and charge flipping. This is known as the phase problem of crystallography.

Nearly every ab initio phase determination technique usually recovers approximate phases, and therefore only an incomplete structure model may be built up from the Fourier map. Once the phases are established at least approximately, they can be combined with the measured $|F(hkl)|$ to calculate a new electron density map or differential electron density map, from which new atoms are probably located. This process may be repeated many times as needed, until all atoms in the structure are determined.

Although the structure model is complete, there are, however, still substantial errors in the position of the atoms. This leads to the fact that the calculated structure factors don't agree well with the observed values. So the next step of structure analysis is to optimize these parameters so as to make the differences as small as possible. This is the process of structure refinement. Least square refinement is by far the most commonly used method to refine a crystal structure.

After refinement, crystal structures generally arrive at a point where the model is good enough to be used for deriving structural parameters of interests to chemists such as bond lengths, bond angles, torsion angles and best planes. Before these calculations, the structure is usually validated and checked to find and solve any potential errors. The structural parameters can be presented either in the form of numbers and tables, or in the form of diagrams and figures. And finally, the structure is published with the supporting of these materials. A molecular structure determined from single crystal X-ray diffraction is shown in Fig.3.22.

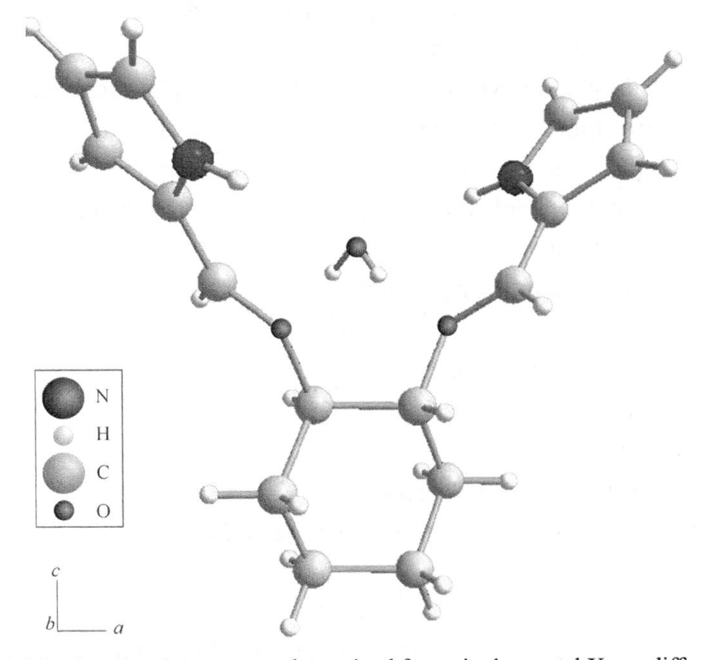

Fig.3.22 A molecular structure determined from single crystal X-ray diffraction

3.3.3 Powder X-ray Diffraction

There is no doubt that single crystal diffraction is the method of choice to obtain high quality structural data. But, in many cases, single crystals of interesting materials of sufficient size and/or quality simply cannot be prepared. Key structural insights of many important material such as high-Tc superconductors, magnetic materials, multiferroics and ionically conducting polymers came from powder diffraction studies since single crystals of these materials were not available. In addition, powder diffraction may provide complementary information like bulk sample composition to sigle crystal studies. And, there are many cases in which crystals need to be investigated under non-ambient conditions such as high temperature, high pressure, high magnetic field. Powder diffraction are often much simpler to perform than single crystal diffraction. The powder diffraction method are invented by Debye, Sherrer (1916) and Hull (1919) in Europe and America respectively.

3.3.3.1 X-ray Diffraction by Powder Crystalline Solids

In a single crystal diffraction experiment, a beam of X-ray is incident on a single crystal, and the positions and intensities of diffracted beams are recoded usually using an area detector. While in a powder diffraction experiment, a collection of randomly oriented polycrystallites is exposed to the incident X-ray beam. When certain geometric requirements are met, the scattered X-rays interfere constructively and produce a diffraction peak (Fig.3.23). In 1912, W. L. Bragg recognized a predictable relationship among several factors.

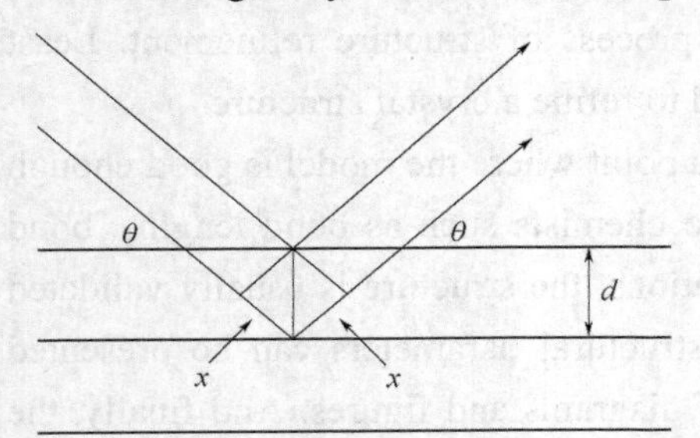

Fig.3.23 Reflection of X-rays from two planes of atoms in a solid.

- The distance between similar atomic planes in a crystal (the interatomic spacing) which we call the *d*-spacing and measure in angstroms.
- The angle of diffraction which we call the theta angle and measure in degrees. For practical reasons the diffractometer measures an angle twice that of the theta angle. Not surprisingly, we call the measured angle '2-theta'.
- The wavelength of the incident X radiation, symbolized by the Greek letter lambda and, in our case, equal to 1.54 angstroms.

These factors are combined in Bragg's Law:

$$n\lambda=2d\sin\theta \tag{3.4}$$

where n=1,2,3,…, usually taken as 1;

λ=wavelength of the X-rays in angstroms;

d=d-spacing;

θ=the diffraction angle in degrees.

We can rearrange this equation for the unknown spacing d:

$$d=\frac{n\lambda}{2\sin\theta} \tag{3.5}$$

Both the Laue equation and Bragg equation, one for single crystal diffraction experiments and the other for powder diffraction experiments, predict the directions of diffracted beams from crystals. It can be proved that they are equivalent to each other although they have different forms.

If the diffracted intensities are recorded with an area detector, just as in the case of single crystal diffraction, one will have rings of diffracted intensities called Debye rings, which are the intersections of cones of diffracted intensity with the detector. Usually the intensities are measured by scanning a point detector or a 1D line detector across a narrow strip of the rings. In either case one can represent the diffraction data as a plot of diffraction intensity against the diffraction angle 2θ, which is a powder diffraction pattern (Fig.3.24). Any powder diffraction pattern is composed of multiple Bragg peaks with different positions, varying intensities and shapes.

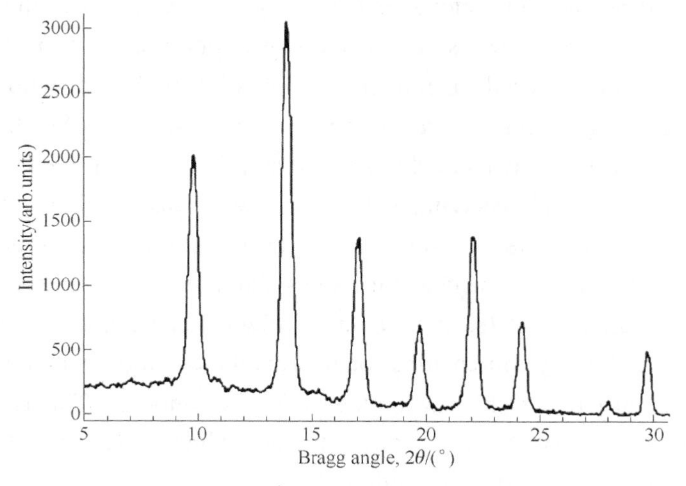

Fig.3.24 A typical powder diffraction pattern

The position of a diffraction peak is a discontinuous function of inter-planar distances, wavelength and miler indices. Therefore, both the unit cell and wavelength are the two major factors that determine the Bragg angle for the same combination of *hkl*. A comparison of the X-ray powder diffraction patterns of NaCl (bottom) and KCl (top) is shown in Fig.3.25. Peaks in the KCl diffraction pattern are labeled with Miller indices, *hkl*, indicating the set of lattice planes responsible for that diffraction peak. The KCl peaks are shifted to lower angles relative to the NaCl pattern due to the larger cubic unit cell of KCl.

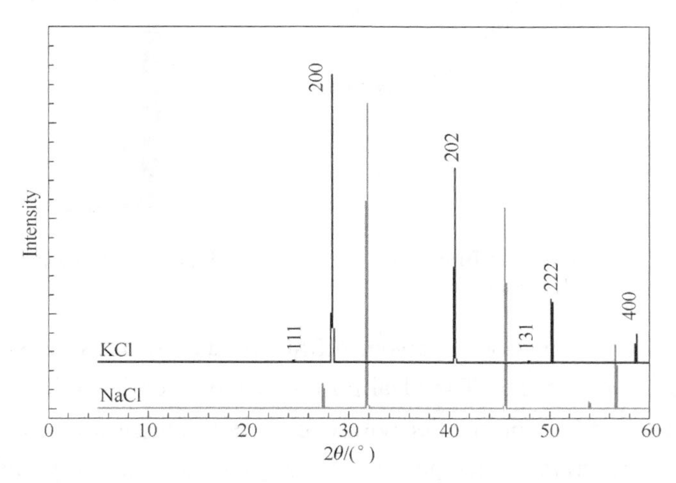

Fig.3.25 X-ray powder diffraction patterns of NaCl (bottom) and KCl (top)

The intensities are a function of structure factors, and it also depends on several sample and instrumental factors. Sample factors include but not limited to preferred orientation, grain size and size distribution. Wavelength, focusing geometry and slit system are several instrumental factors that bias the intensities.

3.3.3.2 X-ray Powder Diffraction Experiments

In the early days, powder diffraction pattern is recorded on a variety of cameras, which are replaced by automated powder diffractometer beginning approximately in the 1970s. There are a large variety of powder diffractometers around the world, but most of them have many common features. The most commonly used instrument configuration with conventional divergent X-ray beam is based on the Bragg-Brentano para-focusing geometry shown in Fig.3.26. Divergent X-ray beam generated by X-ray tube S is monochromatized and focused on F by the monochromator M. Focal point F, flat sample and receiving slit lie on the 'focusing circle', which has a radius dependent on θ. Coherently scattered X-ray from a flat sample then converge on a receiving slit located in front of the detector D. The detector rotates about the goniometer axis through twice the angular rotation of the sample (θ-2θ scans). Powder diffractometer can also be constructed in a way that both the detector and X-ray source arms rotate around a common goniometer axis at the same angular speed in a synchronized fashion while the sample is stationary. This is the θ-θ scan and the geometry commonly used today. A powder diffractometer which takes the vertical Bragg-Brentano parafocusing and θ-θ geometry is shown in Fig.3.27.

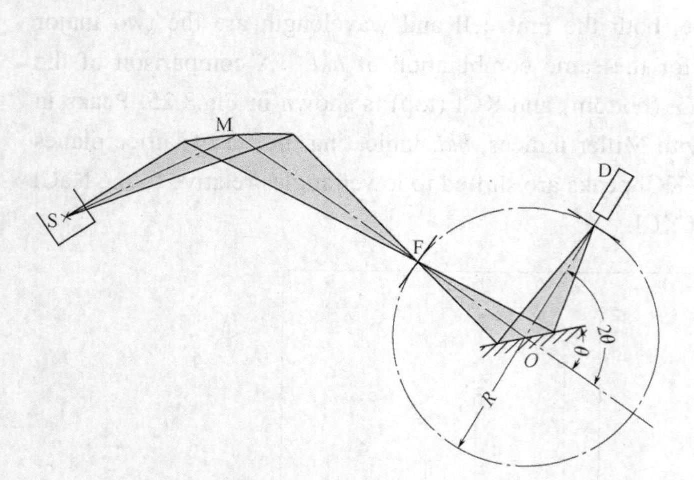

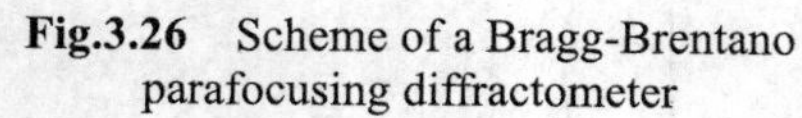

Fig.3.26 Scheme of a Bragg-Brentano parafocusing diffractometer

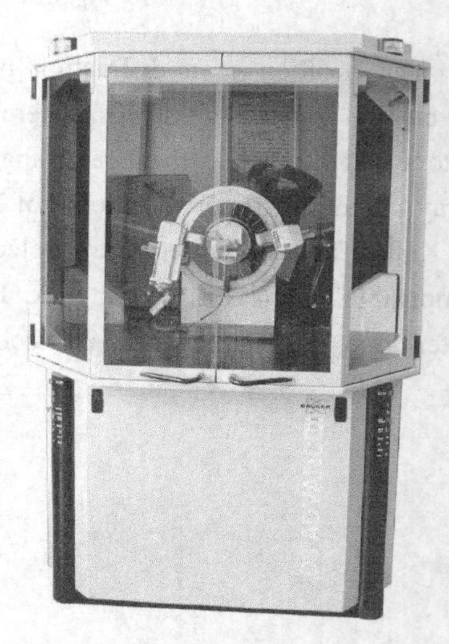

Fig.3.27 A typical powder diffractometer

Many factors affect the quality of powder diffraction data such as the state of the specimen, the wavelength, and the scan speed. The ideal powder sample consists of an enormous number of very tiny crystals of size 50μm or smaller with completely random orientation. Suitable samples may be obtained as fine grained precipitates, or by grinding coarse crystalline materials using mortar and pestle. In many cases, grinding or milling produced particles with very anisotropic shapes and therefore, special precautions must be taken to avoid preferred orientation when mounting powders on sample holder, especially in the case of platelet-like or needle-like particles. The length of the flat sample should be large enough so that at any Bragg angle during the data collection, the projection of the X-ray beam on the sample surface does not exceed the length of the specimen. The thickness of the sample should also be large enough so that the sample is completely opaque to the incident beam. The sample always needs to be properly positioned at the goniometer center. Wavelength of incident X-ray is usually selected based on the materials under investigation and the purpose of the experiments. The most typical radiation is CuK_{α} with a wavelength of 0.154nm. Long wavelength is preferred for accurate determination of lattice constants and short wave length is usually used for the examination of large volume of reciprocal space. Another important consideration is whether or not the materials will produce considerable X-ray fluorescence. For example, nearly all Co-based materials cause severe X-ray fluorescence under CuK_{α}, which is obviously not good for high quality diffraction data. The aperture of the incident beam should be selected to match the diffraction geometry and sample size, and the aperture of the diffracted beam should be selected to maintain both good resolution and good intensity. After all the instrumental parameters have been properly selected, the next step is to determine proper scan mode, scan range, scan step and counting time to obtain high quality data according to the instrument, the sample and what information is going to be obtained from the powder diffraction data.

A typical powder diffraction pattern is collected in the form of scattered intensities as a numerical function of Bragg angles, while what is needed in many applications, such as phase analysis, lattice parameters and structure solution, is a list of integrated intensities, miller indices and Bragg angles. In order to obtain these information, preliminary data processing must be carried out, including peak search, background subtraction, pattern smoothing and K_{α_2} stripping.

3.3.3.3 Interpretation of Powder Diffraction Pattern

Powder diffraction pattern contains lots of information about the sample. Peak positions are determined by the size, shape and symmetry of the unit cell. Peak intensities are determined by the type, number and positions of atom in the unite cell, and peak shape is a convolution of instrumental parameters and microstructure of the sample such as domain size and strain. Although powder diffraction data can be used to determine crystal structures, the goal of the experiment is normally not structure solution, but phase analysis, precise cell parameters, size and strain of crystallites and Rietveld refinement.

Phase composition of a material can be analyzed both qualitatively and quantitatively from powder diffraction data. Each crystalline solid present in the bulk sample has a unique characteristic X-ray powder pattern which may be used as a "fingerprint" for its identification, and the intensity of the peaks are proportional to the amount of this phase. The experimental pattern can be compared to a database of known measured or calculated diffraction patterns to identify the possible phases in the sample. Perhaps the most important application of phase identification in small molecule crystallography is to confirm whether the powder diffraction pattern of a bulk sample corresponds to a structure determined from a single crystal obtained during the same synthesis. It is also possible to obtain quantitative information about the composition of a multiple-phase sample by using various techniques.

Crystal structure of a material may be considered solved if both the size, shape, symmetry of the unit cell and the arrangement of atoms in the unit cell are determined. So, the finding the cell parameters a,b,c,α,β and γ that characterize the size, shape and symmetry of the crystal lattice is unavoidable. Fortunately, powder diffraction data provides this information if suitable single crystals are not available. In addition, very simple and quick phase identification can be made by comparing the cell parameters of a material with a known unit cell. The miller indices of each peak of a diffraction pattern can be calculated from the diffraction data through a process called indexing.

For a powder crystalline material consisting of sufficiently large and strain-free crystallites, theoretical diffraction peaks will be exceedingly sharp. But in actual diffraction experiments, diffraction peaks of such sharpness are never observed because of the convolutional effect of several instrumental and sample factors that broadens the peaks. One of the most important sample factors is the size of crystallites. As the size of the crystallite decreases, the angular spread of the reflection increases. The half height width can be used a measure of the mean particle size of the sample. The first treatment of the particle size broadening was due to Sherrer. And his formula predicts that the peak width and particle size are related by

$$L=K\lambda/(\beta\cos\theta)$$

Where L is the thickness of the crystallite along the direction perpendicular to the planes that gives the diffraction peak, β and θ is the full width at half maximum (FWHM) and Bragg angle of this peak respectively, λ is the wavelength and K is a constant around 1, often taken as 0.89 or 0.94.

For many compounds that are hard to grow suitable single crystals but very good powder diffraction patterns may be obtained easily, it is possible to refine the structure using powder data with the Rietveld method, which is now widely recognized to be uniquely valuable for structural analysis of nearly all classes of crystalline materials in the previous situation. The method is originally designed for netron powder diffraction by H.G. Rietveld in the early 1970s, and is soon followed by its extension to X-ray diffraction. In this method, one can calculate the diffraction intensity y_{calc} at each experimental value of 2θ, using a structure model, a peak shape function, and a back ground function. Then least square refinement are performed to adjust the structural parameters and instrumental parameters so that the difference between y_{calc} and y_{obs} are minimized. Once the agreement is good, the refinement is finished and structure parameters and sample properties are obtained. A example of a Rietveld refinement is given in Fig.3.28.

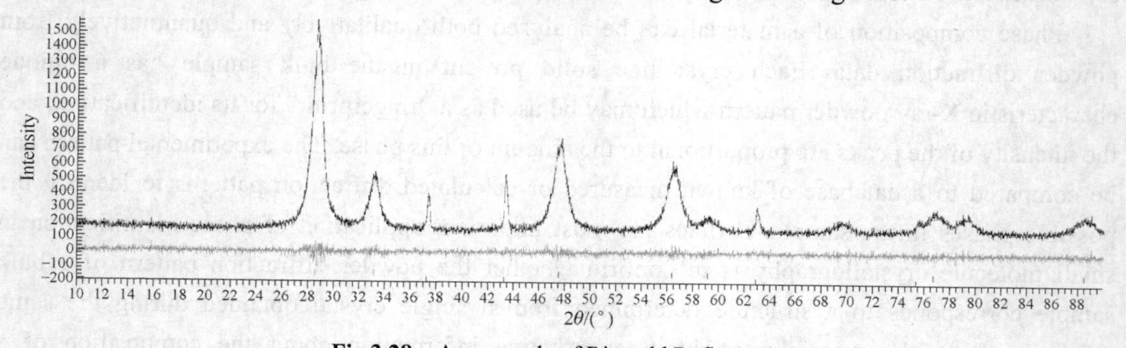

Fig.3.28 An example of Rietveld Refinement

Rietveld method is a powerful refinement and optimization tool which may be used to establish structure details of the sample under investigation. It is important to remember that the Rietveld refinement requires a structure model which the method itself offers no clue on how to create such a model. To build up a model for the refinement, one must do ab initio structure determination from powder diffraction data by using various methods such as Patterson methods, direct methods, charge flipping and simulated annealing. The steps involved in this process are as follows.

- Unit cell determination,
- Decomposition of powder pattern into integrated intensities (optional),
- Assignment of space group from systematic absences (optional),
- Forming an approximate solution using various techniques,
- Refinement of the structure, typically by the Rietveld method.

The ability to determine crystal structures using powder diffraction promises to open up many avenues in structural sciences. However, the determination of structures using powder diffraction data is much more difficult than from single crystal data. This problem arises due to the collapse of the three dimensional crystallographic information into a single dimensional one which is the powder diffraction pattern. This ambiguity creates problems in the determination of the unit cell and accurate intensities. However with the improvements in the instrument and algorithmic developments it is now possible to solve different structures from powder diffraction data alone.

3.4 Photoelectron Spectroscopy

Photoelectron spectroscopy utilizes photo-ionization and energy-dispersive analysis of the emitted photoelectrons to study the composition and electronic state of the surface region of a sample. Traditionally, when the technique has been used for surface studies it has been subdivided according to the source of exciting radiation into:

- X-ray photoelectron spectroscopy (XPS): using soft X-ray (200-2000eV) radiation to examine core-levels.
- Ultraviolet photoelectron spectroscopy (UPS): using vacuum UV (10-45eV) radiation to examine valence levels.

The development of synchrotron radiation sources has enabled high resolution studies to be carried out with radiation spanning a much wider and more complete energy range (5-5000eV) but such work is, and will remain, a very small minority of all photoelectron studies due to the expense, complexity and limited availability of such sources.

Photoelectron spectroscopy is based upon a single photon in/electron out process and from many viewpoints this underlying process is a much simpler phenomenon than the Auger process. The energy of a photon is given by the Einstein equation

$$E=h\nu$$

Where h is Planck constant (6.62×10^{-34}J·s); ν is frequency (Hz) of the radiation. Photoelectron spectroscopy uses monochromatic sources of radiation (i.e. photons of fixed energy).

In XPS the photon is absorbed by an atom in a molecule or solid, leading to ionization and the emission of a core (inner shell) electron. By contrast, in UPS the photon interacts with valence levels of the molecule or solid, leading to ionization by removal of one of these valence electrons. The kinetic energy distribution of the emitted photoelectrons (i.e. the number of emitted photoelectrons as a function of their kinetic energy) can be measured using any appropriate electron energy analyzer and a photoelectron spectrum can thus be recorded.

The process of photoionization can be considered in several ways. one way is to look at the overall process as follows :

$$A+h\nu \longrightarrow A^{+}+e^{-}$$

Conservation of energy then requires that:

$$E(A)+h\nu=E(A^{+})+E(e^{-})$$

Since the electron's energy is present solely as kinetic energy (*KE*), this can be rearranged to give the following expression for the *KE* of the photoelectron:

$$KE=h\nu-[E(A^{+})-E(A)]$$

The final term in brackets, representing the difference in energy between the ionized and neutral atoms, is generally called the binding energy (*BE*) of the electron—this then leads to the following commonly quoted equation:

$$KE=h\nu-BE$$

The *BE* is now taken to be a direct measure of the energy required to just remove the electron concerned from its initial level to the vacuum level.

The basic requirements for a photoemission experiment (XPS or UPS) are:

- A source of fixed-energy radiation (an X-ray source for XPS or, typically, a He discharge lamp for UPS).
- An electron energy analyzer (which can disperse the emitted electrons according to their kinetic energy, and thereby measures the flux of emitted electrons of a particular energy).
- A high vacuum environment (to enable the emitted photoelectrons to be analyzed without interference from gas phase collisions).

Such a system is illustrated schematically in Fig.3.29.

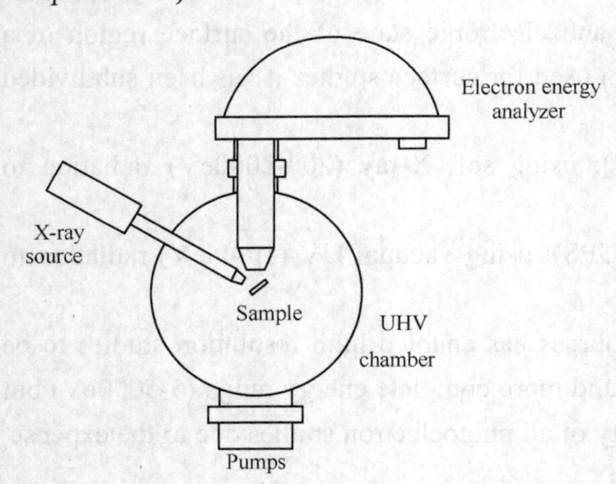

Fig.3.29 The illustration of the experimental system for the photoelectron spectroscopy

There are many different designs of electron energy analyzer but the preferred option for photoemission experiments is a concentric hemispherical analyzer (CHA) that uses an electric field between two hemispherical surfaces to disperse the electrons according to their kinetic energy.

3.4.1 X-ray Photoelectron Spectroscopy (XPS)

For each and every element, there will be a characteristic binding energy associated with each core atomic orbital, i.e. each element will give rise to a characteristic set of peaks in the photoelectron spectrum at kinetic energies determined by the photon energy and the respective binding energies.

The presence of peaks at particular energies therefore indicates the presence of a specific element in the sample under study, furthermore, the intensity of the peaks is related to the concentration of the element within the sampled region. Thus, the technique provides a quantitative analysis of the surface composition and is sometimes known by the alternative acronym, ESCA (Electron Spectroscopy for Chemical Analysis).

The most commonly employed X-ray sources are those giving rise to: Mg-K_α radiation ($h\nu$=1253. 6eV), and Al K_α radiation: $h\nu$=1486. 6eV. The emitted photoelectrons will therefore have kinetic energies in the range of ca. 0-1250eV or 0-1480eV. Since such electrons have very short IMFPs (Inelastic Mean Free Paths) in solids, the technique is necessarily surface sensitive.

Example: the XPS spectrum of Pd metal

The diagram in Fig.3.30 shows a real XPS spectrum obtained from a Pd metal sample using Mg K_α radiation. The main peaks occur at kinetic energies of ca. 330eV, 690eV, 720eV, 910eV and 920eV.

Since the energy of the radiation is known, it is a trivial matter to transform the spectrum so that it is plotted against *BE* as opposed to *KE* (Fig.3.31).

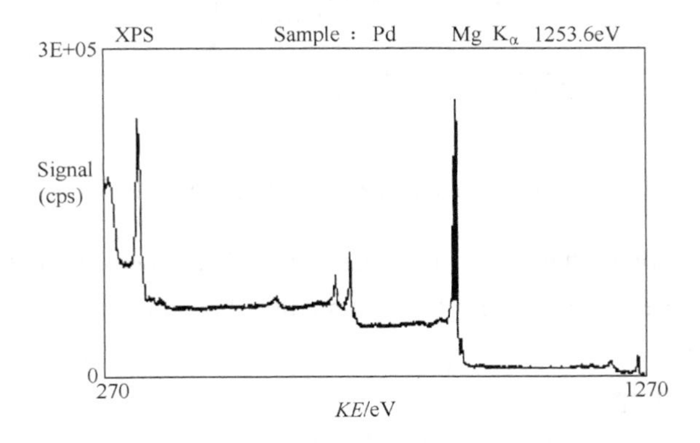

Fig.3.30 XPS spectrum obtained from a Pd metal sample using Mg K_α radiation

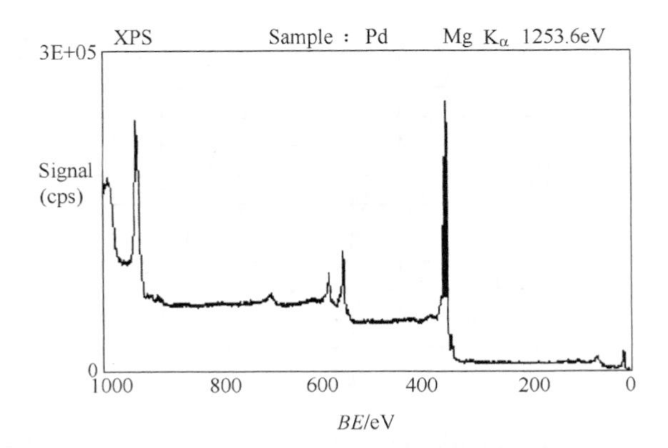

Fig.3.31 XPS spectrum of a Pd metal sample plotted against bond energy (*BE*)

The most intense peak is now seen to occur at a binding energy of ca. 335eV.

Working downwards from the highest energy levels:

- The valence band (4d, 5s) emission occurs at a binding energy of ca. 0-8eV.
- The emission from the 4p and 4s levels gives rise to very weak peaks at 54 and 88eV, respectively.
- The most intense peak at ca. 335eV is due to emission from the 3d levels of the Pd atoms, whilst the 3p and 3s levels give rise to the peaks at ca. 534/561eV and 673eV, respectively.
- The remaining peak is not an XPS peak at all ! It is an Auger peak arising from X-ray induced Auger emission.

These assignments are summarized in the picture (Fig.3.32).

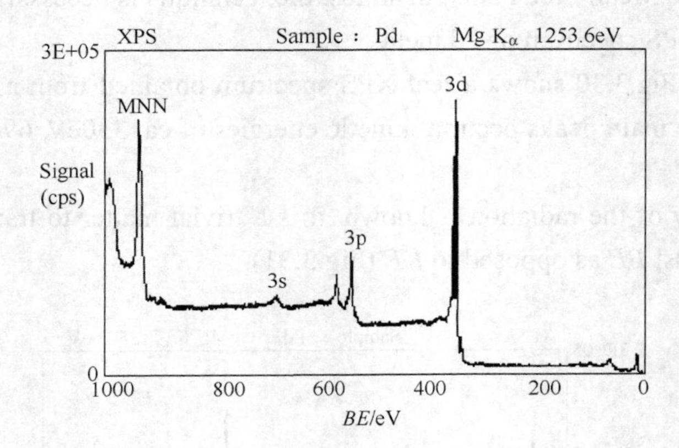

Fig.3.32 The assignments of the XPS peaks of Pd metal

It may be further noted that there are significant differences in the natural widths of the various photoemission peaks and the peak intensities are not simply related to the electron occupancy of the orbitals.

3.4.1.1 Spin-Orbit Splitting

Closer inspection of the spectrum shows that emission from some levels (most obviously 3p and 3d) does not give rise to a single photoemission peak, but a closely spaced doublet (Fig.3.33).

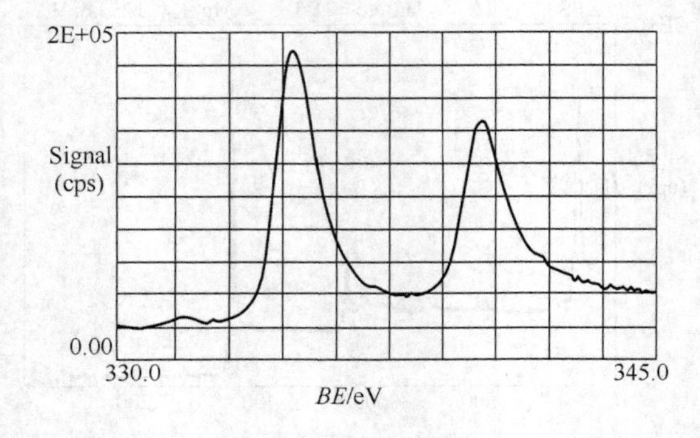

Fig.3.33 The splitting of the XPS peaks of 3d photoemission

We can see this more clearly if, for example, we expand the spectrum in the region of the 3d emission. The 3d photoemission is in fact split between two peaks, one at 334. 9eV *BE* and the other at 340.2eV *BE*, with an intensity ratio of 3 : 2. This arises from spin-orbit coupling effects in the final state.

The inner core electronic configuration of the initial state of the Pd is:

$$(1s)^2(2s)^2(2p)^6(3s)^2(3p)^6(3d)^{10}\cdots$$, with all sub-shells completely full

The removal of an electron from the 3d sub-shell by photo-ionization leads to a $(3d)^9$ configuration for the final state. since the d-orbitals (l=2) have non-zero orbital angular momentum, there will be coupling between the unpaired spin and orbital angular moment. Spin-orbit coupling is generally treated using one of two models which correspond to the two limiting ways in which the coupling can occur—these being the LS (or Russell-Saunders) coupling approximation and the *j-j* coupling approximation.

If we consider the final ionized state of Pd within the LS (Russell-Saunders) coupling approximation, the $(3d)^9$ configuration gives rise to two states (ignoring any coupling with valence levels) that differ slightly in energy and in their degeneracy .

$$^2D_{5/2} \quad g_J=2\times(5/2)+1=6$$
$$^2D_{3/2} \quad g_J=2\times(3/2)+1=4$$

These two states arise from the coupling of the L=2 and S=1/2 vectors to give permitted J values of 3/2 and 5/2 (Fig.3.34). The lowest energy final state is the one with maximum J(since the shell is more than half-full), i.e. J=5/2, hence this gives rise to the "lower binding energy"peak. The relative intensities of the two peaks reflect the degeneracies of the final states (g_J=2J+1), which in turn determine the probability of transition to such a state during photoionization.

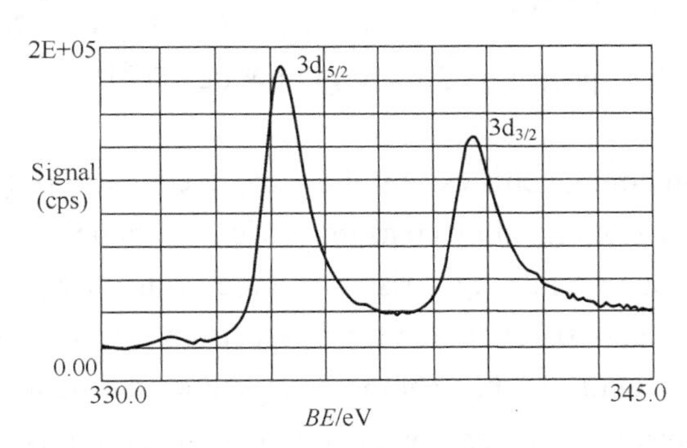

Fig.3.34 The relative intensities of the two peaks in the 3d XPS photoemission

The peaks themselves are conventionally annotated as indicated-note the use of lower case lettering. This spin-orbit splitting is of course not evident with s-levels (L=0), but is seen with p, d and f core-levels which all show characteristic spin-orbit doublets.

3.4.1.2 Chemical Shifts

The exact binding energy of an electron depends not only upon the level from which photoemission is occurring, but also upon ① the formal oxidation state of the atom and ② the local chemical and physical environment.

Changes in either ① or ② give rise to small shifts in the peak positions in the spectrum, so-called chemical shift.

Such shifts are readily observable and interpretable in XPS spectra(unlike in Auger spectra)because the technique is of high intrinsic resolution (as core levels are discrete and generally of a well-defined energy) and an electron process.

Atoms of a higher positive oxidation state exhibit a higher binding energy due to the extra coulombic interaction between the photo-emitted electron and the ion core. This ability to discriminate between different oxidation states and chemical environments is one of the major strengths of the XPS technique.

In practice, the ability to resolve between atoms exhibiting slightly different chemical shifts is limited by the peak widths which are governed by a combination of factors, especially

- the intrinsic width of the initial level and the lifetime of the final state.
- the line-width of the incident radiation—which for traditional X-ray sources can only be improved by using X-ray monochromators.
- the resolving power of the electron-energy analyzer.

In most cases, the second factor is the major contribution to the overall line width.

For example, oxidation states of titanium can be considered. Titanium exhibits very large chemical shifts between different oxidation states of the metal;in Fig.3.35, a Ti 2p spectrum from the pure metal (Ti^0) is compared with a spectrum of titanium dioxide (Ti^{4+}).

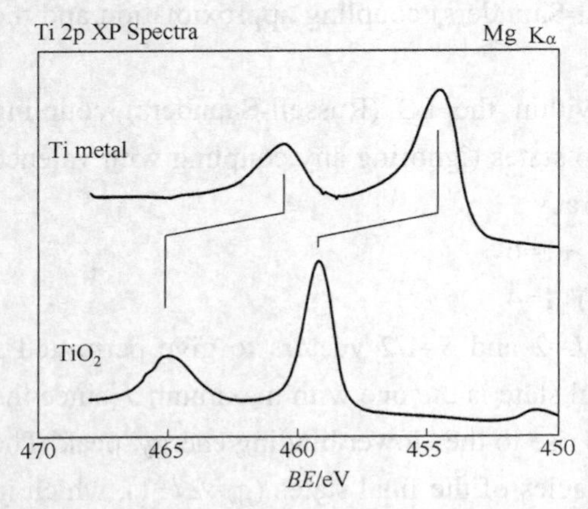

Fig.3.35 The chemical shifts in XPS spectrum of pure metal (Ti^0) and titanium dioxide (Ti^{4+})

Note:

- The two spin orbit components exhibit the same chemical shift (~4.6eV).
- Metals are often characterized by an asymmetric line shape, with the peak tailing to higher binding energy whilst insulating oxides give rise to a more symmetric peak profile.
- The weak peak at ca. 450.7eV in the lower spectrum arises because typical X-ray sources also emit some X-rays of slightly higher photon energy than the main Mg K_α line;this peak is a "ghost"of the main $2p_{3/2}$ peak arising from ionization by these additional X-rays.

3.4.1.3 Angle Dependent Studies

The degree of surface sensitivity of an electron-based technique such as XPS may be varied by photoelectrons emitted at different emission angles to the surface plane. This approach may be used to perform non-destructive analysis of the variation of surface composition with depth (with chemical state specificity).

For example, angle-dependent analysis of a silicon wafer with a native oxide surface layer (Fig 3.36). A series of Si 2p photoelectron spectra are recorded for emission angles of 10°-90° to the surface plane. Note how the Si 2p peak of the oxide (*BE*~103eV) increases markedly in intensity at grazing emission angles whilst the peak from the underlying elemental silicon (*BE*~99eV) dominates the spectrum at near-normal emission angles.

3.4.2 Ultraviolet Photoelectron Spectroscopy (UPS)

In UPS the source of radiation is normally a noble gas discharge lamp; frequently a Hedischarge lamp emitting He-I radiation of energy 21.2eV.

Such radiation is only capable of ionizing electrons from the outermost levels of atoms—the valence levels. The advantage of using such UV radiation over X-ray is the very narrow line width of the radiation and the high flux of photons available from simple discharge sources.

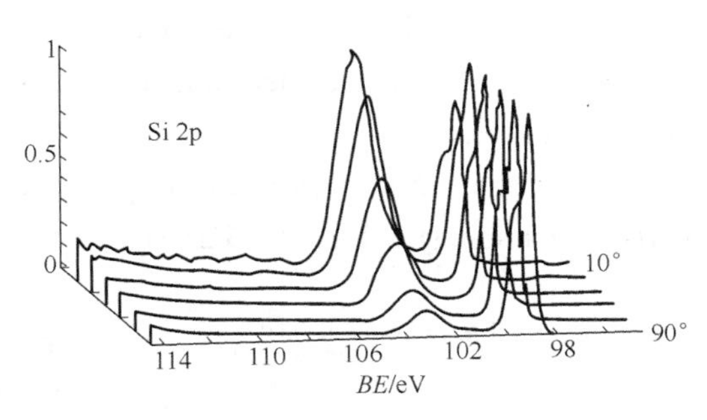

Fig.3.36 Angle-dependent analysis of a silicon wafer with a native oxide surface layer

The main emphasis of work using UPS has been in studying:

- The electronic structure of solids—detailed angle resolved studies permit the complete band structure to be mapped out in k-space.
- The adsorption of relatively simple molecules on metals—by comparison of the molecular orbitals of the adsorbed species with those of both the isolated molecule.

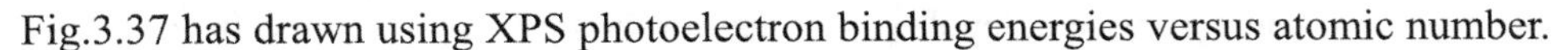

Fig.3.37 has drawn using XPS photoelectron binding energies versus atomic number.

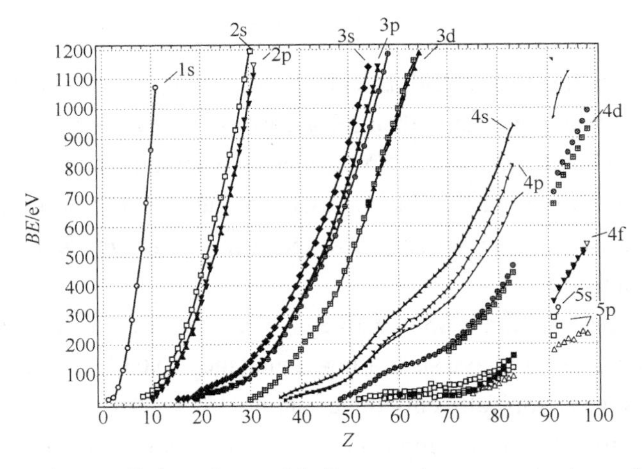

Fig.3.37 XPS photoelectron binding energies versus atomic number

3.5 Nuclear Magnetic Resonance (NMR) Spectroscopy

Nuclear magnetic resonance spectroscopy is a powerful and theoretically complex analytical tool. NMR is performing experiments on the nuclei of atoms, not the electrons. The chemical environment of specific nuclei is deduced from information obtained about the nuclei.

3.5.1 Basic Principle of NMR Spectroscopy

Subatomic particles (electrons, protons and neutrons) can be imagined as spinning on their axes. In many atoms (such as ^{12}C) these spins are paired against each other, such that the nucleus of the atom has no overall spin. However, in some atoms (such as ^{1}H and ^{13}C) the nucleus does possess an overall spin.

The rules for determining the net spin of a nucleus are as follows:

- If the number of neutrons and the number of protons are both even, then the nucleus has no overall spin.
- If the number of neutrons plus the number of protons is odd, then the nucleus has a half-integer spin (i.e. 1/2, 3/2, 5/2···).

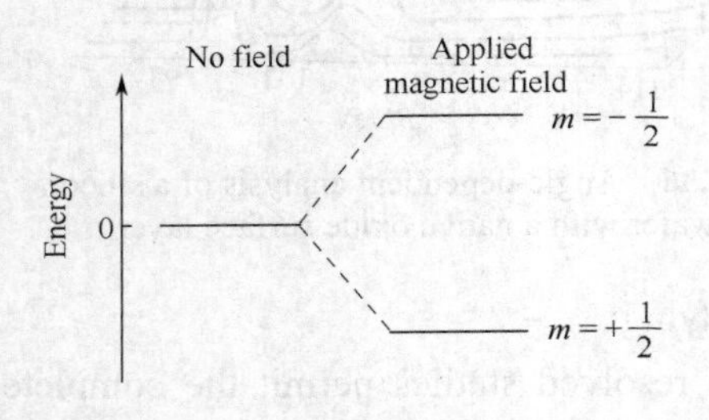

Fig.3.38 Energy level for a nucleus with spin quantum number(1/2) in a applied magnetic field

- If the number of neutrons and the number of protons are both odd, then the nucleus has an integer spin (i.e. 1, 2, 3···).
- The overall spin, I, is important. Quantum mechanics tells us that a nucleus of spin I will have $2I+1$ possible orientations. A nucleus with spin 1/2 will have 2 possible orientations. In the absence of an external magnetic field, these orientations are of equal energy. If a magnetic field is applied, then the energy levels split. Each level is given a magnetic quantum number, m (Fig.3.38).

The difference in energy between levels (the transition energy) can be found from

$$\Delta E=\frac{\gamma hB}{2\pi}$$

This means that if the magnetic field, B, is increased, so is ΔE. It also means that if a nucleus has a relatively large magnetogyric ratio, then ΔE is correspondingly large.

In addition to the ^{1}H and may be ^{13}C NMR, there are actually dozens of other nuclei that are also amenable to study using NMR. Just because a nucleus has a non-zero spin does not automatically mean that we can obtain an NMR spectrum of that nucleus, there are at least four other factors we must consider:

- Isotopic abundance Some spin active nuclei such as ^{19}F are 100% abundant(^{1}H is 99.985%), but others such as ^{17}O have such a low abundance (0.037%) that we can't expect to get much of a signal unless we isotopically enrich the sample. ^{13}C is only 1.1% abundant, that' s one of the reasons we need to use a lot more sample and take more scans to obtain a ^{13}C spectrum versus a ^{1}H spectrum.
- Sensitivity This is typically scaled to ^{1}H=1.0. There is only one (rather uncommon) isotope with a sensitivity higher than 1.0; all other isotopes have sensitivities less than 1.0. The lower the sensitivity, the greater the amount of time and sample it will take to get a signal. Some sensitivities are so low, such as ^{103}Rh (100% abundant but only 0.000031 sensitivity), that obtaining a spectrum for the nucleus is generally impractical. However, the nucleus can still couple to other spin-active nuclei and provide useful information provided it has good abundance.
- Nuclear quadrapole For spins greater than 1/2, the nuclear quadrapole moment is usually larger and the line widths may become excessively large. This can sometimes be overcome by running the sample at low temperature.
- Relaxation time Two factors govern the rate at which the excited nucleus relaxes its spin.

Spin-lattice relaxation, T_1, is the time it takes for a nucleus to relax. Spin-spin relaxation, T_2, refers to how long it takes for a set of aligned nuclei to lose their phase coherence. T_1 is usually much greater than T_2, so we normally only concern ourselves with T_1.

The combination of these four factors governs whether a given nucleus will give a useful NMR spectrum. Some of the more common I=1/2 nuclei used by chemists are listed in Table 3.6.

Table 3.6 The more common I= 1/2 nuclei used in NMR

Nucleus	Natural Abundance/%	Relative Sensitivity	Nucleus	Natural Abundance/%	Relative Sensitivity
^1H	99. 985	1. 0	^{19}F	100	0. 83
^{13}C	1. 108	0. 016	^{31}P	100	0. 07

3.5.2 The Nuclear Absorption of Radiation in Magnetic Field

Imagine a nucleus (of spin 1/2) in a magnetic field. This nucleus is in the lower energy level (i.e. its magnetic moment does not oppose the applied field). The nucleus is spinning on its axis. In the presence of a magnetic field, this axis of rotation will process around the magnetic field (Fig.3.39).

The potential energy of the processing nucleus is given by;

$$E=-\mu B\cos\theta$$

where θ is the angle between the direction of the applied field and the axis of nuclear rotation.

If energy is absorbed by the nucleus, then the angle of precession, θ, will change. For a nucleus of spin 1/2, absorption of radiation“flips”the magnetic moment so that it opposes the applied field (the higher energy state) (Fig.3.40).

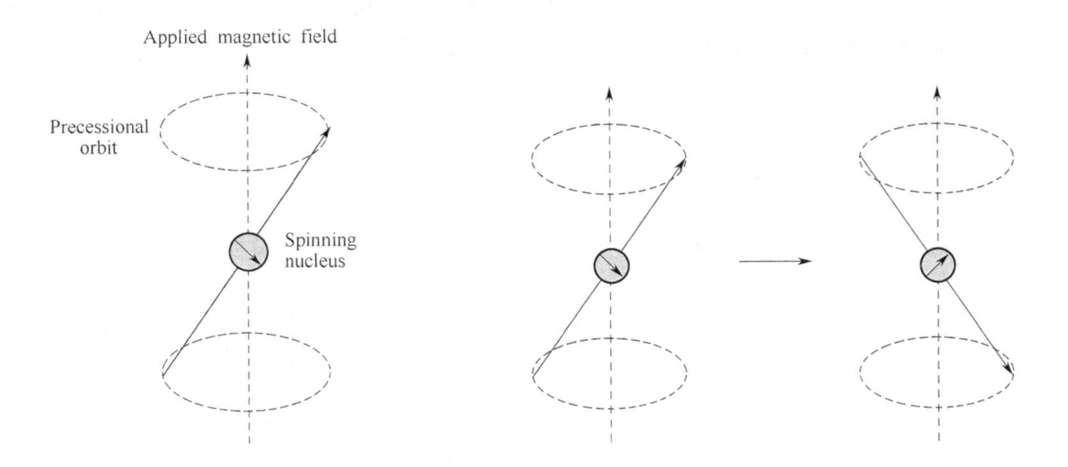

Fig.3.39 The illustration of the axis of rotation processing in the applied magnetic field

Fig.3.40 The magnetic moment opposes the applied field

3.5.3 Chemical Shift

The magnetic field at the nucleus is not equal to the applied magnetic field;electrons around the nucleus shield it from the applied field. The difference between the applied magnetic field and the field at the nucleus is termed the nuclear shielding.

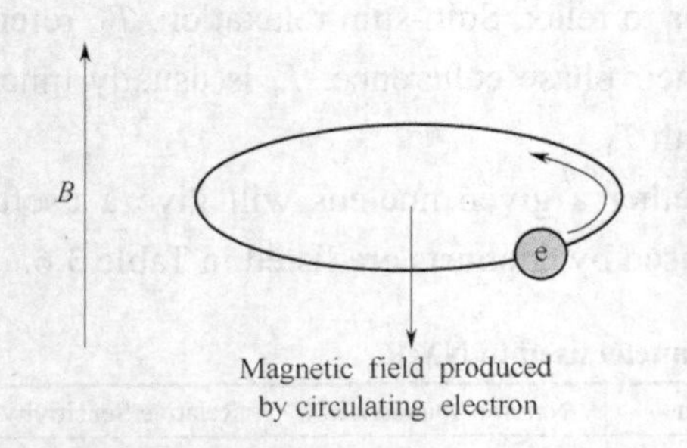

Fig.3.41 Producing a magnetic field by the s-electrons opposing the applied field

Considering the s-electrons in a molecule, They have spherical symmetry and circulate in the applied field, producing a magnetic field that opposes the applied field (Fig.3.41). This means that the applied field strength must be increased for the nucleus to absorb at its transition frequency. This upfield shift is also termed diamagnetic shift.

Chemical shift is defined as nuclear shielding/ applied magnetic field. Chemical shift is a function of the nucleus and its environment. It is measured relative to a reference compound. For ^1H NMR, the reference is usually tetramethylsilane, $Si(CH_3)_4$.

3.5.4 Spin-Spin Coupling

The most important aspect of multinuclear NMR is that all spin active nuclei can couple to each other and that the multiplicity of the coupling is given by $2nI+1$ where n=the number of equivalent nuclei that are being coupled to. For example, if a proton is adjacent to two equivalent protons, the resonance will appear as a triplet because $2nI+1=2\times(2)\times(1/2)+1=3$. If the two adjacent nuclei were fluorine atoms instead of hydrogen atoms the resonance would still be triplet because I=1/2 for fluorine and ^{19}F is 100% abundant. The only difference in these two cases would be the magnitude of the coupling constants, J_{HH} versus J_{HF}.

The ^1H NMR spectrum of ethanol (Fig.3.42) shows the methyl peak has been split into three peaks (a triplet) and the methylene peak has been split into four peaks (a quartet). This occurs because there is a small interaction (coupling) between the two groups of protons.

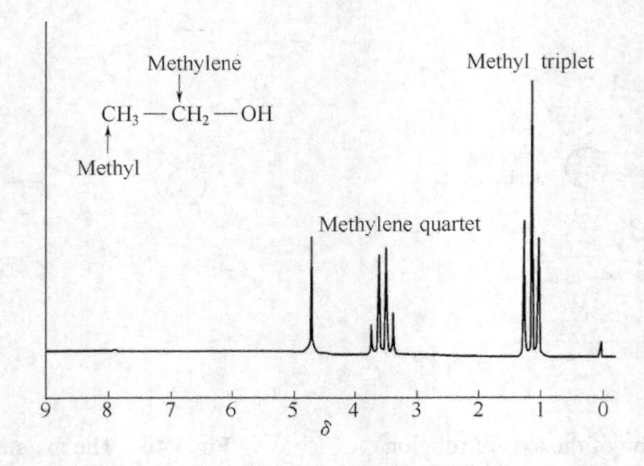

Fig.3.42 The ^1H NMR spectrum of ethanol

In a first-order spectrum (where the chemical shift between interacting groups is much larger than their coupling constant), interpretation of splitting patterns is quite straightforward.

- The multiplicity of a multiplet is given by the number of equivalent protons in neighbouring atoms plus one, i.e. the n+1 rule.

- Equivalent nuclei do not interact with each other. The three methyl protons in ethanol cause splitting of the neighbouring methylene protons, they do not cause splitting among themselves.
- The coupling constant is not dependant on the applied field. Multiplets can be easily distinguished from closely spaced chemical shift peaks.

3.5.5 Some Chemical Shifts in ^1H NMR and ^{13}C NMR

Some important chemical shifts in ^1H NMR and ^{13}C NMR are shown in Table 3.7 and Table 3.8.

Table 3.7 Proton chemical shifts

Type of proton	Chemical shift (δ)	Type of proton	Chemical shift (δ)
Alkyl, RCH_3	0.8—1.0	Ketone, $RCOCH_3$	2.1—2.6
Alkyl, RCH_2CH_3	1.2—1.4	Aldehyde, RCOH	9.5—9.6
Alkyl, R_3CH	1.4—1.7	Vinylic, $R_2C{=}CH_2$	4.6—5.0
Allylic, $R_2C{=}CRCH_3$	1.6—1.9	Vinylic, $R_2C{=}CRH$	5.2—5.7
Benzylic, $ArCH_3$	2.2—2.5	Aromatic, ArH	6.0—9.5
Alkyl chloride, RCH_2Cl	3.6—3.8	Acetylenic, $RC{\equiv}CH$	2.5—3.1
Alkyl bromide, RCH_2Br	3.4—3.6	Alcohol hydroxyl, ROH	0.5—6.0①
Alkyl iodide , RCH_2I	3.1—3.3	Carboxylic, RCOOH	10—13①
Ether, $ROCH_2R$	3.3—3.9	Phenolic, ArOH	4.5—7.7①
Alcohol, $HOCH_2R$	3.3—4.0	Amino, $R{-}NH_2$	1.0—5.0①

① The chemical shifts of these protons vary in different solvents with temperature and concentration.

Table 3.8 Some important chemical shifts in ^{13}C NMR

Type of carbon atom	Chemical shift (δ)	Type of carbon atom	Chemical shift (δ)
Alkyl, RCH_3	0—40	Alkene, $R_2C{=}$	100—170
Alkyl, RCH_2R	10—50	Benzylic carbon	100—170
Alkyl, $RCHR_2$	15—50	Nitriles, $-C{\equiv}N$	120—130
Alkyl halide or amine, $(CH_3)_3C{-}X$ (X=Cl, Br, NR_2)	10—65	Amides, $-CONR_2$	150—180
		Carboxylic acids, esters, $-COOH$	160—185
Alcohol or ether, R_3COR	50—90	Aldehydes, ketones, $-C{=}O$	182—215
Alkyne, $-C{\equiv}$	60—90		

3.6 Electron Paramagnetic Resonance (EPR)

EPR is a spectroscopic technique that detects chemical species that have unpaired electrons.

Substances having a positive magnetic susceptibility are paramagnetic. They are attracted by a magnetic field. A great number of materials contain such paramagnetic entities: electrons in unfilled conduction bands, electrons trapped in radiation damaged sites, free radicals, various transition ions, triplet states, impurities in semi-conductors.

The unpaired electron's spin magnetic moment are very sensitive to local magnetic fields within the sample. These fields often arise from the nuclear magnetic moments of various nuclei that may be present within the bulk medium. EPR provides a unique means of studying the internal structures in great detail.

EPR has been successfully applied in such diverse disciplines as biology, physics, geology, chemistry, medical science, material science. Solids, liquids and gases are all accessible to EPR. By utilizing a variety of specialized techniques (such as spin-trapping, spin-labeling, ESEEM and ENDOR) in conjunction with EPR, researchers are capable of obtaining detailed information about many topics of scientific interest, for example, chemical kinetics, electron exchange, electrochemical processes, crystalline structure, fundamental quantum theory, catalysis, and polymerization reactions.

By application of a strong magnetic field B to material containing paramagnetic species, the individual magnetic moment arising via the electron "spin" of the unpaired electron can be oriented either parallel or anti-parallel to the applied field. This creates distinct energy levels for the unpaired electrons, making it possible for net absorption of electromagnetic radiation (in the form of microwaves) to occur. The situation referred to as the resonance condition takes place when the magnetic field and the microwave frequency are "just right" (Fig.3.43).

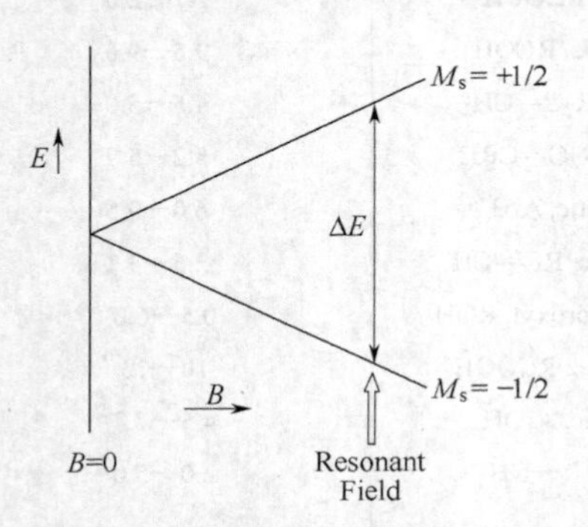

Fig.3.43 Energy-level diagram for two spin states as a function of applied field B

This represents the simplest EPR transition (e.g., free electrons).

When $|\Delta M_s|=1$ (M_s is the magnetic spin quantum number of the spin state), the transition is allowed. The equation describing the absorption (or emission)of microwave energy between two spin states is

$$\Delta E = h\nu = g\beta B$$

where ΔE is the energy difference between the two spin states; h is Planck constant; ν is the microwave frequency; g is the Zeeman splitting factor; β is the Bohr magneton; B is the applied magnetic field.

3.7 Circular Dichroism(CD)

A spectroscopic method that measures the difference in absorbance of left-and right-handed circularly polarized light by a material, as a function of the wavelength.

Most biological molecules, including proteins and nucleic acids are chiral and show circular dichroism (CD) in their ultraviolet absorption bands, which may be used as an indication of secondary structure. Metal centers that are bound to such molecules, even if they have no inherent chirality, usually exhibit CD in absorption bands associated with ligandbased or ligand-metal charge-transfer transitions. CD is frequently used in combination with absorption and MCD studies to assign electronic transitions.

Circular dichroism spectroscopy is particularly good for:

- Determining whether a protein is folded, and if so characterizing its secondary structure, tertiary structure, and the structural family to which it belongs.
- Comparing the structures of a protein obtained from different sources or comparing structures for different mutants of the same protein.
- Demonstrating comparability of solution conformation after changes in manufacturing processes or formulation.
- Studying the conformational stability of a protein under stress—thermal stability, pH stability, and stability to denaturants—and how this stability is altered by buffer composition or addition of stabilizers and excipients.
- CD is excellent for finding solvent conditions that increase the melting temperature and/or the reversibility of thermal unfolding, conditions that generally enhance shelf life.
- Determining whether protein-protein interactions alter the conformation of protein. If there are any conformational changes, this will result in a spectrum that will differ from the sum of the individual components. Small conformational changes have been seen, for example, upon formation of several different receptor/ligand complexes.

The *cis*-[CuNa(GMP)(HGMP)] (**1**) is a transition metal coordination complex of chiral GMP ligand, in which one of the GMP ligands is protonation and is $HGMP^-$. Both of GMP^{2-} ligand and $HGMP^-$ ligand coordinate to Cu(Ⅱ) ion with the N7 site. The coordination geometry of Cu(Ⅱ) ion is square pyramidal with three coordinated water molecules. Two GMP ligands coordinate with the Cu(Ⅱ) center in a face-to-face manner and the glycosyl group of one ligand is located above the subface of the square pyramidal whereas the other is below (Fig.3.44). The dihedral angle of the purine ring-Cu^{2+}-purine ring is 56.6°. Because the GMP ligand is chiral and coordinated with Cu(Ⅱ) ion in an asymmetric mode, the chirality of GMP ligand is preserved in this coordination complex and it can be detected based on liquid-state CD measurements.

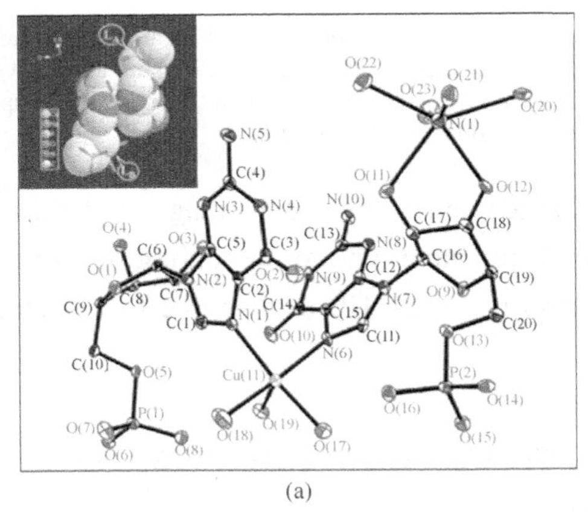

(a)

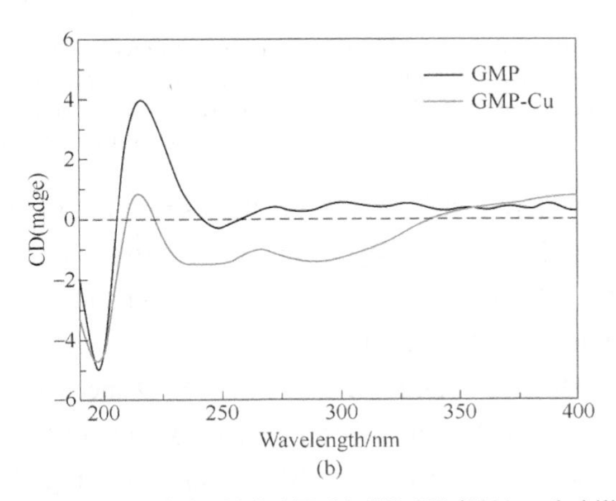

(b)

Fig.3.44 (a) Molecular structure of complex **1**: [CuNa(GMP)(HGMP) $(H_2O)_7$]·$6(H_2O)$·CH_3OH (50% probability). Solvent molecules and hydrogen atoms are not shown for clarity. Inset: Space-filling model of this complex. Ball and stick models are shown for Na, P, O, and Cu atoms for clarity. (b) Liquid-state CD spectra of the GMP ligand and Cu-GMP complex (0.1 mg/mL)

The single crystal structure of *cis*-[CuNa(GMP)(HGMP)] (**1**) shows that the mononuclear coordination complex forms a one-dimensional supramolecular helix chain assembled through

H-bonding interactions between coordinated water molecules and the functional groups of nucleotide ligands [Fig.3.45(a) and (b)]. The Solid-state CD spectra of the GMP ligand and Cu-GMP complex is different from the solution CD spectra. The supramolecular chirality of the H-bonding generated 1D *M*-helix chain can be studied based on single crystal structure analysis and CD measurements [Fig.3.45(c)]. So, the CD signals related to the supramolecular *M*-helix were captured successfully in the solid-state CD spectrum. Therefore, it is an effective way to understand the chirality of coordination complexes based on comprehensive analysis of solution CD and solid-state CD spectroscopy.

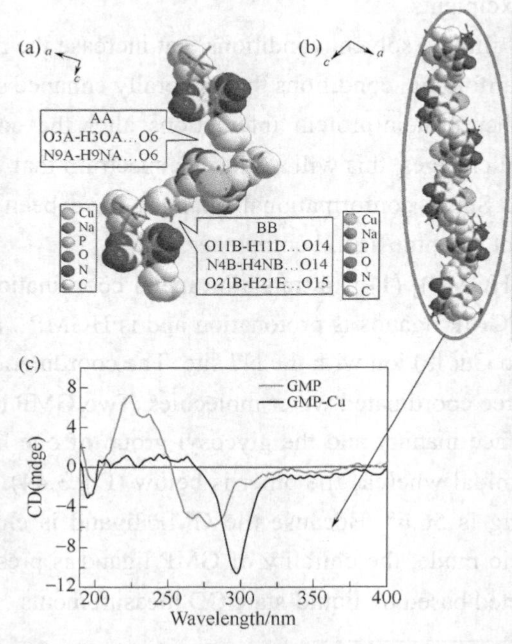

Fig.3.45 (a) Space-filling model of intermolecular hydrogen bonds in the construction of a supramolecular helix. (b) The space-filling model of a helical chain of complex **1**. (c) Solid-state CD spectra of the GMP ligand and Cu-GMP complex (sample:KCl = 1:40). The strong negative Cotton Effect (CE) centered at 297nm due to excitation coupling of the π - π* transitions of aromatic chromophores in the M-helix

Materials that are achiral still exhibit MCD (the Faraday effect), since the magnetic field leads to the lifting of the degeneracy of electronic orbital and spin states and to the mixing of electronic states. MCD is frequently used in combination with absorption and CD studies to effect electronic assignments (Fig.3.46).

The three contributions to the MCD spectrum are:

- the A-term due to Zeeman splitting of the ground and/or excited degenerate states.
- the B-term due to field-induced mixing of states.
- the C-term due to a change in the population of molecules over the Zeeman sublevels of a paramagnetic ground state.

The C-term is observed only for molecules with ground-state paramagnetism, and becomes intense at low temperatures;its variation with field and temperature can be analyzed to provide magnetic parameters of the ground state, such as spin, *g*-factor, and zero-field splitting. Variable-temperature MCD is particularly effective in identifying and assigning electronic transitions originating from paramagnetic chromophores.

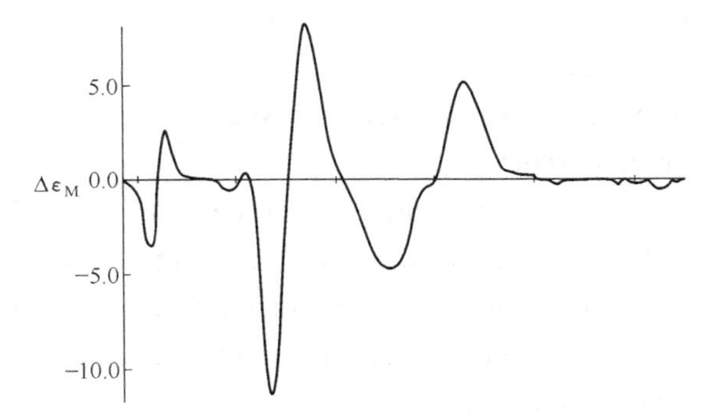

Fig.3.46 MCD spectrum for Pt $[P(t\text{-}Bu)_3]_2$ in acetonitrile

3.8 Advanced Imaging Techniques for Nanosize Coordination Complex

In order to have a deep comprehensive understanding on nano-scaled materials including nano-sized coordination complex, a considerable amount of effort has been devoting to developing characterization techniques or establishing standard evaluation methods. Nowadays, there are many advanced techniques for charactering the materials, each of which has some special advantages. In the following section, we will introduce some advanced imaging techniques including three kinds of electron microscopes (scanning electron microscopy, transmitting electron microscopy, and cryo-electron microscopy) and scanning probe microscopies (scanning tunneling microscopy and atomic force microscopy).

3.8.1 Electron Microscopies

As a kind of optical system, the microscope has been widely used to transform very small objects into magnification images. In order to achieve a visible image, the light microscopy usually detect the light reflected, transmitted, scattered or absorbed by the objects when the light falling on its surface. While the electron microscopy applies an electron beam to impact on the specimen to obtain the surface topography and structural information. An obvious difference with the light (an electromagnetic radiation with a wavelength range of 400-700nm) is that the electron is a kind of radiation with wavelengths between about 0.001 and 0.01nm. This feature allows the electron microscopy to achieve a higher resolution than that of light microscopy, suggesting a very powerful and potential analytical system. Therefore, before going to deal with the electron microscopes, it is useful to consider in more details about possible interactions between high-energy electrons and atoms of the specimen.

Usually, the elastic or inelastic scattering would happen once the primary electrons reach the specimen atoms, resulting in emission of an electron and/or electromagnetic radiation. In an elastic scattering, the primary electrons may pass through the atom close to the nucleus and suffer only a change in direction, which is an important detector for atomic number and contributor to diffraction patterns. For the inelastic scattering, it may causes the primary electron to change the direction and lose a certain amount of energy. Both the elastic and inelastic scattering phenomenon is very useful for imaging and analysis. The various outcomes of the interaction are displayed schematically in Fig.3.47. The representative signals are:

- Secondary electrons (SEs): escaping from a region very close to the surface of specimen with energies less than 50eV; therefore, the yield of SEs is high enough to be used as imaging signals in scanning electron microscopy.
- Backscattered electrons (BSEs): emitting from the specimen with small energy (BSE). It can be used for imaging, diffraction and analysis in the scanning electron microscopy.
- Transmitted electrons: applying in transmission electron microscopy.
- X-ray: emitting if a single outer electron jumps into the inner shell vacancy; this is characteristic of the particular atomic species.
- Auger electrons: emitting as a particular type of interaction between a primary electron and inner shell of electrons around an atomic nucleus in the specimen with an energy up to 1-2 keV. It is useful for elemental analysis of surfaces.

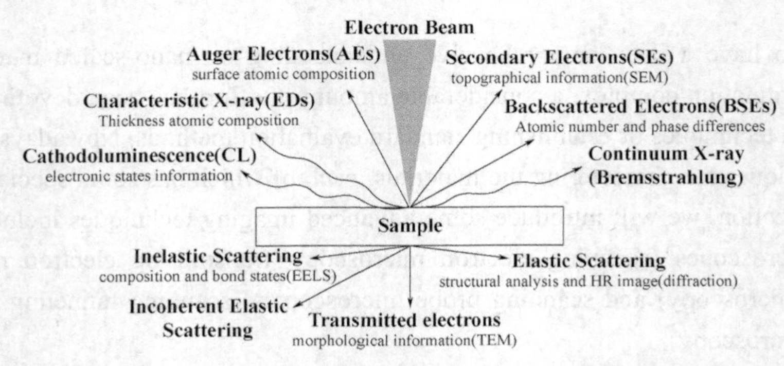

Fig.3.47 Schematic diagram illustrating the signals resulting from the interaction between the electron beam and the sample in an electron microscope

3.8.1.1 Scanning electron microscopy

The scanning electron microscopy (SEM) utilizes a focused high-energy beam of electrons to scan across the specimen. The electrons will interact with the atoms of the specimen, and then produce reflected electrons and other signals to form the image. Fig.3.48 is a schematic diagram showing the essential parts of a simple SEM system. The components are displayed as follows:

- Electron gun: it is made of tungsten thermionic filament (or lanthanum or cerium hexaboride or Schottky field emission gun). A beam of electrons is generated in the electron gun, which delivers a total electron current of 250μA. Besides, the energy is adjustable between less than 1keV and 30keV (or 40keV).
- Condenser lens system: it is usually placed under the electron gun to condense the electrons beam. After passing through one or two condenser lens, the electron beam will be demagnified at 10^4-10^5 times.
- Objective lens (also named as final condenser lens): This lens is used to form the first intermediate image. It can project the diminished image of electron crossover as a spot focused on the surface of the specimen.

● Electron collection system: it is used to detect SE, BSE, and other type of electrons emitted after the primary electrons hitting the specimen atoms.

For SEM imaging, the primary electrons are firstly generated and emitted by thermionic emission in electron gun, which are accelerated to an energy of less than 1keV and 30keV. After the electrons pass through several condenser lens (usually double lens), the electrons beam is demagnified and possesses a diameter of only 2-10nm. Thereafter, the diminished electron beam hit the specimen and produces signals carrying the information about the surface topography, composition and other properties of specimen. The scanning directions and positions of fine electron beam can be controlled by the scan coils, which have already been visualized by manipulating the computer. After detecting and analysis of the signals emitting from the surface of specimen, the resultant image is displayed on a computer screen (see Fig.3.49).

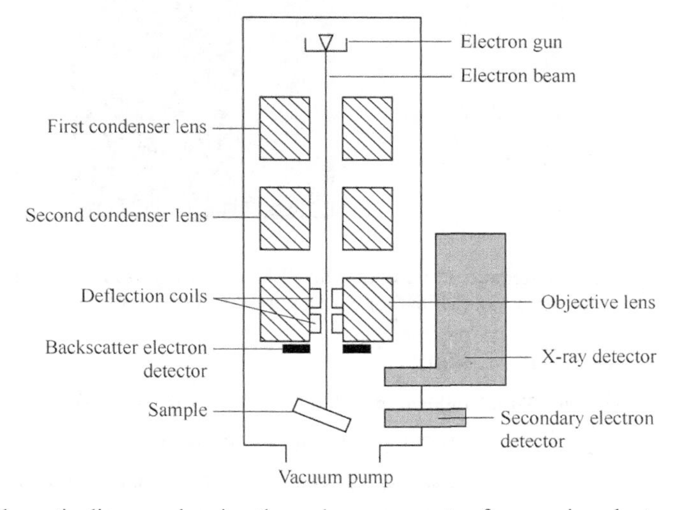

Fig.3.48 Schematic diagram showing the main components of a scanning electron microscope

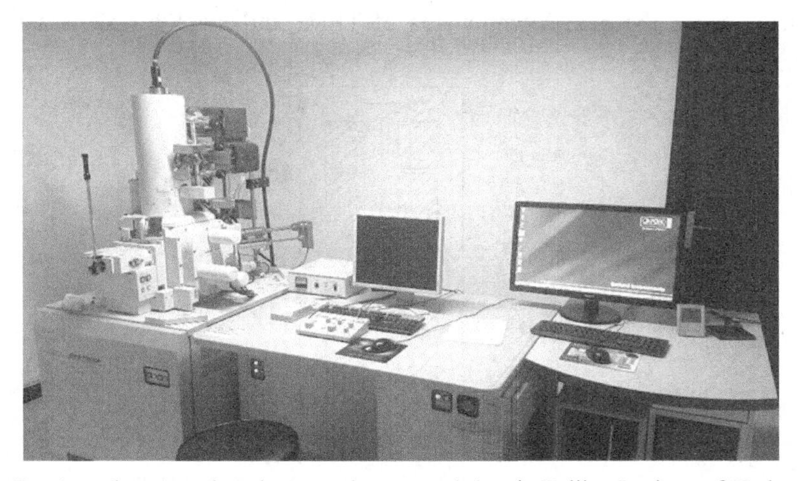

Fig.3.49 A modern scanning electron microscopy (taken in Beijing Institute of Technology)

The basic requirements for the SEM operation and imaging are:

● The column of an electron microscope must be evacuated to a pressure of about 10^{-7} of atmospheric pressure because of the strongly scattering of electrons by gases in the air.

- The specimens must be conductive and dry; those non-conductive specimens must be coated with stable and conductive materials (such as carbon, gold, or silver); the magnetic specimens must be demagnetized before testing.

Usually, SEM provides the information of (sub-)surface structure of bulk specimens. This characterization technology has begun to study the structure of nanosize coordination complex recently. Fig.3.50 is a scanning micrograph showing the structure of MOF-74 with pore apertures of 3.5nm. It is much easier for the eye to interpret this type of image than other indirect analysis.

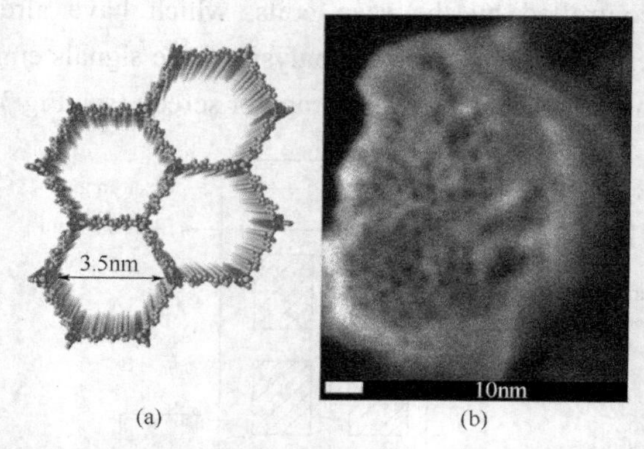

Fig.3.50 Schematic drawing (a) and SEM image (b) of metal organic frameworks (MOF-74) with a pore diameter of 3.5nm

3.8.1.2 Transmitting electron microscopy

Similar to the SEM, the transmission electron microscopy (TEM) also employs a beam of electrons directed at the specimen. It means that the components in their instruments are similar, including electron gun, condenser lenses and vacuum system. Nevertheless, the TEM is used to study the internal arrangement or structure of thin specimens rather than outside of bulk specimens in SEM. Fig.3.51 shows the main components of TEM, and the Fig.3.52 is a photograph of a modern instrument.

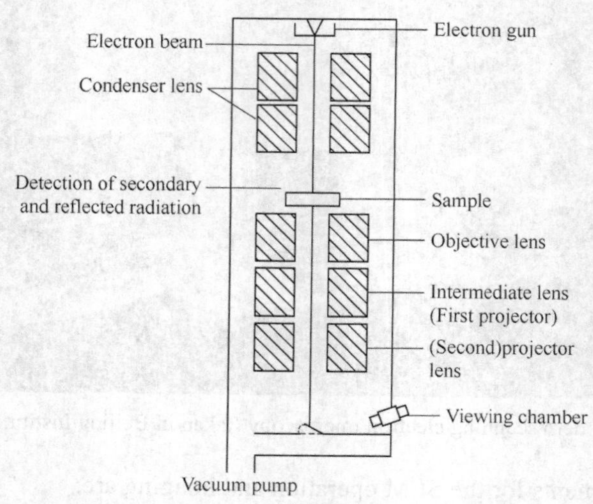

Fig.3.51 Schematic diagram displaying the main parts of transmission electron microscope

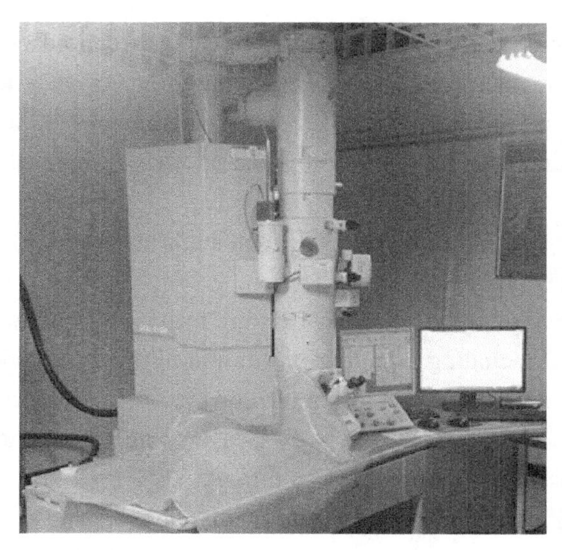

Fig.3.52 A modern transmission electron microscopy (taken in Beijing Institute of Technology).

The basic requirements for the TEM operation and imaging are:

- The electron beam illumination must be generated and used in a high vacuum (10^{-4}mbar or below).
- The specimens must be dry and thin enough to transmit sufficient electrons to form an image; they must be stable under electron bombardment in a high vacuum; the size of specimen must be suitable to fit the specimen holder of microscope.

Nowadays, TEM is a powerful characterization tool for observing local structures of nanomaterials, including spatial distribution of the individual components, crystal structure of a nanocrystalline complex, and identifying surface facets and defects in the crystal structure with electron diffraction or high resolution TEM imaging. An example of crystalline materials under high-resolution TEM is seen in Fig.3.53.

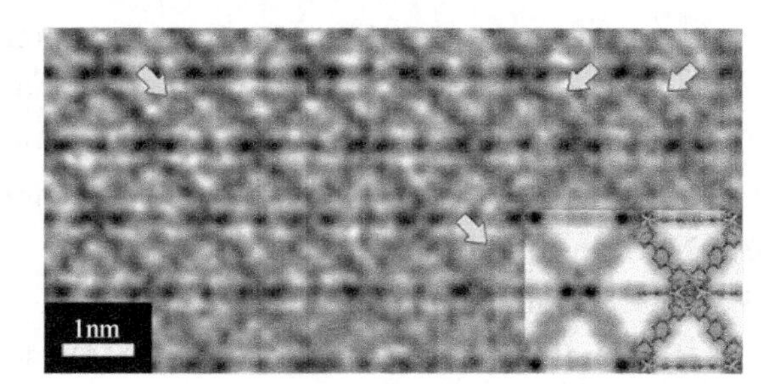

Fig.3.53 High-resolution TEM of MOF UiO-66

3.8.1.3 Cryo-electron microscopy

Since the electrons are strongly scattered by the gases in the air, the specimens for the normal SEM and TEM are required to be dry and observed under a vacuum below 10^{-6}torr. While most of biomolecules, especially for those living organisms live in water environment, once water is removed, these original structures cannot be preserved in these conditions. This challenges the existing characterization technique. The electron microscopes equipped with low temperature specimen stages are regarded as the closest technique to examine fresh biological material.

Cryo-electron microscopy is a form of electron microscopy that is performed on cryogenically cooled samples (the temperature should be as low as generally liquid-nitrogen temperature) which are embedded in an environment of vitreous water. The freezing operation of specimens for cryo-electron microscopy is very important. If the freezing is carried out too slowly or if frozen materials is kept at the temperatures above −80℃, large ice crystals are formed, inducing the structural damage of the specimen and providing artificial structure in the micrographs. Therefore, some principles and requirements have been proposed in order to establish the most suitable way to perform freezing process:

- Anti-freezing agents (such as glycerol, methanol, acetone, etc) are generally introduced into specimen to inhibit the process of crystallization.
- Increasing the rate of freezing (rapid freezing technology) allows water to be in a glassy state at low temperatures, including plunging into liquid-nitrogen-cooled liquid propane to avoid the formation of the ice crystals, spraying liquid propane on to the specimen, and high-pressure freezing (Moor, 1987).

Cryo-electron microscopy is particular good for:

- Directly observing biomolecules, cells, protein, polymers and so on.
- Allowing 3D imaging frozen-hydrated particles or "solution" structures of the investigated molecules with minimal structural distortions.
- Imaging the material specimens that are too volatile in vacuum and sensitive with the electron beam.

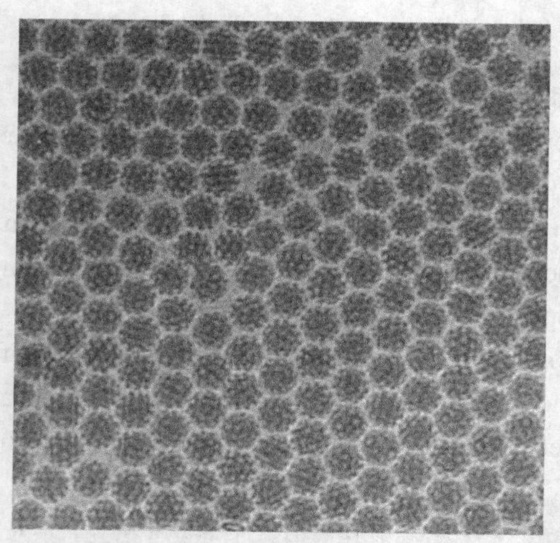

Fig.3.54 The electron cryo-microscopy of Semliki Forest Virus (SFV) floating immobilised in their vitrified aqueous medium

The Nobel Prize in Chemistry 2017 is awarded to Jacques Dubochet, Joachim Frank and Richard Henderson for the development of cryo-electron microscopy, which both simplifies and improves the imaging of biomolecules. This technique has greatly prompted the development of biochemistry and moved into a new era. Fig.3.54 shows a Semliki Forest Virus (SFV) suspension stretched over the 18 mm holes of a grid.

3.8.2 Scanning probe microscopies

3.8.2.1 Scanning tunneling microscopy

Scanning tunneling microscopy (STM) is recognized as one of the great scientific and technological achievements in 1980s. Since the invention of STM by G. Binning and H. Rohrer, it is the first time for human beings to observe the arrangement of individual atoms on the surface of specimen and the physical/chemical properties related to surface electronic behavior. STM is an instrument that utilizes the tunneling effect of quantum theory to detect the surface structure of specimen, which expands a new range of techniques. Therefore, the inventors of this technique was awarded the Nobel Prize for Physics in 1987.

The principle of STM is based on quantum mechanical tunneling effect. In quantum mechanics, even if the energy of particles is less than the threshold energy, many particles rush to the barrier, in which a part of the particles bounce, while some particles pass through the barrier. It seems to be a tunnel (named as quantum tunneling). Specifically, it is believed that the electrons of a solid conductor are usually distributed on the surface and may extend slightly into free space. When a small bias voltage is applied between two solid conductors close to each other, their electron distributions may overlap at certain distance and then a tunneling current can flow between them. The magnitude of tunneling current depends on the distance between the conductors, which is an important detecting parameter in STM.

In the STM operation process (Fig.3.55), a very sharp probe (made of tungsten wire or platinum-iridium alloy wire) is controlled and scanned across the surface of the specimen. The gap between the probe and the specimen surface is equivalent to a barrier; the electrons can pass through this gap to form tunneling current, which will change as the height of the surface changes. Thus, detecting the current strength can provide surface information of specimen.

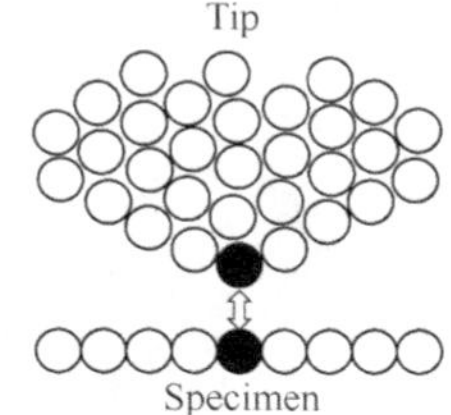

Fig.3.55 Schematic diagram of interatomic interaction for STM

The basic requirements for the STM are:

● In order to avoid dust and other contaminations in the air and obtain clear atomic images, the STM is usually operated and used in a high vacuum environment.

● Piezoelectric crystals are applied to control the movement of the probe in all three directions (x and y for scanning, z for vertical moving).

● It works primarily with conductive materials.

The STM allows the scientists to observe and locate a single atom in practical application, which has a higher resolution than its similar atomic force microscope. Moreover, it is not only an important measurement tool, but also a processing tool in nanotechnology because of its ability on precisely manipulating the atoms. Nowadays, STM has been combined with other existing techniques (such as conventional SEM), which is widely used in common laboratory.

3.8.2.2 Atomic force microscopy

Atomic force microscopy (AFM) is one of high-resolution types of scanning probe microscopy, which performs the detection through sensing and amplifying the force between the cantilever probe and test sample. Since invention of AFM in 1986, this powerful nanoscopic platform has attracted enormous attention and has found widespread use.

Fig.3.56 shows the schematic diagram of the AFM components. In the operation process, a fine tip is attached to a cantilever beam, when the tip is close to the surface of the specimen, the distance between the atoms of the fine tip and the sample surface will be kept constant by their interatomic forces. While, the cantilever beam will move up and down vertically with the surface topography of the sample in response to attraction or repulsion forces with the sample surface. This moving track is detected and recorded by the reflection of a small laser beam from the cantilever and photodetector. Since the sensitivity of AFM on vertical deflections can be as small as 0.1nm, even only one atom high of the specimen surface can be resolved quite easily.

Usually, the AFM technique has three imaging modes in order to meet the various specimens, including contact mode, non-contact mode and tapping mode. There is no special requirements on the specimen; those soft samples (e.g. biological samples and organic thin film) can also be measured with high resolution. In fact, most AFM can be operated in the air, under water or other liquids. Moreover, AFM can not only observe the conductor, but also observe the non-conductor, thus making up for the shortcomings of STM. However, the imaging of AFM requires computer processing is not as straightforward as with an SEM. Nevertheless, it is still a powerful and useful technique for many years. Fig.3.57 shows an example of biointerface under AFM.

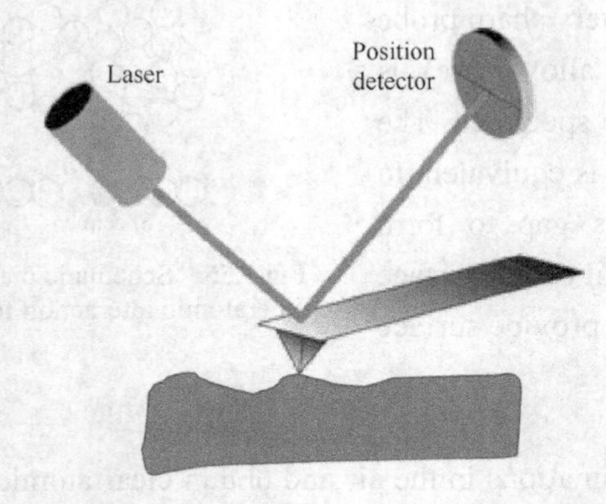

Fig.3.56 The schematic diagram of the AFM components

The movement of the cantilever can be measured by a reflecting laser beam from the cantilever

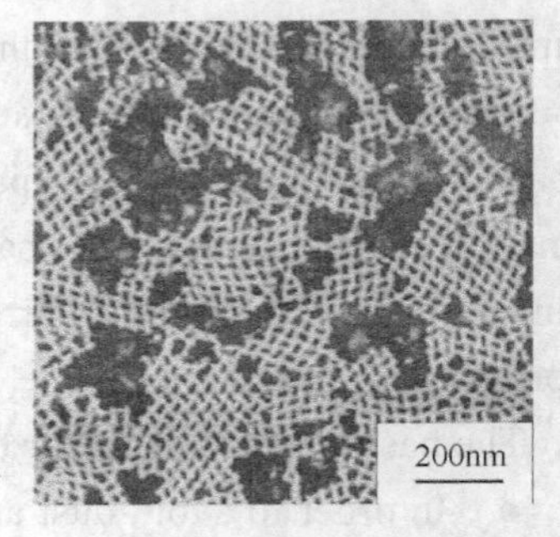

Fig.3.57 The AFM charactering self-assembly of RNA strands

Chapter 4 The Structure and Physicochemical Properties of Coordination Complex

4.1 The Structures of Several Kinds of Coordination Complexes

4.1.1 Organometallic Complex

An organometallic compound is partially characterized by the presence of one or more metal-carbon bonds, in which the carbon involved would be otherwise considered a part of an organic compound. A metal is seen as any element less electronegative than carbon (this includes B, Si and As and excludes P and halogen).

They are normally three types: R—M, R—M—R, R—M—X (R is an organic group, carbon; M is metal ions).

The chemical bonding in the organometallic complexes are shown in the Fig.4.1.

Carbon-metal bonds are polar bonds that can be represented by a resonance hybrid of covalent and ionic structures. Carbon-metal bonds are frequently classified as to how much "ionic character" they have, by using an index called the importance of the ionic resonance contributor relative to covalent structure (Fig.4.2).

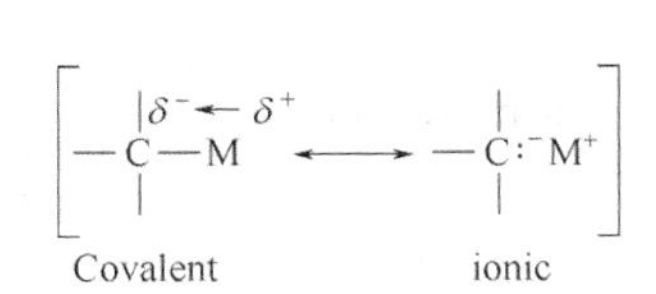

Fig.4.1 The nature of chemical bonding in the organometallic complexes

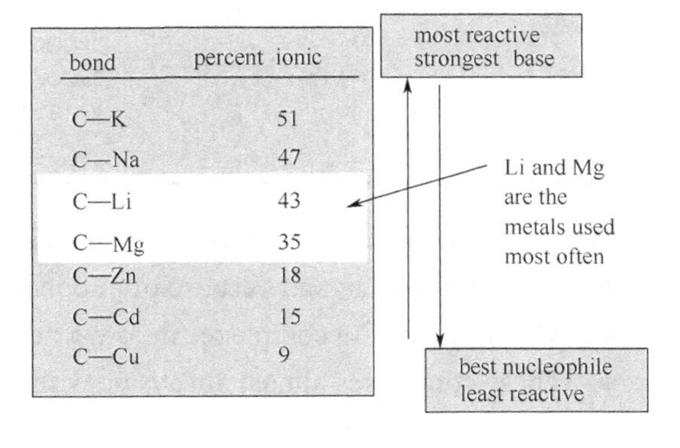

bond	percent ionic
C—K	51
C—Na	47
C—Li	43
C—Mg	35
C—Zn	18
C—Cd	15
C—Cu	9

Fig.4.2 An index of covalent and ionic characters of carbon-metal bonds

(1) 18 electron rule Most π-complexes of the transition metals conform to an emperical rule, know as the "The Eighteen Electron Rule ", which assumes that the metal atom accepts sufficient electrons from the ligands to enable it to attain the electronic configuration of the next noble gas. This means that the valence shell of the metal atom contains 18 electrons. Thus, the sum of the number of metal d-electrons plus the electrons conventionally regarded as being supplied by the ligands is 18. This leads to a constant electronic arrangement for the metal atom.

Take iron complex as examples, counting electrons in organometallic complexes follows the following steps.

- Count number of electrons (e^-) in valence shell of free metal atom. $Fe^0 \rightarrow 8e^-$ ($4s^2 3d^6$ or just count the number of groups in the "periodic system").
- Add or substract electrons depending on the total charge on the metal complex.

$$[Fe(L)_x] \rightarrow 8e^-, [Fe(L)_y]^+ \rightarrow 7e^-, [Fe(L)_z]^- \rightarrow 9e^-$$

- Sum the numbers of electrons that the ligands formally contribute to the metal.

terminal $CO \rightarrow 2e^-$, bridging $CO \rightarrow 1e^-$, metal-metal bond $\rightarrow 1e^-$, organic ligands $\rightarrow$ number of electrons depends on the group.

- Sum the electrons find in the second point and the third point. A stable complex will have 18 electrons.

There are two distinct methods that are used to count electrons, the neutral or covalent method and the effective atomic number or ionic method.

The organometallic chemist considers the transition metal valence electrons to all be d electrons. There are certain cases where the $4s^2 3d^x$ order does occur, but we can neglect these in our first approximation. For zero-valence metals, we see that the electron count simply corresponds to the column it occupies in the periodic table. Ti^0: d^4, not d^2; Fe^0: d^8, not d^6; Re^{3+} is d^4 (seventh column for Re, and then add 3 positive charges⋯or subtract three negative ones).

In the next case the molecule is split in half because of the two iron atoms which both want to attain the electronic configuration of the next noble gas (Table 4.1).

Table 4.1 Sum the numbers of electrons in $[Fe_2(CO)_9]$

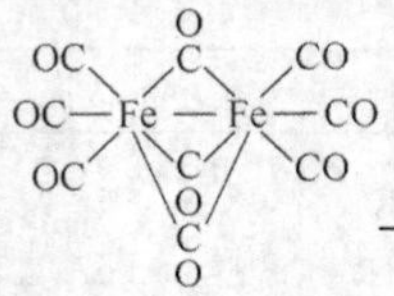

Fe^0:	$8e^-$
$3\times CO_T$:	$6e^-$
$3\times CO_B$:	$3e^-$
Fe—Fe bond:	$1e^-$
Total:	$18e^-$

(2) Carbonyl complex Carbonyl complex is charactrized by:

- CO is a strong ligand because of pi bonding;
- Metals are in CO complexes in low oxidation states (sometimes negative);
- These complexes almost always obey the 18-electron-rule.

$$M + C{\equiv}O \longrightarrow M^- {-} C{\equiv}O^+ \longleftrightarrow M{=}C{=}O$$

Stable carbonyl complexes of the 3d elements: $[V(CO)_6]/[V(CO)_6]^-$, $[Cr(CO)_6]$, $[Fe(CO)_5]$, $[Ni(CO)_4]$, $[Mn_2(CO)_{10}]$, $[Fe_2(CO)_9]$, $[Co_2(CO)_8]$, $[Fe_3(CO)_{12}]$, $[Co_4(CO)_{12}]$, $[Co_6(CO)_{16}]$(Fig.4.3).

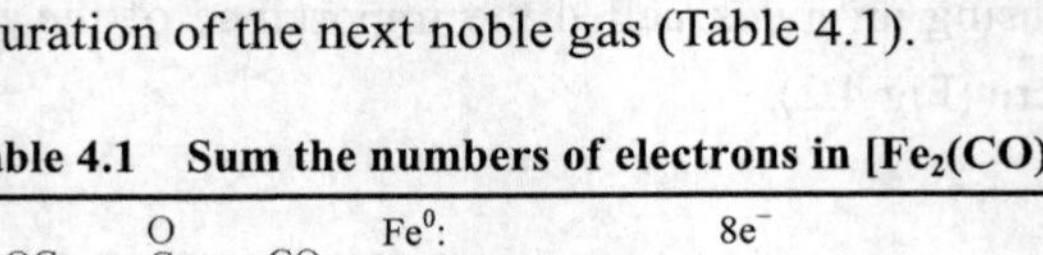

Fig.4.3 The molecular structures of some examples of the 3d element carbonyl complexes

$[V(CO)_6]$: 17-e-complex. Decomposes at 70℃, and can be reduced to the 18-e-complex $[V(CO)_6]^-$ [equation (4.1)]:

$$Na + [V(CO)_6] \longrightarrow Na^+ + [V(CO)_6]^- \quad (4.1)$$

Heavier transition metals: have strong tendency towards multinuclear complexes. Os_3-$(CO)_{12}$is more stable than $Os(CO)_5$, while the opposite is true for the Fe complexes (cf. size, M—M bonding).

We used IR before to determine the C—O bond strength. It can also be used to gain information about the structure/symmetry. Consider $[Fe(CO)_5]$, D_{3h}, two different C—O stretching absorptions (Fig.4.4).

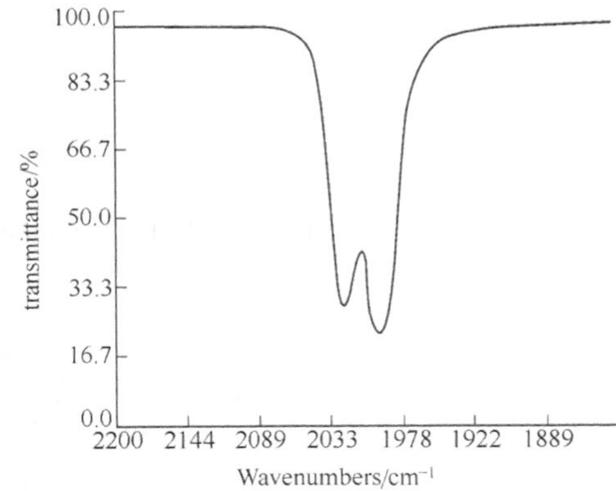

Fig.4.4 The IR spectrum indicating the two different C—O stretching absorptions

(3) Polynuclear carbonyl complex。

Carbonyl complexes with more than one metal atom/ion: e. g. $[Mn_2(CO)_{10}]$, $[Fe_2(CO)_9]$, $[Co_2(CO)_8]$, $[Fe_3(CO)_{12}]$, $[Co_4(CO)_{12}]$, $[Co_6(CO)_{16}]$, $[Os_3(CO)_{12}]$. These may ($[Mn_2(CO)_8]$) or may not ($[Co_2(CO)_{10}]$) include a bridging CO ligand (if not, a M—M bond is needed).

Bridging vs. terminal CO: C—O bond order gets lowered upon bridging, as shown with IR: typical C—O stretching frequencies are 2143cm^{-1}for free CO, 2125-1850cm^{-1} for terminal CO, 1850-1700cm^{-1} for bridging CO, 1715cm^{-1} for saturated ketones. That helps with identifying possible structures (in addition to the 18-electron-rule). Note that the 18-electron-rule cannot help in distinguishing between bridging and terminal CO, for both contribute two electrons (Fig.4.5).

$[Mn_2(CO)_{10}]$ exists in both forms, thus the energy difference between bridging and terminal CO must be small. General trend: complexes with heavier metal atoms have fewer CO bridges (Fig.4.6).

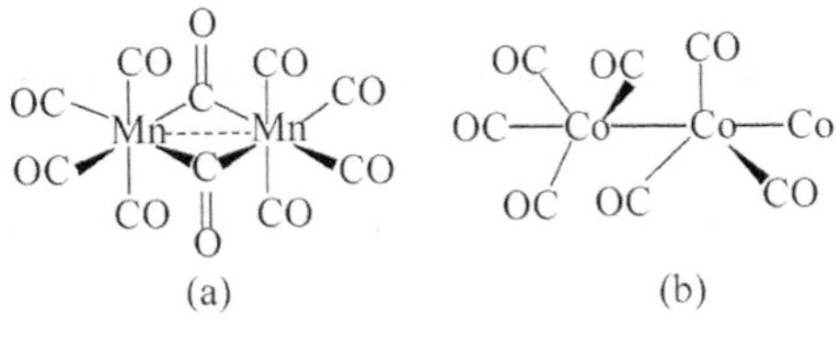

Fig.4.5 The molecular structures of $[Mn_2(CO)_{10}]$ (a) and $[Co_2(CO)_8]$ (b)

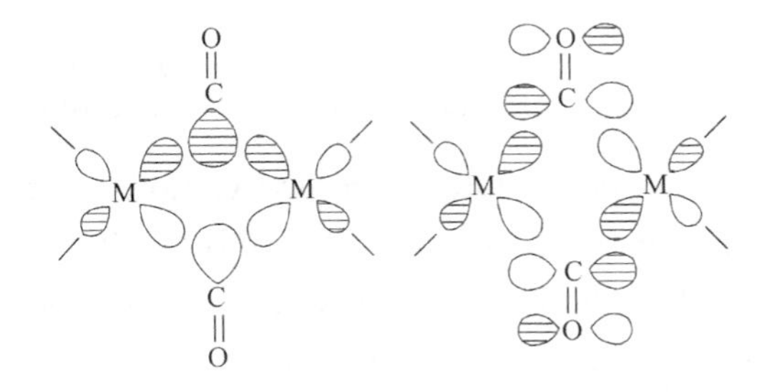

Fig.4.6 The illustration of σ and π or bitals of CO overlap with the d orbitals of M

M—M bonding in polynuclear carbonyls: often needed because of the 18-electron-rule.

Semibridging μ_2-CO(in $[Os_4(CO)_{14}]$): CO is unequally shared between two M centers (Fig.4.7).

μ_3-CO: shown in Fig.4.8.

Heterobimetallic carbonyl complex: this term describes a complex in which there are two different metal centers, for example, ruthenium (Ru) and molybdenum (Mo) [Fig.4.9(a)].

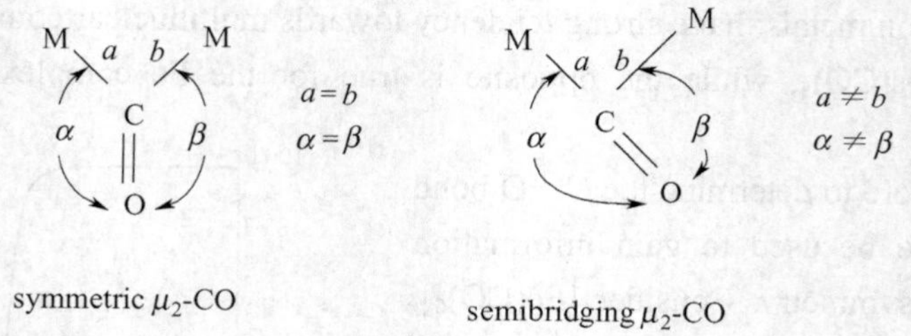

symmetric μ_2-CO　　semibridging μ_2-CO

Fig.4.7　The illustration of the symmetric and semibridging μ_2-CO

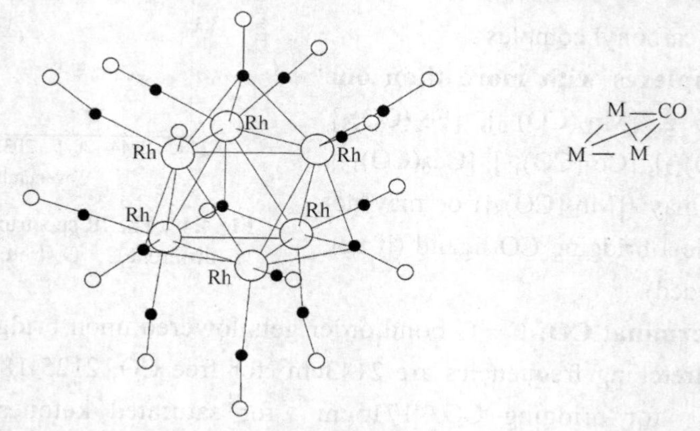

Fig.4.8　The illustration of the molecular structure of $Rh_6(CO)_{16}$

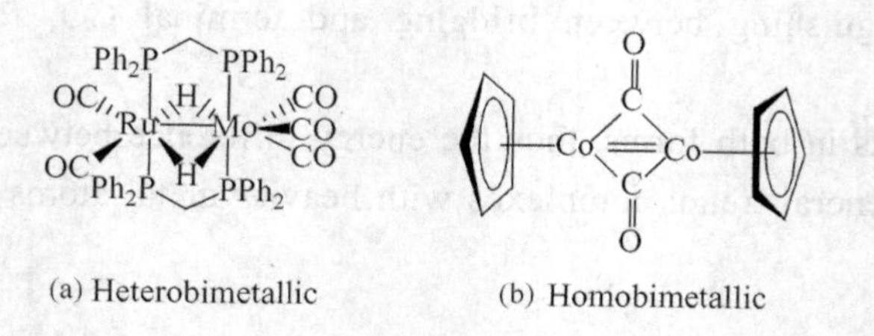

(a) Heterobimetallic　　(b) Homobimetallic

Fig.4.9　The molecular structures of the $Ru(CO)_2(\mu\text{-}H)_2(\mu\text{-}CH_2(PPh_2)_2)_2Mo(CO)_3$(left) and $Co(CO)_2(C_5H_5)_2$(right)

Homobimetallic carbonyl complexes: these complexes have two metal centers that are the same element such as cobalt (Co) [Fig.4.9(b)].

The two metals centers do not need to have identical ligands or coordination number, but are usually found as symmetric dimers.

(4) Carbonylate ions　Anionic carbonyl complexes obey 18-e-rule. Electronically and structurally related to the neutral complexes, anionic carbonyl complexes are as following.

$[Ti(CO)_6]^{2-}$-$[V(CO)_6]^{-}$-$[Cr(CO)_6]$ (IR: $1748cm^{-1}$-$1860cm^{-1}$-$2000cm^{-1}$)

$[V(CO)_5]^{3-}$-$[Cr(CO)_5]^{2-}$-$[Mn(CO)_5]^{-}$-$[Fe(CO)_5]$

$[Cr(CO)_4]^{4-}$-$[Mn(CO)_4]^{3-}$-$[Fe(CO)_4]^{2-}$-$[Co(CO)_4]^{-}$-$[Ni(CO)_4]$

$[Co(CO)_3]^{3-}$

$[Cr_2(CO)_{10}]^{2-}$-$[Mn_2(CO)_{10}]$

$[Fe_2(CO)_8]^{2-}$-$[Co_2(CO)_8]$

(5) Dihydrogen complexes Vaska's complex reacts reversibly with H_2 (oxidative addition) [equation (4.2)]. The process proceeds via a concerted mechanism [equation (4.3)].

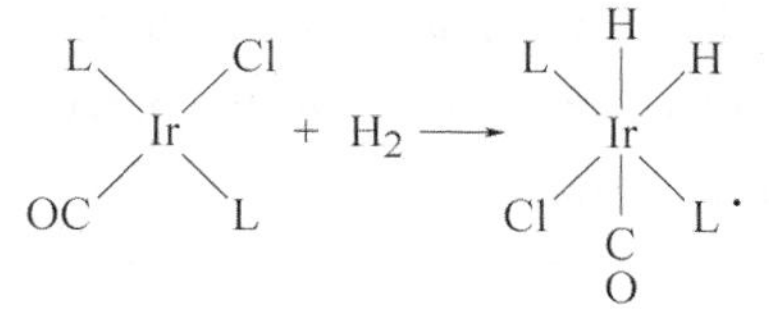

$$\text{Ir(L)}_2\text{Cl(CO)} + H_2 \longrightarrow \text{Ir(H)}_2\text{(L)}_2\text{Cl(CO)} \qquad (4.2)$$

$$M + H_2 \longrightarrow \left[M\begin{smallmatrix} H \\ \vdots \\ H \end{smallmatrix} \right] \longrightarrow M\begin{smallmatrix} H \\ \\ H \end{smallmatrix} \qquad (4.3)$$

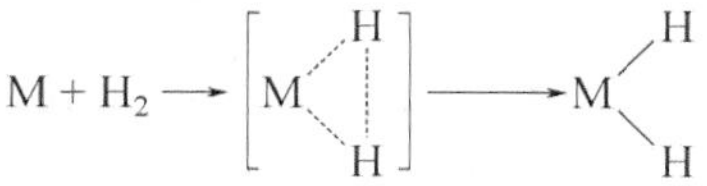

$$\text{Mo(CO)}_3\text{(cht)} + 2\text{PPr}^i_3 \longrightarrow \text{Mo(CO)}_3\text{(PPr}^i_3)_2 + \text{cht}$$

$$\text{Mo(CO)}_3\text{(PPr}^i_3)_2 + H_2 \longrightarrow \text{Mo(CO)}_3\text{(PPr}^i_3)_2(H_2)$$

(cht = cycloheptariene)

The intermediate contains an intake H—H bond, i.e. an η_2-H_2 ligand in $M(H_2)$ [instead of two bonded H atoms/ions in *cis*-$M(H)_2$], which could be isolated in the following complexes (Fig.4.10).

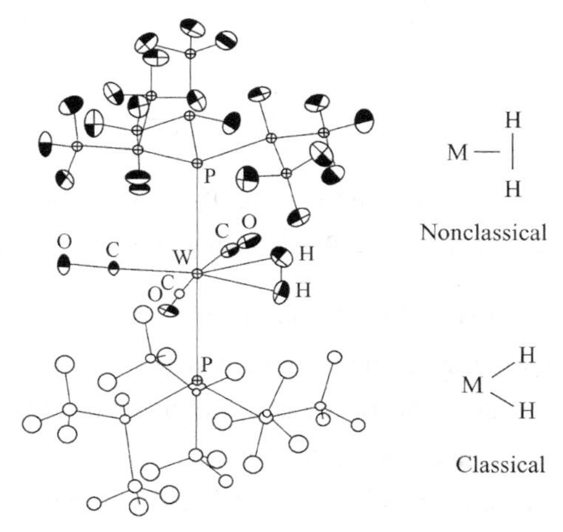

Fig.4.10 The molecular structure of the $W[P(i\text{-}Pr)_3]_2(CO)_3(H_2)$

The $M—H_2$ bond involves (like M—CO) L—M σ donor and L—M π acceptor interactions, the latter introducing electrons into the σ^* orbital of H_2, thereby weaking the H—H bond which might lead to breaking it.

(6) μ-H-complexes $[(OC)_5Cr—H—Cr(CO)_5]^-$ is comparable to B_2H_6 in the donation of a bonding electron pair to a Lewis acid [equation (4.4)].

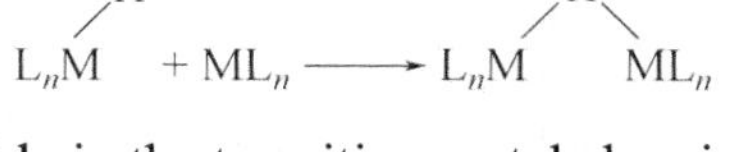

$$L_nM\text{—H} + ML_n \longrightarrow L_nM\text{—H—}ML_n \qquad (4.4)$$

Double hydrido bridges are possible in the transition metal chemistry as well (Fig.4.11).

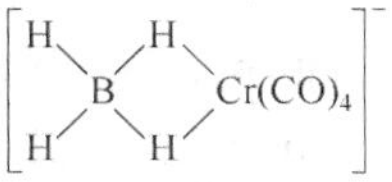

$$\left[H_2B(\mu\text{-}H)_2Cr(CO)_4 \right]^-$$

Fig.4.11 The illustration of the molecular structure of $Cr(CO)_4(\mu\text{-}H)_2BH_2$

4.1.2 Cluster

Compounds containing metal-metal bond is referred as cluster. Certainly, the clusters are polynuclear compounds (Fig.4.12). One of the prerequisites for the formation of metal-metal bond is a low formal oxidation state. The second important factor for low oxidation state is the nature of the d orbitals, since effective overlap of d orbitals appears necessary to stabilize cluster. In general, the first transition series are unfavorable to from cluster based on their d orbitals with relatively small sizes even at their low oxidative state.

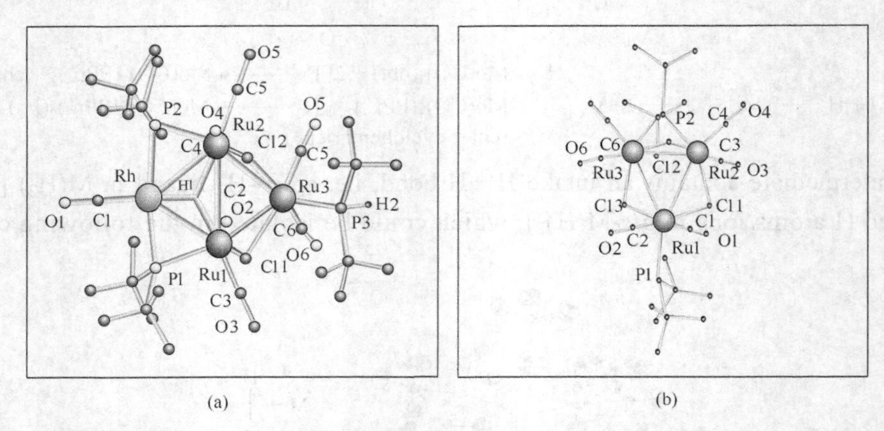

Fig.4.12 The molecular structure of $[Ru_3Rh(CO)_7(\mu_3\text{-}H)(\mu\text{-}PBu^t_2)_2(Bu^t_2PH)(\mu\text{-}Cl)_2]$ (a) and molecular structure of $[Ru_3(CO)_6(\mu\text{-}Cl)_3(\mu\text{-}PBu^t_2)(Bu^t_2PH)]$(b)

A way to determine the number of M—M bonds in a polynuclear complex is to determine the total number of electrons in the entire complex and subtract this number from (n×18), where n is the number of metals in the system. The result is the number of electrons necessary to obtain an 18-electron configuration, and if it is assumed that these electrons are obtained by the formation of metal-metal bonds, division by two will give the predicted number of bonds. Not all polynuclear species fit this localized eighteen-electron bonding model.

The development of cluster chemistry has been achieved by the inspiration of supramolecular chemistry. For example, the heterothiometallic clusters (Fig.4.13).

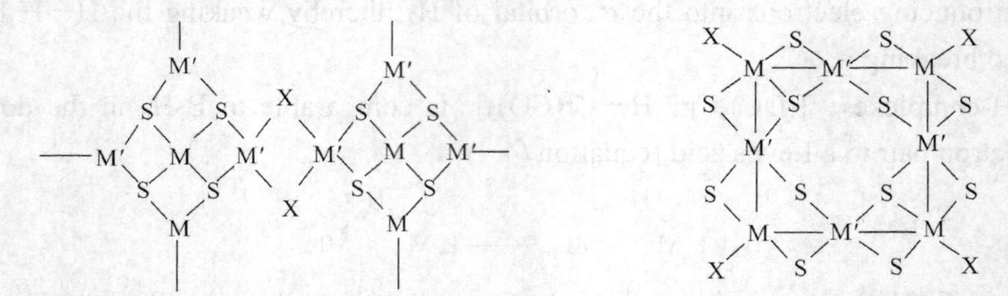

Fig.4.13 The molecular structures of molecular motifs in some heterothiometallic clusters

4.1.3 Macrocyclic Complex and Bioinorganic Complex

(1) Macrocyclic ligand (Fig.4.14) A new type of the macrocyclic ligands has been design and synthesized especially for polynuclear complexes (Fig.4.15). The dinuclear mixed-valence Mn(Ⅱ) Mn(Ⅲ) complex with this kind of ligand are reported recently (Fig.4.16)。

Some bis- and trismacrocyclic ligands have been developed for synthesis of the mimic complexes of active metal center of enzyme or protein (Fig.4.17 and Fig.4.18).

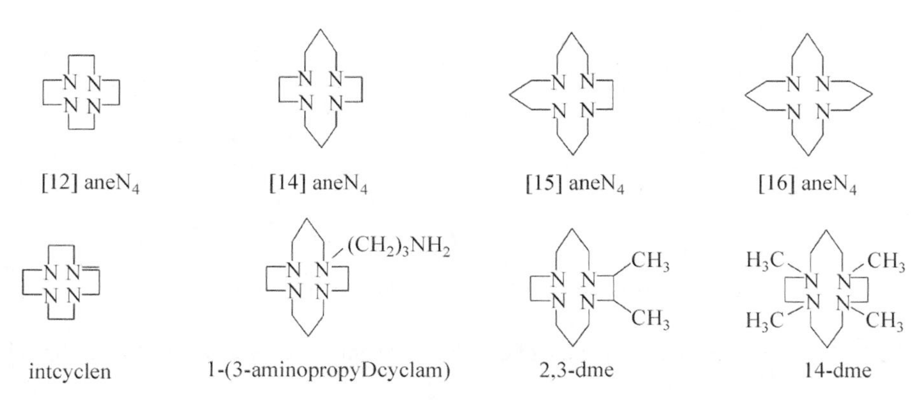

Fig.4.14 Some examples of Tetraaza-macrocyclic ligands

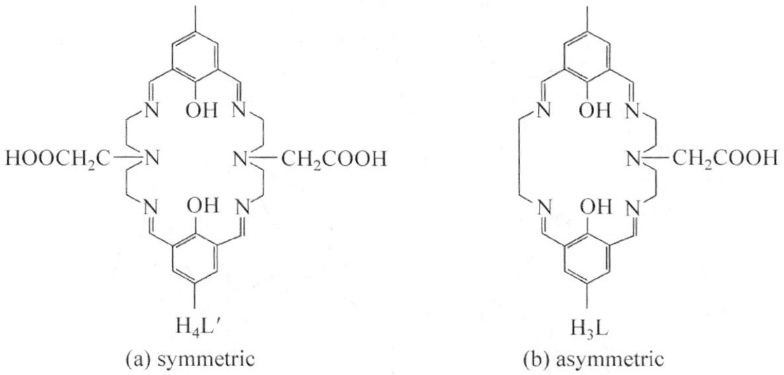

HOOCH$_2$C N OH N CH$_2$COOH

H_4L'

(a) symmetric

N OH N CH$_2$COOH

H_3L

(b) asymmetric

Fig.4.15 The illustration of the structure of macrocyclic ligands for dinuclear complex

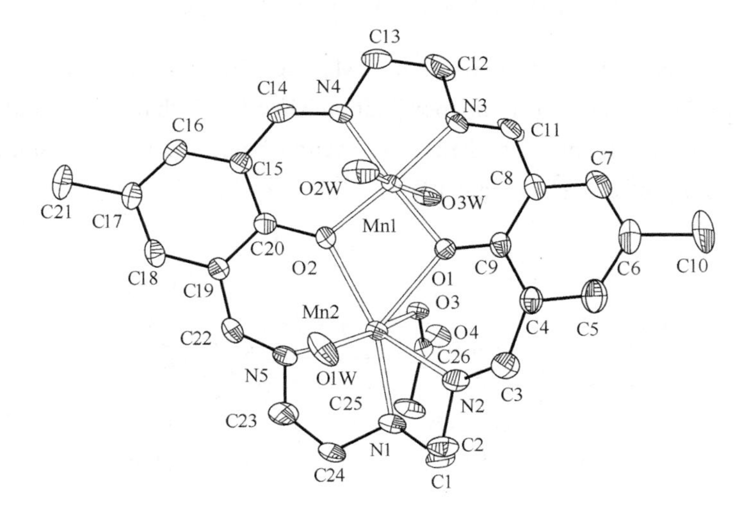

Fig.4.16 An ORTEP view of the complex cation $[Mn(II)Mn(III)L(H_2O)_3]^{2+}$[L is shown in Fig 4.15 (b)]

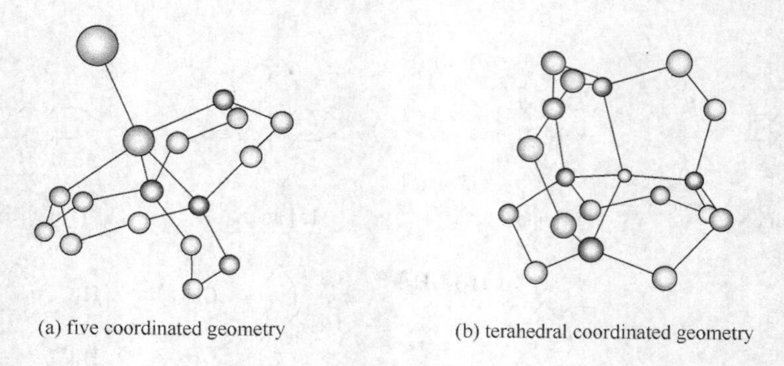

Fig.4.17 The illustration of structure of coordination compounds with bi- and tricycli ligand

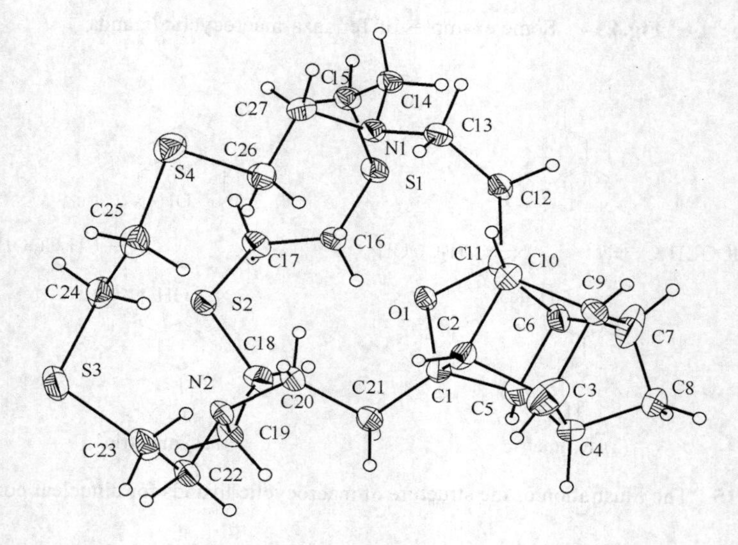

Fig.4.18 An ORTEP view of the tricyclic ligand, $C_{27}H_{41}S_4N_2O$

(2) Macrocyclic coordination compound Macrocyclic coordination compounds mainly belong to bioinorganic chemistry (or biocoordination chemistry) that is a rapidly growing field. Example is the family of cytochromes, their active center being the heme group, a porphyrin ring surrounding a Fe atom [Fe^{2+} or Fe^{3+}] (Fig.4.19).

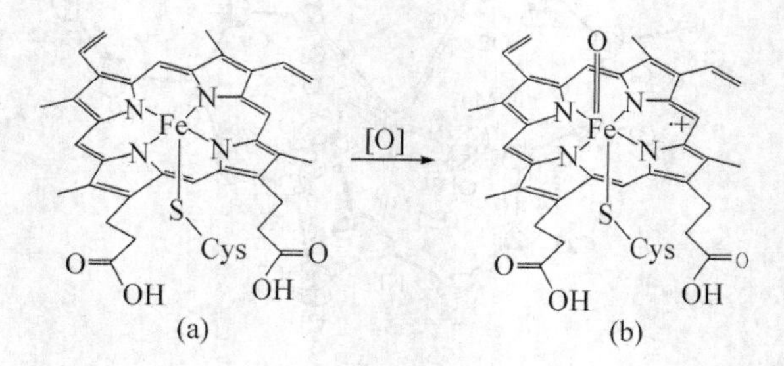

Fig.4.19 Iron(Ⅲ) protoporphyrin with a cysteinate as the axial ligand (a), which is typical of cytochrome P450, CPO, and NOS enzymes. The active oxygen species of these proteins and related heme enzymes is an oxoiron(Ⅳ) porphyrin cation radical (b)

The scheme of the catalytic cycle for oxygen activation and transfer is shown in Fig.4.20.

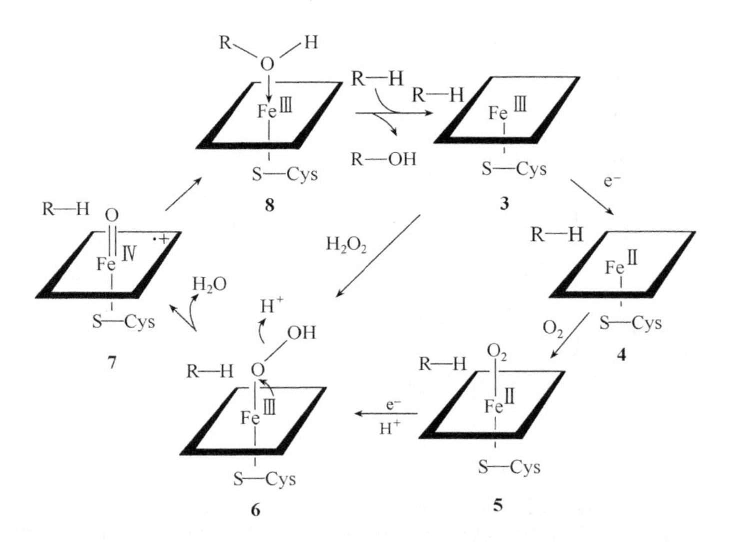

Fig.4.20 The catalytic cycle for oxygen activation and transfer by cytochrome P450

The order of stability of porphyrin complexes is $Ni^{2+}>Cu^{2+}>Co^{2+}>Fe^{2+}>Zn^{2+}$. Kinetics of formation follow this order: $Cu^{2+}>Co^{2+}>Fe^{2+}>Ni^{2+}$.

So why Fe? Natural abundance is large and Fe^{II}/Fe^{III} redox chemistry is rich. Redox potentials for Fe^{II}/Fe^{III} pairs is shown in Table 4.2.

Table 4.2 Redox potentials for Fe^{II}/Fe^{III} pairs

compound	E_0/mV	compound	E_0/mV
hemoglobin	170	$[Fe(H_2O)_6]^{2+/3+}$	771
myoglobin	46	$[Fe(bpy)_3]^{2+/3+}$	960
cytochrome a_3	400	$[Fe(C_2O_4)_3]^{4-/3-}$	20
cytochrome c	260	$[Fe(CN)_6]^{4-/3-}$	358
cytochrome b_5	20		
cytochrome P450	−400		

$$HFeO_4^- \xrightarrow{+2.07} Fe^{3+} \xrightarrow{+0.77} Fe^{2+} \xrightarrow{-0.44} Fe$$

The structures of heme A, B and C are shown in Fig 4.21. The redox potential of cytochrome a is very high, it can reduces O_2 to H_2O in the cytochrome c oxidase complex. Since the M atom in cytochrome c is six-coordinated, cytochrome c can only react in electron transfer reactions (e. g. , in photosynthesis and respiration).

Porphyrin ligand is macrocyclic tetrapyrrole systems with conjugated double bonds (Fig.4.22). Probably, the most important class of metal complexes in biological systems are the transition metal complexes with the porphyrin ligands.

The chlorophyll system (vitamin B_{12}) is a porphyrin ring with a long alkyl chain at the bottom. A double bond in one of the pyrrole rings has been reduced (Fig.4.23).

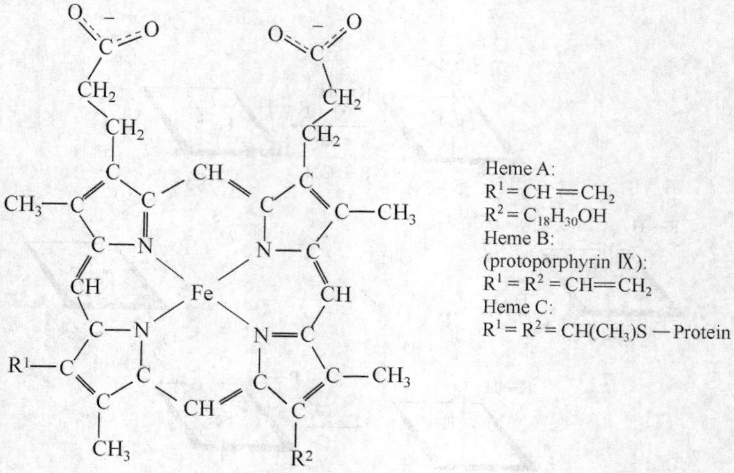

Heme A:
$R^1 = CH{=}CH_2$
$R^2 = C_{18}H_{30}OH$
Heme B:
(protoporphyrin Ⅸ):
$R^1 = R^2 = CH{=}CH_2$
Heme C:
$R^1 = R^2 = CH(CH_3)S$ — Protein

Fig.4.21 The structural scheme of Heme A, B, C

Heme A: cytochrome a; heme B: hemoglobin, myoglobin, peroxidase, cytochrome b; heme C: cytochrome c

(a)
(b)

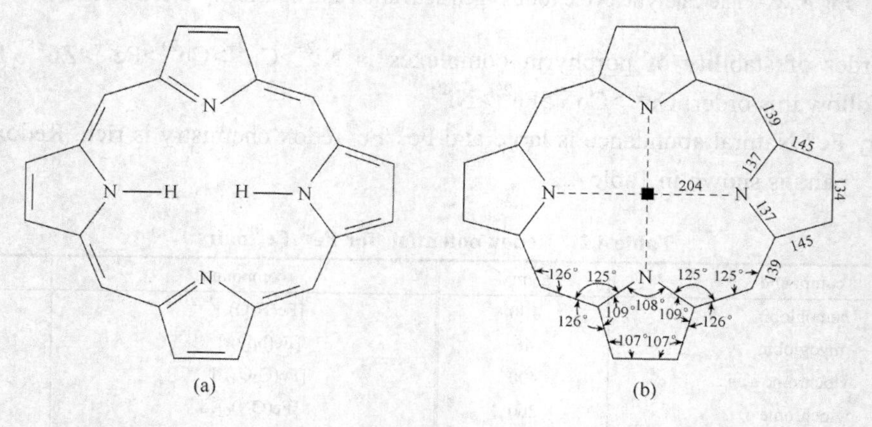

Fig.4.22 The molecular structure of the porphyrin ligand (a) and a set of rational parameters for the porphyrin skeleton (b)

Chlorophyll a (R=CH_3)
Chlorophyll b (R=CHO)

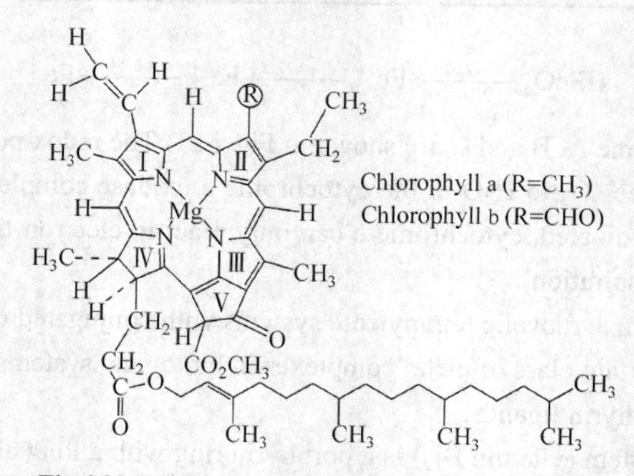

Fig.4.23 The molecular structure of the chlorophyll

(3) O_2 and N_2 complexes in bioinorganic chemistry Dioxygen and dinitrogen ligands are soft ligands with some (weak) π bonding. O_2 binds well to Fe^{II} in the porphyrin ring because of the symbiotic action of the ring system. Since CO binds much better, it will make the hemoglobin unavailable for O_2.

N_2 is isoelectronic with CO, but it took a long time to prepare the first dinitrogen complex in 1965. N_2 complexes are important as they reduce the N—N bond order compared to gaseous N_2, rendering bond breaking possible (fixation of N_2 from the atmosphere). Synthesis is possbile using N_2 gas, where nucleophilic N_2 replaces the weak ligand H_2O. Formation of bridging N_2 is shown in equation (4.5).

$$[Ru(NH_3)_5Cl]^{2+} \xrightarrow[H_2O]{Zn(Hg)} [Ru(NH_3)_5(H_2O)]^{2+}$$
$$[Ru(NH_3)_5(H_2O)]^{2+} + N_2 \longrightarrow [Ru(NH_3)_5N_2]^{2+} \tag{4.5}$$

Structural possibilities for N_2 as ligand is shown in Fig.4.24.

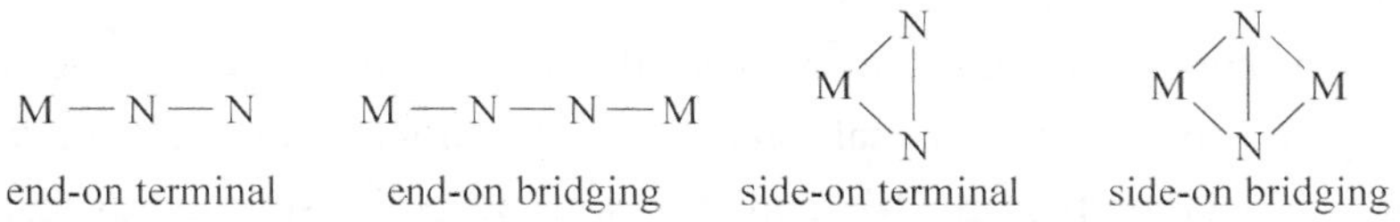

M—N—N (end-on terminal) M—N—N—M (end-on bridging) M(η²-N₂) (side-on terminal) M(μ-η²:η²-N₂)M (side-on bridging)

Fig.4.24 Possible coordination modes of N_2 molecule as ligand

Bonding of $N_2$is comparable to CO, i.e. there is π back bonding as identified based on short M—N and elongated N—N bonds and lower N—N stretching frequency, compared to the gas N_2 from (only Raman active) $2331cm^{-1}$ to $2105cm^{-1}$ in $[Ru(NH_3)_5N_2]Cl_2$.

However, CO is a better σ donor as well as a much better π acceptor. Because of the superior π accepting ability of CO, carbonyl dinitrogen complexes are rare and somewhat unstable. $[Cr(CO)_5N_2]$ and $[Cr(CO)_4(N_2)_2]$ exist only well below room temperature, while $[Cr(CO)_6]$ melts at 154℃.

The formation of hematin follows these steps (Fig.4.25). These reactions have to be suppressed in the living systems, i.e. by steric hindrance.

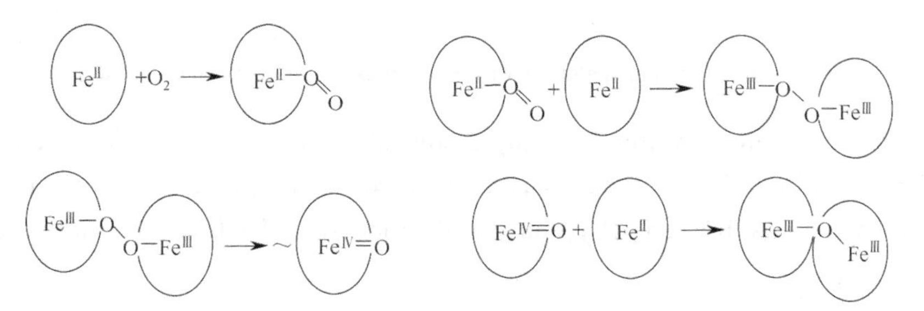

Fig.4.25 The formation steps of hematin

4.1.4 Supramolecular Assemblies Containing Transition Metal Ions (Polynuclear Complex)

Supramolecular chemistry is a new rapidly progressing field on the crossroads among chemistry, biochemistry, physics and technology. Its core theme is characterization, understanding and application of novel non-covalent bonding types.

Progress in supramolecular chemistry is driven by the human quest for sophisticated, high-symmetry structures and by the possible applications these structures have in fields like materials science, chemical technology, and medicine. Structures containing transition metal ions are of particular interest because of special magnetic and electric properties conferred upon the supramolecular assemblies by the presence of multiple metal paramagnetic centers. Metal ions have been used for "programmed" reading of the information contained in ligands with multiple coordination sites, which is essential for directing the organization of elaborate structures in self-assembly processes.

One of the most interesting structures pursued in supramolecular chemistry is that of helix. DNA is the classical example of biological helix formed by using a network of hydrogen bonds and base stacking interactions. One of the goal in this area is to create novel nanosize structures that contain transition metal ions and have potential nanotechnology applications. One strategy for creating such structures is based on the use of the molecular recognition properties of transition metal ions and peptide nucleic acids (PNA).

The spatial arrangement and dimensions of the nanostructures is determined by the coordination properties of transition metal ions and by the structure of PNA, a synthetic analogue of DNA. Metal ion incorporation in PNA duplexes is achieved by substituting ligands for natural nucleobases. This procedure allows the precise control of the number and position of metal ions in the PNA duplexes. Formation of duplexes or higher complexity structures is driven by hydrogen bonds between the natural nucleobases and coordinative bonds between the ligands incorporated in PNA strands and the metal ions bridging these strands (Fig.4.26).

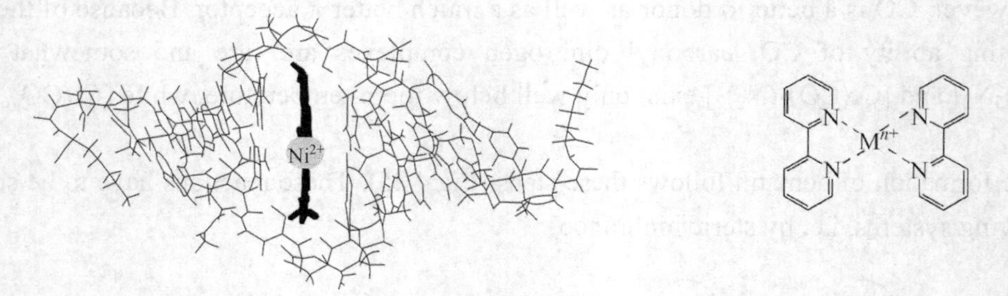

Fig.4.26 The representation of PNA duplex that contains one Ni^{2+} site

Fig.4.27 gives another example of supramolecular assembly with one-dimensional helix structure. The formation of the structure has achieved by two organic long-chain molecules, then induced by the metal ions that can offer four coordination sites. Each of metal ions has tetrahedron coordination geometry.

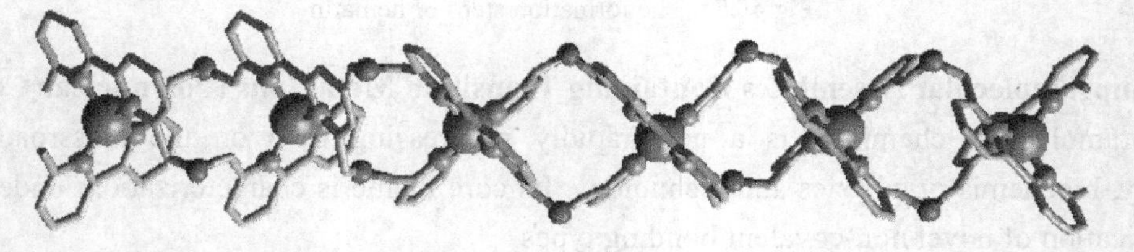

Fig.4.27 One-dimensional supramolecular assembly with helix structure

4.2 Thermodynamic Properties

Metal ion in solution does not exist in isolation, but in combination with ligands (such as solvent molecules or simple ions) or chelating groups, giving rise to complex ions or coordination compounds.

What happens when SCN^- is added to an aqueous solution containing $Fe(NO_3)_3$?

In aqueous solution, $[Fe(H_2O)_6]^{3+}$ ions react with SCN^- (not free Fe^{3+} ions). In the reaction shown in equation (4.6), the SCN^- ion replaces a water molecule in the coordination sphere to produce the complex ion, $[Fe(H_2O)_5(SCN)]^{2+}$.

$$\begin{array}{c} Fe(NO_3)_3(s) \xrightarrow{H_2O} Fe^{3+}(aq) + NO_3^-(aq) \\ \quad\quad\quad\quad ||| \\ \quad\quad\quad\quad [Fe(H_2O)_6]^{3+} \\ [Fe(H_2O)_6]^{3+} + SCN^- \longrightarrow [Fe(H_2O)_5(SCN)]^{2+} + H_2O \end{array} \quad (4.6)$$

4.2.1 Thermodynamic Stability

It will have been pointed out that the "stability of a complex in solution" refers to the degree of association between the two species involved in the state of equilibrium. Qualitatively, the greater the association, the greater the stability of the compounds. The magnitude of the (stability or formation) equilibrium constant for the association, quantitatively expresses the stability. Thus, if we have a reaction of the type [equation (4.7)]:

$$M + 4L \longrightarrow ML_4 \quad (4.7)$$

then the larger the stability constant, the higher the proportion of ML_4 that exists in the solution. Free metal ions rarely exist in solution so that M, will usually be surrounded by solvent molecules which will compete with the ligand molecules, L, and be successively replaced by them. For simplicity, we generally ignore these solvent molecules and write four stability constants as follows [equation (4.8)]:

$$\begin{array}{ll} M+L \longrightarrow ML & K_1=[ML]/\{[M][L]\} \\ ML+L \longrightarrow ML_2 & K_2=[ML_2]/\{[ML][L]\} \\ ML_2+L \longrightarrow ML_3 & K_3=[ML_3]/\{[ML_2][L]\} \\ ML_3+L \longrightarrow ML_4 & K_4=[ML_4]/\{[ML_3][L]\} \end{array} \quad (4.8)$$

where K_1, K_2 etc. are referred to as "stepwise stability constants".

Alternatively, we can write the "Overall Stability Constant" [equation (4.9)].

$$M + 4L \longrightarrow ML_4 \qquad \beta_4=[ML_4]/\{[M][L]^4\} \quad (4.9)$$

The stepwise and overall stability constants are therefore related as follows [equation (4.10)].

$$\beta_4=K_1K_2K_3K_4$$

or more generally

$$\beta_n=K_1K_2K_3K_4\cdots K_n \quad (4.10)$$

If we take as an example, the steps involved in the formation of the cuprammonium ion, we have the following [equation (4.11)].

$$\begin{array}{ll} Cu^{2+}+NH_3 \longrightarrow Cu(NH_3)^{2+} & K_1=[Cu(NH_3)^{2+}]/[Cu^{2+}][NH_3] \\ CuNH_3^{2+} +NH_3 \longrightarrow Cu(NH_3)_2^{2+} & K_2=[Cu(NH_3)_2^{2+}]/\{[Cu(NH_3)^{2+}][NH_3]\} \\ & \beta_4=[Cu(NH_3)_4^{2+}]/\{[Cu^{2+}][NH_3]^4\} \end{array}$$

$$\lg K_1=4.0, \lg K_2=3.2, \lg K_3=2.7, \lg K_4=2.0 \text{ or } \lg\beta_4=11.9 \tag{4.11}$$

So, it is usual to represent the metal-binding process by a series of stepwise equilibria which lead to stability constants that may vary numerically from hundreds to enormous values such as 10^{35} and more. For this reason, they are commonly reported as logarithms. It is additionally useful to use logarithms, since $\lg K$ is directly proportional to the free energy of the reaction [equation (4.12)].

$$\Delta G^{\ominus}=-RT\ln b, \Delta G^{\ominus}=-2.303\,RT\lg 10^{\beta}, \Delta G^{\ominus}=\Delta H^{\ominus}-T\Delta S^{\ominus} \tag{4.12}$$

For instance, the stability constant in real numbers for zinc acetate is a bit over 10, while the stability constant for copper gluconate is over 10^{18}. Consequently, the logrithim of the actual stability of the compound offers an easy way to write and understand how stable a metal-ligand is. Obviously copper gluconate is not biologically available. Some substances, like EDTA are so strong in their chelating, sequestering, binding power that they are used to bind metals, particularly zinc.

4.2.2 Stability of Complexes

(1) Metal effects

- The smaller the ionic radius, the more stable the complex.
- The larger the oxidation state, the more stable the compound.
- Crystal field effects. Natural order of stability:

$$Ca^{2+}<Sc^{2+}<Ti^{2+}<V^{2+}<Cr^{2+}<Mn^{2+}<Fe^{2+}<Co^{2+}<Ni^{2+}<Cu^{2+}<Zn^{2+}$$

- Hard acids form most stable complexes with hard bases, soft acids with soft bases.

(2) Ligand effects

- Ligands with high base strengths (affinities for H^+) will form more stable complexes.
- Chelate effect: multidentate ligands have equal probability of forming a coordination bond as do monodentate ions. Once the bond is formed, however, there is a high likelihood that further bonding will occur.
- Chelate ring size: 5-membered saturated ligands and 6-membered unsaturated chelate rings form stable complexes.
- Steric strain: large ligands form less stable complexes.

(3) The chelate effect The chelate effect can be seen by comparing the reaction of a chelating ligand and a metal ion with the corresponding reaction involving comparable monodentate ligands, for example, comparison of the binding of 2,2′-bipyridine with pyridine or 1,2-ethylenediamine (en) with ammonia.

It has been known for many years that a comparison of this type always shows that the complex resulting from coordination with the chelating ligand is much more thermodynamically stable.

4.2.3 Calculation of Species Concentrations

Formation constant express the stability of complex ion. Stepwise formation constant represents the bonding of individual ligands to a metal ion, K [equation (4.13)].

$$Zn^{2+}+NH_3 \rightleftharpoons Zn(NH_3)^{2+} \qquad K_1=\frac{[Zn(NH_3)^{2+}]}{[NH_3][Zn^{2+}]}=3.9\times10^2$$

$$Zn(NH_3)^{2+} + NH_3 \Longrightarrow Zn(NH_3)_2^{2+} \qquad K_2 = \frac{[Zn(NH_3)_2^{2+}]}{[ZnNH_3^{2+}][NH_3]} = 2.1\times10^2$$

$$Zn^{2+} + 2NH_3 \Longrightarrow Zn(NH_3)_2^{2+} \qquad \beta_2 = \frac{[Zn(NH_3)_2^{2+}]}{[NH_3]^2[Zn^{2+}]} = K_1K_2 = 8.2\times10^4$$

$$Zn^{2+} + 3NH_3 \Longrightarrow Zn(NH_3)_3^{2+} \qquad \beta_3 = K_1K_2K_3 \tag{4.13}$$

In complexation competition exists between H^+, metal ions, ligands, precipitation of metal ions by various precipitants.

Solubilization of metal complexes from solids depends on stability of metal chelates, concentration of complexing agent, pH, nature of insoluble metal deposit.

4.3 Molecular Electronic Devices—Redox-active Coordination Complex

4.3.1 Concept of Molecular Electronics

Molecular electronics is defined as the use of single molecule based devices or single molecular wires to perform signal and information processing. It is an interdisciplinary field combining the efforts of biologists, chemists, physicists, mathematicians, computer scientists and genetic engineers from all over the world. Molecular electronics will make room for new technological advances such as molecular wires, molecular switches and molecular systems. Such systems might include biocomputers that could be injected into a biological system (the human body) and have the ability to perform repair functions on the cellular level.

Molecular electronics is the idea of building electronic circuits using molecules and molecular structures as building blocks. Molecules have the capability of conducting and transferring energy between one another. If this process can somehow be manipulated and controlled, it would be possible to have these molecules and molecular structures perform tasks required for information processing. This would include basic tasks similar to circuits in computers today.

4.3.2 Molecular Wires

One of the applications of molecular electronics is molecular wire. The term molecular wire (or nanowire) has been used to describe many disparate substances, including carbon nanotubes, other nanotubes, polyalkynes, other doped polymers, and still other entities. The broad concept underlying all uses of the term envisions an elongated array of atoms, the thickness and composition of which may vary greatly, which may be assumed to provide a conductive path for electrons.

Generally, the definition of molecular wire: a pre-one dimension linear molecular materials that can transmit charge carrier (electrons or cavities) along the molecular chain.

A molecular wire must posses several different properties to be of any use.

- It must be able to conduct electrons along its length. This is completed by the movement of either a hole or electrons (electron tunneling, resonant and nonresonant) in electronic molecular wires and by excitation in a photonic wire as well as other types of electron movement on the molecular level.

- The wire must also be easily oxidized or reduced.
- A third feature that a molecular wire must have is an insulating sheath to prevent the current from leaking to the surroundings. One example of this in the porphyrin wire is to attach a *tert*-butyl group to the backbone of the molecule.
- Finally, molecular wires must have a defined and fixed length.

One-dimensional redox molecular wires with simple electro-active termini present a simplified conduction mechanism compared with the photonic wire (Fig.4.28).

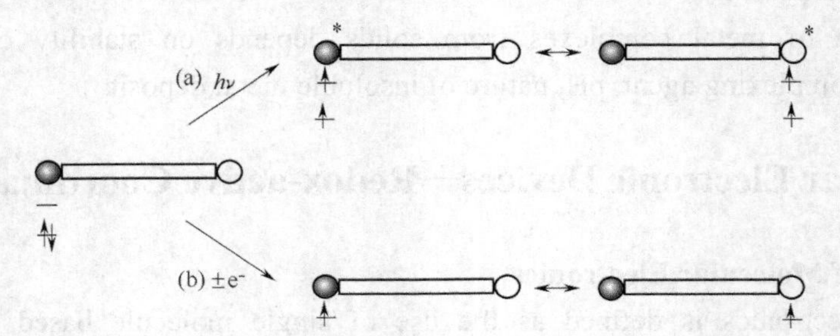

Fig.4.28 Schematic presentation of electron conduction in a photonic (a) and a redox (b) molecular wire

Triad molecular systematization is an example of molecular wire (Fig.4.29). It is consist of three parts, ① donor: redox-active termini; ② conducting space: unsaturated organic units which have conjugated π-electron (Fig.4.30); ③ acceptor: redox-active termini.

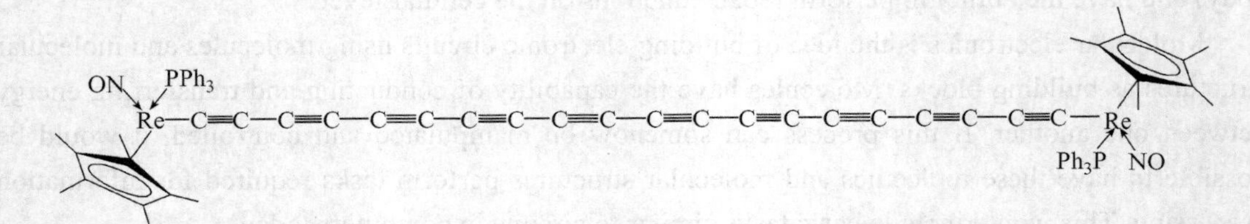

Fig.4.29 The molecular wire with metallofragment as chromophore and unsaturated organic units as conducting space

$M{-}(C{\equiv}C)_n{-}M$ $\quad M{=}(C)_n{=}M$ $\quad M{\equiv}C{-}(C{\equiv}C)_n{-}C{\equiv}M$ $\quad M{=}C{-}(C{\equiv}C)_n{-}M$ $\quad M{=}C{-}(C{\equiv}C)_n{-}C{=}M$

Fig.4.30 Some structural motifs of the conducting space

4.3.3 Molecular Switching

Another device in molecular electronics is the switch. Molecular switches are molecules or aggregate made of a few molecules that can reversibly change their situations by an external stimulus, for example light, electronic, magnetic, or chemical influences. These molecules are able to exist in two different states to having two different properties: an ON state (allowing a complete electron transfer to occur) and an OFF state (electron transfer is blocked).

The types of molecular switches are:

- Torsion process—twists the molecule around a single bond to achieve decoupling in the OFF state although no easy way has been discovered to twist and lock the molecule in place.
- Saturation/unsaturation—a double bond is transformed into a single one therefore

breaking conjugation. This is accomplished by reduction or molecular reorganization. ON and OFF states are obtained by deprotonation or protonation of the intervalence bands.

- Quantum interfaces—based on electronic effects and uses the wave nature of electrons.
- Tunneling switch—works on excitation of the molecule. Switch can be turned OFF by changing either the barrier height or the depth of the potential well.

Switches are needed in all types of circuits where charge flow is needed at some time intervals and not at others. They may also be used for data registers at some point in the future, allowing for memory storage.

Chiropticene molecules are switchable between two distinct states that are spatial mirror images of each other. These mirror images are electronically and optically distinct enabling sharp and stable switching properties.

The chiropticene switch is a single molecule designed to be switchable between left-handed and right-handed forms thus capable of forming a molecular-based binary code. This molecular code can be read by its chiral effect, called chiroptical properties, on a passing light beam (Fig.4.31).

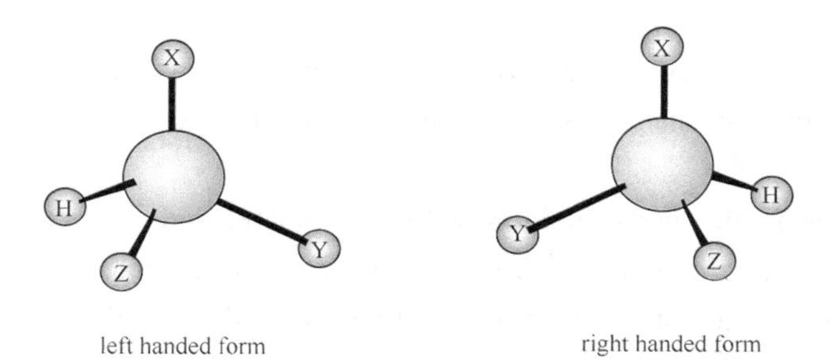

Fig.4.31 The structural illustration of chiral center with left and right hand forms

The chiral switching process can be represented as Fig.4.32.

Each of the two crossed arrows represents one of two states of the molecule, constituting the binary pair of the switch(Fig.4.33).

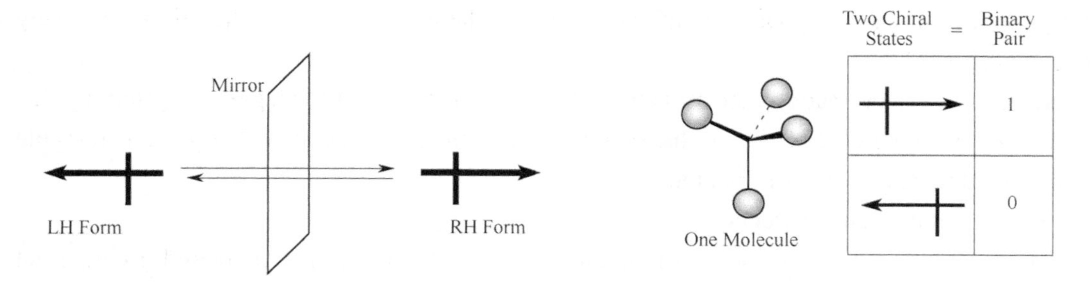

Fig.4.32 The representation of chiral switch process **Fig.4.33** The binary pair of one chiropticene molecule

The switch action of chiropticene molecule is schemed in Fig.4.34.

The switch may also be flipped by an electrical field that flips the dipole and its state may be read with a field detector that detects the molecular capacitance induced by the dipole.

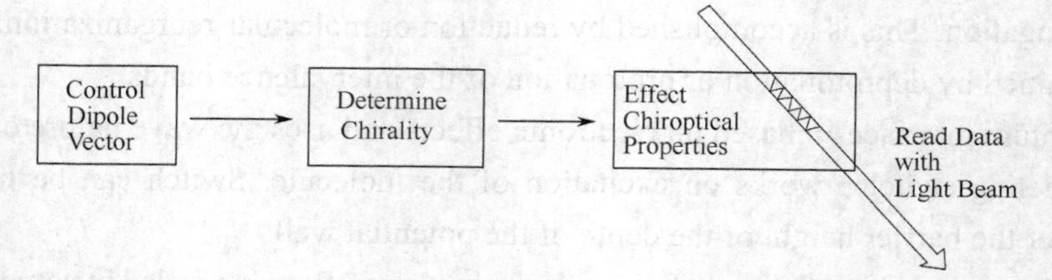

Fig.4.34 The scheme of the switch action of chiropticene molecule

4.4 Magnetic Properties of Coordination Complex

Magnetism arises from moving charges, such as an electric current in a coil of wire. In a material that does not have a current present, there are still magnetic interactions. Atoms are made of charged particles (protons and electrons) that are moving constantly. The processes that create magnetic fields in an atom are:

- Nuclear spin Some nuclei, such as a hydrogen atom, have a net spin that creates a magnetic field.
- Electron spin An electron has two intrinsic spin states that we call up and down.
- Electron orbital motion There is a magnetic field due to the electron moving around the nucleus.

Each of these magnetic fields interacts with one another and with external magnetic fields. However, some of these interactions are strong and others are negligible.

Measurement of interactions with nuclear spins is used to analyze compounds in nuclear magnetic resonance (NMR) and electron spin resonance (ESR) spectroscopy. In most other situations, interaction with nuclear spins is a very minor effect.

Interactions between the intrinsic spin of one electron and the intrinsic spin of another electron are strongest for very heavy elements such as the actinides. This is called spin-spin coupling. For these elements this coupling can shift the electron orbital energy levels.

The interaction between an electron's intrinsic spin and it's orbital motion is called spin-orbit coupling. Spin-orbit coupling has a significant effect on the energy levels of the orbitals in many inorganic compounds.

Macroscopic effects, such as the attraction of a piece of iron to a bar magnet are primarily due to the number of unpaired electrons in the compound and their arrangement. The various possible cases are called magnetic states of matter.

4.4.1 Magnetic States of Material

(1) Diamagnetic A diamagnetic compound has all of it's electron spins paired giving a net spin of zero. Diamagnetic compounds are weakly repelled by magnet. Diamagnetism is a very weak form of magnetism that is only exhibited in the presence of an external magnetic field. It is the result of changes in the orbital motion of electrons due to the external magnetic field. The induced magnetic moment is very small and in a direction opposite to that of the applied field.

(2) Paramagnetic A paramagnetic compound will have some electrons with unpaired spins.

Paramagnetic compounds are attracted by magnet. Paramagnetism is the tendency of the atomic magnetic dipoles, due to quantum-mechanical spin as well as electron orbital angular momentum, to align with an external magnetic field.

(3) Ferromagnet In a ferromagnetic substance there are unpaired electron spins, which are held in alignment by a process known as ferromagnetic coupling. Ferromagnetic compounds, such as iron, are strongly attracted to magnets. Ferromagnetism is a phenomenon by which a material can exhibit a spontaneous magnetization, and is one of the strongest forms of magnetism. A phenomenon can be exhibited by materials like iron (nickel or cobalt) that become magnetized in a magnetic field and retain their magnetism when the field is removed.

(4) Ferrimagnet Ferrimagnetic compounds have unpaired electron spins, which are held in an pattern with some up and some down. This is known as ferrimagnetic coupling. In a ferrimagnetic compound, there are more spins held in one direction, so the compound is attracted to a magnet.

(5) Antiferromagnetic When unpaired electrons are held in an alignment with an equal number of spins in each direction, the substance is strongly repelled by a magnet. This is referred to as an antiferromagnet.

(6) Superconductor Superconductors are repelled by magnetic fields because the magnetic field is excluded from passing through them. This property of superconductors, called the Meissner effect, is used to test for the presence of a superconducting state. The underlying theory of how superconductivity arises is still a matter of much research. It does appear that the mechanism behind the magnetic properties of superconductors is significantly different from the other classes of compounds discussed here. For these reasons, superconductors will not be discussed further here.

4.4.2 Interaction with an External Magnetic Field

When a material is placed in a magnetic field, the magnetic field inside the material will be the sum of the external magnetic field and the magnetic field generated by the material itself. The magnetic field in a material is called the magnetic induction and given the symbol "*B*". The formula for this is [equation (4.14)]:

$$B=H+4\pi M \tag{4.14}$$

where B=magnetic induction; H=external magnetic field; π=3.14159; M=magnetization (a property of the material).

The magnitude of the magnetic field is usually given in units of gauss (G) or tesla (T), where 1 tesla=10000gauss.

For mathematical and experimental convenience this equation if often written as [equation (4.15)]

$$B=1+4\pi M=1+4\pi\chi_V \tag{4.15}$$

where $\chi_V=M/H$=volume magnetic susceptibility.

(1) Magnetic susceptibility The magnetic susceptibility is the degree of magnetization of a material in response to a magnetic field. The mass magnetic susceptibility is represented by the symbol χ [equation (4.16)].

$$\chi=M/J \tag{4.16}$$

where J is the magnetic dipole moment per unit mass (A/m) of the material.

If χ is positive the material is called paramagnetic, and the magnetic field is strengthened by the presence of the material. If χ is negative then the material is diamagnetic and the magnetic field is weakened in the presence of the material (Table 4.3).

Table 4.3 Magnetic susceptibilities of some materials

Material	$\chi_V/10^{-5}$	Material	$\chi_V/10^{-5}$
Aluminium	+2.2	Hydrogen	−0.00022
Ammonia	−1.06	Oxygen	+0.19
Bismuth	−16.7	Silicon	−0.37
Copper	−0.92	Water	−0.90

The volume magnetic susceptibility (χ_V) is so named because B, H and M are defin-ed per unit volume. However this results in χ_V being unitless. It is convenient to use the magnetic susceptibility instead of the magnetization because the magnetic susceptibility is independent of the magnitude of the external magnetic field, H, for diamagnetic and para-magnetic materials.

There are variant forms of susceptibility, the main two of which are mass susceptibility and molar susceptibility [equation (4.17)].

$$\chi_\rho=\chi_V/\rho \ (\mathrm{m^3/kg});$$
$$\chi_m=\chi_V M_a/\rho \ (\mathrm{m^3/mol}) \qquad (4.17)$$

where ρ is the density of the substance in $\mathrm{kg/m^3}$ and M_a is the molar mass in kg/mol.

Magnetic mass susceplibilities of some common substances are shown in Table 4.4 and Table 4.5.

Table 4.4 Magnetic mass susceptibilities

Material	$\chi_\rho/(10^{-8}\mathrm{m^3/kg})$	Material	$\chi_\rho/(10^{-8}\mathrm{m^3/kg})$
Aluminium	+0.82	Hydrogen	−2.49
Ammonia	−1.38	Oxygen	+133.6
Bismuth	−1.70	Silicon	−0.16
Copper	−0.107	Water	−0.90

Table 4.5 Materials classified by their magnetic properties

Class	χ dependant on B	Dependant on temperature	Hysteresis	Example	χ
Diamagnetic	No	No	No	Water	-9.0×10^{-6}
Paramagnetic	No	Yes	No	Aluminium	2.2×10^{-5}
Ferromagnetic	Yes	Yes	Yes	Iron	3000
Antiferromagnetic	Yes	Yes	Yes	Terbium	9.51×10^{-2}
Ferrimagnetic	Yes	Yes	Yes	$MnZn(Fe_2O_4)_2$	2500

(2) Effective magnetic moment Another measure of magnetic interaction that is often used is an effective magnetic moment, μ, where

$$\mu=2.828\ (\chi_m T)^{1/2} \qquad (4.18)$$

where μ=effective magnetic moment; χ_m=molar magnetic susceptibility; T=temperature.

The numeric factor puts μ in units of Bohr magnetons (BM), where one BM equals 9.274×

10^{-24} joules per tesla. The effective magnetic moment is a convenient measure of a material's magnetic properties because it is independent of temperature as well as external field strength for diamagnetic and paramagnetic materials.

This said, we would now like to examine how the magnetization, magnetic susceptibility and effective magnetic moment depend on molecular structure.

The magnetic properties of coordination compounds arise from the spin and orbital angular moment of electrons in unfilled shells. Along with the paramagnetism from this source, a smaller diamagnetic effect from the precession of electron orbitals in the appli-ed magnetic filed is always present. The Pascal constants give an empirical method for estimation of the diamagnetic susceptibilities of atoms in transition metal complex molecules (Table 4.6).

Table 4.6 Diamagnetic Corrections of some atoms and groups

Ion	Na^+	K^+	NH_4^+	Hg^{2+}	Fe^{2+}	Fe^{3+}	Cu^{2+}	Br^-	I^-	NO_3^-	ClO_4^-	IO_4^-	CN^-	NCS^-	H_2O	$EDTA^{4-}$
DC	6.8	14.9	13.3	40	12.8	12.8	12.8	34.6	50.6	18.9	32	51.9	13	26.2	13	~150
Ion	Co^{2+}	Co^{3+}	Ni^{2+}	VO^{2+}	Mn^{3+}	Cr^{3+}	Cl^-	SO_4^{2-}	OH^-	$C_2O_4^{2-}$	OAc^-	pyr	Me-pyr	$Acac^-$	en	urea
DC	12.8	12.8	12.8	12.5	12.5	12.5	23.4	40.1	12	34	31.5	49.2	60	62.5	46.3	33.4

4.4.3 Diamagnetism

Diamagnetism can be described by electrons forming circular currents, orbiting the nucleus, in the presence of a magnetic field. As such, a diamagnetic contribution can be calculated for any atom. However, the magnitude of the diamagnetic contribution is so much smaller than the magnitude of paramagnetic and other effects that it is usually ignored for any other type of materials.

4.4.4 Paramagnetism

The structural feature most prominent in determining paramagnetic behavior is the number of unpaired electrons in the compound. A spin only formula for the magnetic moment of a paramagnetic compound is equation (4.19).

$$\mu=g[S(S+1)]^{1/2} \tag{4.19}$$

where μ=effective magnetic moment; g=2.0023; S=1/2 for one unpaired electron, 1 for two unpaired electrons, 3/2 for three unpaired electrons, etc.

This equation is sometimes written with g=2. This does not introduce a significant error since this simple spin only treatment is a decent approximation but is often not accurate.

The equation (4.20) takes into account both spin and orbital motion of the electrons.

$$\mu=[4S(S+1)+L(L+1)]^{1/2} \tag{4.20}$$

where μ=effective magnetic moment; S=1/2 for one unpaired electron, 1 for two, etc; L=total orbital angular momentum.

This equation is applicable only to molecules with very high symmetry where the energies of the orbitals containing unpaired electrons are degenerate. A discussion of the calculation of "L" can be found in any introductory quantum mechanics text or in the chapter on quantum mechanics in many physical chemistry texts.

For all of the cases of paramagnetic behavior the spin only formula is often used as a first rough approximation. If the only purpose for measuring the magnetic susceptibility is to determine the number of unpaired electrons this is often all that is done.

4.4.5 Ferromagnetism, Antiferromagnetism and Ferrimagnetism

The advantage of using effective magnetic moments for describing paramagnetic beha-vior is that it is a measure of the materials magnetic behavior that is not dependent upon either the temperature or the magnitude of the external field. It is not possible to set up such a convention for ferromagnetic, antiferromagnetic and ferrimagnetic materials.

All three of these classes of materials can be considered a special case of paramagnetic behavior. The description of paramagnetic behavior is based on the assumption that every molecule behaves independently. The materials discussed here result from a situation in which the direction of the magnetic field produced by one molecule is affected by the direction of the magnetic field produced by an adjacent molecule, in other words their behavior is coupled. If this occurs in a way in which the magnetic fields all tend to align in the same direction, a ferromagnetic material results and the phenomenon is called ferromagnetic coupling. Antiferromagnetic coupling gives an equal number of magnetic fields in opposite directions. Ferrimagnetic coupling gives magnetic fields in two opposite orientations with more in one direction than in the other.

With a few exceptions, the magnetic moments are not aligned through out the entire material. Typically regions, called domains, will form with different orientations. The existence of domains of coupled molecules gives rise to a number of types of behavior as described in the following paragraphs.

The tendency of molecules to align themselves to one another enhances the magnetization of the material due to the presence of an external magnetic field. This is why ferromagnetic and ferrimagnetic materials can have magnetic susceptibilities several orders of magnitude large than paramagnetic materials. This also gives rise to the fact that the magnetic susceptibility of these materials is not independent of the magnitude of the external magnetic field as was the case for diamagnetic and paramagnetic materials.

For a ferromagnetic material, the actual field acting on a given magnetic dipole (unpaired electron) is designated H_t and given by an equation [equation (4.21)] similar to the equation for magnetic induction given above.

$$H_t = H + N_w M \tag{4.21}$$

where H_t=magnetic field felt by an electron; H=external magnetic field; N_w=molecular field constant, approximately 10000; M=magnetization.

This equation is used because it allows a mathematical treatment of a ferromagnetic substance similar to that used for paramagnetic substances. In this form the molecular field constant, N_w, is typically defined empirically in order to take the ferromagnetic coupling into account. To obtain the molecular field constant in a rigorous way would require a quantum mechanical calculation that takes into account the elements, their arrangement in the solid, kinetic energy of the electrons, Coulombic attraction of electrons to the nucleus and repulsion with other electrons as well as spin interactions.

Vibrational motion of the molecules, which increases with temperature, can disrupt the domain structure. Thus the magnetic properties of all three of these types of materials are strongest at low temperatures. At sufficiently high temperatures, no domain structure is able to form so all of these materials become paramagnetic at high temperatures. The temperature at which paramagnetic behavior is seen called the Curie temperature for ferromagnetic and ferrimagnetic materials and called the Neel temperature for antiferromagnetic materials. This is why a temperature independent effective magnetic moment cannot be defined for these materials.

The alignment of the magnetic moments of the domains may give the material a net magnetic moment even in the absence of an external field. This gives a permanent magnet, such as a bar magnet. A material with no net moment prior to being exposed to an external magnetic field may retain a net moment after being exposed to an external magnetic field. This is how cassette and video tapes and computer disks store information. The magnitude of this memory effect can be quantified by plotting magnetization vs field strength as the external field intensity is varied from one polarity to the other and back again. A strong memory effect will be indicated by a wide hysteresis loop.

4.4.6 Magnetic Behaviour of Variation with Temperature

The source of variation of magnetic properties with temperature is the disruption of the alignment of molecular magnetic moments due to the thermal motion of the atoms. As such, it should come as no surprise that diamagnetic behavior shows no variation with temperature.

(1) Paramagnetism As temperature increases, the magnetic susceptibility of a paramagnetic substance decreases.

In some paramagnetic compounds the magnetic susceptibility is inversely proportional to the temperature. These are called "normal paramagnets" and have magnetic properties arising primarily due to the presence of permanent magnetic dipoles. This is referred to as the Curie Law and is expressed in mathematical form as [equation (4.22)].

$$\chi=C/T \tag{4.22}$$

$$C=N_{A}g^{2}b^{2}/4k$$

where χ=magnetic susceptibility; C=the Curie constant; T=temperature; N_A=Avogadro's number; g=the electron g factor; b=the Bohr magneton; k=the Boltzman constant.

In most paramagnetic compounds, an inverse relationship is observed, but the extrapolation to zero temperature does not obey the Curie Law. These compounds obey the Curie-Weiss Law [equation (4.23)].

$$\chi=C/(T-\theta) \tag{4. 23}$$

where χ=the magnetic susceptibility;C=a material-specific Curie constant;T=absolute temperature, measured in Kelvins;θ=the Curie temperature, measured in Kelvins.

(2) Ferromagnetism and ferrimagnetism Ferromagnetic and ferrimagnetic compounds also show a decrease in magnetic susceptibility with increasing temperature. However, a plot of magnetic susceptibility vs. temperature shows a different line shape for these compounds than for paramagnetic compounds. This plot would have a positive curvature for paramagnetic compounds and a negative curvature for ferromagnetic compounds. A rough sketch of the shapes of these curves is as follows (Fig.4.35).

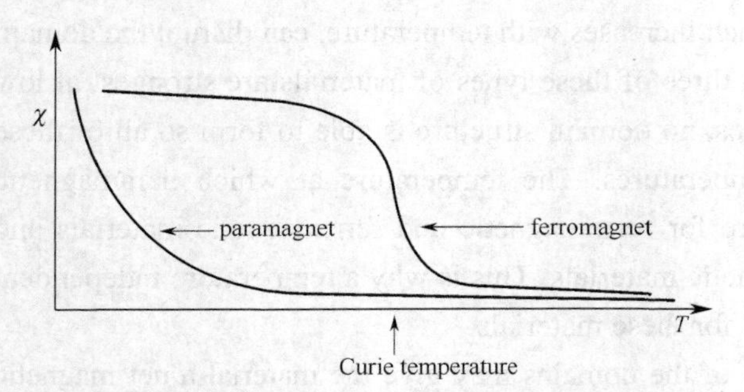

Fig.4.35 A rough sketch of the shapes of χ-T curve for paramagnetic and ferromagnetic compounds

When the Curie temperature is reached, the curvature of the plot changes. At the Curie temperature, ferromagnetic and ferrimagnetic compounds become paramagnetic. Curie temperatures range from 16℃ for Gd to 1131℃ for Co.

(3) Antiferromagnetism Antiferromagnetic compounds show an increase in magnetic susceptibility until their critical temperature, called the Neel temperature, is reached. Above the Neel temperature these compounds also become paramagnetic. Neel temperature range from 1. 66K for $MnCl_2 \cdot 4H_2O$ to 953 K for α-Fe_2O_3.

4.5 Photochemical Properties of Coordination Complex

Photoreactivity is a widespread phenomenon in organic and inorganic chemistry. A compound may change its structural and electronic properties under light irradiation. Light-induced charge separation or *cis-trans* isomerization processes under light are well-known examples.

As a consequence, a phase transformation may occur, with or without hysteresis. If the light-induced new phase possesses sufficiently long lifetime and differs in optical and/or magnetic properties such that it can be detected easily by optical or magnetic means, the material bears the potential for possible technical applications in switching or display devices.

4.5.1 Fundamental Properties of a Photochemical Process

In a molecule, photon absorption is followed by a rapid vibrational relaxation (vr), which causes the molecule to reach an equilibrium geometric configuration corresponding to its electronic excited state. In every excited state there is a competition between physical, radiative (f=fluorescence, p=phosphorescence), nonradiative (ic=internal conversion, isc=intersystem crossing), and chemical reaction modes of deactivation (Fig.4.36).

There are some advantages of employing transition metals to build supramolecular systems.

- Involvement of d orbitals which offer more bonding modes and geometric symmetries than simple organic molecules;
- A range of electronic and steric properties which can be fine-tuned by employing various ancillary ligands;
- Easily modified size of the desired supramolecules by utilizing various lengths of bridging ligands;
- Incorporation of their distinct spectral, magnetic, redox, photophysical, and photochemical properties.

Moreover, the diverse bonding angles imported by the transition-metal centers and the high directionality of the bonding between the ligands and metals also provide superior features over weak electrostatic, van der Waals, and π-π interactions. Another interesting aspect is that thermodynamically driven spontaneous self-assembly of individual molecular components into well-defined molecular structures in solution is expected to be rather similar for both coordination chemistry and biology, and this enables transition metal complexes to be valuable mimics of the more complicated biological systems.

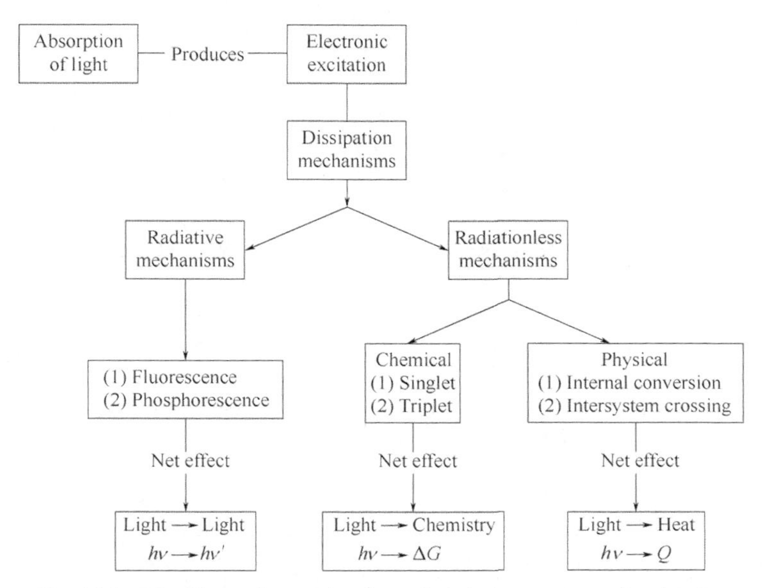

Fig.4.36 The illustration of the photochemical processes of molecules

4.5.2 Artificial Photosynthesis

An increasing demand for cheap and clean energy has stimulated development of novel chemical systems capable of efficient solar energy conversion. Until now natural photosynthesis has been the cleanest and most efficient process. It involves a large array of chromophores (antennae) to harvest visiblelight, a charge-separation system, and redox centers. All the elements (antenna, charge separation, and reaction centers) may involve transition metals. Application of metal complexes facilitates mimicking of this complex chemical system due to rich and versatile photochemical processes typical for transition-metal complexes.

(1) Light-harvesting antennae An antenna for light harvesting is an organized multicomponent system in which an array of chromophoric molecules absorbs the incident light and channels the excitation energy to the common acceptor component. The antenna effect can only be obtained in supramolecular arrays suitably organized in space, time, and energy domains. Each molecular component must absorb the incident light, and the excited state obtained this way must transfer electronic energy to the nearby component before undergoing radiative or nonradiative deactivation. One of the obvious candidates for light-harvesting antennae chromophores are porphyrin and phthalocyanine arrays, synthetic analogues of natural chlorophyll arrays present in photosystems of green plants (Fig.4.37).

Porphyrins is characterized by high molar absorption coefficients and fast energy/electron transfer to other components of the system. They can be easily used as building blocks in large supramolecular systems.

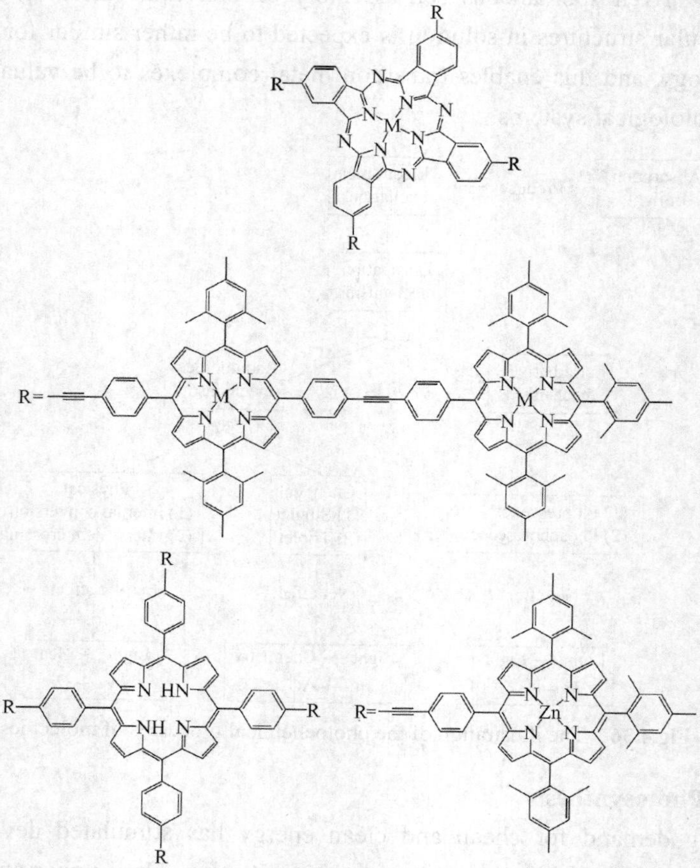

Fig.4.37 Porphyrin and phthalocyanine arrays for light-harvesting antennae

Another class of extensively studied artificial antennae encompasses metallodendrimers and multimetal arrays based on polypyridine ligands. Especially promising, due to high molar absorptivity, photostability, and excited-state properties, are polypyridine complexes of ruthenium (Ⅱ) and osmium (Ⅱ) (Fig.4.38).

(2) Charge-separation system Once the solar energy is harvested and focused in the reaction center it must be converted into a more useful form, i.e. , chemical energy. This goal can be achieved in the systems that allow formation of the charge-separated state with a significant lifetime. It is obvious that in such a structure charge recombination will occur readily. To slow the recombination it is necessary to introduce an additional reaction step in which the charges are moved far apart. The simplest way to do so is to introduce the secondary donor (D) or secondary acceptor (A) (Fig.4.39).

Structure of the ferrocene-porphyrin-fullerene tetrad is capable of efficient light-induced charge separation, for example, the tetrad Fe-ZnPH_2P-C_{60} with the long lifetime (380ms) (Fig.4.40).

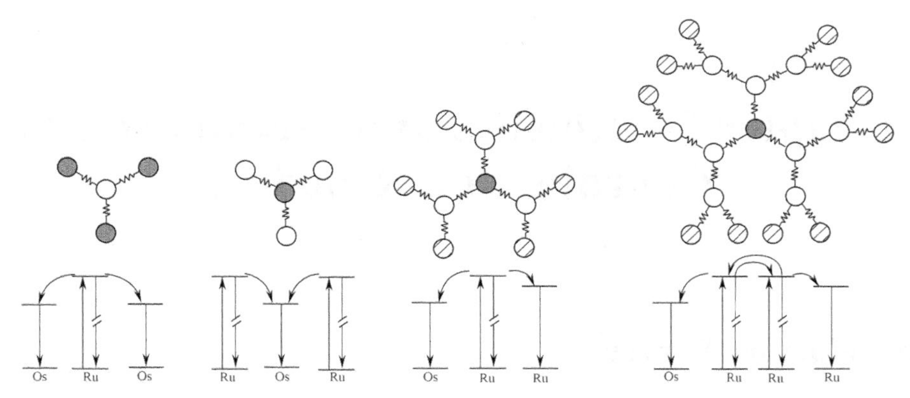

Fig.4.38 Schematic structures of several Ru^{II} and Os^{II} based polypyridine dendrimers together with schematic energy diagrams indicating the energy-transfer processes

Ruthenium centers are marked in ○ and ⊘, while osmium centers are in ●. The terminal ligands are omitted for clarity

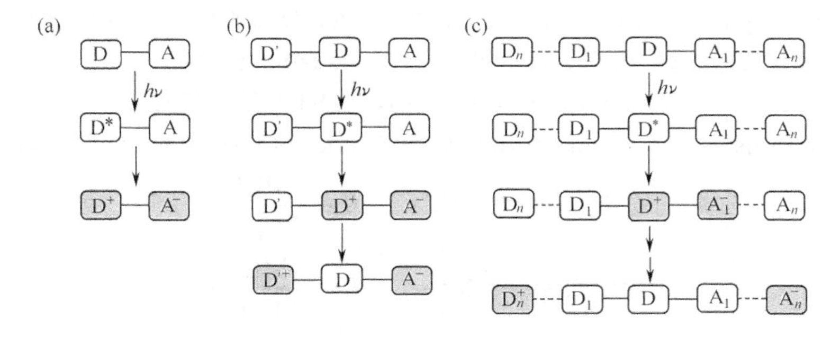

Fig.4.39 The illustration of the path of charge separation including the secondary donor (D′) or acceptor (A′)

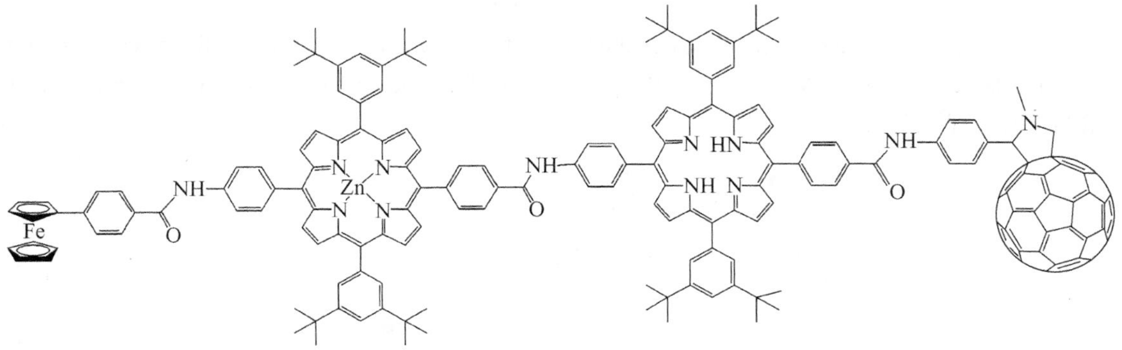

Fig.4.40 The molecular structure of the Fe-ZnPH$_2$P-C$_{60}$

Chapter 5 Kinetics and Mechanisms of Coordination Reactions

5.1 Introductory Survey

(1) Formation of coordination complexes Typically coordination compounds are more labile or fluxional than other molecules [equation (5.1)].

$$MX+Y \rightleftharpoons MY+X \tag{5.1}$$

X is leaving group and Y is entering group. One example is the competition of a ligand, L for a coordination site with a solvent molecule such as H_2O [equation (5.2)].

$$[Co(OH_2)_6]^{2+}+Cl^- \longrightarrow [Co(OH_2)_5Cl]^{+}+H_2O \tag{5.2}$$

(2) Formation constants Consider formation as a series of formation equilibria [equation (5.3)]:

$$\begin{aligned} M+L &\rightleftharpoons ML \qquad K_1=\frac{[ML]}{[M][L]} \\ ML+L &\rightleftharpoons ML_2 \qquad K_2=\frac{[ML_2]}{[ML][L]} \end{aligned} \tag{5.3}$$

Summarized as [equation (5.4)]:

$$M+nL \rightleftharpoons ML_n \qquad \beta_n=\frac{[ML_n]}{[M][L]^n}=K_1K_2K_3\cdots K_n \tag{5.4}$$

Generally, the value of K_{n-1} is greater than that of K_n expected statistically, fewer coordination sites available to form ML_n than ML_{n-1}. The Fig.5.1 shows the sequential formation of $[Al(H_2O)_{6-x}F_x]^{(3-x)+}$.

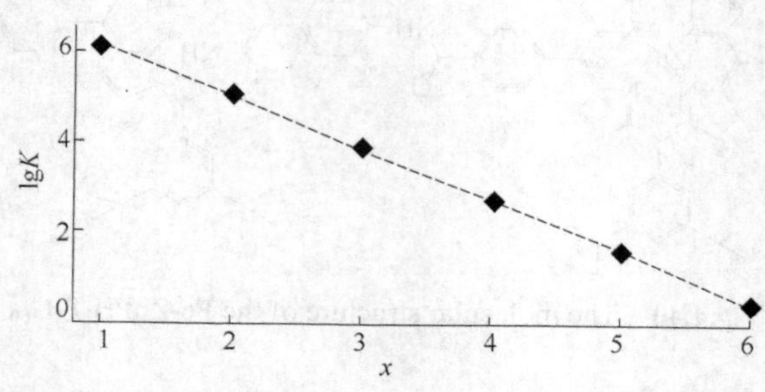

Fig.5.1 The relationship between the lg$K$ with x in the sequential formation of $[Al(OH_2)_{6-x}(F)_x]^{(3-x)+}$

(3) Chelate effect Compare K_1 to β_2 ($M=Cu^{2+}$) [equation (5.5)]:

$$\begin{aligned} M(OH_2)_2^{2+} +en &\rightleftharpoons M(en)^{2+}+2H_2O \\ M(OH_2)_2^{2+} +2NH_3 &\rightleftharpoons M(NH_3)_2^{2+} +2H_2O \end{aligned} \tag{5.5}$$

$\lg K_1$=10.6, $\lg\beta_2$=7.7. Chelate complex is three orders of magnitude more stable. The enhanced stability of a chelated complex over its non-chelating analog. Attributed to the change in entropy, chelation trades two restricted solvent molecules for one bound ligand.

(4) Ring formation and electron delocalization Ability to form rings with metal center improves stability, particularly five or six membered rings. Additionally, ligands with aromatic rings can behave as π acceptors and form back bonding complexes (Fig.5.2).

(5) The nature of the metal ions on the stability of the complexes for a given ligand The Irving-Williams series can tell us some information, which has discussed in the chapter 2. For example, Irving-Williams series gives Cu^{2+} more stable than Ni^{2+}. The figure (Fig.5.3) is shown the stepwise K_f for displacement of H_2O by NH_3 ligands from aquated Ni^{2+} and Cu^{2+}.

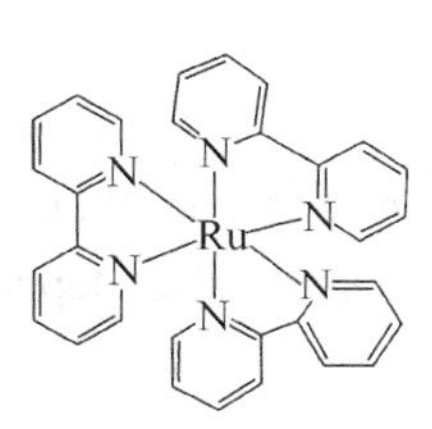

Fig.5.2 The molecular structure of the $Ru(bpy)_3^{2+}$ chelated cation

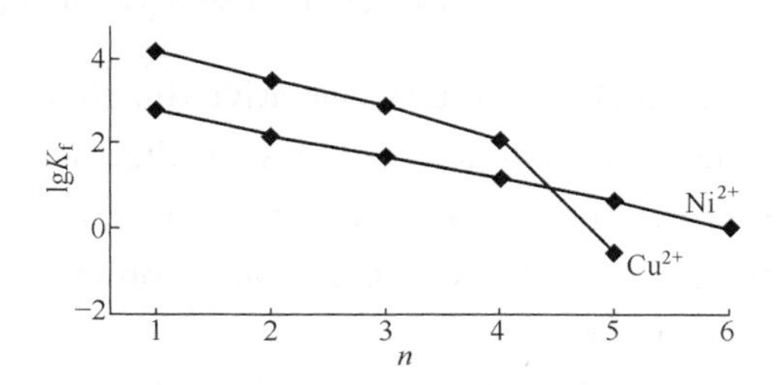

Fig.5.3 The stepwise K_f for displacement of H_2O by NH_3 ligands from aquated Ni^{2+} and Cu^{2+}

5.2 Reaction Mechanisms of d-block Metal Complex and Mechanisms of Organometallic Reaction

Rate is important for understanding coordination complexes chemistry.

Inert: species that are unstable but survive for minutes or more.

Labile: species that react more rapidly than inert complexes.

General rules:

- For 2+ ion, d metals are moderately labile, particularly $d^{10}(Hg^{2+}, Zn^{2+})$.
- Strong field d^3 and d^6 octahedral complexes are inert. i.e. Cr(Ⅲ) and Co(Ⅲ).
- Increasing ligand field stabilization energy improves inertness.
- 2nd and 3rd row metals are generally more inert.

Ligand field stabilization energy (LFSE) for O_h-geometry is listed in Table 5.1. The relationship of the electron configuration and stabilization is shown in Fig.5.4.

Table 5.1 LFSE for O_h Geometry

d^n	High Spin		Low Spin		d^n	High Spin		Low Spin	
	config	LFSE(Δ_o)	config	LFSE(Δ_o)		config	LFSE(Δ_o)	config	LFSE(Δ_o)
d^1	$t_{2g}^1e_g^0$	−0.4			d^6	$t_{2g}^4e_g^2$	−0.4	$t_{2g}^6e_g^0$	−2.4
d^2	$t_{2g}^2e_g^0$	−0.8			d^7	$t_{2g}^5e_g^2$	−0.8	$t_{2g}^6e_g^1$	−1.8
d^3	$t_{2g}^3e_g^0$	−1.2			d^8	$t_{2g}^6e_g^2$	−1.2		
d^4	$t_{2g}^3e_g^1$	−0.6	$t_{2g}^4e_g^0$	−1.6	d^9	$t_{2g}^6e_g^3$	−0.6		
d^5	$t_{2g}^3e_g^2$	0	$t_{2g}^5e_g^0$	−2.0	d^{10}	$t_{2g}^6e_g^2$	0		

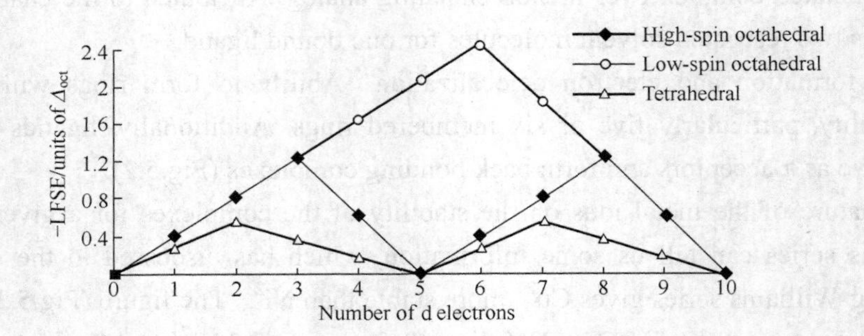

Fig.5.4 The relationship of the electron configuration and stabilization

5.2.1 Associative and Dissociative Reactions

Ligand substitution reactions are either associative or dissociative.

Associative: Reactions intermediate has higher coordination number than reactants or products. Two characters are lower coordination number complexes and rates depending on the entering group.

Dissociative: Reactions intermediate has lower coordination number than reactants or products. Two characters are octahedral complexes and smaller metal centers, rates depending on leaving group.

5.2.2 Measurements of Rates

"Inert" species: $t_{1/2}$>1min. We can use classical static techniques, e. g. light absorption, pH measurements.

"Labile" species: $t_{1/2}$=1min-1ms, Use stop flow measurements, rapid mixing, fast spectroscopy.

"Rapid" reactions: Relaxation techniques and fast spectrophotometry.

We cannot conclude a mechanism from a rate law! e. g. Rate=k[A][B], or Rate=k[A] (Fig.5.5).

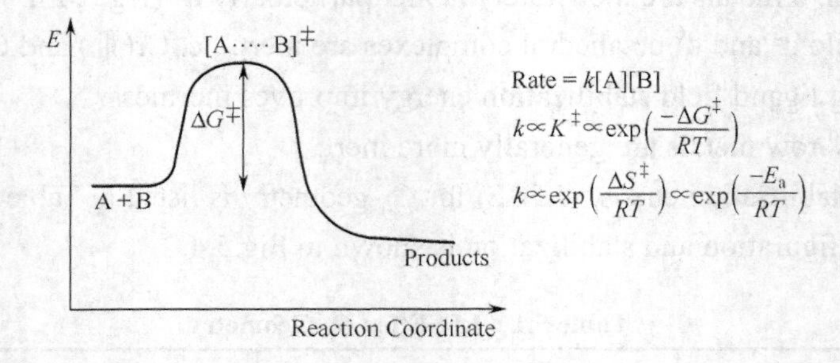

Fig.5.5 Plots of energy versus reaction coordinate for reactions of A+B ⟶ product

Some points that can be drawn out from Fig.5.5.

- involvement of solvent (pseudo-first order behaviour).
- complex reactions with only one rate limiting step.

Finally, you can only disprove a mechanism you can never prove a mechanism. A rate law can at best only be consistent with a mechanistic scheme, it can never prove it.

Distinguish some definitions:

- "Lability" and "inert" are kinetic terms!!
- "Stable" and "unstable" are thermodynamic statements.
- "Intimate" mechanism refers to the details of the mechanism on the molecular scale.

The rates of the substitution reactions can span from 1ms to 10^8s.

Since the rate is exponentially dependent on both $\Delta S^{\ddagger}$ and ΔE_{act}, small rate changes are "not significant" for the interpretation of mechanism with these very crude theories.

5.2.3 Typical Reaction Coordinates

The 1st order dependence of the rate of the reaction on: ① the concentration of the substrate and ② the concentration entering reagent, indicates that these complexes undergo substitution of ligands by a bimolecular mechanism.

Some important points concerning the "intimate" mechanism are shown in Fig.5.6.

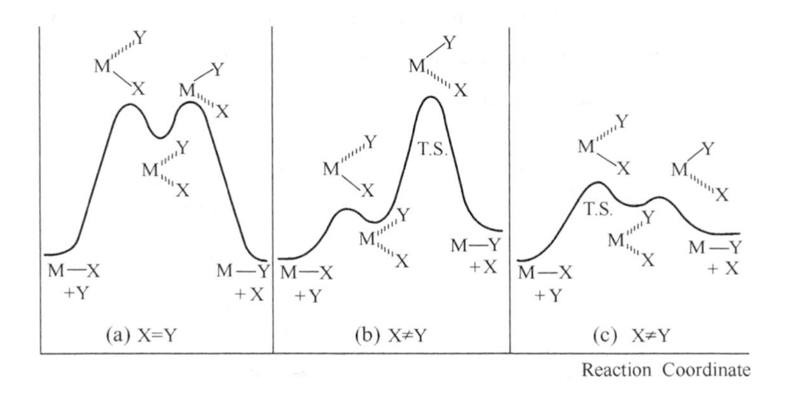

Fig.5.6 Plots of the energy versus reactions coordinate for the three kinds of the intimate mechanism

- Profile (a) applies when the energies of the reactants and products are almost equal, and in which the energies of the transition states are almost equal.
- Profile (b) is appropriate to a reaction for which the bond breaking is rate determining (d mechanism from the intermediate).
- Profile (c) depicts the situation when bond forming is rate determining, a situation in which the entering group (Y) lies higher in the *trans* effect series than the leaving group.

In a few cases the intermediate is reasonable stable and so lies at lower energy than the products. For example, a reaction profile for a reaction where a quasi-stable intermediate is formed before the transition state: bond breaking in the five-coordinate intermediate is more important than bond making to form the intermediate (Fig.5.7).

5.2.4 Mechanisms of Organometallic Reactions

(1) Introduction

Some basic concepts for this part are included: Green-Davies-Mingos Rules, review of ligand types and 18 Electron Rule (in chapter 4).

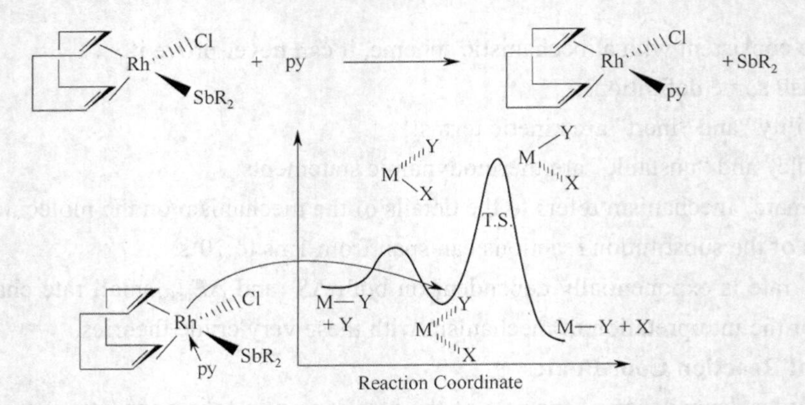

Fig.5.7 A reaction profile for a reaction where a quasi-stable intermediate is formed before the transition state

① Davies-Green-Mingos Rules Nucleophilic addition to cationic-complexes is one of the most common and important reactions in organometallic chemistry. Allyl- and enyl- ligands as well as unsaturated hydrocarbons like ethene and butadiene or benzene that do not normally undergo nucleophilic addition reactions, are readily attacked by nucleophiles like H^-, CN^- or MeO^-. For example [equation (5.6) and equation (5.7)]:

(5.6)

(5.7)

The DGM Rules (1978) have proposed three general rules that allow predictions of the direction of kinetically controlled nucleophilic attack at 18-electron cationic metal complexes.

Rule 1 Nucleophilic attack occurs preferentially at even coordinated polyenes.

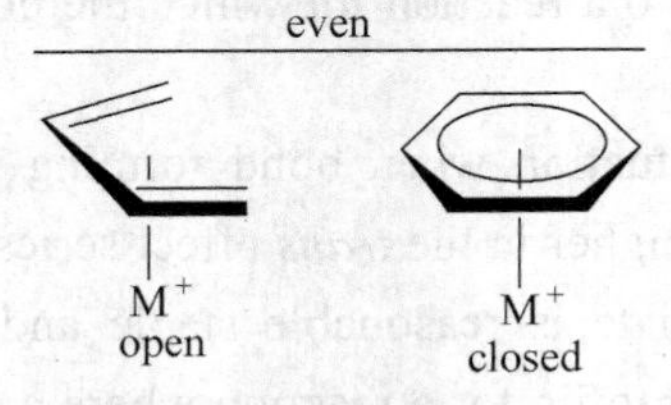

Rule 2 For odd coordinated polyenes, nucleophilic attack to open coordinated polyenes is preferred to closed polyene ligands.

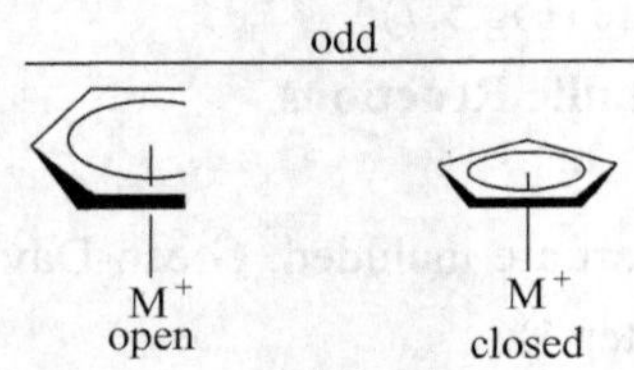

Rule 3 In the case of even open polyenes, nucleophillic attack always occurs at the terminal carbon atom;for odd open polyenes, attack at the terminal carbon occurs only if L_nM^+ fragment is strongly electron withdrawing.

② Review of ligand types shown in Table 5.2.

Table 5.2 Some common ligands in organometallic compounds

Number of electrons donated	Name	Example	Number of electrons donated	Name	Example
1	yl	methyl,ethyl,phenyl = • 	5	dienyl	cyclopentadienyl = 
2	ene	ethene 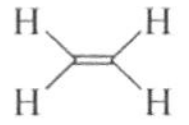 = H H H H	6	triene	benzene  = cyclopentatriene  =
3	enyl	η^3-allyl enyl propenyl 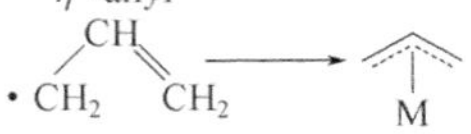 = CH • CH_2 CH_2 → M	7	trienyl	cycloheptatrienyl  =
4	diene	butadiene  = cyclopentadiene  =	8	tetraene	cyclootatetraene  =

The ionic (charged) model The basic premise of this method is that we remove all of the ligands from the metal and, if necessary, add the proper number of electrons to each ligand to bring it to a closed valence shell state. For example, if we remove ammonia from our metal complex, NH_3 has a completed octet and acts as a neutral molecule. When it bonds to the metal center it does so through its lone pair (in a classic Lewis acid-base) and there is no need to change the oxidation state of the metal to balance charge. We call ammonia a neutral two-electron donor.

The covalent (neutral) model The major premise of this method is that we remove all of the ligands from the metal, but rather than take them to a closed shell state, we do whatever is necessary to make them neutral. Let's consider ammonia once again. When we remove it from the metal, it is a neutral molecule with one lone pair of electrons. Therefore, as with the ionic model, ammonia is a neutral two electron donor.

The most critical point is that like oxidation state assignments, electron counting is a formalism and does not necessarily reflect the distribution of electrons in the molecule. However, these formalisms are very useful to us, and both will give us the same final answer (Table 5.3).

Table 5.3 Some examples of the two methods compared

	ionic		covalent			ionic		covalent	
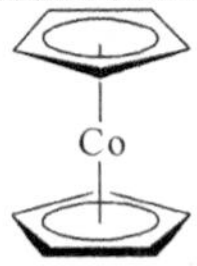 Co	Co^{II}	$7e^-$	Co	$9e^-$	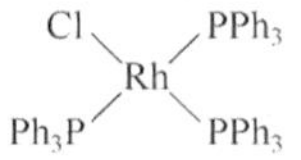 Cl, PPh₃, Rh, Ph₃P, PPh₃	Rh^{I}	$8e^-$	Rh	$9e^-$
	$2Cp^-$	$12e^-$	$2Cp\cdot$	$10e^-$		Cl^-	$2e^-$	$Cl\cdot$	$1e^-$
	Total	$19e^-$	Total	$19e^-$		$3PPh_3$	$6e^-$	$3PPh_3$	$6e^-$
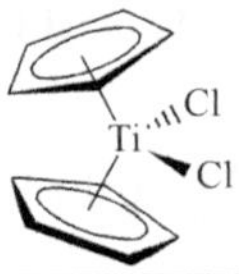 Ti, Cl, Cl	Ti^{IV}	$0e^-$	Ti	$4e^-$		Total	$16e^-$	Total	$16e^-$
	$2Cl^-$	$4e^-$	$2Cl\cdot$	$2e^-$	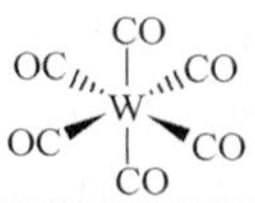 CO, OC, CO, W, OC, CO, CO	W^0	$6e^-$	W	$6e^-$
	$2Cp^-$	$12e^-$	$2Cp\cdot$	$10e^-$		6CO	$12e^-$	6CO	$12e^-$
	Total	$16e^-$	Total	$16e^-$		Total	$18e^-$	Total	$18e^-$

(2) Fundamental Reaction Types

① Oxidative addition reactions　Formal definition: (a) Increase of 2 in formal oxidation state. (b) Increase of 2 in coordination number [equation (5.8)].

$$L_nM + X—Y \longrightarrow L_nM(X)(Y) \tag{5.8}$$

$$Mn^+ \longrightarrow M^{(n+2)+} \qquad d^n \longrightarrow d^{n-2}$$

Requirements for the metal complex: (a) Metal centre must be d^2 or greater. (b) Metal centre must be coordinatively unsaturated. (c) Metal must have appropriate energy for both empty and vacant orbitals [equation (5.9)].

Empty Orbital
L_nM + X—Y ⟶ L_nM(Y)(X)　(5.9)
Filled Orbital

② Intramolecular oxidative addition [equation (5.10)]

$IrCl(PPh_3)_3$ —△→ $IrHCl(PPh_3)_2(Ph_2P\text{-}C_6H_4)$　(5.10)

It is an example of an orthometallation reaction.

③ Bimolecular reaction [equation (5.11)]

$$L_nM—ML'_n + X—Y \longrightarrow L_nM—X + L'_nM—Y$$

$(OC)_5Mn—Mn(CO)_5 \xrightarrow{Br_2} 2\,Mn(CO)_5Br$　(5.11)

④ Carbon-hydrogen bond activation [equation (5.12)]

$Cp^*Rh(PMe_3)H_2 \xrightarrow[-H_2]{h\nu} [Cp^*Rh(PMe_3)]$; $+C_6H_6$ → $Cp^*Rh(PMe_3)(H)(C_6H_5)$; $+CH_4$ → $Cp^*Rh(PMe_3)(H)(CH_3)$　(5.12)

Note: reaction proceeds faster for C_6H_6 versus CH_4 but Ph—H bond is stronger than CH_3—H bond.

5.2.5 Kinetic Rate Laws for Oxidative Addition Reaction

(1) d^8 Square planar complexes　In the oxidative addition of XY to $IrCl(CO)\text{-}(PPh_3)_2$(Vaska's complex), It is observed that rate is k_2[Ir(I)][XY].

It is a simple 2nd order kinetics and characteristic of coordinatively unsaturated systems [equation (5.13)].

$$Ir(Cl)(CO)(PPh_3)_2 \xrightarrow{X-Y} Ir(X)(Y)(Cl)(CO)(PPh_3)_2 \quad (5.13)$$

(2) d^8 Trigonal bipyramidal complexes Oxidative addition of XY to $IrH(CO)(PPh_3)_3$ [equation (5.14)].

$$IrH(Cl)(PPh_3)_3 \xrightarrow{X-Y} Ir(X)(Y)(H)(CO)(PPh_3)_2 \quad (5.14)$$

It is observed

$$\text{Rate}=\frac{k_1k_2[\text{Ir(I)}][\text{XY}]}{k_{-1}[\text{PPh}_3]+k_2[\text{XY}]} \quad (5.15)$$

It has character of coordinatively saturated complexes and is consistent with the following scheme:

$$IrH(CO)(PPh_3)_3 \underset{k_{-1}}{\overset{k_1}{\rightleftharpoons}} IrH(CO)(PPh_3)_2+PPh_3$$

$$IrH(CO)(PPh_3)_2+XY \overset{k_2}{\rightleftharpoons} IrH(X)(Y)(CO)(PPh_3)_2 \quad (5.16)$$

Note: reaction is retarded by extra phosphine.

5.2.6 Mechanisms of Oxidative Addition Reaction

No single mechanism holds for all oxidative-addition reactions. Three main mechanisms have been observed depending on the reactants, and conditions. Three types are:

- Concerted;
- (i) Nucleophillic attack (S_N2), (ii) Ionic mechanism;
- Radical reaction.

(1) Concerted mechanism There are three characteristics:

- Involves a 3-centre transition state, with side on approach of the incoming molecule;
- Molecule X—Y attaches to the reaction centre and simultaneously or subsequently undergoes electron rearrangement from X—Y bond to M—X and M—Y bonds [equation (5.17)];
- In transition state, there is synergetic electron-flow:

$$X—Y(\sigma) \longrightarrow M(\sigma);\ M(\pi) \longrightarrow X—Y(\sigma^*).$$

$$L_nM + X—Y \longrightarrow [L_nM\cdots(X,Y)]^{\ddagger} \longrightarrow L_nM(X)(Y) \quad (5.17)$$

Features for equation (5.18): (a)Coordinatively unsaturated molecule;(b)Common when X—Y group has low bond polarity, e. g. H_2, $R_3Si—H$;(c)Stereospecific and *cis*-addition [equation (5.18)].

$$Ir(Cl)(CO)(PPh_3)_2 \xrightarrow{H_2} Ir(H)_2(Cl)(CO)(PPh_3)_2 \quad (5.18)$$

Cis-addition can be shown by infrared spectroscopy and observed two M—H stretching bands (symmetric and asymmetric).

Note: only one asymmetric stretching band if *trans*. If X—Y is chiral then reaction proceeds with retention, for example:

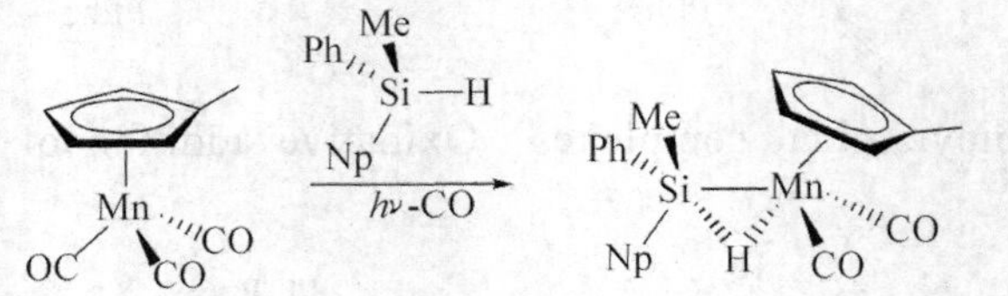

(5.19)

Side addition of H_2 has low activation energy; $\Delta G^{\ddagger}$ around 40kJ/mol (Fig.5.8).

The strength of H—H bond (420kJ/mol) suggests "early T. S. " with little H—H stretching. Thus kinetic isotope effect is small:

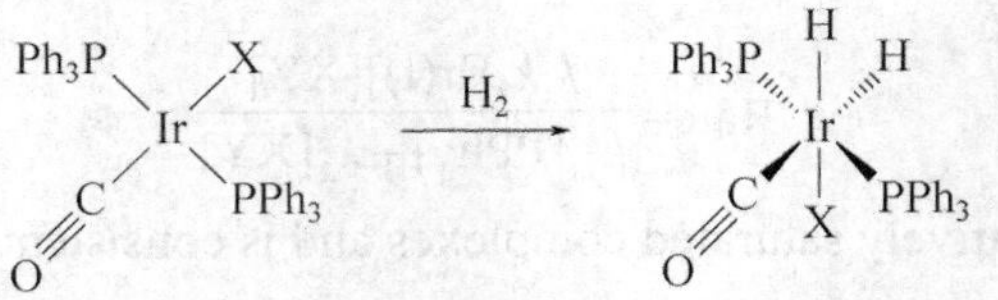

(5.20)

Rates for addition of H_2 in Vaska's complex X: I > Br > Cl

Ratio's of rate: 100 4 0.9

Electron density on the metal centre is crucial in determining ΔG. Donation into H_2 (σ^*) appears to be the most important factor. If replaced CO by CS (better π-acceptor), no reaction with H_2 get.

(2) Nucleophillic attack It is favoured when X—Y bond is polar, for example: RX, HX, $HgCl_2$. Dipolar transition state is expected (Fig.5.9).

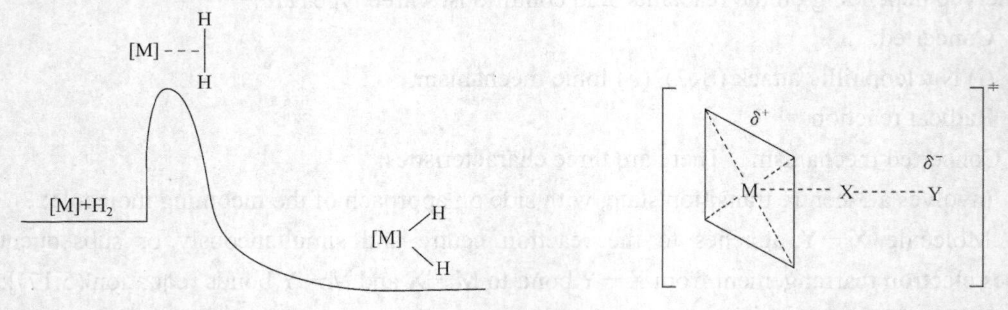

Fig.5.8 The energy versus reactions coordinate for the side on addition of H_2 mechanism

Fig.5.9 The illustration of dipolar transition state for the mechanism of nucleophillic attack

Metal centre acts as electron pair donor (nucleoplile). For reactions involving R—X bond, inversion of sterochemistry at the carbon, analogous to S_N2 reactions is appear [equation (5.21)].

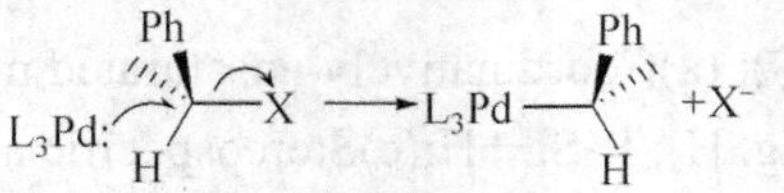

(5.21)

Features of nucleophilic attack: (a) *Trans* addition; (b) Rates faster in polar solvents; (c) Large $-\Delta S^{\ddagger}$, due to solvent organization around the dipolar transition state; (d) Rates sensitive to nature of other ligands (L). Example [equation (5.22)]:

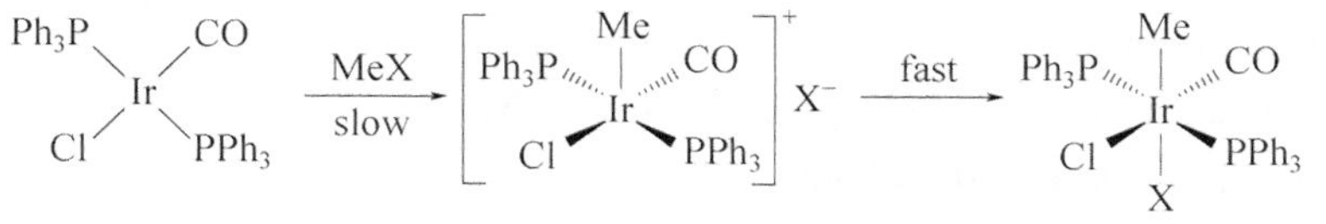

$$trans\text{-}[Ir(CO)Cl(PPh_3)_2] \xrightarrow[\text{slow}]{MeX} [Ir(Me)(CO)Cl(PPh_3)_2]^+ X^- \xrightarrow{\text{fast}} [Ir(Me)(X)(CO)Cl(PPh_3)_2] \tag{5.22}$$

A few features of reaction [equation (5.22)] are: (a) More electron donating the phosphine, faster the rate of addition; (b) Hammett plots indicate increased ligand basicity leads to increased nucleophillicity of Ir-centre; (c) Steric effects, increasing size of ligands decreases the rate of oxidative addition;due to crowding of transition state.

(3) Ionic mechanism In polar solvents (eg. MeOH), sometimes addition is less stereospecific. The lifetime of the positive charged 5-coordinate intermediate for the nucleophilic mechanism is enhanced. Thus, in equation (5.23):

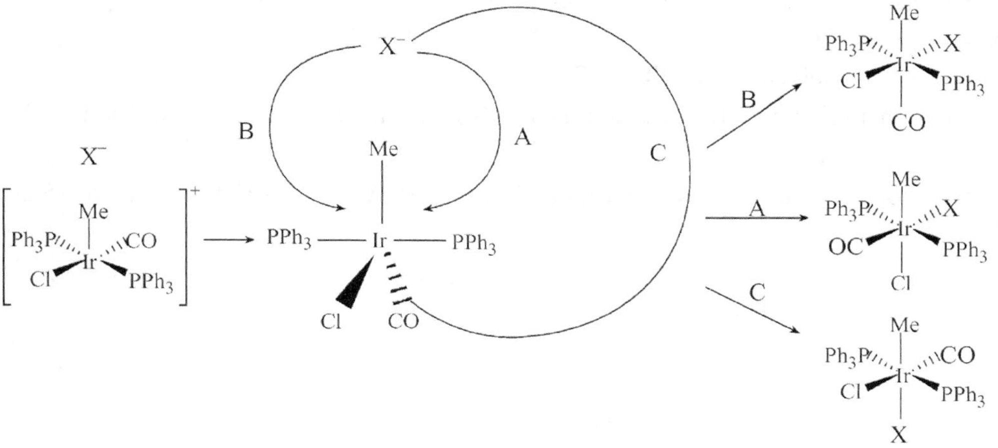

$$[Ir(Me)(CO)Cl(PPh_3)_2]^+ X^- \longrightarrow PPh_3\text{—}Ir(Me)(Cl)(CO)\text{—}PPh_3 \xrightarrow{A,\ B,\ C} \text{products} \tag{5.23}$$

(Path B: Me trans to CO, X trans to Cl; path A: Me trans to Cl, X trans to CO; path C: Me trans to X.)

Polar solvents(DMF, MeOH, H_2O, MeCN) ⟶ *cis*+*trans*

Non-polar solvents(C_6H_6, $CHCl_3$) ⟶ *cis* only

Gas phase ⟶ *cis* only

(4) Radical reactions Alkyl (not methyl), vinyl, aryl halides react with Vaska's complex by radical chain reaction. Rate depends on (a) initiators; (b) inhibitors; (c) radical spin traps. Typical initiators include O_2 and Ir(Ⅱ) (Fig.5.10).

$$\left.\begin{array}{l}[Ir^{II}] + O\cdot \text{ (trace radical)} \longrightarrow [Ir^{II}]\text{—}O\cdot \\ [Ir^{II}]\text{—}O\cdot + R\text{—}X \longrightarrow X\text{—}[Ir^{III}]\text{—}O + R\cdot\end{array}\right\} \text{initiation steps}$$

$$\left.\begin{array}{l}[Ir^{II}] + R\cdot \longrightarrow R\text{—}[Ir^{II}]\cdot \\ \qquad \longrightarrow R\text{—}[Ir^{III}]\text{—}X + R\cdot\end{array}\right\} \text{propagation steps}$$

$$[Ir^{II}] + R\text{—}X \longrightarrow R\text{—}[Ir^{III}]\text{—}X \quad \text{net reaction}$$

Fig.5.10 Typical initiators include O_2 and Ir(Ⅱ)in radical reactions

Features of chain mechanism:

- Complete loss of stereochemistry of R in R—X;
- 17-electron ML_5 complexes such as Co(Ⅱ)-d^7 undergo stepwise free radical addition.

Features of stepwise reaction:

- 2nd order kinetics;
- Reactants extracts a radical fragment from X—Y;
- Racemisation at C-based radicals.

5.2.7 Migration Reaction ("Migratory Insertions")

(1) Formal definition Insertion of an unsaturated ligand(Y)into an adjacent M—X bond is shown in equation (5.24).

$$\underset{\substack{M^{n+}\\ d^{n}}}{\overset{X}{\overset{|}{M}}} \leftarrow Y \rightleftharpoons \underset{d^{[n-2]}}{[M—Y—X]} \rightleftharpoons \underset{\substack{M^{n+}\\ d^{n}}}{\overset{L}{\overset{\downarrow}{M}}}—Y—X \qquad (5.24)$$

Y=CO, C_2H_4, C_2H_2, benzene, cyclopentadienyl

X=H, R, Ar, C(O)R

L=Lewis base, solvent, PR_3, NR_3, R_2O

(2) Two main types 1,1-Insertion and 1,2-insertion are shown in equation (5.25) and equation (5.26) respectively.

For example: X=Me;AB=CO in equation (5.25). X=H; AB=H_2C═CH_2 in equation (5.26).

$$\overset{X}{\overset{|}{M}}—A\equiv B \longrightarrow M—A(X)═B \qquad (5.25)$$

$$\overset{X}{\overset{|}{M}}—\underset{\underset{B}{\|}}{A} \longrightarrow M—A—B—X \qquad (5.26)$$

(3) General features of insertion reactions

- No change in formal oxidation state of metal centre, except if inserting ligand (Y) is M═C or MC.
- Groups must be adjacent (*cis*) to each other.
- Vacant coordination site is created during the forward reaction.
- When the X ligand is chiral, the reaction generally proceeds with retention of configuration.
- Examples of X-migration and Y-insertion are known.
- Position of equilibrium depends upon the strength of M—X, M—Y, and M—(XY) bonds.
- 1-electron oxidation often speeds up reaction.

(4) Specific examples of migration reactions

Hydride migrations [equation (5.27) and equation (5.28)]:

$$Cp^{*}Rh(PMe_3)(H)(C_2H_4) \rightleftharpoons [Cp^{*}Rh(PMe_3)(H\cdots CH_2CH_2)]^{\ddagger} \underset{-PMe_3}{\overset{+PMe_3}{\rightleftharpoons}} Cp^{*}Rh(PMe_3)_2Et \qquad (5.27)$$

[Mo–D complex]$^+$ $\xrightarrow{+PMe_3}$ [Mo–PMe$_3$ complex]$^+$ (H, D) (5.28)

η^6-cycloheptatriene η^5-cycloheptatrienyl

Alkyl migration [equation (5.29) and equation (5.30)]:

Ph$_3$P–Ru(CH$_3$)(CH$_3$) $\xrightarrow[-H^-]{Ph_3C^+}$ [Ph$_3$P–Ru=CH$_2$(CH$_3$)]$^+$ ⟶ [Ph$_3$P–Ru–Et] ⟶ [Ph$_3$P–Ru(H)(=)]$^+$ (5.29)

Mo(Et)(Cl) $\xrightarrow{PR_3}$ Mo(PR$_3$)(Cl) (Et) $\xrightarrow{TlPF_6}$ [Mo(Et)(PR$_3$)]$^+$ PF$_6$ (5.30)

Other examples [equation (5.31)]:

Co(ON)(Me) $\underset{k_{-1}}{\overset{k_1}{\rightleftharpoons}}$ [Co–N(=O)Me]‡ $\xrightarrow[L]{k_2}$ Ph$_3$P–Co–N(=O)Me

ν_{NO}=1780cm^{-1}

k_{-3} k_3,L

O=N–Co(L)(Me)

ν_{NO}=1530cm^{-1}

L=PEt$_3$

ν_{NO}=1310cm^{-1}

L=PPh$_3$PEt$_3$ (5.31)

(5) Combined oxidative addition and alkyl migration reactions

Alkyl ⟶ Acyl [equation (5.32)]:

$$\text{Cp*M(CO)L} \xrightarrow[\text{Oxidative Addition}]{\text{MeI}} [\text{Cp*M(Me)(CO)L}]^+\text{I}^- \xrightarrow{\text{Methyl Migration}} \text{Cp*M(I)(L)(C(=O)Me)} \quad \text{M=Co,Rh} \tag{5.32}$$

Acyl ⟶ Alkyl [equation (5.33)]:

$$\text{Rh(PMe}_2\text{Ph)}_3\text{Cl} \xrightarrow[\text{Oxidative Addition}]{\text{RC(O)Cl}} \text{Rh(C(O)R)(PMe}_2\text{Ph)}_3\text{Cl}_2 \xrightarrow[\text{Alkyl Migration}]{-\text{PPhMe}_2} \text{Rh(R)(CO)(PMe}_2\text{Ph)}_2\text{Cl}_2 \tag{5.33}$$

Carbonylation reaction [equation (5.34)]

$$(OC)_5MnCH_3 + CO \longrightarrow (OC)_5Mn\overset{\overset{\displaystyle O}{\|}}{C}CH_3 \tag{5.34}$$

There are three possible mechanisms: (a)direct insertion; (b) "knock-on"insertion;(c)alkyl migration. The mechanism of this reaction can be elucidated by using ^{13}CO to examine stereochemistry of products.

Infrared spectroscopy: Acetyl derivative has separate band $\nu(CO)=1664cm^{-1}$. Position of entry/exit of ^{13}CO can be deduced from intensity changes and band positions(5.35).

$$Me—Mn(CO)_5 + {}^{13}CO \longrightarrow \textit{cis-}(OC)_4({}^{13}CO)Mn—C(=O)Me \tag{5.35}$$

I. R. Data: Intensity at $1970cm^{-1}$, $1963cm^{-1}$, $1625cm^{-1}$. Thus, acetyl group is *cis* to ^{13}CO; ^{13}CO cannot be directly inserted; no scrambling takes place in the product.

For equation (5.36), it is proved using intensities of $1976cm^{-1}$ (*cis*) and $1949cm^{-1}$ (*trans*) in I. R. spectrum; Only migration can give correct product distribution; It cannot get *trans* isomer by de-insertion of CO.

$$\textit{cis-}(OC)_4({}^{13}CO)Mn—C(=O)Me \xrightarrow{\triangle} \textit{trans-}MeMn(CO)_4({}^{13}CO) + \textit{cis-}MeMn(CO)_4({}^{13}CO) \tag{5.36}$$

Rate law for carbonylation reaction Under normal conditions the following rates are observed [equation (5.37)].

$$RMn(CO)_5 \underset{k_{-1}}{\overset{k_1}{\rightleftharpoons}} (OC)_4Mn—C(=O)R \underset{k_{-2}(-L)}{\overset{k_2(+L)}{\rightleftharpoons}} (OC)_4Mn(L)—C(=O)R \tag{5.37}$$

k_1 is rate determining;i.e. 1st order in alkyl-Mn. zero order in L (incoming ligand), if $k_2 >> k_{-2}$, then [equation (5.38)]:

$$\text{Rate} = \frac{k_1 k_2 [\text{Mn(CO)}_5\text{R}][\text{L}]}{k_{-1} + k_2[\text{L}]} \tag{5.38}$$

If [L] is high, it is first order kinetics in starting material; If [L] is low, it is mixed order kinetics. Rate constants are found to be identical for the substitution of CO in Me—CO—Mn(CO)$_5$ and for the decarbonylation of the acetyl compound [equation (5.39)].

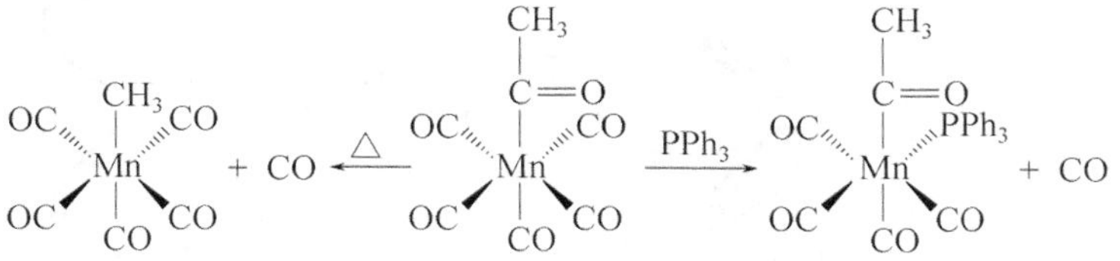

(5.39)

Rate can be accelerated by Lewis acid catalysis, e.g. H^+, $AlCl_3$, BF_3. Lewis acids are presumed to stabilize transition state as shown below [equation (5.40)]:

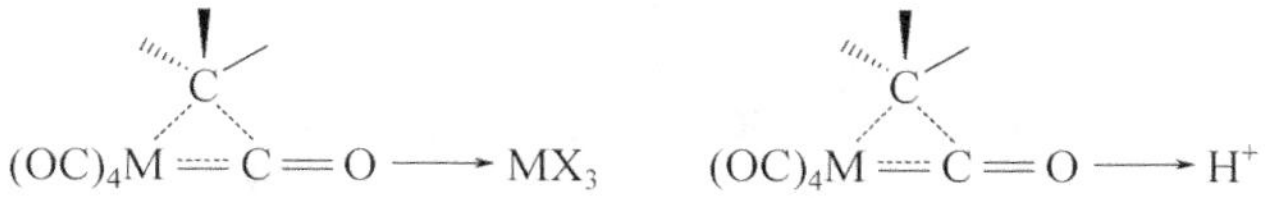

(5.40)

Some sterospecificity: Metal centre; carbon center [equation (5.41)]; adjacent sites [equation (5.42)].

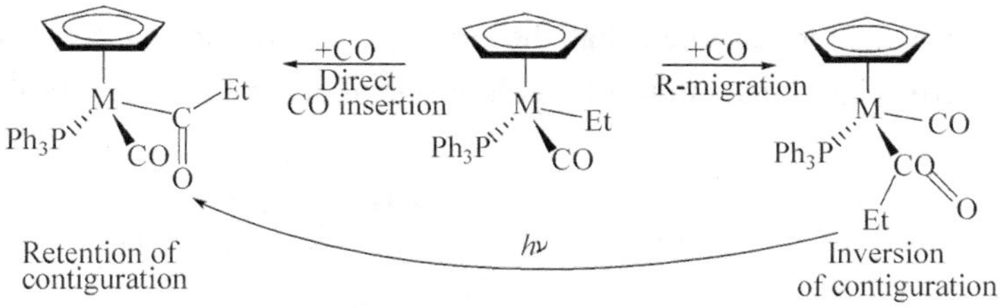
+CO
Direct
CO insertion
+CO
R-migration
hν
Retention of
configuration
Inversion
of configuration

(5.41)

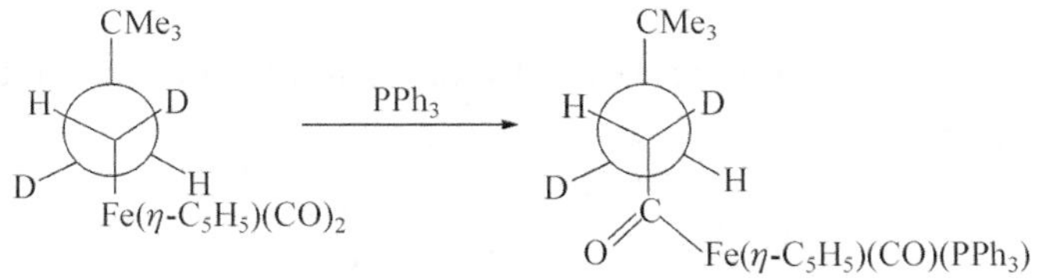

(5.42)

5.2.8 Elimination Reactions

(1) Hydrogen Elimination Reaction

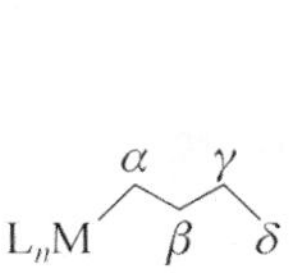

① α-Hydrogen elimination [equation (5.43)]

$d^n \longrightarrow d^{n-2}$; $M^{m+} \longrightarrow M^{(m+2)+}$ (5.43)

Previously α-elimination reactions were thought to be much less common than β-elimination. Probably due to the fact that M (carbene) hydrides are very reactive [equation (5.44)].

$$Cp_2Mo(CH_3)Cl \xrightarrow{PR_3} [Cp_2Mo(CH_3)(PR_3)]^+ \text{ (Kinetic Product)} \xrightarrow{-PR_3} [Cp_2Mo(CH_3)]^+ \rightleftharpoons [Cp_2Mo(=CH_2)H]^+ \underset{PR_3}{\rightleftharpoons} Cp_2Mo(H)(CH_2-\overset{+}{P}R_3) \text{ (Thermodynamic Product)} \tag{5.44}$$

② β-Hydrogen elimination [equation (5.45)]

$$L_nM-CH_2-CH_2-R \rightleftharpoons L_nM(H)(\eta^2\text{-}CH_2{=}CHR) \tag{5.45}$$

$$d^n \longrightarrow d^n;\ M^{m+} \longrightarrow M^{m+}$$

Traditionally, It has been assumed that the presence or absence of β-hydrogens is the most important factor determining the stability of an alkyl ligand coordinated to a transition metal center [equation (5.46)].

$$(PEt_3)_2Pt(CH_2CH_3)_2 \rightleftharpoons (PEt_3)Pt(CH_2CH_3)_2 \rightleftharpoons (PEt_3)Pt(CH_2CH_3)(H)(H_2C{=}CH_2) \tag{5.46}$$

Ligands containing no β-hydrogens exhibt extra stability, for example: —Me, —CH_2Ph, —CH_2Bu-*t* [equation (5.47)].

$$TaCl_5 \xrightarrow{Zn(CH_3)_2} TaMe_5 \tag{5.47}$$

③ γ-Hydrogen elimination [equation (5.48)]

$$L_nM-CH_2-C(CH_3)_3 \rightleftharpoons L_nM(H)(CH_2C(CH_3)_2CH_2) \tag{5.48}$$

$$d^n \longrightarrow d^{n-2};\ M^{m+} \longrightarrow M^{(m+2)+}$$

Neopentyl platinum alkyls compounds tend to decompose by γ-elimination, in contrast to α-elimination that is favoured by Ta. It may imply a different mechanism operating in each case. In Ta case, one alkyl may be deprotonated at α-carbon by another alkyl group. Whereas oxidative addition mechanism is more likely for Pt(Ⅱ) [equation (5.49)].

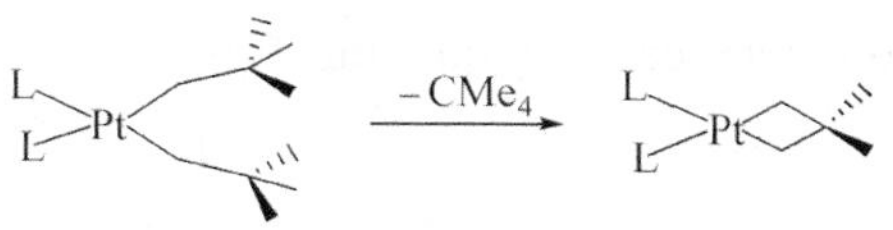

$$\xrightarrow{-CMe_4} \tag{5.49}$$

④ Other hydrogen eliminations [equation (5.50)]

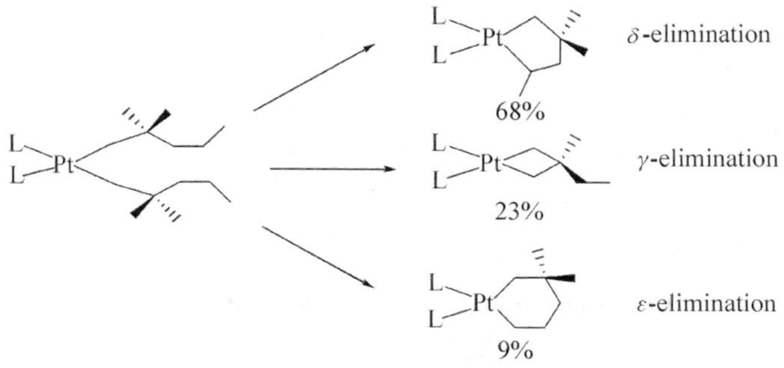

δ-elimination 68%

γ-elimination 23%

ε-elimination 9%

(5.50)

(2) Reductive Hydrogen Elimination Reaction

$$L_nM\begin{matrix}Y\\X\end{matrix} \underset{\text{Oxidtive Addition}}{\overset{\text{Reductive Elimination}}{\rightleftharpoons}} L_nM + X—Y \tag{5.51}$$

$$d^n \longrightarrow d^{n+2};\ M^{m+} \longrightarrow M^{(m-2)+}$$

A few points:

- Reverse of oxidative addition;
- Leads to the formation of new bond between X and Y groups;
- Examples using a wide variety of X and Y ligands are known;
- Concerted reaction.

Examples [equation (5.52)]:

$$L_nM(R)(H) \longrightarrow L_nM + R—H \qquad L_nM(C(=O)R)(H) \longrightarrow L_nM + RCHO$$

$$L_nM(R)(R) \longrightarrow L_nM + R—R \qquad L_nM(X)(R) \longrightarrow L_nM + R—X \tag{5.52}$$

To enhance and/or allow intramolecular reductive elimination:

- Require *cis* disposition of X and Y ligands around metal centre [equation (5.53)];

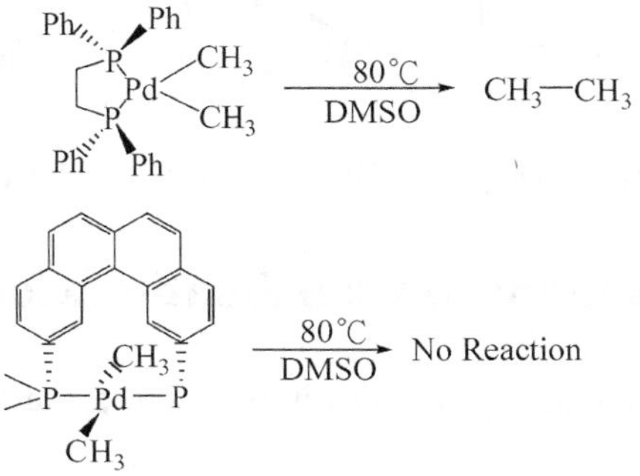

$$(Ph_2P\frown PPh_2)Pd(CH_3)_2 \xrightarrow[\text{DMSO}]{80^\circ C} CH_3—CH_3$$

$$trans\text{-}(P)Pd(CH_3)_2(P) \xrightarrow[\text{DMSO}]{80^\circ C} \text{No Reaction} \tag{5.53}$$

- High formal charge on metal centre [equation (5.54)].

$$[(bipy)_2FeEt_2] \rightleftharpoons [(bipy)_2FeEt_2] \longrightarrow (bipy)_2FeH(C_2H_4)(Et) \longrightarrow C_2H_4 + C_2H_6$$

$$[(bipy)_2FeEt_2] \xrightarrow{-1e^-} [(bipy)_2FeEt_2]^+ \longrightarrow Et\cdot \longrightarrow C_2H_4,\ C_2H_6,\ n\text{-}C_4H_{10}$$

$$[(bipy)_2FeEt_2]^+ \xrightarrow{-1e^-} [(bipy)_2FeEt_2]^+ \longrightarrow n\text{-}C_4H_{10}\ \text{only} \tag{5.54}$$

In equation(5.54), thermolysis of neutral complex gives ethylene and ethane as a result of initial β-hydrogen elimination. Electrochemical oxidation to the monocation gives ethylene, ethane, and butane as a result of homolysis of Fe—Et bond, giving Et· radicals. Further oxidation to the dication gives butane rapidly by reductive elimination.

- If the metal centre is too electron-rich, reductive elimination will be inhibited [equation (5.55)].

$$Ir(H)_2(Cl)(CO)(PPh_3)_2 \xrightarrow[-CO]{h\nu} Ir(H)_2(Cl)(PPh_3)_2 \xrightarrow{-H_2} IrCl(PPh_3)_2$$

$$Ir(H)_2(Cl)(PPh_3)_3 \xrightarrow[-PPh_3]{h\nu} Ir(H)_2(Cl)(PPh_3)_2 \tag{5.55}$$

The reaction is inhibited by addition of PPh_3, implying a pre-equilibrium involving loss of PPh_3. Three PPh_3 ligands make the metal centre too electron rich for elimination.

Labelling experiments show it is an intramolecular reaction [equation (5.56)] and no crossover happens [equation (5.57)].

$$Rh(H)(CH_2COCH_3)Cl(PMe_3)_3 \rightleftharpoons Rh(H)(CH_2COCH_3)Cl(PMe_3)_2 \xrightarrow[+PMe_3]{-CH_3CCH_3\ (\text{O})} RhCl(PMe_3)_3 \tag{5.56}$$

$$(PPh_3)_2IrClH_2 + (PPh_3)_3IrClD_2 \longrightarrow 2(PPh_3)_2IrCl + H_2 + D_2 \tag{5.57}$$

5.3 Substitution Reactions of Coordination Complex

In the ligand labels for nucleophillic substitution, there are three types of ligands that are important.

- Entering ligand: Y;
- Leaving ligand: X;
- Spectator ligand: species that neither enters nor leaves, which are particularly important when located in a *trans* position, designated “T”.

5.3.1 The Three Patterns of the Reaction Mechanisms

Associatve A (2 steps) [equation (5.58)]

$$ML_nX+Y \longrightarrow ML_nXY \longrightarrow ML_nY+X \qquad (5.58)$$

Dissociative D (2 steps) [equation (5.59)]

$$ML_nX+Y \longrightarrow ML_n+X+Y \longrightarrow ML_nY+X \qquad (5.59)$$

Interchange I (a continuous process) [equation (5.60)]

$$ML_nX+Y \longrightarrow Y\cdots ML_n\cdots X \longrightarrow ML_nY+X \qquad (5.60)$$

5.3.2 Substitution of Square Planar Metal Complex

The metal ions that are possible to form the square planar metal complexes are shown in Fig. 5.11. The element labeled in “” places are the common square planar metal complexes. The element labled in “” is the four coordinate planar complexes with special ligands only.

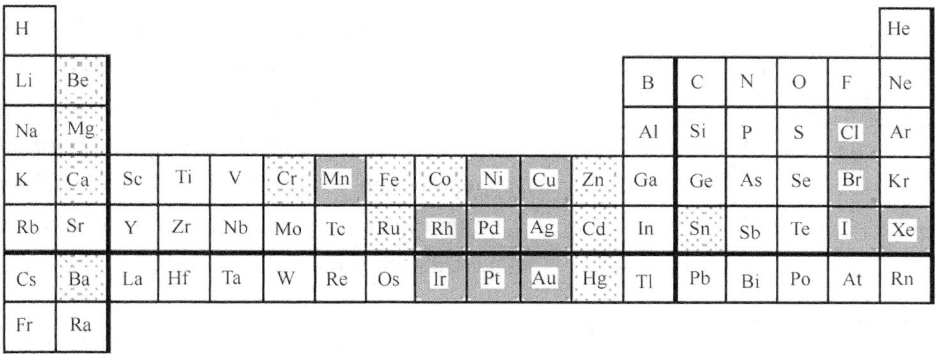

H																	He
Li	Be											B	C	N	O	F	Ne
Na	Mg											Al	Si	P	S	Cl	Ar
K	Ca	Sc	Ti	V	Cr	Mn	Fe	Co	Ni	Cu	Zn	Ga	Ge	As	Se	Br	Kr
Rb	Sr	Y	Zr	Nb	Mo	Tc	Ru	Rh	Pd	Ag	Cd	In	Sn	Sb	Te	I	Xe
Cs	Ba	La	Hf	Ta	W	Re	Os	Ir	Pt	Au	Hg	Tl	Pb	Bi	Po	At	Rn
Fr	Ra																

Fig.5.11 The possible elements to form the square planar metal complexes in period table

The examples of square planar transition metal complexes are often Ni(Ⅱ) (mainly d^8), Rh(Ⅰ), Pd(Ⅱ), Ir(Ⅰ), Pt(Ⅱ), Au(Ⅲ) complexes.

Substitution of square planar complexes is almost always A mechanism. The leaving ligand presumably moves down as the entering ligand approaches, so that the intermediate has a trigonal bipyramidal configuration. The rate depends on the entering group and the rate-determining step is the M—X bond formation.

General Rate Law [equation (5.61)]:

$$ML_3X+Y \xrightarrow{\text{solvent}} ML_3Y+X$$

$$\frac{-d[ML_3X]}{dt}=(k_s+K_Y[Y])[ML_3X] \qquad (5.61)$$

Factors that affect the rate of substitution are

- The *trans* effect: the ligand, T, *trans* to the leaving ligand (X) can alter the reaction rate.
- The role of the entering group and the leaving group.

- Steric effects: bulky *cis* ligand reduce Y nucleophillic attack.
- Stereochemistry: *cis/trans* conserved for A mechanism unless activated complex is long lived.
- $\Delta V^{\ddagger}$ and $\Delta S^{\ddagger}$ are both negative for A mechanism.

(1) The *trans* effect The *trans* ligands: The *trans* effect is best defined as the effect of a coordinated ligand upon the rate of substitution of ligands opposite to it.

When the ligand T is *trans* to the leaving group in square planar complexes, it will effect the rate of substitution. If T is a strong σ donor or π acceptor, the rate of substitute is dramatically increased. The reasons are: (a) if T contributes a lot of e^- density (is a good σ donor), the metal has less ability to accept electron density from X (the leaving ligand). (b) if T is a good π acceptor, electron density on the metal is decreased and nucleophillic attack by Y is encouraged.

Trans effect is more pronounced for σ donor as follows:

$$H_2O\sim OH^-\sim NH_3\sim \text{amines}\sim Cl^-<Br^-<SCN^-\sim I^-<CH_3^-<\text{Phosphines}\sim H^-<\text{Olefins}<CO\sim CN^-$$

Trans effect is more pronounced for π acceptor as follows:

$$Br^-<Cl^-<NCS^-<NO_2^-<CN^-<CO$$

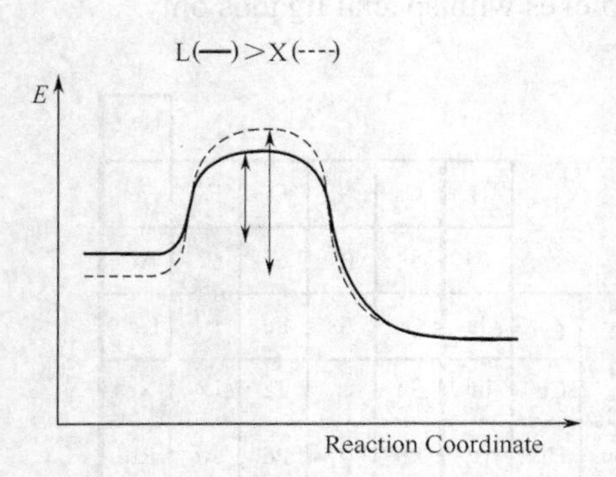

Fig.5.12 The illustration of the labilizing effect based on the destabilization of the ground state and/or a stabilization of the transition state

Note that the "labilizing effect" is used to emphasize the fact that this is a kinetic phenomenon. This labilization may arise because of destabilization (a thermodynamic term) of the ground state and/or a stabilization of the transition state (Fig.5.12).

The *trans* influence is purely a thermodynamic phenomenon. That is, ligands can influence the ground state properties of groups to which they are *trans*. Such properties include:

- Metal-ligand bond lengths;
- Vibration frequency or force constants;
- NMR coupling constants.

The *trans* influence series based on structural data, has been given as:

$$R^-\sim H^- > {=}PR_3 > CO\sim C{=}C\sim Cl^-\sim NH_3 \text{ (Fig.5.13)}$$

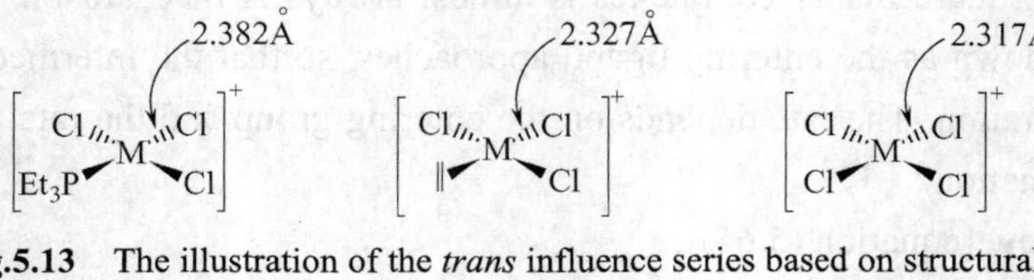

2.382Å [Cl, Cl, M, Et_3P, Cl]$^+$ 2.327Å [Cl, Cl, M, ‖, Cl]$^+$ 2.317Å [Cl, Cl, M, Cl, Cl]$^+$

Fig.5.13 The illustration of the *trans* influence series based on structural data

(2) The role of the entering group and the leaving group The *cis* ligands: in cases where a relatively poor nucleophile acts as the entering group [equation (5.62)].

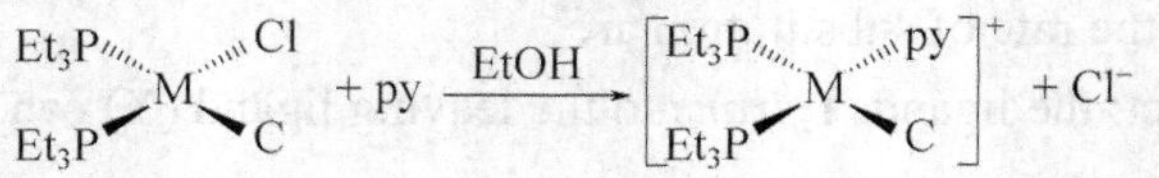

$$Et_3P, Et_3P, Cl, C\text{-}M + py \xrightarrow{EtOH} [Et_3P, Et_3P, py, C\text{-}M]^+ + Cl^- \qquad (5.62)$$

$$C = CH_3^- > Ph^- > Cl^-$$

$$\frac{k_y(C)}{k_y(Cl^-)} = 3.6 \quad 2.3 \quad 1.0$$

Compare with the *trans* series [equation (5.63)], it acts in the same way:

$$[M(Et_3P)(Cl)(T)(PEt_3)] + py \xrightarrow{EtOH} [M(Et_3P)(py)(T)(PEt_3)]^+ + Cl^- \tag{5.63}$$

$$T = H^- > CH_3^- > Ph^- > Cl^-$$

$$\frac{k_y(T)}{k_y(Cl^-)} = 10^4 \quad 1700 \quad 400 \quad 1.0$$

Although the entering and leaving-ligand series have a close similarity to the *trans*-effect series, the distinct differences can be seen. Some of these differences are probably due to the fact that the entering and leaving group series are affected by differences in the solvation of ligands, whereas the *trans*-effect series is independent of such solvation effects (Fig.5.14).

The associative pathways is utilized in the substitution of one ligand for another at a square planar reaction centre [equation (5.64)].

$$ML_2X_2 + L' \xrightarrow{solvent} MLL'X_2 + L \tag{5.64}$$

(3) Steric effects Steric crowding reduces the rate of A mechanisms and increases D mechanisms. Simply a spatial phenomenon: less room around the metal means that a higher coordination number transition state is higher energy, for example: *cis*-$[PtXL(PEt_3)_2]$ (Fig.5.15).

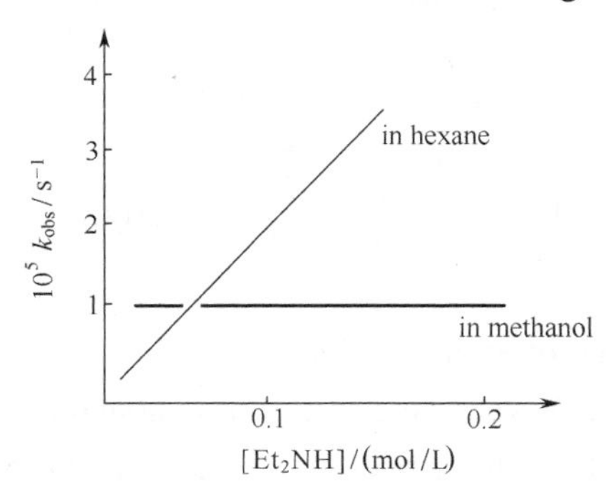

Fig.5.14 An illustration of the importance of solvent on the substitution pathways for square planar reaction centres

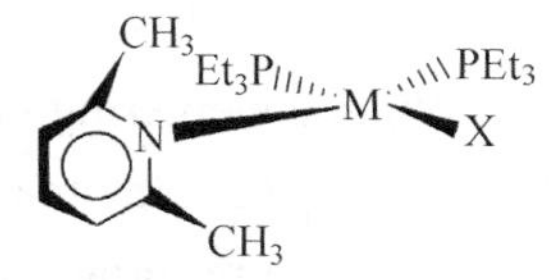

Fig.5.15 The molecular structure of the *cis*-$[PtX(2,6\text{-dimethylpyridine})(PEt_3)_2]$

Rate varies with L: pyridine>2-methyl py>2, 6-dimethyl py.

Stereochemistry of square planar substitution is also important. Observing the final product stereochemistry can provide information on the mechanism and intermediate life-times (Fig.5.16).

(4) $\Delta V^{\ddagger}$ and $\Delta S^{\ddagger}$

- Changes in volume along a reaction pathway can be determined by observing reaction rate as a function of pressure usually. A negative $\Delta V^{\ddagger}$ suggests an associative complex.
- The change in entropy from the reactants to the activated complex is $\Delta S^{\ddagger}$, which can be determined by the temperature dependence of the rate. Associative mechanism has a negative $\Delta S^{\ddagger}$

as expected from increasing order of the system by loss of freedom for the entering group without release of the leaving group.

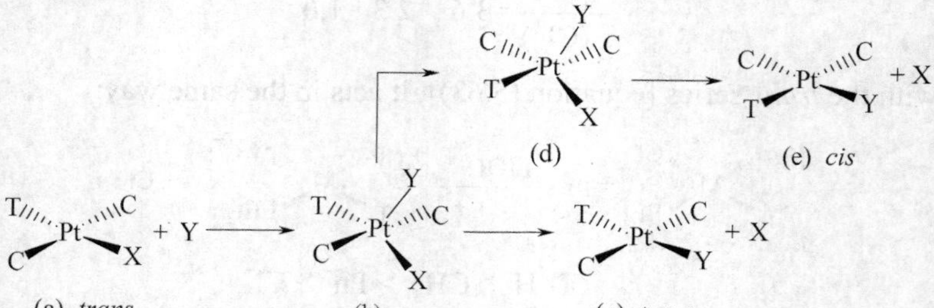

(a) *trans* (b) (c) *trans* (d) (e) *cis*

Fig.5.16 The illustration of the stereochemistry of square planar substitution

5.3.3 Substitution of Octahedral Complex

Interchange mechanism (I) is the most important reaction mechanism for substitution of O_h complexes. But, It can be I_a or I_d. It depends on the rate determining step being Y—M formation vs M—X breaking. For associative, the rate depends heavily on the entering group;for dissociative, the rate is independent of the entering group.

Studies on octahedral complexes have largely been limited to two types of reaction:

(1) Replacement of coordinated solvent Perhaps the most thoroughly studies replacement reactions of this type is the formation of a complex ion from a hydrated metal ion in solution [equation (5.65) and equation (5.66)].

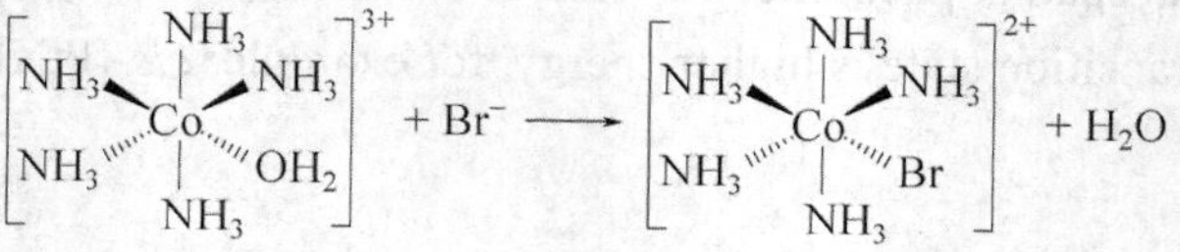

$$[Co(NH_3)_5(OH_2)]^{3+} + Br^- \longrightarrow [Co(NH_3)_5Br]^{2+} + H_2O \quad (5.65)$$

$$[Ni(OH_2)_6]^{2+} + \text{phen} \longrightarrow [Ni(OH_2)_4(\text{phen})]^{2+} + 2H_2O \quad (5.66)$$

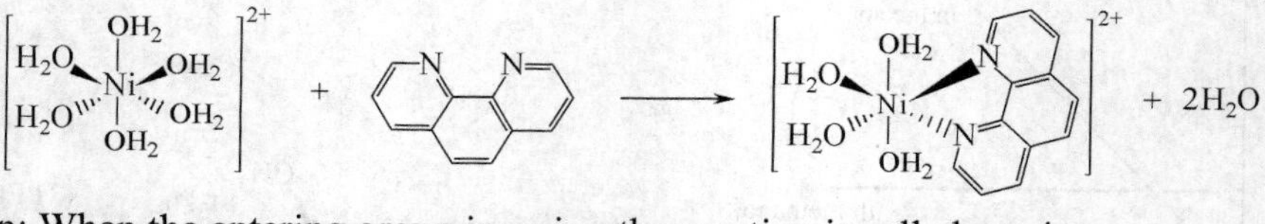

Anation: When the entering group is an ion the reaction is called anation.

(2) Solvolysis Since the majority of such reactions have been carried out in aqueous solution, hydrolysis is a more appropriate term. Hydrolysis reactions have been done under acidic or basic conditions [equation (5.67)].

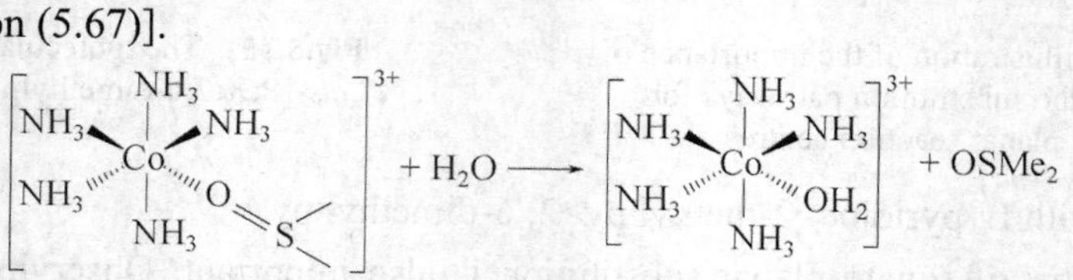

$$[Co(NH_3)_5(OSMe_2)]^{3+} + H_2O \longrightarrow [Co(NH_3)_5(OH_2)]^{3+} + OSMe_2 \quad (5.67)$$

The standard mechanism for O_h I substitutions reactions is Eigen-Wilkins mechanism, which is based on the formation of an "encounter complexes". The first pre-equilibrium [equation (5.68)]:

$$ML_6+Y \rightleftharpoons \{ML_6, Y\} \quad K_E=\frac{[\{ML_6, Y\}]}{[ML_6][Y]} \quad (5.68)$$

Followed by the product formation[equation (5.69)]:

$$\{ML_6, Y\} \longrightarrow \text{Product} \quad \text{Rate}=k[\{ML_6, Y\}] \quad (5.69)$$

The rate expression can be written in terms of the K_E so that [equation (5.70)]:

$$\text{Rate}=\frac{kK_E[C]_{tot}[Y]}{1+K_E[Y]} \tag{5.70}$$

Where $[C]_{tot}$ is the total of all of the complex species, If $K_E[Y]\ll 1$ then the rate becomes [equation (5.71)]:

$$\text{Rate}=k_{obs}[C]_{tot}[Y] \tag{5.71}$$

Why the Eigen-Wilkins is so important. It is very useful to help us to predict reaction mechanism. When $k_{obs}=kK_E$, so we can get k. Now testing k to see if it varies with Y or not so, we can assign I_a or I_d.

General rules for O_h substitution:

- Most 3d metals undergo I_d substitutions, i.e. the rate determining step is independent of the entering group and primarily is the breaking of the M—X bond.
- Larger metals (4d, 5d) lean towards I_a characteristics.
- Leaving group nature of X is important as expected for I_d as bond breaking of M—X is the rate determining step.
- Spectator ligands(*cis-trans* effect) no clear *trans* effects for O_h complexes. In general, good spectator sigma donors will stabilize the complex after the departure of the leaving group.
- Steric effects (a) Steric crowding around the metal center favors dissociative activation. (b) Dissociative activation relieves crowding around the complex. (c) Steric crowding has been qualitatively and quantitatively explored—Tolman Cone Angle (Fig.5.17).
- The $\Delta V^{\ddagger}$ of octahedral substitution $\Delta V^{\ddagger}$ is not large for I mechanism, but I_a tends to be negative value, I_d tends to be positive values. Decreasing d number shows tendancy towards I_a mechanism.
- O_h stereochemistry of substitution It is more complicated than for T_d complexes. For example: *cis*- or *trans*-$[CoAX(en)_2]^{2+}$, *cis* complexes tends to retain *cis*; *trans* complexes can isomerize depending on the spectator ligand, depends on geometry of the activated complex. Trigonal bipyramidal results in isomerization depending on where Y enters. Square planar leads to retention of stereochemistry (Fig.5.18).

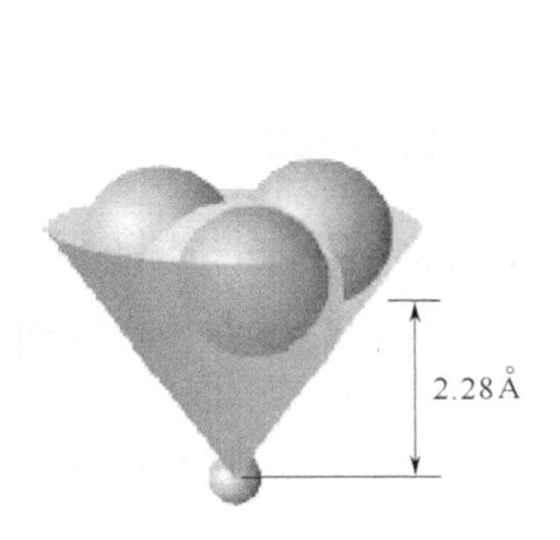

Fig.5.17 Tolman Cone Angle—qualitatively and quantitatively explored steric effects

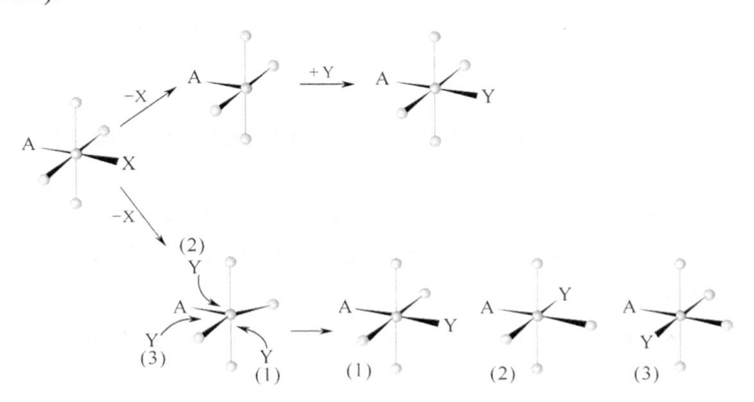

Fig.5.18 Stereochemistry of substitution O_h complex for *cis*- or *trans*-$[CoAX(en)_2]^{2+}$

5.3.4 Isomerization Reactions

This kinds of reaction is similar to substitution reactions. Berry pseudorotation mixes axial and equatorial positions in a five-coordinated TBP species. Both square planar complexes that undergo D or I_d mechanisms involve a five-coordinated state. So, isomerization is possible (Fig.5.19).

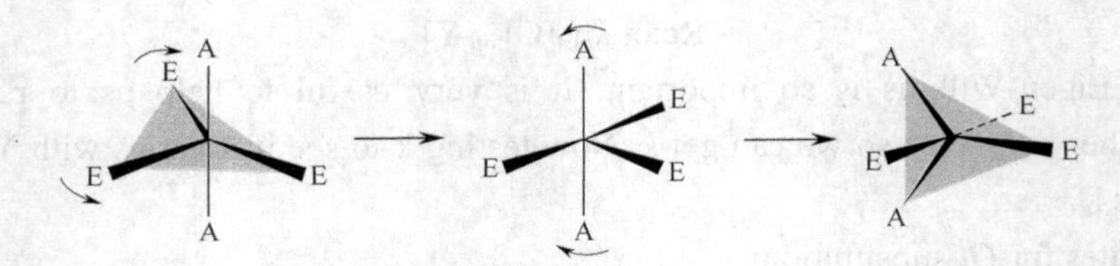

Fig.5.19 Pseudorotation mixing axial and equatorial positions in a five-coordinated TBP species

O_h complexes may also isomerize via "twist" mechanisms. It does not require loss of ligands or breaking bonds, just depends on energy barriers between conformations: Bailar Twist (a) and Ray Dutt Twist (b). Both of them occur via trigonal prismatic confirmation (Fig.5.20).

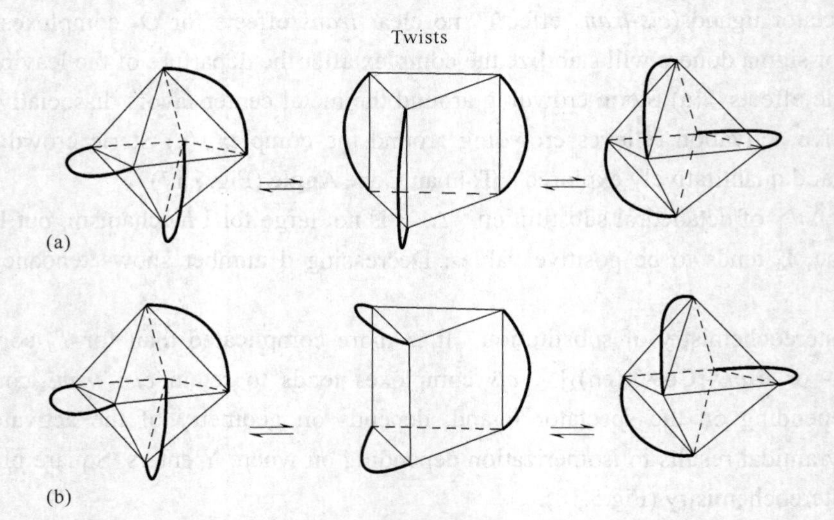

Fig.5.20 Bailar Twist (a) and Ray Dutt Twist (b) in O_h complex isomerization

5.4 Electron Transfer Reactions of Coordination Complex

Electron transfer reactions may occur by either or both of two mechanisms.

- Outer sphere: No bridging ligand involved. Direct transfer of electrons between the metal centers.
- Inner sphere: It requires formation of bridging bimetallic species. It results in ligand transfer at the same time.

5.4.1 Outer Sphere Electron Transfer

In principle all outer sphere mechanism involves electron transfer from reductant to oxidant with the coordination shells or spheres of each staying intact. That is one reactant becomes involved in the outer or second coordination sphere of the other reactant and an electron flows from the reductant to oxidant. Such a mechanism is established when rapid electron transfer occurs between two substitution-inert complexes [equation (5.72)].

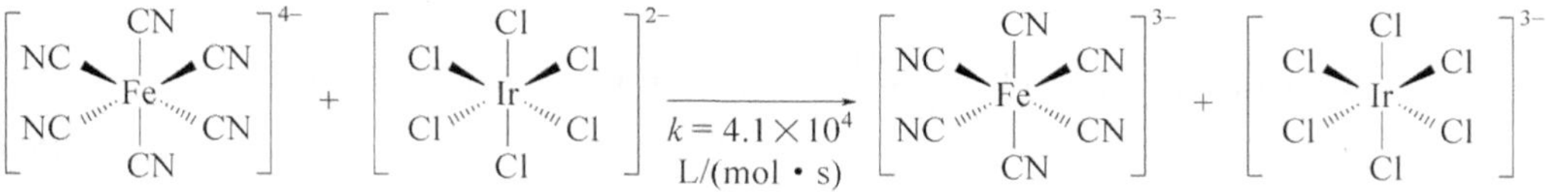

$$[Fe(CN)_6]^{4-} + [IrCl_6]^{2-} \xrightarrow[L/(mol \cdot s)]{k=4.1\times 10^4} [Fe(CN)_6]^{3-} + [IrCl_6]^{3-} \quad (5.72)$$

This mechanism is readily identified when no ligand transfer occurs between the species. It is easier to identify when complexes are inert with respect to ligand substitution. Born Oppenheimer Approximation can be used in this case because electron move faster than nuclei and complexes reorganization can be considered in a separate step from electron transfer. Another model—Marcus Equation is suitable for the mechanism. Electron transfer requires vibrational excited states, shape of potential energy well determines rate of transfer.

5.4.2 Inner Sphere Electron Transfer

An inner sphere mechanism is one in which the reactant and oxidant share a ligand in their inner or primary coordination spheres the electron being transferred across a bridging group.

The bridging ligands required in the inner sphere reactions should be with multiple pairs of electrons to donate, for example (Fig.5.21):

:C̈l⁻:   :S—C≡N⁻:   :N≡N: :C≡N⁻:

Fig.5.21 Some coordination ligands with multiple pairs of electrons

The rate of electron transfer is dependent on the ligands that are present.

Normally, inner sphere reactions has three steps:

- Formation of bridged complex [equation (5.73)]

$$M^{II}L_6 + XM^{III}L'_5 \longrightarrow L_5M^{II}—XM^{III}L'_5 + L \quad (5.73)$$

- Electron transfer [equation (5.74)]

$$L_5M^{II}—XM^{III}L'_5 \longrightarrow L_5M^{III}—XM^{II}L'_5 \quad (5.74)$$

- Decomposition into final products [equation (5.75)].

$$L_5M^{III}—XM^{II}L'_5 \longrightarrow L_5M^{III}—X + M^{II}L'_5 \quad (5.75)$$

As we expected, the rate-determining step is usually the electron transfer step. However, the formation of bridging complex or the decomposition could also limit the rate. Good conjugation could provide a simple path for the electron. Therefore, how to construction of bridging ligand systems as models is important.

The reduction of hexaamminecobalt (Ⅲ) by hexaaquochromium (Ⅱ) occurs slowly [k=10^{-3}L/(mol·s)] by an outer sphere mechanism. However, if one ammonia ligand on Co(Ⅲ) is substituted by NCS^- or Cl^-, reaction now occurs with a substantially greater rate [k=6×10^5L/(mol·s)] [equation (5.76)- equation (5.78)].

$$[Co(NH_3)_5NCS]^{2+} + [Cr(H_2O)_6]^{2+} \longrightarrow \left[(NH_3)_5Co{-}NCS{-}Cr(OH_2)_5 \right]^{4+} \longrightarrow [Co(H_2O)_6]^{2+} + 5NH_4^+ + [Cr(H_2O)_5SCN]^{2+} \quad (5.76)$$

$$\underset{t_{2g}^6}{[Co(NH_3)_5Cl]^{2+}} + \underset{t_{2g}^3e_g^1}{[Cr(H_2O)_6]^{2+}} \overset{H^+}{\rightleftharpoons} \underset{t_{2g}^5e_g^2}{[Co(H_2O)_6]^{2+}} + \underset{t_{2g}^3}{[Cr(H_2O)_6]^{3+}} + 5NH_4^+ \quad (5.77)$$

$$\underset{t_{2g}^6}{[Co(NH_3)_6]^{3+}} + \underset{t_{2g}^3e_g^1}{[Cr(H_2O)_6]^{2+}} \overset{H^+}{\rightleftharpoons} \underset{t_{2g}^5e_g^2}{[Co(H_2O)_6]^{2+}} + \underset{t_{2g}^3}{[Cr(H_2O)_6]^{3+}} + 6NH_4^+ \quad (5.78)$$

5.5 Homogeneous Catalysis

The research interest in the organometallic chemistry is largely fuelled by potential application of organometallic compounds as catalysts in industrial chemistry. Catalysis was responsible for billion dollars in revenue on the world. A catalyst is defined as a substance that alters the rate of a reaction without appearing in any of the products of that reaction; it may speed up or slow down a reaction. For a reversible reaction, a catalyst alters the rate at which equilibrium is attained; it does not alter the position of equilibrium (Fig.5.22).

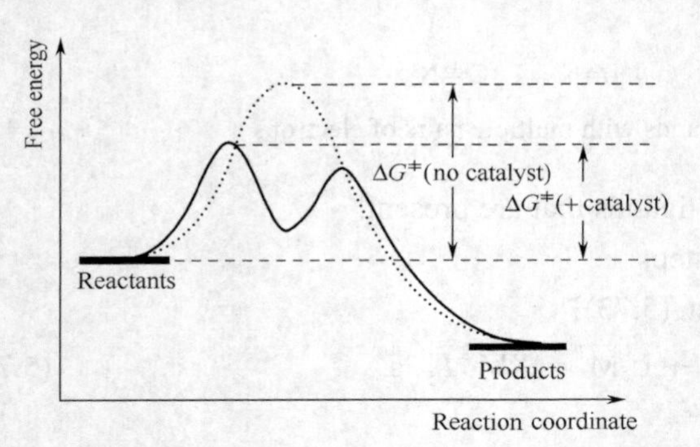

Fig.5.22 The comparison of the transition states of reaction(dotted)and the reaction with catalyst

Catalytic processes can be broadly defined into two categories.

- Homogeneous catalysis A process where the catalyst and the components of the reaction are in the same phase. Homogeneously catalyze reactions tend to be very specific and are typically carried out in solution. Therefore, homogeneous catalysts can be studied spectroscopically. But it is difficult to separate the catalyst from the product mixture. Homogenous catalysis is dominated by organometallic complexes. Some common examples of homogeneous catalyst are listed in Table 5.4.
- Heterogeneous catalysis A process where the catalyst and the components of the reaction are in different phases. In most heterogeneous catalysis systems the catalysts is in the solid phase and the reactants are liquids or gases. Therefore, it is very easy to separate the products from the catalyst.

Homogeneous catalysts must have easily generated open coordination sites, in general must be $ML_4(D_{4h})$ or have readily dissociable ligands. Homogeneous catalysts are generally much more selective and tuneable.

Table 5.4 Examples of homogeneous catalyst

Homogeneous catalyst	Application
$RhCl(PPh_3)_3$(Wilkinson's catalyst)	Alkene hydrogenation
cis-$[Rh(CO)_2I_2]^-$	Monsanto acetic acid synthesis
$HCo(CO)_4$	Hydroformylation
$HRh(CO)_4$	Hydroformylation(branched alkenes)
$[HFe(CO)_4]^-$	Water-gas shift reaction
Cp_2TiMe_2	Alkene polymerization
Cp_2ZrH_2	Hydrogenation of alkenes/alkynes
$Pd(PPh_3)_4$	Heck, Suzuki, Stille, Negishi coupling reactions(very synthetically important)

5.5.1 Alkene Hydrogenation

The most successful homogeneous organometallic catalyst to date is Wilkinson's catalyst, $RhCl(PPh_3)_3$, which catalyzes the hydrogenation of olefins [equation (5.79)].

$$RCH{=}CH_2 + H_2 \longrightarrow RCH_2CH_3 \tag{5.79}$$

Interestingly, the catalytic discovery were made independently and nearly simultaneously by Wilkinson and Coffey catalyzes the hydrogenation of olefins in the 1960s and the compound is now commonly referred to as"Wilkinson's catalyst". The major mechanistic feature of the reaction sequence can be shown by using what is known as a catalytic cycle of a Tolman loop (Fig.5.23).

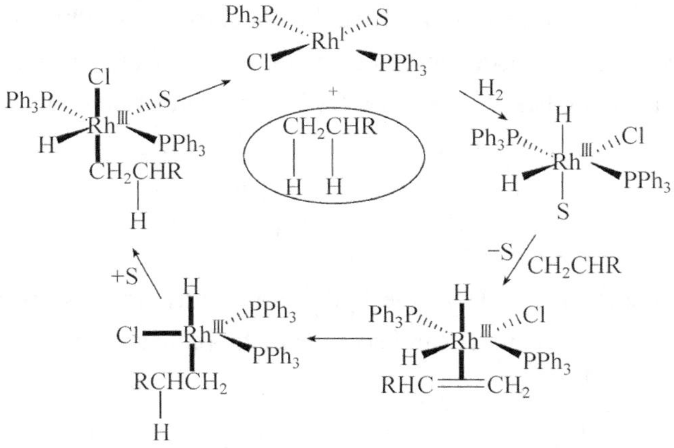

Fig.5.23 Wilkinson's catalyst—alkene hydrogenation

In the catalytic cycle there are several important step central to many organometallic reaction mechanisms. A key example is the reaction of dihydrogen with the solvent complex to form a *cis*-dihydride species [equation (5.80)].

$$\underset{A}{RhCl[P(C_6H_5)_3]_2(S)} + H_2 \longrightarrow \underset{B}{cis\text{-}RhCl[P(C_6H_5)_3]_2(S)(H)_2} \tag{5.80}$$

This reaction is known as an oxidative-addition reaction. Note that in this chemical transformation, A is bound to only four ligands, called species A a four-coordinate"coordinately unsaturated"compound while B is bound to six that is"coordinately saturated". Wilkinson's catalyst, $RhCl(PPh_3)_3$, is a Rh(I) complex with 16 total valence electrons square planar geometry. Therefore, we can add two electrons to the metal centers [equation (5.81)].

$$[RhCl(PPh_3)_3] \xrightarrow{H_2} [RhH_2Cl(PPh_3)_3] \tag{5.81}$$

Species B is a Rh(Ⅲ) complex with 18 valence electrons and there is no vacant coordination sites. Thus, in an oxidative-addition reaction the coordination number of the metal changes from four to six and the oxidation state of the metal increase by two. Therefore, before any other reaction can occur, a ligand needs to dissociate [equation (5.82)].

$$[RhH_2Cl(PPh_3)_3] \longrightarrow [RhH_2Cl(PPh_3)_2] + PPh_3 \tag{5.82}$$

Now the 16 electron $[Rh(PPh_3)_2Cl(H_2)]$ complex can add a 2 electron ligand such as $H_2C{=}CH_2$ [equation (5.83)].

$$[RhH_2Cl(PPh_3)_2] \xrightarrow{H_2C=CH_2} [RhH_2Cl(PPh_3)_2(H_2C{=}CH_2)] \tag{5.83}$$

The 18-electrons $[Rh(PPh_3)_2Cl(H_2)\ (\eta^2\text{-}H_2C{=}CH_2)]$ complex can now undergo some interesting chemistry that is unique to organometallic complexes: the ethylene inserts into the Rh—H bond. This migratory insertion reaction results in the formation of a C—H bond and a Rh—C σ bond. Consequently, the $[Rh(PPh_3)_2Cl(H_2)(\eta^1\text{-}H_2C{-}CH_3)]$ is an 18 electron complex [equation (5.84)].

$$[RhH_2Cl(PPh_3)_2(H_2C{=}CH_2)] \longrightarrow [RhHCl(CH_2{-}CH_3)(PPh_3)_2] \tag{5.84}$$

In$[Rh(PPh_3)_2Cl(H_2)(\eta^1\text{-}CH_2{-}CH_3)]$, the hydride and ethyl groups are in a *cis* arrangement, which means that the $CH_2{-}CH_3$ and H can leave the metal [equation (5.85)].

$$[RhHCl(CH_2{-}CH_3)(PPh_3)_2] \longrightarrow [RhHCl(PPh_3)_2] + CH_3{-}CH_3 \tag{5.85}$$

Notice that the coordination number about the metal has decreased by 2 and the electron count has also decreased by two. In other words, the reaction is the opposite of the oxidative addition in the first step. Therefore, the reverse of an oxidative-addition reaction is also common and is termed a reductive-elimination reaction. A color change can be observed in this reaction.

The final step in the catalytic cycle involves the addition of PPh_3 in order to return to the starting point [equation (5.86)].

$$[Rh(PPh_3)_2Cl]+PPh_3 \longrightarrow [Rh(PPh_3)_3Cl] \tag{5.86}$$

Once Wilkinson's catalyst has been re-generated, the entire cycle can repeat. The cycle can repeat up to around 1000 times before the catalyst decomposes. Most organometallic catalytic reactions occur by between through 16- and 18-electron intermediates.

Summary of catalytic cycle:

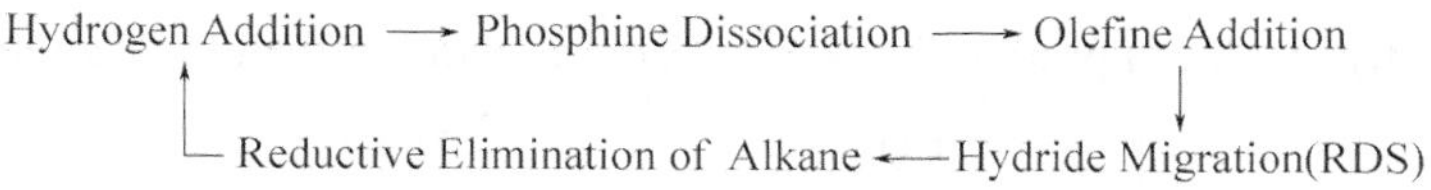

Hydrogen Addition ⟶ Phosphine Dissociation ⟶ Olefine Addition
↑ ↓
Reductive Elimination of Alkane ⟵ Hydride Migration(RDS)

Chirality can be introduced into prochiral olefins. Chiral diphosphines are used. William Knowles and Ryoji Noyori won the Noble Prize (2001) for the asymmetric hydrogenation with DIPMAP Rh complexes (Fig.5.24) and BINAP Ru complexes respectively.

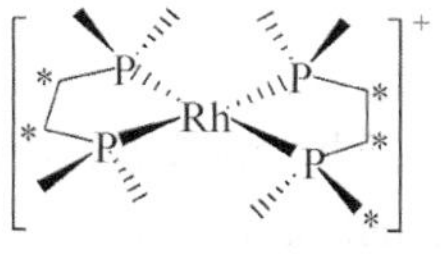

Fig.5.24 The molecular structure of chiral Wilkinsontype catalyst

5.5.2 Monsanto Acetic Acid Synthesis

Acetic acid is the chemical compound responsible for the characteristic odor and sour taste of vinegar. Typically, vinegar is about 4% to 8% acetic acid. As the defining ingredient of vinegar, acetic acid has been produced and used by humans since before the dawn of recorded history. In fact, its name comes from the Latin for vinegar, acetum. Vinegar is formed from dilute solutions of alcohol, such as wine, by the action of certain bacteria in the presence of oxygen. These bacteria require oxygen, and the overall chemical change is the reaction of ethanol with oxygen to form acetic acid and water [equation (5.87)].

$$CH_3CH_2OH + O_2 \longrightarrow CH_3COOH + H_2O \tag{5.87}$$

Vinegar may also be obtained from other fermented beverages such as malt or cider. For many years, the bulk of commercial acetic acid was produced by the oxidation of ethanol.

Today, most industrial production of acetic acid is by the Monsanto process, in which carbon monoxide reacts with methanol under the influence of a rhodium complex catalyst at 180℃ and pressures of 30-60atm. About 3.85million tons of acetic acid is produced each year. Approximately 60% of the acetic acid produced involves the conversion of methanol to acetic acid with Rh catalyst. The mechanism is shown in Fig.5.25.

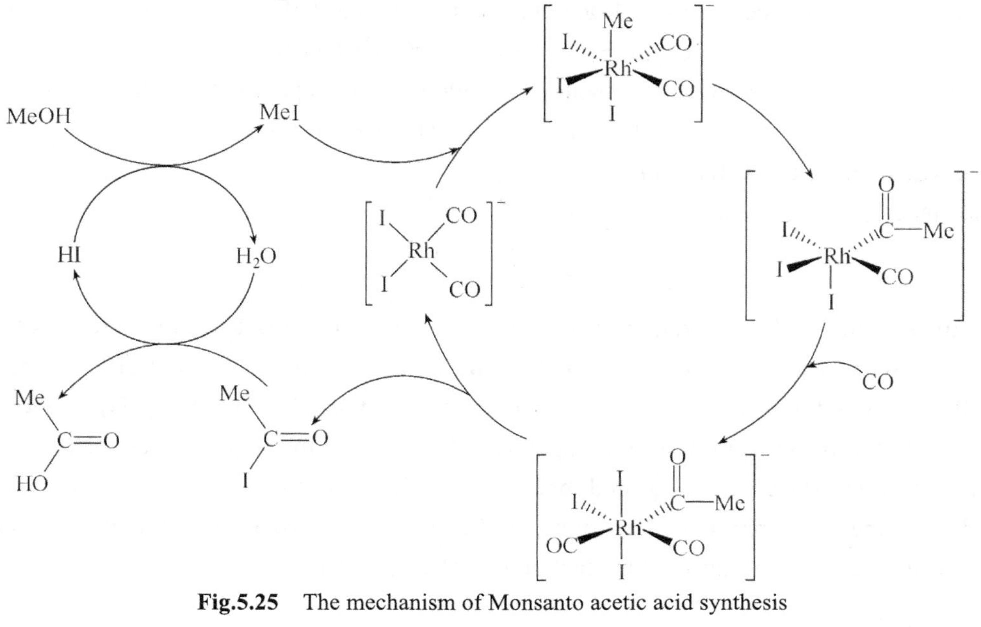

Fig.5.25 The mechanism of Monsanto acetic acid synthesis

The catalyst for this reaction is *cis*-$[Rh(CO)_2I_2]^-$, which is a 16 electron complex and can undergo oxidative addition. In this case, the molecule that is oxidatively added is CH_3I. Since CH_3I is a polar molecule, the oxidative addition product is the *trans* complex rather than the *cis* [equation (5.88)].

$$[RhI_2(CO)_2]^- \xrightarrow{Me-I} [RhMeI_3(CO)_2]^- \qquad (5.88)$$

The methyl group can migrate to one of the coordinated CO ligand, which results in the formation of a Rh—COMe group (acyl ligand). The product, $[Rh(COMe)I_3(CO)]^-$, is a 16 electron complex, which can add a 2 electron ligand, such as CO [equation (5.89)].

$$[RhMeI_3(CO)_2]^- \longrightarrow [Rh(COMe)I_3(CO)]^- \qquad (5.89)$$

The COMe and I are in an arrangement that is appropriate for reductive elimination of I—COMe [equation (5.90) and equation (5.91)].

$$[Rh(COMe)I_3(CO)]^- \xrightarrow{CO} [Rh(COMe)I_3(CO)_2]^- \qquad (5.90)$$

$$[Rh(COMe)I_3(CO)_2]^- \longrightarrow [RhI_2(CO)_2]^- + I-C(=O)-Me \qquad (5.91)$$

Notice that we have already regenerated our catalyst, but we do not have the required product. Acetic acid can be produced from ICOMe with water [equation (5.92)].

$$ICOMe + H_2O \longrightarrow MeCOOH + HI \qquad (5.92)$$

The HI can be used to convert methanol into iodomethane [equation (5.93)].

$$HI + MeOH \longrightarrow MeI + H_2O \qquad (5.93)$$

5.5.3 Hydroformylation Reaction

(1) General reaction [equation (5.94)]

$$C=C + CO + H_2 \xrightarrow{[CAT]} R-CHO \qquad (5.94)$$

Homogeneous hydroformylation catalysis is one of the largest volume processes in the chemical industry with a worldwide oxo-aldehyde production of 7.8×10^6t in 1997, which was originally developed by Roelen in the late 1930's. The origin catalyst was $Co_2(CO)_8$, which is still widely used. Hydroformalation is an important C—C bond forming reaction. There are two possible isomeric products, linear and branched products. Usually the more desirable linear aldehydes are the major products. Depending on the catalyst and conditions, the aldehydes can be directly reduced to alcohols during the reaction [equation (5.95)].

$$R-CH=CH_2 \xrightarrow[120\text{-}170^\circ C,\ 200\text{-}300atm]{Co_2(CO)_8,\ H_2,CO} RCH_2CH_2CHO + RCH(CH_3)CHO \xrightarrow{reduction} RCH_2CH_2CH_2OH + RCH(CH_3)CH_2OH \qquad (5.95)$$

Three main classes of homogeneous catalysts:

- Hydrido-cobalt-carbonyl;
- Hydrido-cobalt-phosphines;
- Hydrido-rhodium-phosphine-carbonyls.

(2) Hydrido-cobalt-carbonyl Although the exact nature of some steps has not been conclusively determined, the catalytic cycle is as follows (Fig.5.26).

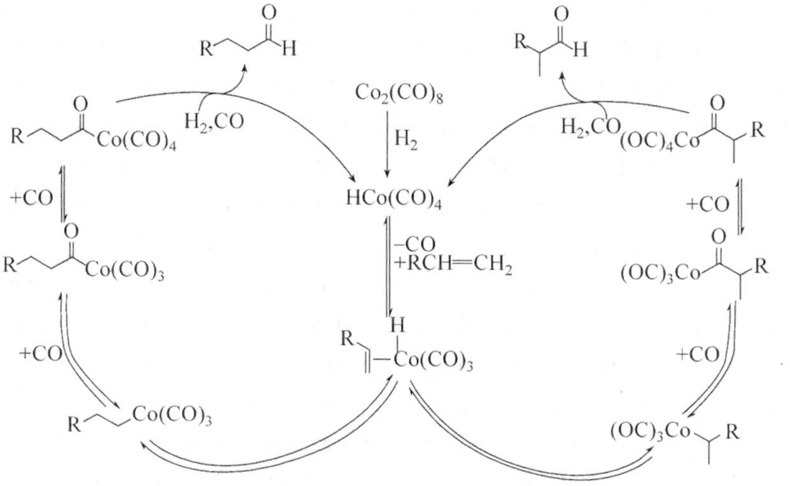

Fig.5.26 The catalytic cycle of $Co_2(CO)_8$ hydrogenation

$Co_2(CO)_8$ undergoes hydrogenation to give the acidic "hydride" complex $HCo(CO)_4$, which can also be prepared directly and used as the catalyst. The first step is a dissociative substitution of alkene for CO. Migratory insertion can result in either a primary or secondary metal alkyl. Although this is the step that sets the regiochemistry of the products, it is rapidly reversible. This rapid reversibility results in alkene isomerization and H/D exchange. β-H elimination requires an open coordination site, isomerization and isotope exchange are inhibited by increased CO pressures.

The next step is a second migratory insertion to form the coordinatively unsaturated acyl that can coordinate another CO to give the 18 electron acyl complex. Under standard catalytic conditions with 1-octene, this is the only species observed by IR.

One possible mechanism is the reaction with H_2 to give the aldehyde and $HCo(CO)_4$, which could occur by oxidative addition/reductive elimination. Another possible mechanism would involve a dinuclear reaction with $HCo(CO)_4$. Evidence for both mechanisms has been observed [equation (5.96)].

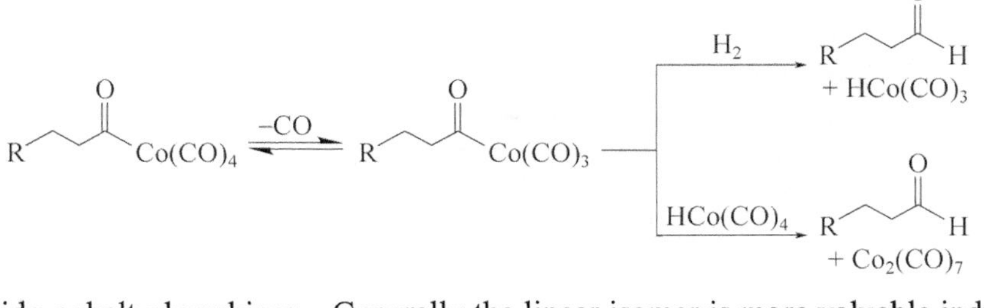

$$RCH_2CH_2C(O)Co(CO)_4 \underset{}{\overset{-CO}{\rightleftharpoons}} RCH_2CH_2C(O)Co(CO)_3 \begin{cases} \xrightarrow{H_2} RCH_2CH_2CHO + HCo(CO)_3 \\ \xrightarrow{HCo(CO)_4} RCH_2CH_2CHO + Co_2(CO)_7 \end{cases} \tag{5.96}$$

(3) Hydrido-cobalt-phosphines Generally the linear isomer is more valuable industrially than

the branched isomer. Since the catalytic cycle is irreversible, the observed regioselectivity represents the kinetic product ratio rather than the thermodynamic preference. Normally the regioselectivity is not determined by the rate at which alkene insertion occurs to give primary and secondary Co-alkyls. Rather the regioselectivity is determined by the relative rate of migratory insertion of the primary and secondary alkyl to CO. Less hindered alkyl groups migrate more rapidly than larger groups.

Adding phosphine to the Co-catalyzed hydroformylation system can modify the regioselectivity. Using $HCo(CO)_3PBu_3$ as catalyst, higher linear to branched ratios are achieved. Probably, it is due to a greater degree of regiocontrol in the initial migratory insertion of the alkene into the Co—H. There are two aspects. ①steric; substituting PBu_3 for CO increases the steric bulk of the metal center, which would be expected to favour formation of a primary Co-alkyl. ②electronic; substituting PBu_3 for CO significantly decreases the acidity of the cobalt-hydride. So, there is a lower preference for addition of H—Co to the alkene.

(4) Hydrido-rhodium-phosphine-carbonyls Since simple rhodium carbonyls tend to form a stable, catalytically inactive Rh-carbonyl cluster. They are inactive for the hydroformylation reaction. $HRh(CO)_4$ can be formed under high H_2 and CO pressures to give a very active catalyst, but it also isomerizes alkenes and gives low linear to branched ratios.

Addition of phosphines to these rhodium systems dramatically improves both the chemoand regioselectivity, while maintaining high reaction rates of the initial Rh-carbonyls [equation (5.97)].

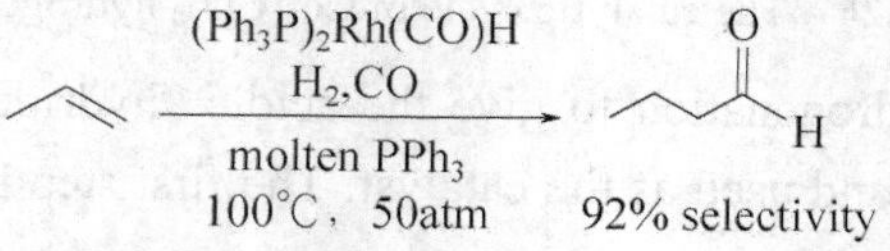

$$\text{CH}_2\text{=CHCH}_3 \xrightarrow[\text{molten PPh}_3,\ 100^\circ\text{C},\ 50\text{atm}]{(\text{Ph}_3\text{P})_2\text{Rh(CO)H},\ \text{H}_2,\text{CO}} \text{CH}_3\text{CH}_2\text{CH}_2\text{CHO}\quad 92\%\ \text{selectivity} \tag{5.97}$$

The generally accepted mechanism for the Rh systems, which was first proposed by Sir Geoffrey Wilkinson in 1968, is similar to that of the Co catalyst (Fig.5.27).

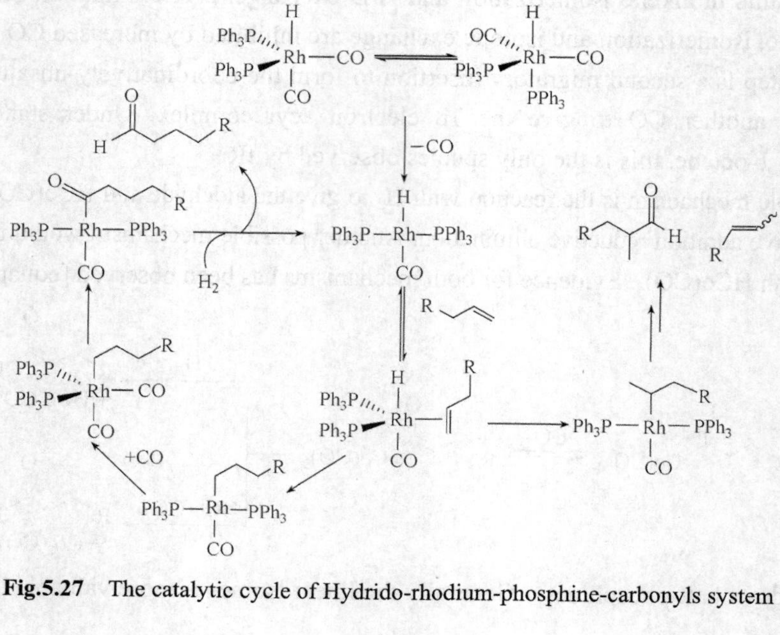

Fig.5.27 The catalytic cycle of Hydrido-rhodium-phosphine-carbonyls system

The key Rh complex is a 5-coordinate 18 electron species. Due to the high *trans*-effect ligands (CO, H), the large phosphines occupy *cis* coordination sites. Both phosphines can be equatorial, or one can occupy an axial position.

These catalysts that have a square planar intermediate have very high linear to branched ratios. Alkene migratory insertion largely determines the regioisomeric ratio because the square planar Rh-alkyl is rapidly trapped by CO in an irreversible reaction under the reaction conditions. For a primary Rh-alkyls this is followed by CO in migratory insertion and then hydrogen cleavage. Secondary Rh-alkyl undergos CO insertion slowly, so alkene isomerization products are also observed. The strong kinetic preference for CO insertion to occur for primary Rh-alkyls can be very useful.

Finally, the problem of identifying the true active catalyst in catalytic systems is exceedingly difficult. Only through detailed mechanistic studied can an experimentalist gain any certainty of the active catalyst. There exist many reports in the scientific literature of "catalysts" that in reality are not catalysts at all. Often impurities or decomposition products catalyze the reaction of interest.

NMR is very useful in the monitoring of reacting systems to gain insight in mechanistic and kinetic details, which is the key for understanding and improving catalysts and reaction conditions. Although application of NMR methods to problems of catalysis, homogeneous as well as heterogeneous, is an established technology, it is still a fascinating field of research because of the unique ability of NMR to deliver information on structure and dynamics, qualitative and quantitative, at the same time.

第 1 章　配位化学简介

1.1　配位化学的发展历史

1.1.1　配位化学的起源

20 世纪最多产的研究领域之一，就是根据 Alfred Werner 的理论发展起来的配位化学的研究。今天，在 Werner 开展最初研究的百年之后，配位化合物的数量、种类以及复杂性仍在不断增加，这足以证明它对无机化学领域的影响是非常之大的。

第一个配位化合物是由法国化学家 Tassaert 在 18 世纪末期制备的。他观察到氨水与钴矿石反应产生了一种红褐色的产物（这可能就是我们已知的第一种配位化合物）。在后来的一个世纪中，合成和表征了许多的配合物，不过，有关配合物结构方面的进展却很慢（图 1.1）。对于配合物的发现和理解应该以原子结构、周期表、分子中化学键的巨大进步为背景。

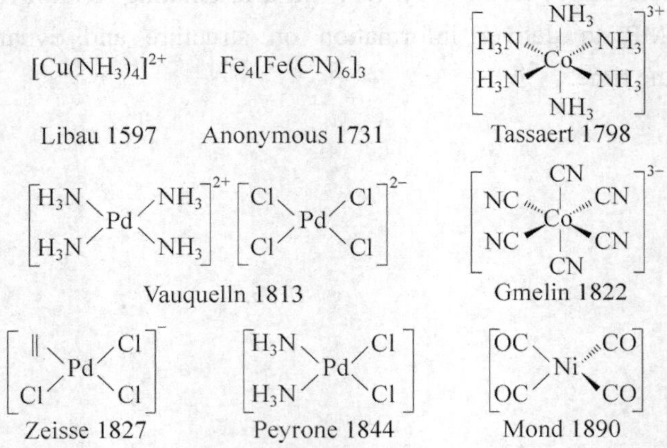

$[Cu(NH_3)_4]^{2+}$　　$Fe_4[Fe(CN)_6]_3$　　$[Co(NH_3)_6]^{3+}$

Libau 1597　　Anonymous 1731　　Tassaert 1798

$[Pd(NH_3)_4]^{2+}[PdCl_4]^{2-}$　　$[Co(CN)_6]^{3-}$

Vauquelln 1813　　Gmelin 1822

$[PdCl_3(C_2H_4)]^-$　　$[PdCl_2(NH_3)_2]$　　$[Ni(CO)_4]$

Zeisse 1827　　Peyrone 1844　　Mond 1890

图 1.1　无机化学中具有标志性的几种重要化合物

Proust、Lavoisier 的贡献促进了道尔顿简明原子理论的诞生（1808 年）。Mendeleev 在 1869 年发表了他的第一张元素周期表。20 世纪，伴随着 X 射线、放射性、电子、原子核等的发现，原子的现代量子力学模型开始呈现，这一模型可以在理论上解释原子的线性光谱和现代元素周期表。但是在那时，还没有一种可以合理解释这些特殊化合物的理论基础。

考虑到有机化学家在描述碳基化合物的结构单元和原子的化合价方面所取得的成功，这些理论被应用于氨合物是很自然的。但是，结果令人失望。例如，氨合钴氯化物的一些典型数据（表 1.1）。在 19 世纪最后几十年里所使用的化学式只表示了氨和钴的物质的量之比，而没有考虑它们之间的成键作用。这种不确定性可以从化学式中用来连接氯化钴和氨的“点”反映出来（由于氨/钴比为 3∶1 的化合物难以制备，所以表中用铱的化合物来代替）。

表 1.1　氨合钴氯化物的导电率

分　子　式	导　电　率	被沉淀的 Cl^- 的数目
$CoCl_3 \cdot 6NH_3$	高	3
$CoCl_3 \cdot 5NH_3$	中	2
$CoCl_3 \cdot 4NH_3$	低	1
$IrCl_3 \cdot 3NH_3$	零	0

电导率用于测量这些化合物溶于水后溶液中离子的数量。“被沉淀的氯离子的数目”通过加入硝酸银来测定，如方程式（1.1）所示：

$$AgNO_3\ (aq) + Cl^-\ (aq) \longrightarrow AgCl\ (s) + NO_3^-\ (aq) \tag{1.1}$$

现在如何解释表中的这些数据呢？1869 年，Christian Wilhelm Blomstrand 首先阐述了他的理论，并解释了氨合钴氯化物以及其他氨合物。他画出的 $CoCl_3 \cdot 6NH_3$ 的结构如图 1.2（a）所示。

基于当时的主流思想，这是一个非常合理的结构。他认为二价的氨与写成 H—NH_3—Cl 形式的氯化铵是一致的，钴为 3 价，氮原子连接在一起，与有机化合物中的碳非常相似，而三个一价的氯与金属离子钴具有足够远的距离，所以它们可以离开金属离子，与硝酸银反应生成氯化银。

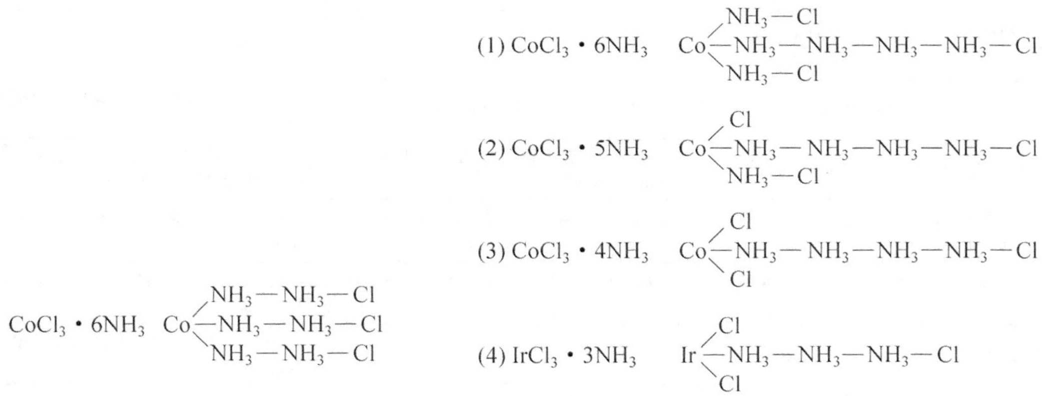

$CoCl_3 \cdot 6NH_3$　Co(—NH_3—NH_3—Cl)(—NH_3—NH_3—Cl)(—NH_3—NH_3—Cl)

(a) 氨合钴氯化物的Blomstrand表达式

(1) $CoCl_3 \cdot 6NH_3$　Co(—NH_3—Cl)(—NH_3—NH_3—NH_3—NH_3—Cl)(—NH_3—Cl)

(2) $CoCl_3 \cdot 5NH_3$　Co(—Cl)(—NH_3—NH_3—NH_3—NH_3—Cl)(—NH_3—Cl)

(3) $CoCl_3 \cdot 4NH_3$　Co(—Cl)(—NH_3—NH_3—NH_3—NH_3—Cl)(—Cl)

(4) $IrCl_3 \cdot 3NH_3$　Ir(—Cl)(—NH_3—NH_3—NH_3—Cl)(—Cl)

(b) 其他四种氨合钴氯化物的Jørgensen表达式，(4)为预期的氨合钴氯化物的铱的替代物

图 1.2　氨合钴氯化物的 Blomstrand 和 Jørgensen 表达式

1884 年，S.M.Jørgensen 对他导师的结构图提出了一些修改（表 1.2）。首先，他得到新的证据正确地表明了这些化合物是单体。其次，他调整了氯离子到钴的距离，来解释氯具有不同沉淀速率的问题。第一个氯比其他氯离子沉淀快得多，所以它的位置离钴更远，因此受到钴原子的作用较少。他的前三个氨合钴氯化物见图 1.2（b）。注意，在第二个化合物中，一个氯是直接与钴相连的，所以，Jørgensen 假设它不能被硝酸银沉淀。在第三个化合物中，两个氯的位置相似，这个改变是很重要的，这表明 Blomstrand-Jørgensen 的理论似乎是在向正确的方向发展。

但是，是否存在只含有三个氨分子的化合物呢？如图 1.2（b）（4）所示，该理论预言这种化合物是应该存在的，而且应该有一个可电离的氯。但是这一关键化合物始终没有得到。

在付出相当长的时间和努力之后，得到了该类氨合钴氯化物的替代物氨合铱氯化物。但这个化合物是一个中性化合物，没有可电离的氯。该理论遇到了麻烦。

表 1.2 配位化合物发展的历史背景

原子结构和周期表	分子结构和化学键	配 位 化 学
1750		
1774 年物质守恒定律：Lavoisier		1798 年观察到第一个氨合钴化合物：Tassaert
1799 年定组分定律：Proust		
1800		
1808 年道尔顿原子理论发表在化学哲学新体系上	1830 年结构的自由基理论：Liebig，Wöhler，Berzelius，Dumas（有机化合物由甲基、乙基等自由基组成）	1822 年制备出氨合钴的草酸盐：Gmelin
1859 年光谱学的发展：Bunsen 和 Kirchhoff	1852 年化合价的概念：Frankland（所有原子有一固定的化合价）	1851 年制备出 $CoCl_3 \cdot 6NH_3$、$CoCl_3 \cdot 5NH_3$ 和其他的钴氨化合物：Genth，Claudet，Fremy
1869 年门捷列夫第一个周期表列出 63 个已知元素	1854 年四价碳原子的概念：Kekulé	1869 年氨合物的链理论：Blomstrand
1885 年可见氢光谱的 Balmer 公式	1874 年四面体碳原子的概念：Le Bel 和 Van't Hoff	1884 年链理论的修正：Jørgensen
1894 年第一个“惰性”气体被发现	1884 年电解液离解理论：Arrhenius	1892 年 Werner 关于配合物的设想
1895 年 X 射线被发现：Roentgen		
1896 年放射性被发现：Becquerel		
1900		
1902 年电子的发现：Thomson	1923 年电子-点图形：Lewis	1902 年预测的配位理论的三个假设：Werner
1905 年光的波粒二象性：Einstein	1931 年价键理论：Pauling，Heitler，London，Slater	1911 年顺式$[CoCl(NH_3)(en)_2]X_2$ 的光学异构体的拆分：Werner
1911 年 α 粒子和金箔使用，原子核模型：Rutherford	19 世纪 30 年代分子轨道理论：Hund，Bloch，Mulliken，Hückel	1914 年不含碳的光学异构体的拆分：Werner
1913 年原子的玻尔模型（电子能量的量子化）	1940 年价层电子对互斥（VSEPR）理论：Sidgwick	1927 年 Lewis 概念应用于配合物：Sidgwick
1923 年电子的波粒二象性：De Broglie		1933 年晶体场理论：Bethe 和 Van Vleck
1926 年薛定锷的量子力学模型（关于原子核轨道的电子、电子光谱被认为是轨道间跃迁）		
包括性质变化趋势的现代元素周期表	化学键的现代概念	现代配位理论

1.1.2 现代配位化学——Werner 配位理论

Alfred Werner（1866—1919），作为有机化学专业无薪水的讲师，打破了有机和无机化学领域的界线。他的第一个贡献（立体化学，或原子在氮化合物中的空间排布）是有机领域的，但是那段时期有那么多有趣的无机化学问题。他注意到了无机化学家在解释配位化合物时遇到的困难。已经建立的有机化学理论可能会把配位化学引入歧途。1892 年，他提出了配位理论。但是他的新理论受到早期传统理论的抨击，而且他也没有实验证据来支持自己的理论。他的理论被认为是不切实际的空想。Werner 用他一生中剩下的时间进行了系统、彻底的实验，证明自己的直觉是正确的。

Werner 坚信单一固定化合价理论不能应用于钴和其他类似的金属。通过对氨合钴及其他相关体系（如铬和铂）的研究，他提出这些金属应该有两种价态，一个主要价态和一个次要价态。主要价态或电离价态即为我们今天所说的氧化态；对于钴来说，是+3 价。次要价态一般称为配位数；对于钴来说，是 6。Werner 主张这个次要价态是由空间稳定的几何位置决定的。

图 1.3 表示了 Werner 早期对于氨合钴成键的一些设想方案。他认为钴的主要价态和次要价态必须同时满足。图中实线代表基团为满足主要价态而形成的键，虚线代表次要价态。在化合物（1）中，三个氯只用来满足主要价态，六个氨用来满足次要价态。在化合物（2）中，一个氯必须具有双重作用，同时满足主要、次要两种价态。这个用于满足次要价态的氯（同时直接连接 Co^{3+}）被断定不能被硝酸银沉淀。化合物（3）有两个具有双重作用的氯，所以只有一个氯可以被沉淀。Werner 认为，化合物（4）应该是一个中性化合物，没有可电离的氯。这一点与 Jørgensen 在铱的化合物中的发现完全相同。

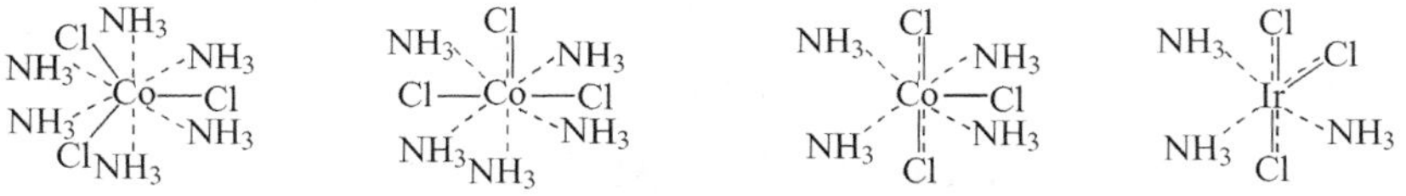

(1) $CoCl_3 \cdot 6NH_3$　(2) $CoCl_3 \cdot 5NH_3$　(3) $CoCl_3 \cdot 4NH_3$　(4) $IrCl_3 \cdot 3NH_3$

图 1.3　氨合钴氯化物的 Werner 表达式，实线表示满足钴主要价态或（+3）氧化态的基团；虚线表示满足钴次要价态或配位数（6）的基团；次要价态的基团在空间中有固定的位置

随后，Werner 开始研究次要价态（或配位数）的几何构型。如表 1.3 所示，六个氨围绕一个中心金属原子或离子可以有几种不同的几何构型，包括平面六边形、三角锥形和八面体形。表 1.3 比较了预测的和实际的配合物异构体的数目。

表 1.3　三种构型的预测的和实际的异构体的数目

分子式	配位数为 6 的几何构型 平面六边形	 三角双锥	 八面体	实际异构体的数目
	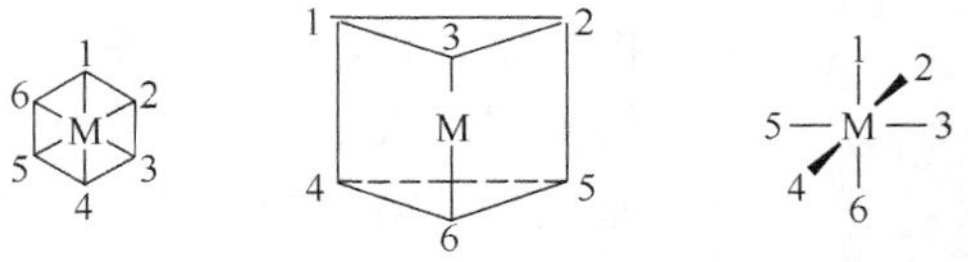			
	预测异构体的数目（括号中的数字表示 B 配体的位置）			
MA_5B	1	1	1	1
MA_4B_2	3	3	2	2
	（1，2）	（1，2）	（1，2）	
	（1，3）	（1，4）	（1，6）	
	（1，4）	（1，6）		
MA_3B_3	3	3	2	2
	（1，2，3）	（1，2，3）	（1，2，3）	
	（1，2，4）	（1，2，4）	（1，2，6）	
	（1，3，5）	（1，2，6）		

在讨论之前，需要对表格中的数据作一些说明。首先，使用符号 M 代表中心金属，A 和 B 代表不同配体。其次，括号中的数字表示每个异构体中配体 B 的相对位置。

异构体在这里定义为：具有相同数目和种类的化学键，但是键的空间排布不同的化合物（在以后的章节里将有异构体更详细的讨论）。

对于 MA_5B（表 1.3），通过实验只可以制备一种异构体，这一结果与三种几何构型的预测值都一致。但是，对于 MA_4B_2，Werner 只制得两种异构体（图 1.4）。这个实际数目与八面体情况下的可能数目相吻合，而在平面六边形和三角锥形的情况下，应有 3 种异构体。假如 Werner 没有遗漏一种异构体的制备，那么数据表明对于六个配体的情况，“空间中最稳定的形式”是八面体形。对于 MA_3B_3 进行同样的分析，得到类似的结果。只有采取八面体形的结构才可以得到与实际相符的异构体数目。

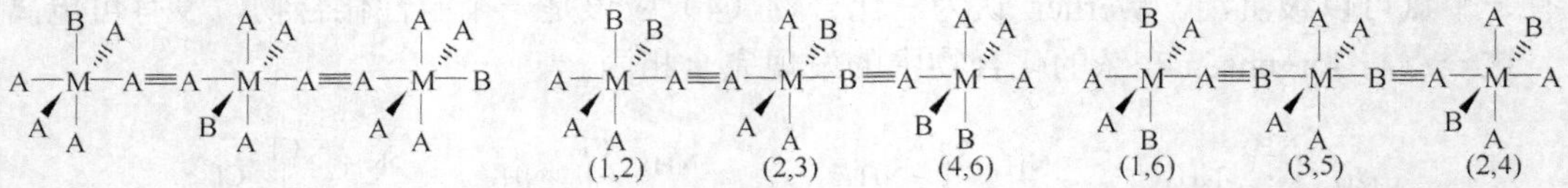

(a) MA_5B (b) MA_4B_2(异构体 1) (c) MA_4B_2(异构体 2)

图 1.4 几种八面体构型的等价构型

得到这些结论后，Werner 预测 $CoCl_3 \cdot 4NH_3$ 可有两种异构体。虽然难以制备，但 1907 年，Werner 成功地证明了他的预测。他发现了两种异构体，一个是浅绿色的，另一个是鲜艳的紫罗兰色。通过与实际存在的异构体的数目相比较，Werner 得出结论：这六个配体应该是八面体构型的。从而，配位理论逐渐强大。

George B.Kauffman
Alfred Werner
Founder of Coordination Chemistry
Alfred Werner(1866—1919)
Springer-Verlag,Berlin Heidelberg,New York,1966

图 1.5 《阿尔弗雷德·沃纳——配位化学之父》一书的封面的复印件

所有这些都证实，在科学界，有时需要冒险。有时需要跟随自己的直觉，提出一种新的有时可能是没有人支持的方式，思考一个现象，从而取得革命性的进步。Blomstrand 和 Jørgensen 努力扩展已经建立的有机化学理论，来解释更新的配位化合物。事实上，他们阻碍了人们对于这个化学分支的认识和发展。那么，避免这种情况发生的诀窍是，我们应该知道什么时候该抓住已有的理论，什么时候又该从中脱离出来。Werner 选择了后者，在 20 年后的 1913 年，他获得了诺贝尔化学奖。

在 20 世纪初，无机化学不是一个主要的领域，直到 Werner 开始金属-氨化合物如$[Co(NH_3)_6Cl_3]$的研究——Werner 化合物，他被誉为配位化学之父。图 1.5 是《阿尔弗雷德·沃纳——配位化学之父》一书的封面。

1.1.3 广义配位化学——超分子化学

超分子化学是一个研究分子间非共价相互作用的化学领域。传统化学关注的是共价键作用，那么超分子化学关注的则是分子间弱的并且可逆的非共价相互作用，这些作用力包括氢键、π-π 堆积作用、静电作用、疏水作用和范德华力等。近年来，金属配合物中的配位键也被认为是超分子化学中的重要相互作用之一，从这一点来讲，可以认为超分子化学是拓展的配位化学。分子自组装、分子识别、预组装和大环效应等重要概念都在超分子化学中得到了阐释。生物体系通常是超分子化学研究灵感的源泉，研究非共价相互作用对于我们理解许多生命过程是至关重要的，其中包括了从细胞的结构到特定的结构和功能对这些力的依赖性的理解。

超分子化学的重要性在 1987 年得以确立，在这一年 Donald J. Cram、Jean-Marie Lehn 和 Charles J. Pedersen 由于在该领域的工作被授予诺贝尔化学奖。具体而言，在选择性主客体复合物的研究方面取得了重大进展——主体分子可以识别并选择性地与特定的客体分子键合。

超分子化学处于化学、生物化学、物理学和工艺技术的交叉点，是一个高速发展的新领域。超分子化学实现了在传统有机化学中令人难以置信的激动现象（如，碱金属阴离子），不仅为许多工业改革提供了基础，同时也促进了我们对于生命体和生命起源的认识。

在 20 世纪 90 年代，在众多学者的努力和推动下，超分子化学日趋成熟。James Fraser Stoddart 促进了分子机器和高度复杂的自组装结构的研究，Itamar Willner 开创了传感器与电子学和生物学相互结合的研究。同时，由于富勒烯、量子点和枝状聚合物等均可以作为构筑单元参与超分子体系的合成，因此，纳米技术这一新兴领域也对超分子化学产生了重大影响。

（1）超分子化学中的概念

① 分子自组装　分子自组装是指在缺少外力引导和监管下构造化学体系（仅提供一个适宜的环境）。分子是通过非共价相互作用的诱导进行组装。自组装可分为分子间自组装和分子内自组装。

② 分子识别　分子识别是指客体分子通过特定的键合作用与互补的主体分子形成主客体复合物。然而对于主体分子和客体分子的界定是主观的。通过非共价相互作用分子可以进行识别。

③ 超分子构筑单元　超分子体系极少是按照“第一性原理”设计的。相反，化学家有一系列结构和性质都经过彻底研究的构筑单元用于建造更大的体系。这些构筑单元通常是相似的单位组成的一个系列，因而通过与所追求的性质的对比就可从中选择适合的构筑单元。

④ 合成识别组分　合成识别组分包括羧酸二聚物和其他简单的氢键键合作用；冠醚和铵阳离子的键合；联吡啶和二氧芳烃或二氨基芳烃间的 π-π 电荷转移作用等。

⑤ 大环　大环在超分子化学中是非常有用的，它们可以提供一个空穴将客体分子完全包围，也可以通过化学修饰来对它们的性质进行调控。

⑥ 结构单元　许多超分子体系要求它们的各组成部分在空间和形态上要相互匹配，因此易于利用的结构单元就是必需的了。

⑦ 动态共价化学　在动态共价化学中，共价键的断裂和形成平衡通过热力学控制。尽管在这个过程中共价键是关键，然而非共价相互作用将最终引导体系形成能量最低的构型。

⑧ 仿生学　许多合成的超分子体系都是设计用来模拟生物体系中的某些功能。这些仿生学结构对于我们学习生物模型和合成工具都是有用的。

⑨ 印迹　分子印迹描述了通过合适的分子模板将小分子构造成主体分子的过程。在主体分子构筑完成之后，移除模板剩下的就是主体分子。用于构造主体分子的模板也许和完成键合的客体分子有微小的差异。最简单情形就是印迹只使用了空间的相互作用，但是对于更为复杂的体系，则包括氢键等相互作用来增加键的强度和特异性。

⑩ 分子机器　分子机器是指可以进行诸如线性运动或旋转、开关和圈套等功能的分子或分子组装体。这些器件处于超分子化学和纳米技术的交叉领域，它们的原型已经在超分子

化学中进行了阐述。

（2）超分子化学的运用

设计具有特定功能的超分子化学体系可以使化学工业更清洁和安全，通过开发由单个分子或数个分子组成的器件将使电子学微型化。当然我们使用能源的方式也将彻底改变。它也可以为制药工业和医学带来改变，如开发药品管理的新途径，合成可用于牙科、外科等医学分支的生物相容性的材料。

● 材料　超分子化学中的分子自组装过程已经用于一些新材料的研制。按照自下而上的合成将小分子构造成更大的结构是可行的。因此可以说纳米技术中大部分的自下而上（的合成方式）是基于超分子化学。

● 催化　超分子化学的一个主要应用是合成催化剂和研究催化作用。非共价相互作用在催化过程中极为重要，它们将反应物连接成适合反应的构型，从而降低反应中过渡态的能量。模板引导的合成是起分子化学催化的一个特殊例子。

● 医药　超分子化学对新的药物治疗的发展很重要，它有助于我们了解反应中药物键合的位点。同时，由于超分子化学促进了药物输送中胶囊和药物靶点释放机制的研究，它也带来了药物输送领域的重大进步。

● 绿色化学　在非共价键诱导的固态反应得到发展时，超分子化学在绿色化学中也有了应用。这个过程是极为理想的，因为这将减少化学过程中溶剂的用量。

● 信息存储和处理　超分子化学将计算的功能在分子尺度上进行展示。在许多情况下，光和化学信号已经在这些部件中得到了应用，但是这些单元的电子界面也在超分子单个转换器件中呈现出来。信息存储通过使用包含光致变色和可光致异构的分子开关、电致变色和氧化还原开关单元，甚至是分子的移动得以实现。

● 其它功能器件　超分子化学常追求于开发那些单个分子不能呈现的功能。这些功能包括磁学性质、光反应、自愈聚合物、分子传感器等。超分子的研究工作已经应用到高科技传感器的开发上。

1.2　配合物的基本特征

1.2.1　配合物的概念

今天，配合物的分子式可以更加清楚地表示它的含义，哪个部分在配位层中，哪个不在。金属原子或金属离子与它的配体被括在括号中。氨合钴氯化物可以如下表示。

（1）$CoCl_3 \cdot 6NH_3$　　$[Co(NH_3)_6]Cl_3$

（2）$CoCl_3 \cdot 5NH_3$　　$[Co(NH_3)_5Cl]Cl_2$

（3）$CoCl_3 \cdot 4NH_3$　　$[Co(NH_3)_4Cl_2]Cl$

（4）$CoCl_3 \cdot 3NH_3$　　$[Co(NH_3)_3Cl_3]$

括号中的氨分子和氯离子满足了钴的配位数。配位层中的氯离子同时还用来满足钴的+3价氧化态。括号外面的氯离子，有时称为抗衡离子（counterions），只用来满足氧化态。只有这些氯离子才可以被硝酸银沉淀。例如，如果化合物（2）在水溶液中与银离子反应，其反应方程式应如式（1.2）所示：

$$[Co(NH_3)_5Cl]Cl_2(s) + 2Ag^+(aq) \longrightarrow 2AgCl(s) + [Co(NH_3)_5Cl]^{2+}(aq) \qquad (1.2)$$

考虑下面一系列用现代分子式表示的氨合铂的体系，体系可以扩展到形成带负电荷的配离子，如化合物（10）和（11），其抗衡离子为钾离子。

（5）$[Pt(NH_3)_6]Cl_4$

（6）$[Pt(NH_3)_5Cl]Cl_3$

（7）$[Pt(NH_3)_4Cl_2]Cl_2$

（8）$[Pt(NH_3)_3Cl_3]Cl$

（9）$[Pt(NH_3)_2Cl_4]$

（10）$K[Pt(NH_3)Cl_5]$

（11）$K_2[PtCl_6]$

在此基础上，有关 配位化学和配合物的定义如下。

配位化学是研究金属离子和其他中性或带负电荷的分子间所形成的化合物的化学。

配位化合物是一种在路易斯酸碱反应中，中性分子或阴离子（称为配体）与金属原子（或离子）通过配位键而形成的产物。

氨分子是一种非常著名的配体。这种只与金属或金属离子共用一对电子的配体称为单齿配体（monodentate ligand）。“monodentate”一词来源于希腊语“monos”和拉丁语“dentis”。形象地说，就像一个配体只用一对电子“咬住”金属或金属离子。另外一些常见的配体列于表 1.4 中（关于这些配体的命名将在下一部分中讨论）。

表 1.4　一些常见的单齿、多齿、桥联和两可配体

<table>
<tr><th colspan="4">常用的单齿配体</th></tr>
<tr><td>F^-</td><td>fluoro</td><td></td><td></td></tr>
<tr><td>Br^-</td><td>bromo</td><td></td><td></td></tr>
<tr><td>I^-</td><td>iodo</td><td></td><td></td></tr>
<tr><td>CO_3^{2-}</td><td>carbonato</td><td></td><td></td></tr>
<tr><td>NO_3^-</td><td>nitrato</td><td></td><td></td></tr>
<tr><td>SO_3^{2-}</td><td>sulfito</td><td></td><td></td></tr>
<tr><td>$S_2O_3^{2-}$</td><td>thiosulfato</td><td></td><td></td></tr>
<tr><td>SO_4^{2-}</td><td>sulfato</td><td rowspan="10">常见的桥联配体</td><td></td></tr>
<tr><td>CO</td><td>carbonyl</td><td></td></tr>
<tr><td>Cl^-</td><td>chloro</td><td></td></tr>
<tr><td>O^{2-}</td><td>oxo</td><td></td></tr>
<tr><td>O_2^{2-}</td><td>peroxo</td><td></td></tr>
<tr><td>OH^-</td><td>hydroxo</td><td></td></tr>
<tr><td>NH_2^-</td><td>amino</td><td></td></tr>
<tr><td>CN^-</td><td>cyano</td><td></td></tr>
<tr><td>SCN^-</td><td>thiocyanato</td><td rowspan="2">两可配体</td></tr>
<tr><td>NO_2^-</td><td>nitro</td></tr>
<tr><td>H_2O</td><td>aqua</td><td></td><td></td></tr>
<tr><td>NH_3</td><td>ammine</td><td></td><td></td></tr>
<tr><td>CH_3NH_2</td><td>methylamine</td><td></td><td></td></tr>
<tr><td>$P(C_6H_5)_3$</td><td>triphenylphosphine</td><td></td><td></td></tr>
<tr><td>$As(C_6H_5)_3$</td><td>triphenylarsine</td><td></td><td></td></tr>
<tr><td>N_2</td><td>dinitrogen</td><td></td><td></td></tr>
<tr><td>O_2</td><td>dioxygen</td><td></td><td></td></tr>
</table>

续表

常用的单齿配体		
NO	nitrosyl	
C_2H_4	ethylene	
C_5H_5N	pyridine	
多齿配体		
$NH_2CH_2CH_2NH_2$	乙二胺（en）	（2）
$CH_3\underset{\underset{O}{\|}}{C}\bar{C}H\underset{\underset{O}{\|}}{C}CH_3$	乙酰丙酮（acac）	（2）
$C_2O_4^{2-}$	草酸（ox）	（2）
$NH_2CH_2COO^-$	甘氨酸（gly）	（2）
$NH_2CH_2CH_2NHCH_2CH_2NH_2$	二亚乙基三胺（dien）	（3）
$N(CH_2COO)_3^{3-}$	次氮基三乙酸(NTA)	（4）
$(OOCCH_2)_2NCH_2CH_2N(CH_2COO)_2^{4-}$	乙二胺四乙酸（EDTA）	（6）

毫不奇怪，还有双齿（bidentate）、三齿（tridentate）以及多齿（multidentate）配体。一般来讲，配体的齿数定义为配体与金属或金属离子共用的电子对数目。表 1.4 中还给出了一些常见的多齿配体，配体的齿数在括号中给出。例如，乙二胺的齿数为 2（图 1.6）。

$H_2\ddot{N}—CH_2—CH_2—\ddot{N}H_2$

H H H H H H H H N N C C M

(a) (b)

图 1.6 双齿乙二胺配体

1.2.2 配体的分类

（1）单齿配体　仅有一个点与金属中心配位，只有一个配位位置，例如：NH_3、H_2O、PMe_3、CO 和其他中性的两电子给体。

（2）双齿配体　如图 1.6 所示，乙二胺在 Werner 和 Jørgensen 的研究中是一个非常重要的双齿配体。这个化合物中的两个氮原子各有一对孤对电子可以与金属共用。当双齿配体中的一端并没有与金属离子相连时，则作为单齿配体，此时又称为两可配体（见下一部分）。2,2′-联吡啶（bipy）、乙酰丙酮（acac）、二苯基膦乙烷（dppe）和草酸（ox）都是双齿配体的例子（图 1.7）。

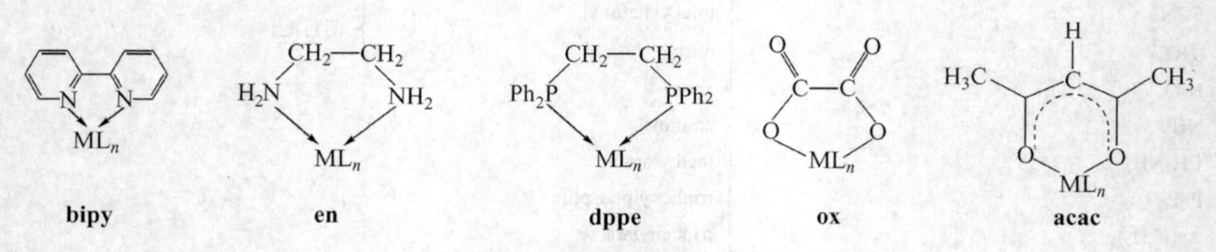

图 1.7 一些常见的双齿配体

（3）两可配体　在表 1.4 中还有两种常见的配体，在这里简单介绍一下。第一种是桥联配体，定义为含有两对共用电子对，且分别与两个不同金属原子共用的配体。配体与金属之间的相互作用可表示为 M←:L:→M。桥联配体包括氨基化合物（NH_2^-）、一氧化碳（CO）、氯离子（Cl^-）、氰基（CN^-）、羟基（OH^-）、硝基（NO_2^-）、氧化物（O^{2-}）、过氧化物（O_2^{2-}）、硫酸盐（SO_4^{2-}）和硫氰酸盐（SCN^-）等。另一种配体就是两可配体（ambidentate ligand），这种配体，根据实验条件和金属的不同，可以使用其中两个不同原子中的任意一个与金属离子共用电子对。如果把这种配体表示为:AB:，那么它与金属形成的配位键有两种可能的形式，M←:AB:或:AB:→M。一般的两可配体包括氰化物、硫氰酸盐和亚硝酸盐。最经典的例子就是 NO_2^- 配体（图 1.8）。又如：SCN^-配体可以通过 S 原子或 N 原子配位。

$[(H_3N)_5Co—NO_2]^{2+}$　　$[(H_3N)_5Co—O—\ddot{N}{=}O]^{2+}$

Nitro　　Nitrito

五氨硝基钴(Ⅲ)　　五氨亚硝酸钴(Ⅲ)

图 1.8　NO_2^- 配体的配位模式

（4）多齿配体　有多个配位点并可占据多个配位位置的配体。双齿配体和两可配体都是多齿配体中的一种。下面所示的具有六个配位点的负 4 价阴离子是乙二胺四乙酸（EDTA）的未配位和配位状态（为清楚起见，氢离子略去）。salen 配体是一种灵活的负 2 价四齿配体。三齿的三咪唑硼配体是一个三齿的负 1 价配体的例子，常用来作为 Cp 的替代物（图 1.9）。

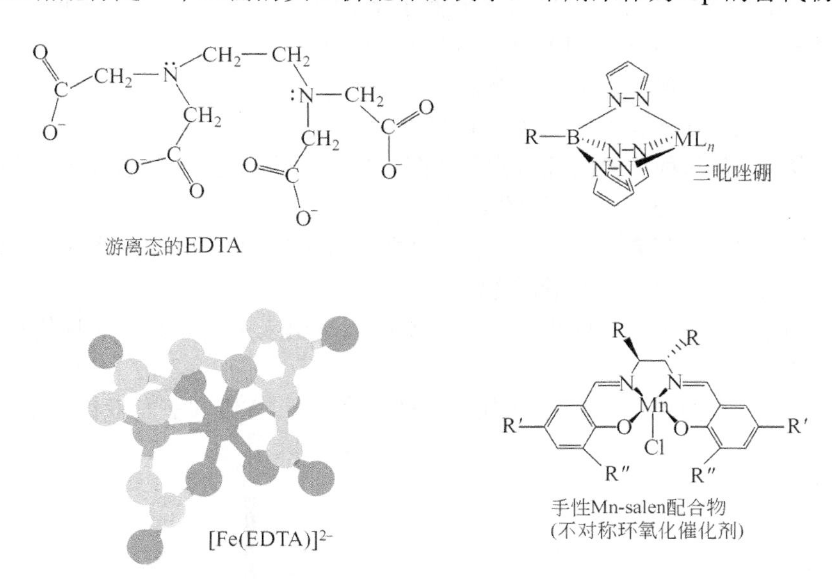

图 1.9　一些常见的多齿配体

（5）螯合配体　当多齿配体中的两对电子同时作用在一个金属原子或离子上时，形成的结构就像一只螃蟹抓着它的猎物一样。通过这种方式与金属原子形成的一个或多个环的多齿配体，被称为螯合剂（chelates or chelating agents，源自希腊语 chele，为“爪”的意思）。上面所示的双齿配体是螯合配体的一个例子。

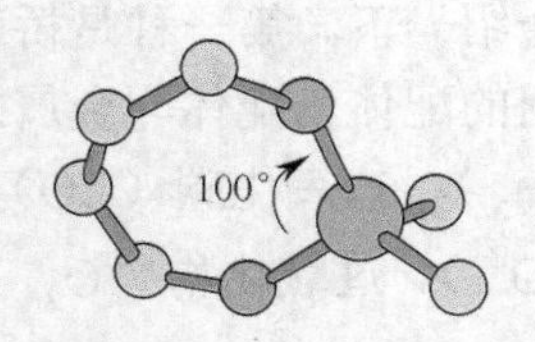

图 1.10 偏离理想四面体角的螯合角

（6）螯合角　它是由双齿或多齿配体配位到金属中心上形成的配体-金属-配体间的角。其值可用来表征理想几何构型的扭曲或测量环张力角度。由图 1.10 可见，四面体金属中心的螯合角偏离了理想的四面体角 109.5°。

1.2.3　配位数与配位几何构型

配位数定义为配体与中心元素接触的点数。它的范围可以从 2 到 16，但 6 较为常见。

（1）配位数 2　这种排布对第一过渡系列的金属配合物而言很少见，最著名的例子是 Ag^+。例如，用于检测氯离子的方法中就包括了线形二氨合银的配合物。第一步是

$$Ag^+ + Cl^- \longrightarrow AgCl(\downarrow)$$

为了证明沉淀的是氯离子，还必须进一步做两个实验：

$$AgCl + 2NH_3 \cdot H_2O \longrightarrow [Ag(NH_3)_2]^+ + Cl^- + 2H_2O$$

$$\Big\| \ HNO_3$$

$$AgCl(\text{沉淀再析出}) + NH_4^+ + NO_3^- + 2H_2O$$

像乙二胺这样的双齿配体与 Ag^+反应是不能形成螯合环的，相反会形成线形双金属配合物。一个原因是该配体不能以 180°的跨度反式配位。

（2）配位数 3　同样，这种配位数对于第一过渡系的金属来讲很不寻常。已知这一配位数有三种不同的几何构型（图 1.11）。

图 1.11　配位数为 3 的三种配位几何构型

- 平面三角形：对于主族元素，著名的例子是 CO_3^{2-}。所有的四个原子在同一平面上，且配体间的键角为 120°。
- 三角锥形：对于主族离子较为普遍。
- T-形分子：第一个 T-形分子是在 1977 年发现的。

（3）配位数 4　有两种可能的几何构型（图 1.12）。四面体是最为普遍的一种；而平面四边形的构型几乎都在具有 d^8 电子构型的离子中。

- 四面体：以四面体碳为中心的分子的化学是有机化学。原则上，将 C 变成 Co 是可以的。目前，已有大量的四面体钴的配合物。
- 平面四边形：这种构型较少，这里之所以列举出来是因为具有这种构型的分子相当重要。

（4）配位数 5　有两种可能的几何构型（图 1.13）：四方锥和三角双锥。配合物$[Cr(en)_3]$

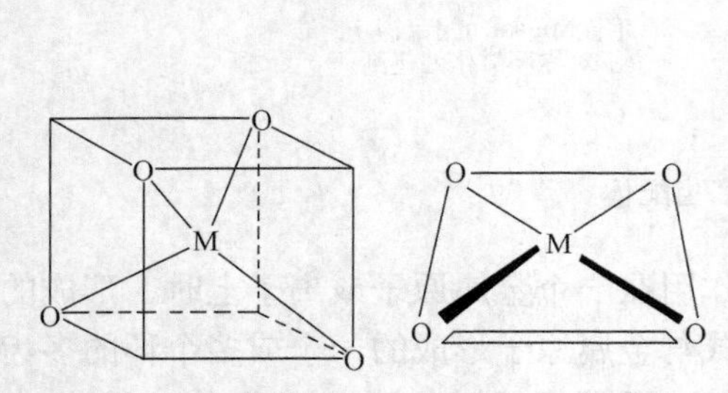

图 1.12　配位数为 4 的两种配位几何构型

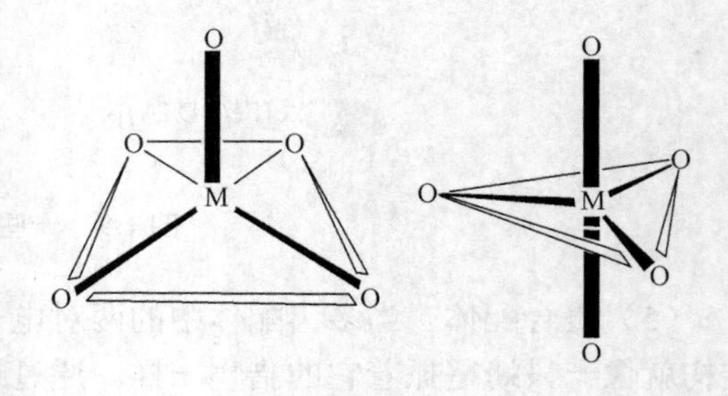

图 1.13　配位数为 5 的两种配位几何构型

$[Ni(CN)_5]\cdot 1.5H_2O$的结构在1968年被报道，这是在同一晶体中同时具有上述两种构型的例子。氰根离子与Ni^{2+}的反应可以通过以下几步来完成：

$$Ni^{2+} + 2CN^{-} \longrightarrow Ni(CN)_2$$

$$Ni(CN)_2 + 2CN^{-} \longrightarrow [Ni(CN)_4]^{2-}\text{（橘红），}\lg\beta_4=30.1$$

$$[Ni(CN)_4]^{2-} + CN^{-} \longrightarrow [Ni(CN)_5]^{3-}\text{（深红）}$$

钒氧盐（VO^{2+}）常常表现为四方锥构型，例如，VO(acac)$_2$。注意，V（Ⅳ）是配位不饱和的，吡啶的加入将导致八面体的形成。

（5）配位数 6（图 1.14）

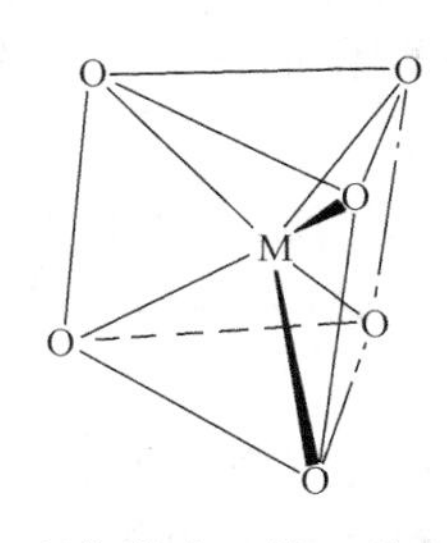

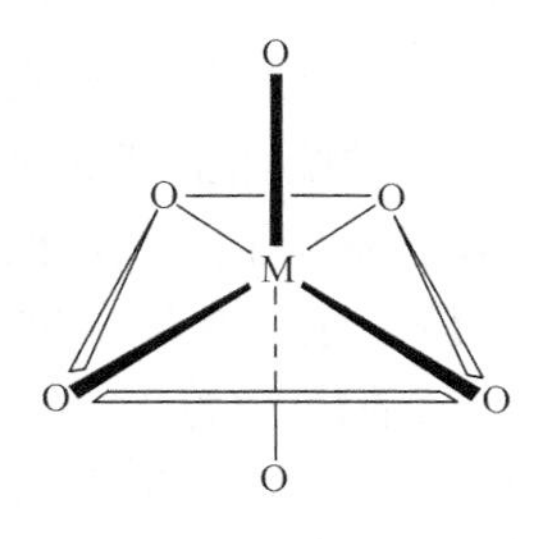

图 1.14　配位数为 6 的三种配位几何构型

- 平面六边形：虽然六个基团在同一平面上排布的情况在高配位数的构型中存在，但这种配位构型对第一过渡系金属离子来讲还未发现。
- 三角棱柱：大多数三角棱柱形化合物是含有三个像草酸这样的双齿配体。第一过渡系金属的配合物具有这样几何构型的例子很少。
- 八面体：对于第一过渡系的金属离子，包括水合离子，这是最常见的配位几何构型。有时有畸变，可用 John-Teller 理论来解释。

（6）配位数 7　有三种可能的几何构型：帽形八面体，帽形三角棱柱，五角双锥（图 1.15）。

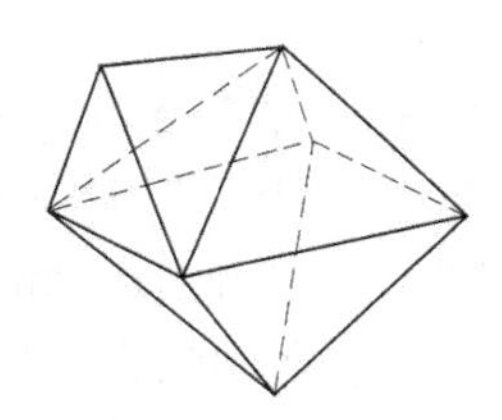

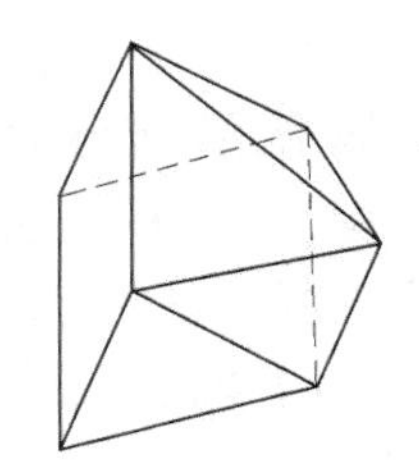

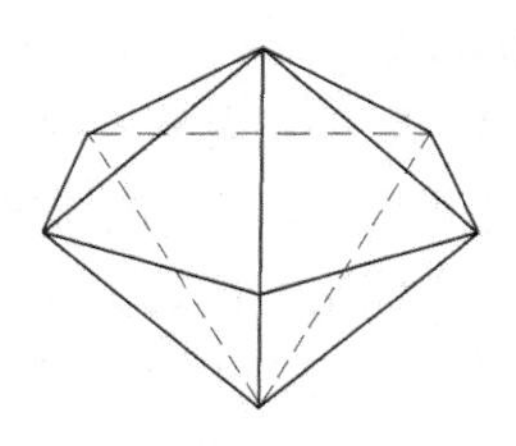

图 1.15　配位数为 7 的三种配位几何构型

（7）配位数 8　有三种可能的几何构型：十二面体（D_{2d}），立方体（O_h），四方反三棱柱（D_{4d}）（图 1.16）。

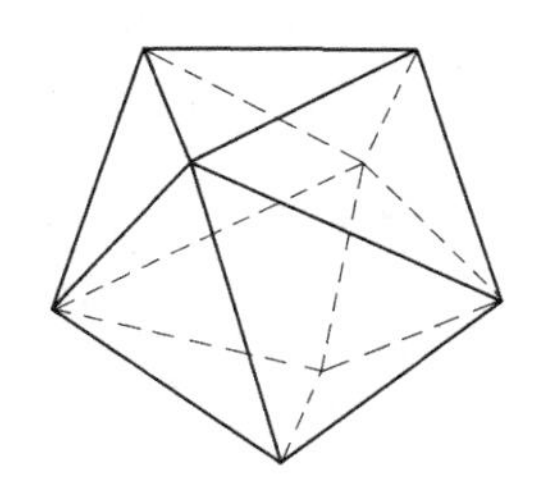

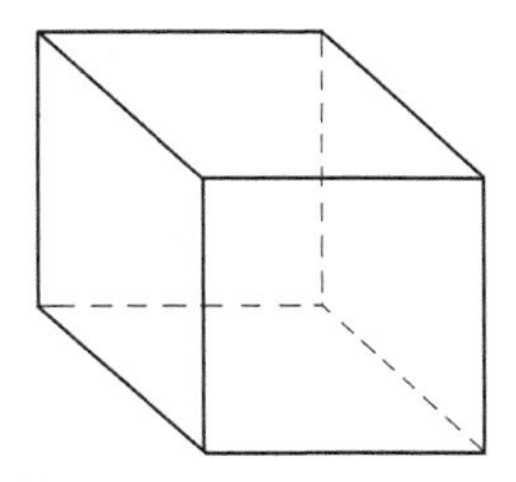

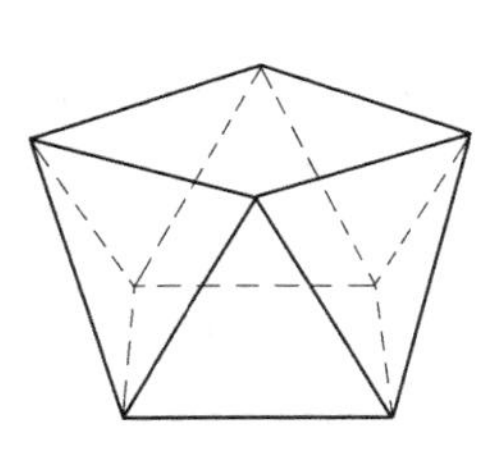

图 1.16　配位数为 8 的三种配位几何构型

（8）配位数 9　三面心三角棱体（D_{3h}）。

（9）配位数 10　双帽四方反棱柱（D_{4d}）。

（10）配位数 11　全帽三角棱体（D_{3h}）。

（11）配位数 12　二十面体(O_h)。

有许多关于配位数的定义，但没有一个简单明确的定义可适用于所有的情况。对于单齿配体和多齿配体，配位数可以定义为直接键合到金属原子上的原子或配体的数目。如$[Fe(NH_3)_6]^{3+}$和$[Fe(en)_3]^{3+}$都是 6 配位的配合物。但当键合到烯烃上如乙烯和环戊二烯上时就变得复杂了。

例如，在茂铁中十个环戊二烯碳等同地键合到铁中心上，但很少称其为 10 配位配合物。已知茂铁的反应性和成键性，称它为 2 配位的配合物似乎也不正确。普遍认为茂铁是 6 配位的，因为参与成键的配体上有 6 对电子。

这种方法似乎减少了麻烦，但进一步增加了复杂性。如下例所示：

- 醇盐、酰亚胺、含氧配体能为金属中心提供一对、两对或三对电子，但在所有例子中每一个仅占一个配位位置。
- 大多数化学家都认为 Cp_2TiCl_2 不是 8 配位配合物。
- 烯烃配合物能看作中性双电子供体但有时被认为双阴离子双齿配体。此外，烯烃经常被认为仅占据一个配位位置。

1.2.4　不饱和配位

我们用不饱和配位来描述这样一类配合物，即具有一个或多个开放配位位置的配合物。虽有很多不同的情况，但少于 6 个 CN 配体的配合物是这一类配合物的典型例子。不饱和配位在过渡金属有机化学中是一重要概念。例如，要将烯烃插入、调换或聚合，烯烃必须有空间去接近或键合到金属上。构造这种必要缺位的常用方法是失去一个配体或改变配体的键合方式（如三配位丙烯基配体转化为单配位形式）。这种改变伴随着电子数的减少。因此，尽管不饱和配位和饱和电子是不同的，但它们经常共同作为催化性循环中的一部分。

1.2.5　第一配位层和第二配位层

金属的第一配位层是直接连接到金属上最接近金属的配体［图 1.17(a)］。中心金属原子和有机配体之间相互作用的本质是共价键。第一配位层具有固定的配位几何构型（见 1.2.3）。第二配位层（或外层）的边界比第一配位层来说变得比较模糊，它常常由运动的阳离子、抗衡离子或溶剂分子构成［图 1.17(b)］。中心金属离子和第二配位层之间的相互作用是非共价键，包括氢键、离子-偶极、偶极-偶极作用等。

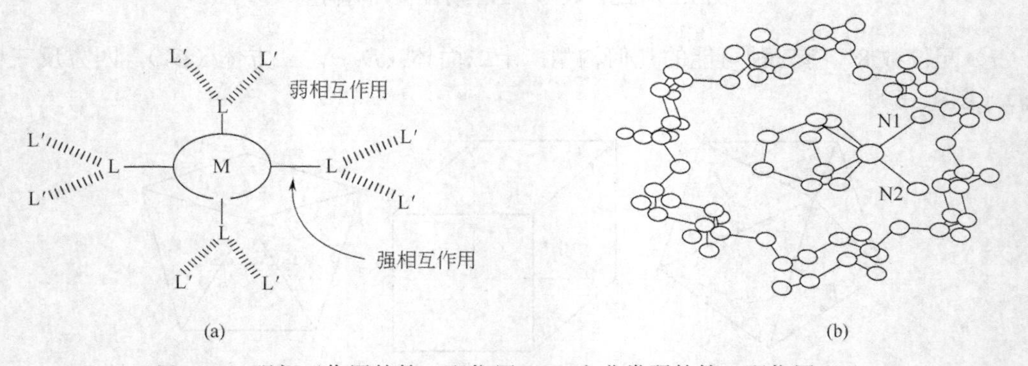

图 1.17　强相互作用的第一配位层（a）和非常弱的第二配位层（b）

第一配位层和第二配位层在参与生命体系很多重要过程的金属蛋白中是很重要的。金属蛋白是一类活性中心中含有一个或多个金属离子作为辅因子的蛋白质，可看作是过渡金属配合物。金属蛋白的中心金属的配体环境通常表现为低对称性、不寻常的配位构型和独特的配位键。

细胞色素 P450（P450）是关于这个问题的一个很好例子。P450 活性部位的最显著特点是通过近端半胱氨酸连接到亚铁血红素形成的亚铁血红素-硫醇盐键［图 1.18（a）］。在亚铁血红素-硫醇盐的边端及其末端底物结合的部位所形成的第二配位层中的氨基酸所发挥的微妙作用是很重要的。它们允许酶有效地发挥单加氧酶的作用，用于调控它们的选择性。具体地说，在 P450 末端一侧的第二配位层的相互作用主要是形成了氢键网络结构和提供质子用于氧气活化并结合底物。末端未配位的组氨酸变为缬氨酸或异亮氨酸而形成了带有 Fe（Ⅲ）的类似 P450 的蛋白质［图 1.18（b）］。这种效应（被作者称为反式效应），可以模拟细胞色素 P450 的更加疏水的环境，非常有助于轴向配体的结合和稳定。同样，在环胍氨酸肽中，轴向的组氨酸变为半胱氨酸就形成了一个非常不稳定的配体，它很快被氧化为半胱磺酸。由此可知，在 P450 蛋白质中靠近半胱氨酸的非极性残基被保留，然而在环胍氨酸肽中，氨基酸是天冬氨酸［图 1.18（c）］。设计可稳定配位的羰基铁衍生物的半胱氨酸硫醇盐，可通过把细胞色素 b562 轴向的组氨酸和蛋氨酸分别变为半胱氨酸和甘氨酸［图 1.18（d）］。

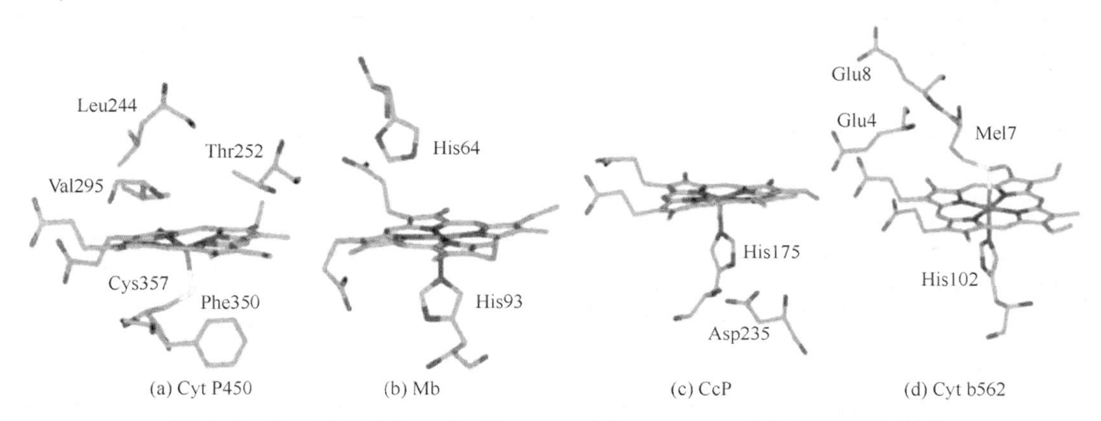

图 1.18 Cyt P450（a）；Mb（b）；CcP（c）；Cyt b562（d）关键结构差别

1.3 配合物的命名法

配位化合物的结构可能很复杂，其名称可能很长，因为配体的名称可能就很长。了解命名法则不仅使我们可以理解配合物是什么，也可以帮助我们给配合物一个适当的名称。

配合物的命名以配体（包括多齿、两可和桥形）和简单配合物的命名为基础。

表 1.5 给出了一些配体和简单配合物命名的规则。归纳如下。

- 命名配合物的时候，阳离子在前，阴离子在后（就像一般的盐一样——例如，氯化钠），而不管阳离子或阴离子是否是配离子（注：中文的命名规则与英文是不同的）。
- 在配离子中，配体的命名总是在金属离子的前面（其顺序正好与写分子式的顺序相反）。
- 在配体的命名中，带负电的配体名称常用“o”结尾（chloro ⟶ Cl^-；cyano ⟶ CN^-，hydrido ⟶ H^-），中性配体一般不做修改（methylamine→$MeNH_2$）。

表 1.5 一些配体和简单配合物命名的规则

配 体

1．以“-o”结尾的阴离子配体

F^-	fluoro	NO_2^-	nitro	SO_3^{2-}	sulfito	OH^-	hydroxo
Cl^-	chloro	ONO^-	nitrito	SO_4^{2-}	sulfato	CN^-	cyano
Br^-	bromo	NO_3^-	nitrato	$S_2O_3^{2-}$	thiosulfato	NC^-	isocyano
I^-	iodo	CO_3^{2-}	carbonato	ClO_3^-	chlorato	SCN^-	thiocyanato
O^{2-}	oxo	$C_2O_4^{2-}$	oxalato	CH_3COO^-	acetato	NCS^-	isothiocyanato

2．以中性分子命名的中性配体

C_2H_4	ethylene	$(C_6H_5)_3P$	triphenylphosphine
$NH_2CH_2CH_2NH_2$	ethylenediamine	CH_3NH_2	methylamine

3．四种特殊名称的中性配体

H_2O	aqua	NH_3	ammine	CO	carbonyl	NO	nitrosyl

4．以“-ium”结尾的阳离子配体

$NH_2NH_3^+$ hydrazinium

5．两可配体

（a）两种形式有各自的名称，如：— NO_2^- 为 nitro、—ONO^-为 nitrito

（b）在配体的名称前加上配位原子的符号，如：—SCN^-为 *S*-thiocyanato 和—NCS^-为 *N*-thiocyanato

6．桥联配体在配体的名称前加 μ-

简单配合物

1．先命名阳离子，再命名阴离子

2．配体按字母顺序排列

3．配体个数（2，3，4，5，6）的使用规则

（a）前缀 di-、tri-、tetra-、penta-、hexa-用于：

① 所有单原子配体；

② 具有较短名称的多原子配体；

③ 具有特殊名称的中性配体。

（b）前缀 bis-、tris-、tetrakis-、pentakis-、hexakis-用于：

① 名称中已含有第一种类型前缀（di-，tri-等）的配体；

② 不具有特殊名称的中性配体；

③ 具有相当长名称的离子配体。

4．如果配合物是阴离子，金属的名称要加后缀-ate（有时正常名称中的-ium 或其他的后缀要去掉，再加-ate 后缀。有些金属，如：铜，铁，金和银等，由金属的拉丁语名称分别变成：cuprate，ferrate，aurate 和 argentate）

5．中心金属的氧化态以括号中的罗马字母表示

特殊配体的命名如下：

水（water）→aqua

氨（ammonia）→ammine

一氧化碳（carbon monoxide）→羰基（carbonyl）

一氧化氮（nitrogen oxide）→亚硝酰基（nitrosyl）

负离子的名称以“e”结尾的，在形成配离子时，需要“e”→“o”，这样的阴离子配体的例子如下：

溴离子（Br^-）→bromo

氯离子（Cl^-）→chloro

氢氧根离子（OH^-）→hydroxo

氧离子（O^{2-}）→oxo

过氧离子（O_2^{2-}）→peroxo

氰基阴离子（CN^-）→cyano

叠氮阴离子（N_3^-）→axido

氨阴离子（NH_2^-）→amido

碳酸根阴离子（CO_3^{2-}）→carbonato

亚硝酸盐或硝基阴离子（NO_2^-）→nitrito（M—O bond）or nitro（M—N bond）

硝酸根阴离子（NO_3^-）→nitrato （when bonded through N）

硫阴离子（S^{2-}）→sulfido

硫氰根阴离子（SCN^-）→*S*-thiocyanato

硫氰根阴离子（NCS^-）→*N*-thiocyanato

草酸根阴离子（$C_2O_4^{2-}$）→oxalato

硫酸根阴离子（SO_4^{2-}）→sulfato

硫代硫酸根阴离子（$S_2O_3^{2-}$）→thiosulfato

四乙酸乙二胺阴离子$[CH_2—N(CH_2COO^-)_2]_2$（EDTA）

- 希腊语中的前缀（mono，di，tri，tetra，penta，hexa 等）常常表示每一种配体的数目（对于一种给定的单一配体而言，“mono”常被省略）。如果配体名称本身包含有 mono、di、tri 等，例如，三苯基膦（triphenylphosphine），则该名称应放在括号中，并用另一种前缀（bis，tris，tetrakis）来表示配体的数目。

例如，$Ni(PPh_3)_2Cl_2$ 被命名为二氯• 二（三苯基膦）合镍（Ⅱ）。希腊语的前缀 mono-，di-（或 bis），tri-，tetra-，penta-，hexa，hepta-，octa-，nona-，（ennea-），deca 对应于 1，2，3，…，10 等。

- 括号中的罗马数字或零用来表示中心离子的氧化态。
- 如果配合物是一个阴离子，在金属名字后加后缀-ate。例如：

钪，Sc=scandate

钛，Ti=titanate

钒，V=vanadate

铬，Cr=chromate

锰，Mn=manganate

铁，Fe=ferrate

钴，Co=cobalatate

镍，Ni=nickelate

铜，Cu=cuprate

锌，Zn=zincate

- 若在配合物或配离子中有多于一种的配体，命名时将根据配体名称的字母顺序而不考虑配体的数量，例如，氨 NH_3（ammine）将在氯 Cl^-（chloro）的前面（这是 1971 年的规则，不同于 1951 年的规则，有些教科书中仍然以中性配体在前、配阴离子在后的顺序来命名）。
- 一些附加说明

（a）一些金属在配阴离子中具有特殊的名称，如 B→Borate，Au→Aurate，Ag→Argentate，Pb→Plumbate，Sn→Stannate。

（b）用方括号来标注配离子或中性配合物，如$[Co(en)_3]Cl_3$，$[Co(NH_3)_3(NO_2)_3]$，$K_2[CoCl_4]$。需要注意的是，卤离子不需要用括号括起来。还有一些名称是不变的，如

C_5H_5N，pyridine

$NH_2CH_2CH_2NH_2$，ethylenediamine

$C_5H_4N—C_5H_4N$，dipyridyl

$P(C_6H_5)_3$，triphenylphosphine

$NH_2CH_2CH_2NHCH_2CH_2NH_2$，diethylenetriamine

（c）对于桥联配体，通常在配体名称前面加希腊字母“μ”。所以桥联的氢氧化物（OH^-）、氨基化合物（NH_2^-）和过氧化物（O^{2-}）配体分别写成μ-hydroxo，μ-amido 或 μ-peroxo。如果含有一个以上的相同桥联配体，那么表明配体数的前缀加在μ后面。例如，如果含有两个桥联氯配体，则表示为μ-dichloro。例如，$(H_3N)_3Co(OH)_3Co(NH_3)_3$被命名为$\mu_3$-羟基•三氨合钴(Ⅲ)。

（d）对于两可配体的命名有两种方法。一是在配体名字中有微小的区别，这个区别取决于贡献电子对给金属的那个原子。二是在配体名字前面标出贡献电子的原子。所以，—SCN可以被称为 thiocyanato 或 *S*-thiocyanato，而—NCS 称为 isothiocyanato 或 *N*-thiocyanato。但是，—NO_2^-和—ONO 通常分别被称为硝基（nitro）和亚硝酸盐（nitrito）。

顺便说一下，化学式的书写规则是由国际纯粹及应用化学联合会（IUPAC）规定的。该规则中配合物的配体书写顺序是相当复杂的，一般教科书中的做法都达不到这种水平。

1.4 配合物的同分异构体

1.4.1 异构体的定义

分子式相同而原子的空间排列不同的化合物称为异构体。

Werner 曾预言配合物$[Co(en)_2ClNH_3]^{2+}$存在两种形式，它们彼此互为镜像。Werner 分离出了这两种形式固体，并且结构研究证实了他的结论。

一般流行的观点认为光学活性是对碳原子而言的。Werner 认为有必要制备出无机配合物（即非碳原子）来证明“非碳的无机化合物也存在镜像异构体”（图 1.19）。

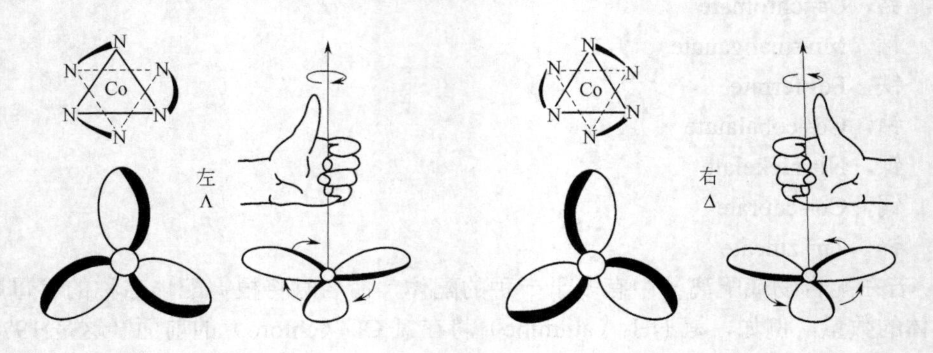

图 1.19 $[Co(en)_3]^{3+}$的绝对构型

异构体的类型主要有两种，二者又都可进一步细分。

（1）结构异构现象 a．配位异构；b．电离异构；c．水合异构；d．连接异构。

（2）立体异构现象　a．几何异构；b．光学异构。

1.4.2　结构异构体

这类异构体的差别在于原子连接的方式不同（具有不同的键）。下面说明几种在配位化学中经常遇到的这类异构体。

（1）配位异构化　配合物含有配合的阴离子和阳离子，这种异构体可以认为是一些从阴离子到阳离子间随机相互交换配体而形成的。

一个异构体　　$[Co(NH_3)_6][Cr(C_2O_4)_3]$

另一个异构体　　$[Co(C_2O_4)_3][Cr(NH_3)_6]$

（2）离子化异构体　与金属键合的阴离子不同，这类异构体可以认为是溶液中不同的离子随机地与金属配位而形成的。

一个异构体　　$[Co(NH_3)_5Br]Cl$　　Cl^-是溶液中的阴离子

另一个异构体　　$[Co(NH_3)_5Cl]Br$　　Br^-是溶液中的阴离子

注意：对于溶液中的电荷平衡而言，两种阴离子都是十分重要的。它们的区别在于一种离子是直接与中心金属配位，而另一种则不是。另一种非常类似的异构化现象是源于溶剂分子对配位基团的置换（溶剂异构化）。若溶剂是水分子的话，则称作水合异构化。

（3）水合异构化　这种异构化作用中最著名的离子是氯化铬“$CrCl_3 \cdot 6H_2O$”，它可能含有 4 个、5 个或 6 个配位的水分子。

$[Cr(H_2O)_4Cl_2]Cl \cdot 2H_2O$　　鲜绿色

$[Cr(H_2O)_5Cl]Cl_2 \cdot H_2O$　　灰绿色

$[Cr(H_2O)_6]Cl_3$　　紫罗兰色

这些异构体的化学性质差别很大，在与硝酸银反应测试氯离子的实验中，可发现溶液中分别有 1 个、2 个和 3 个氯离子。

（4）连接异构体　这类异构体中，与金属键合的配体的配位原子不同。连接异构化现象发现在两可型配体的配合物中。这些两可型配体能够以一种以上的方式配位。最为著名的例子是含有单齿的配体 SCN^-/NCS^-和 NO_2^-/ONO^-。

例如：$[Co(NH_3)_5ONO]Cl$（氧配位硝基异构体）和$[Co(NH_3)_5NO_2]Cl$（氮配位硝基异构体）。

1.4.3　立体异构体

立体异构体有着同样的原子、同样的化学键，但是这些化学键在空间的相对取向是不同的。忽略某些包含非常复杂的配体的配合物的特殊情况，那么有两种异构现象。

（1）几何异构体（构型异构体）　这种构型异构体对于平面四边形配合物和八面体形配合物是可能的，但对于四面体形配合物是不可能的。例如，顺式和反式$[Pt(NH_3)_2Cl_2]$。顺式和反式是指两个基团相对位置的不同。在顺式异构体中，它们是彼此紧挨的，也就是说相对于中心金属离子互成 90°角；然而在反式结构中它们是“背道而驰”的，也就是说相对于中心金属离子互成 180°。

```
a
|
M····· b        a ····· M ····· b
顺式              反式
```

第一次所报道的三种几何异构体是由 Il’ya Chernyaev 在 1928 年合成和表征的[Mabcd]类型的配合物。图 1.20 所示的例子是由 Anna Gel’man 在 1948 年报道的。

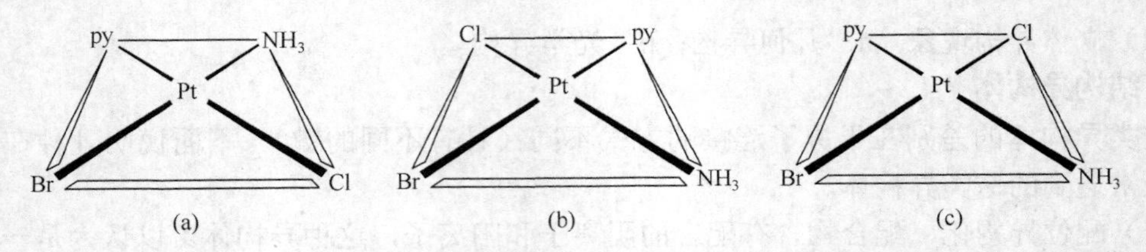

图 1.20 平面四边形配合物[$PtNH_3BrClpy$]的三种构型异构体

预期可能的平面四边形配合物的几何异构体的数目如下：

化合物类型	异构体数目
Ma_2b_2	2（顺式或反式）
Mabcd	3（用顺式和反式关系）

（这里 a、b、c、d 是指单齿配体）

已经合成和表征了一些这种类型的异构体，它们都表现出不同的化学和生物学性质。例如，顺式 $Pt(NH_3)_2Cl_2$ 是一个抗癌物质，然而反式的异构体却对癌细胞毫无活性（它是有毒的物质），在化学疗法中毫无用处（图 1.21）。

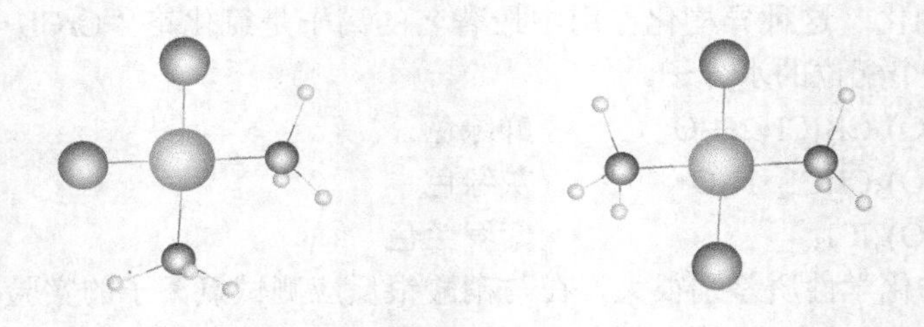

图 1.21 顺式、反式[$Pt(NH_3)_2Cl_2$]的几何异构体的结构示意图

预期可能的八面体形配合物的几何异构体的数目如下：

化合物类型	异构体数目
Ma_4b_2	2（顺式-和反式-）
Ma_3b_3	2（面向-和径向-）
$M(AA)_2b_2$	3（两种顺式-和一种反式-）

（这里 a、b 是单齿配体；AA 是双齿配体）

在第二个例子中，引用了新的标准来反映八面体结构中配体的相对位置。这样，将三个配体放在八面体的一个面上将给出面向异构体，将三个配体围绕中心放置将给出径向（meridinal）异构体。

预期[Mabcdef]将有 15 种异构体。1956 年，Anna Gel'man 合成并表征了[$Pt(NH_3)BrClpyNO_2$]可能有的 15 种异构体中的几个，这 15 种异构体都可能有光学异构体，因此，总共应有 30 种异构体。

（2）镜像异构体（光学异构体） 镜像异构体涉及非重复（non-superimposable）的镜像，它们对偏振光的旋转方向是不同的。这些异构体互称为镜像对映异构体，它们的非重复（non-superimposable）的镜像是不对称的。

光学异构体对于四面体形配合物和八面体形配合物是可能的，但对于平面四边形配合物是不可能的。虽然大多数的研究主要集中在 $M(AA)_3$ 型的光学异构体上，但顺式的 $M(AA)_2b_2$

型配合物也可能表现出光学异构现象，例如：*cis*-$[Co(en)_2Cl_2]^+$（图 1.22）。1911 年，第一个被 Werner 和 King 报道的光学异构体就是顺式$[CoNH_3(en)_2X]^{2+}$（X=Cl^-或 Br^-）。

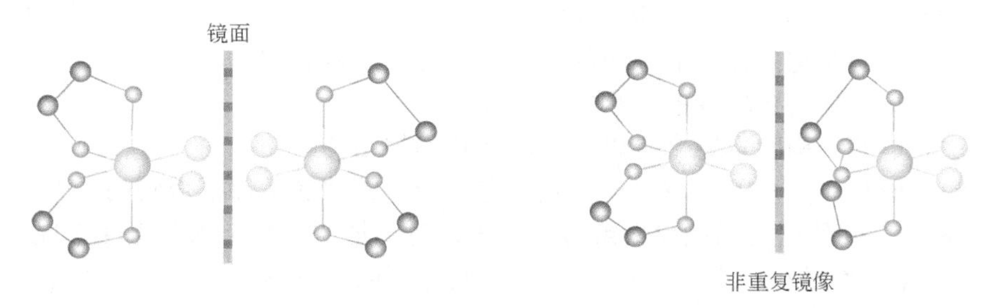

图 1.22　顺式$[Co(en)_2Cl_2]^+$的光学异构体

有许多方法已经被用来表示光学异构体的绝对构型，比如：D 或 L，*R* 或 *S*。1970 年国际纯粹与应用化学联合会定义了螺旋形化合物的概念，并指定希腊字母的第十一个字母 L 代表左旋 D，希腊字母第四个字母 D 代表右旋 L。非常类似地，我们或许可以用相同的规则来看待三叶螺旋桨类的配合物。以此来决定它们是左手或右手螺旋。

异构体具有相同的化学性质，仅仅根据它们的绝对构型不能给出任何有关它们在旋转偏振光方面的任何信息。它们是左旋还是右旋只有通过测量得到，并根据测量结果给它们加以前缀左旋的（−或 L）、右旋的（+或 D）。不提倡使用“*l*”和“*d*”，因为它们可能与 D 和 L 造成混淆。L(D)-$[Co(en)_3]^{3+}$型异构体能右旋偏振光，所以相应是（+）异构体。

注意，尽管预期具有四种不同的配体的四面体形的配合物应该有光学异构体（比较碳化学），但因它们特别易变而难以分离。

几何异构体具有不同的化学和物理性质：不同的颜色、熔点、极性、溶解度、化学反应性等。但两种光学异构体的大多数性质是相同的，如：溶解度、熔点、沸点、化学反应性（与非手性的化学试剂）等。两种光学异构体只有在与手性化学试剂和偏振光相互作用时才不同（图 1.23）。

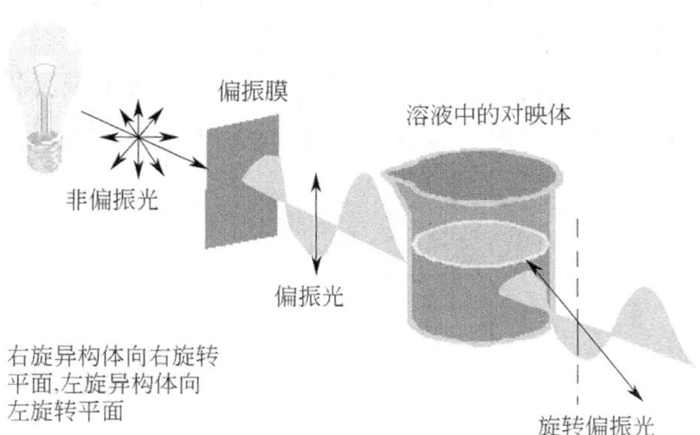

图 1.23　光学异构体与偏振光相互作用的示意图

1.4.4　超分子异构

（1）超分子的结构异构　超分子的结构异构对含有最简单的结构单元和线形桥连配体的配位聚合物的结构和性质有重要的影响。这些构筑单元可以认为是基于 T 形节点的自组装。

在该领域已经有丰富多彩的结构存在。以下指出对于这些结构需要注意的几点。

- 这些复合物并不是真正意义上的聚合物，因为在晶格内存在着客体分子和溶剂分子。然而，它们又都不是传统意义上的溶剂化物。
- 网状物结构及与之伴随的性质的丰富多彩是极其显著的。
- 这些结构在任何自然界的矿物中都不存在。
- 通过简单的结构因素完全可以预测出网状物的结构。
- 一些结构是在同样的构筑单元在几乎完全一样的结晶条件下得到的。

总之，研究表明使用合适的模板或客体分子，对于把一个给定的节点稳定地合成出各种可能的超分子异构物是有帮助的。因此，我们或许可以得到这样的结论，对于一个给定的分子基团，它能构筑的超分子结构是有限的，并且最终可以判断出每一种结构所需要的结晶条件。

（2）超分子的构象异构　像二(4-吡啶基)乙烷这样的柔性配体由于其柔性可以导致或小或大的结构变化，并由此产生一个不同却又相关的网状结构，如1,2-二(吡啶)乙烷、联吡啶就有顺式和反式两种构象。T-形节点通过线性双功能外齿配体如一维梯子形、三维林肯堆木形、二维人字形、二维双层结构、二维围墙形和三维骨架连接成的超分子异构物已经有报道了。

对于构象变化如何影响有机复合物的晶体堆积，5-甲基-2-[(2-硝基苯基)氨基]-3-噻吩三苯肼做出了一种极具戏剧性的解释。这种复合物存在至少六种形态。而这六种构象的本质区别在于噻吩和*o*-硝基苯胺片段间的扭转角在21.7°到104.7°之间变化。

（3）索烃　网状物相互贯通或交织的方式和程度的不同都可导致整个结构和功能的重大变化，当然这种变化取决于分子构筑单元。相互贯通和非相互贯通结构由于它们性质上的重大差异被认定是两种有本质区别的复合物。

令人惊奇的是，如果在网状物内存在相对大的空穴，那么独立的相互贯通的网状结构的存在也是普遍的。相互贯通结构的存在被认为是严重妨碍稳定的开放式框架产生的因素。然而，有一点是清楚的，通过使用适宜的模板可以由同一个网状物得到开放式结构或相互贯通结构。典型的钻石形和四方形刚性网状结构$Cd(CN)_2$和M（bipy）$_2X_2$是其典型例子。这两种复合物都可以得到相互贯通和非贯通式结构。而且，如果相互贯通不能为密堆积提供必要的条件，那么这种贯通结构也被认为是开放式框架结构。

（4）光学　网状物可以具备固有的手性，因此得到的晶体属于手性的（对映的）空间群。那么，对于纯手性的复合物就可以进行类推。这种超分子异构现象正是手性固体自动分解问题的核心。

第 2 章　配合物的对称性和化学键

2.1　化学中的对称性——群论

群论是量子化学和光谱学中最有用的一种数学工具。它可以让使用者预测、解释以及使理论合理化，并且常会简化复杂的理论和数据。它的中心思想是：与分子对称元素相关的一系列操作构成了数学上的一个集合，叫做群。这样，就可以将与这些群有关的数学原理应用于对称操作。

2.1.1　对称元素

与一个分子有关的对称元素有以下几个。

（1）对称轴（C_n）　n=1，2，…，说明绕该轴具有 n 重旋转对称性。

（2）反映镜面（σ）　关于平面左右对称。这种平面可进一步分为：σ_h 是垂直于主轴（轴的最高值为 n）；如果没有主轴存在，σ_h 可定义为分子所在平面。σ_v 或 σ_d 是包含了旋转主轴并且垂直于 σ_h 平面的垂直平面。当 σ_v 或 σ_d 平面都存在时，σ_v 平面包含了最多的原子。σ_d 平面包含了角平分线。如果仅有一种类型的垂直平面存在，σ_v 或 σ_d 的运用将取决于分子的总对称性。

（3）反演中心（I）　所有的对称轴和对称面都一定通过的中心点。如果没有这样的共同点存在就没有对称中心。

（4）反映轴（非真轴）（S_n）　这由两部分组成：C_n 和 σ_h，它们可能是也可能不是分子的真正对称元素。若 C_n 和 σ_h 都存在，那么 S_n 一定存在。以下的关系在这一点上是有帮助的：

- 若 n 为偶数，则 S_n^m =E（当 m=n 时，上标 m 表示操作的次数），同时存在 $C_{n/2}$ 轴；
- 若 n 为奇数，则 S_n^m =s（m=n 时）；S_n^{2m} =E（m=n 时）；同时，C_n 和垂直于 C_n 的 σ 都存在。

（5）恒等操作（E）　每一个分子都至少拥有一个对称元素：恒等操作。恒等操作意味着对分子没做任何改变。

2.1.2　对称操作

所有与独立分子相关的对称操作都具有旋转的特征。

（1）真旋转　C_n^k；k=1，…，n。当 k=n 时，C_n^k =E，恒等操作，旋转了 360°/n，n=1，…。

（2）非真旋转　S_n^k；k=1，…，n。当 k=1、n=1 时，S_n^k =s，反映操作；当 k=1、n=2 时，S_n^k =i，反演操作。

通常情况下，我们要区分以下五种操作：

- E，恒等操作；
- C_n^k，绕某真轴的转动；
- σ，镜面反射；
- i，中心衍射；
- S_n^k，先绕某真轴转动后再通过垂直于该轴的平面进行反映。

每一个对称操作都与一个对称元素有关，该对称元素可以是一个点、一条线或一个平面。操作前后，分子的方向和位置都没有变化。

2.1.3 分子点群

分子可分为以下几类：线形分子、平面形分子和非平面形分子。知道了分子的对称元素，我们可以用图 2.1 来确定分子的点群。

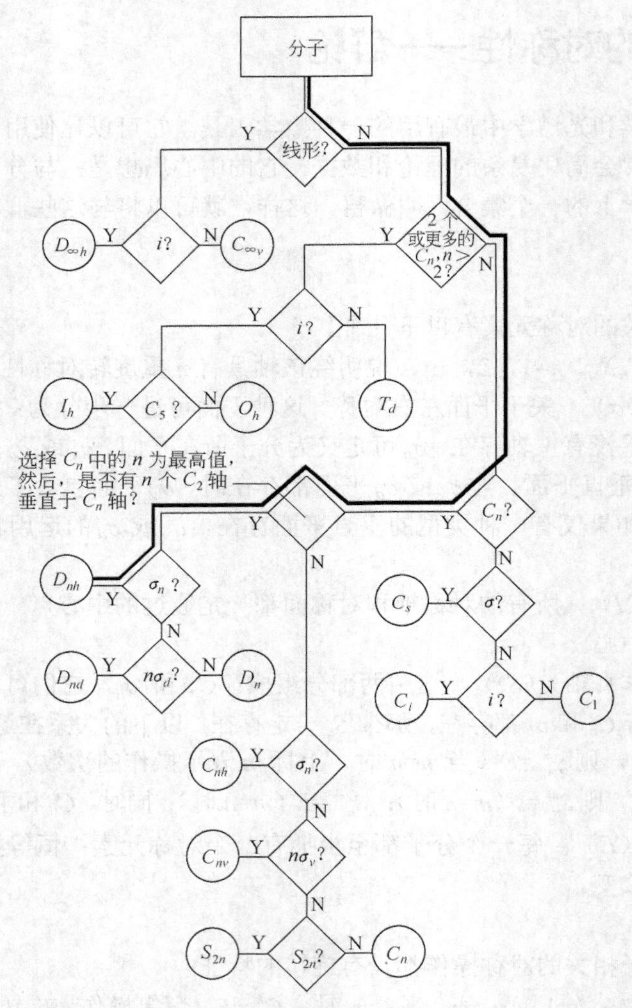

图 2.1 确定分子点群的流程图

（1）线形分子　我们讨论两种线形分子：一氯乙炔和乙炔。

● 所有的线形分子都拥有一条与所有原子中心形成一线的轴，记为 C_∞，因为不管绕轴旋转多少度，都会产生相同的结构（图 2.2）。

● 除了上述对称元素之外，线形分子可能还有一个对称中心 i，如果分子有对称中心，它还会有：

通过这个中心且垂直于主轴（C_∞）的无数的 C_2 轴。

一条非真轴 S_∞，先绕 C_∞ 轴旋转，然后再作通过对称中心且垂直于 C_∞ 轴的平面的反映。

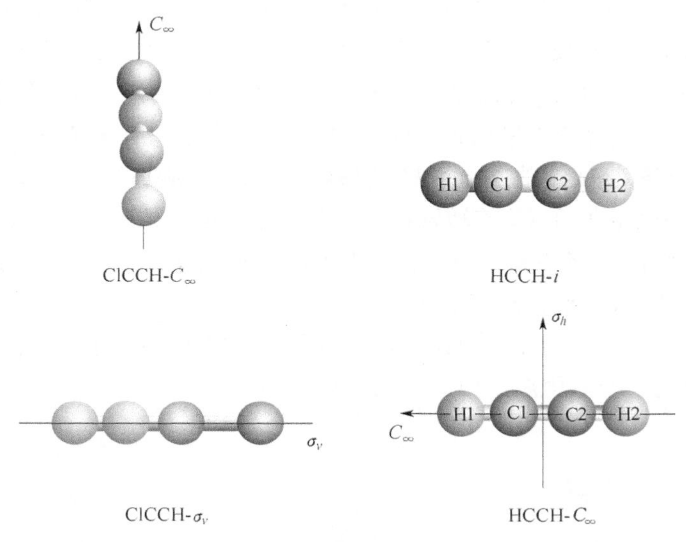

图 2.2　线形分子的真转和非真轴

（2）平面形分子　这类分子的一个重要特征是所有原子在一个共同的平面上。我们主要讨论以下四种分子：甲醛，反式乙二醛，乙烯，苯（图 2.3）。

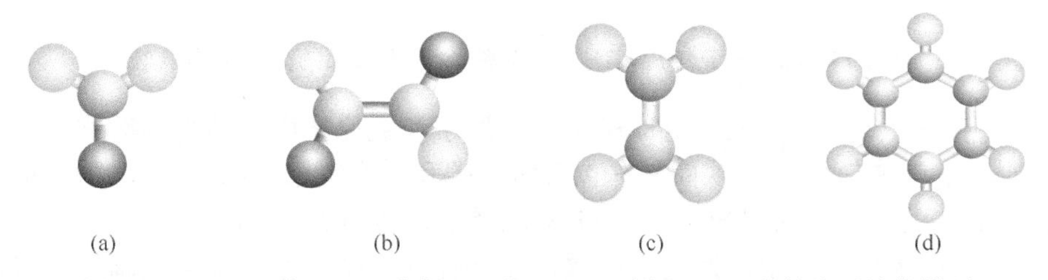

图 2.3　（a）甲醛，（b）反式乙二醛，（c）乙烯和（d）苯的分子结构模型

- 甲醛　首先分子中存在一个二次旋转轴 C_2。它通过碳原子和氧原子且平分 HCH 键角。另外还包含一个 C_2 轴的对称平面，即 σ_v平面；一个对称面——垂直于分子平面、平分 HCH 键角且包含 C_2 轴，也是一个 σ_v平面。甲醛分子没有对称中心，没有反映轴。所以甲醛分子的对称元素有：一条 C_2 轴和两个 σ_v平面。
- 反式乙二醛　分子拥有对称中心 i，它还有一条穿过对称中心且垂直于分子平面的 C_2 轴。既然分子平面垂直于 C_2 轴（主轴），它就拥有一个 σ_h 平面（水平），没有反映轴。反式乙二醛的对称元素有：一条 C_2 轴，一个 σ_h 平面和一个反演中心 i。
- 乙烯　它也有一个对称中心 i；有三条通过对称中心且互相垂直的 C_2 轴。与这三条轴相对应的是三个互相垂直的 σ_h 平面（每一个都垂直于一条 C_2 轴），没有反映轴。乙烯的对称元素是：三条 C_2 轴，三个 σ_h 平面和一个反演中心 i。
- 苯　苯具有对称中心 i，它还具有通过对称中心且垂直于分子平面的旋转轴。这些轴包括一条 C_6 轴、一条 C_3 轴和一条 C_2 轴。在分子平面上有两组 C_2 轴。

（a）三条 C_2' 轴通过对称中心且包含两个 C—H 键。

（b）三条 C_2'' 轴通过对称中心和六边形中相对的两个键的中心。苯还有一个垂直于 C_6 旋转轴的对称平面即一个 σ_h 平面。它还有对应于三条 C_2' 轴的三个包含 C_6 旋转轴和对称中心 i 的 σ_v 平面，另外还有三个对应于三条 C_2'' 轴的包含 C_6 旋转轴和对称中心 i 的 σ_v 平面。总之，因为存在垂直于分子平面的 C_6 旋转轴、C_3 旋转轴和 C_2 旋转轴，所以也存在 S_6 轴、S_3 轴和一个 S_2 轴，不过因为 S_2 轴等价于早已确定的对称中心 i，所以它有无数条。苯的对称元素有：一个对称中心 i；一条 C_6 旋转轴；一条 C_3 旋转轴；一条 C_2 旋转轴；三条在分子平面内的 C_2' 旋转轴；三条在分子平面内的 C_2'' 旋转轴；一个与分子平面一致的 σ_h 反映面；三个包含三条 C_6 旋转轴的 σ_v 反映面；三个包含三条 C_6 旋转轴的 σ_d 反映面；一个与 C_6 旋转轴一致的 S_6 反映轴；一个与 C_3 旋转轴一致的 S_3 反映轴。

（3）非平面分子　这里的非平面只是意味着所有的原子不在同一个平面上，即分子是三维的。我们主要讨论：一氯甲烷，甲烷，丙二烯（图 2.4）。

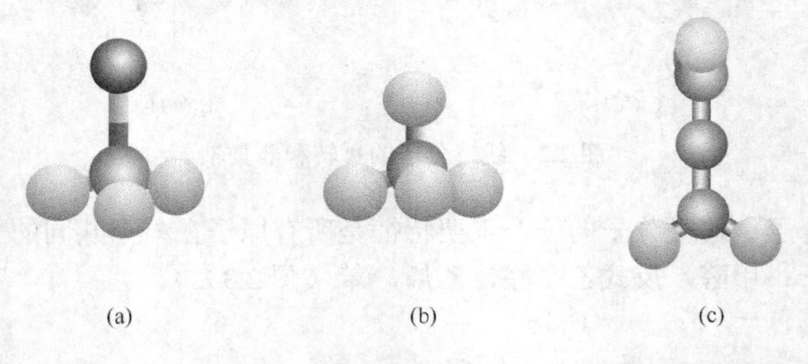

图 2.4　（a）一氯甲烷，（b）甲烷和（c）丙二烯的分子结构模型

- 一氯甲烷　此分子中三个氢原子关于碳原子和氯原子所在的直线对称分布，即存在 C_3 轴。另外，还存在三个 σ_v 反映面——包含氯、碳和一个氢原子以及 C_3 轴的平面。一氯甲烷的对称元素有：一个 C_3 旋转轴和三个包含 C_3 旋转轴的 σ_v 反映面。
- 甲烷　在甲烷分子中，通过中心碳原子和任何一个氢原子都可以画一条直线，那么剩下的三个氢原子就会关于这条直线采取对称分布，即存在一个 C_3 轴，所以它会有四个 C_3 轴。包含碳原子和任两个氢原子的平面会平分其他两个氢原子与碳原子的 HCH 键的键角。这样就会得到以下四个平面：CH1H2，CH1H3，CH1H4，CH2H3，CH2H4，CH3H4。每一个平面都包含了 C_3 轴，都是 σ_d 反映面[图 2.5（a）]。如果将甲烷分子画在一个立方体内，且每个氢原子都在一个顶点上[图 2.5（b）]，就会确定三条互相垂直的 C_2 轴。

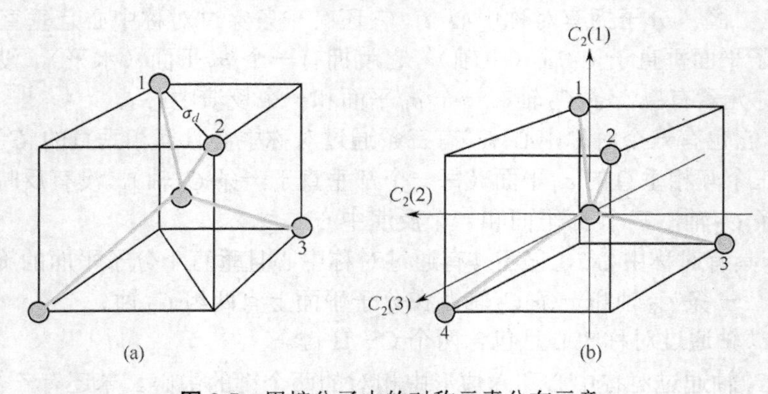

图 2.5　甲烷分子中的对称元素分布示意

用同样的模型，我们会看到先绕任一平面的法线旋转 90°，然后再对垂直于旋转轴的平面做反映操作，就会得到相同的结构，即有三个 S_4 反映轴。在它们中旋转轴和反映面都不是对称元素。甲烷的对称元素有：四条 C_3 旋转轴，三条 C_2 旋转轴，三条 S_4 反映轴和六个 σ_d 反映面（图 2.5）。

- 丙二烯　这种分子画在一个长方形中会很容易研究，其中四个氢原子位于长方形的四个顶点。经过三个碳原子的直线就是一条 C_2 轴。它有两条垂直于 C_2 轴且通过中心碳原子的 C_2'轴。它有两个对称平面——包含两个氢原子且平分剩下的一对氢原子与碳原子的键角。这两个平面都包含了 C_2 轴，都是 σ_d 反映面。总之，它有对应于 C_2 轴的 S_4 反映轴，它的反映平面垂直于 C_2 轴且通过中心碳原子。丙二烯的对称元素有：一条 C_2 旋转轴，两条垂直于 C_2 轴的 C_2'旋转轴，两个对称的 σ_d 平面，一条对应于 C_2 轴的 S_4 反映轴。一些重要的分子点群见表 2.1。

表 2.1　一些重要的分子点群

非轴群	C_1	C_s	C_i	—	—	—	—
C_n 群	C_2	C_3	C_4	C_5	C_6	C_7	C_8
D_n 群	D_2	D_3	D_4	D_5	D_6	D_7	D_8
C_{nv}群	C_{2v}	C_{3v}	C_{4v}	C_{5v}	C_{6v}	C_{7v}	C_{8v}
C_{nh} 群	C_{2h}	C_{3h}	C_{4h}	C_{5h}	C_{6h}	—	—
D_{nh} 群	D_{2h}	D_{3h}	D_{4h}	D_{5h}	D_{6h}	D_{7h}	D_{8h}
D_{nd} 群	D_{2d}	D_{3d}	D_{4d}	D_{5d}	D_{6d}	D_{7d}	D_{8d}
S_n 群	S_2	S_4	S_6	S_8	S_{10}	S_{12}	—
立体群	T	T_h	T_d	O	O_h	I	I_h
线性群	$C_{\infty v}$	$D_{\infty h}$	—	—	—	—	—

2.1.4　特征标表

所有的特征标表都具有下列格式：

- 首行和第一列分别是对称操作和不可约表示；
- 表的内容就是所有的特征标；
- 最后两栏是关于笛卡尔坐标的一阶和二阶组合。

在点群中经常发生相同类型的操作不是等价操作。例如，在 D_{2d} 点群中的三个 C_2 操作，它们中的两个与第三个就不是等价操作。在这种例子中，可以用“撇”或者标记一些笛卡尔坐标（像 D_2 中与 C_2 操作有关的 x，y 和 z）来区别（表 2.2，表 2.3）。

表 2.2　C_2 点群的特征标表

C_2	E	C_2	线性函数，旋转	二次函数	立方函数
A	+1	+1	z，R_z	x^2，y^2，z^2，xy	z^3，xyz，y^2z，x^2z
B	+1	−1	x，y，R_x，R_y	yz，xz	xz^2，yz^2，x^2y，xy^2，x^3，y^3

表 2.3　D_2 点群的特征标表

D_2	E	C_2（z）	C_2（y）	C_2（x）	线性函数，旋转	二次函数	立方函数
A	+1	+1	+1	+1	—	x^2，y^2，z^2	xyz
B_1	+1	+1	−1	−1	z，R_z	xy	z^3，y^2z，x^2z
B_2	+1	−1	+1	−1	y，R_y	xz	yz^2，x^2y，y^3
B_3	+1	−1	−1	+1	x，R_x	yz	xz^2，xy^2，x^3

2.2 价键理论

一定数量的原子轨道的混合会产生相同数量的杂化轨道，运用杂化轨道的概念来解释化学成键、分子的形状和结构是化学界近年来的事情。20 世纪前半段，最重大的进步是人类具备了认识原子和分子结构的能力。量化计算使我们能够看到原子和分子轨道的层面。另一方面，用常见的材料也可以制造出漂亮的原子杂化轨道的模型。这节内容包括以下几个方面：

- 用化学成键的方法描述共价键；
- 用杂化原子轨道描述共价键；
- 分子形状与中心原子的杂化轨道的关系；
- 结合杂化轨道、共价理论、价层电子对互斥理论、共振结构和八隅体定则说明一些常见分子的形状和结构。

共价键的方法认为参与成键的原子通过原子轨道的重叠来形成化学键。由于轨道的重叠，电子被限定在成键区域。重叠的原子轨道可分为不同的类型，例如，σ 键可以通过重叠下面的原子轨道来形成（表 2.4）。

表 2.4　原子轨道重叠形成的化学键

s-s	s-p	s-d	p-p	p-d	d-d
H—H	H—C	H—Pd 在	C—C	F—S	Fe—Fe
Li—H	H—N	Pd 的氢化物中	P—P	在 SF_6 中	
	H—F		S—S		

然而，成键的原子轨道可能不是直接解薛定谔方程得到的“纯”的原子轨道。通常，成键的原子轨道具有几种可能轨道类型的特征。这种组合出具有合适成键特征的原子轨道的方法叫杂化。得到的原子轨道叫做杂化原子轨道或简称杂化轨道。我们首先来看一些杂化轨道的形状，因为它们的形状决定了分子的形状。

2.2.1 原子轨道的杂化

通过解薛定锷方程可以得到下面原子轨道的波函数：1s，2s，2p，3s，3p，3d，4s，4p，4d，4f 等。

当原子具有两个或更多的电子时，它的能级相对于氢原子中的能级会发生位移。原子轨道实际上是束缚在原子核周围的电子的能量状态。当一个原子与另一个键合时，它的能态会改变。

量子力学的方法通过组合波函数给出新的波函数——杂化原子轨道。杂化轨道有一个确定的波函数，但在这儿详细地列出这一波函数显得有点复杂。因此，我们仅讨论下面一组杂化原子轨道（或杂化轨道）：sp；sp^2；sp^3；sp^3d；sp^3d^2。

（1）sp 杂化轨道　sp 杂化轨道是原子中电子的可能状态，尤其是与其他原子键合时。这些电子的状态具有一半 2s 和一半 2p 的特征。从数学的观点来看，2s 和 2p 轨道有两种可能的组合方式：

$$sp_1=2s+2p；\quad sp_2=2s-2p$$

这些能量状态（sp_1 和 sp_2）都是电子出现概率很高的区域，这两个原子轨道以头对头的方式定位在两个原子的中间。sp 杂化轨道可用图 2.6 来描述。

例如，H—Be—H 分子就是两个氢原子的 1s 轨道和 Be 原子的两个 sp 杂化轨道叠加形成的。所以，H—Be—H 是线状的。

Be 原子基态的电子构型是 $1s^22s^2$，我们可以认为成键前电子构型是 $1s^2[sp]^2$，sp 杂化轨道中的两个电子具有相同的能量。成键的杂化原子轨道的概念可能仅仅是为了解释 Cl—Be—Cl 分子的结构提出的一个假说。通常，当两个且只有两个原子与第三个原子键合时，第三个原子会利用 sp 杂化轨道，这三个原子在一条直线上。

（2）sp^2 杂化轨道　第二周期中原子的价电子的能态在 2s 和 2p 轨道上。如果我们混合两个 2p 轨道和一个 2s 轨道，就能得到三个 sp^2 杂化轨道。这三个轨道在同一平面上，如图 2.7 所示，它们指向等边三角形的三个顶点。

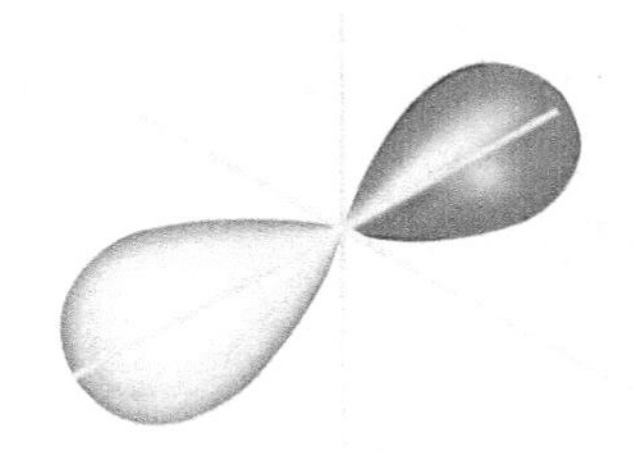

图 2.6　sp 杂化轨道的示意图

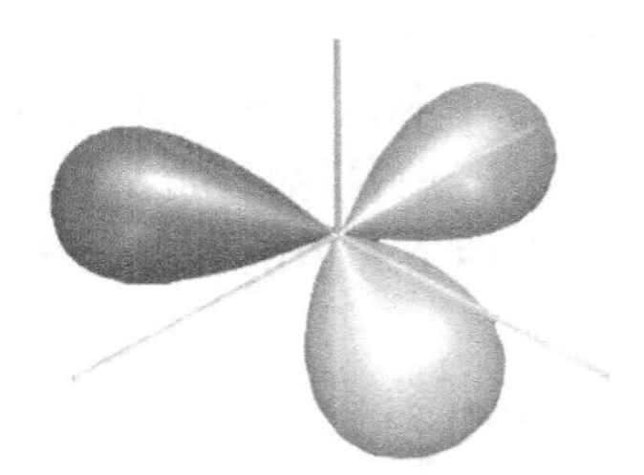

图 2.7　sp^2 杂化轨道的示意图

当中心原子利用 sp^2 杂化轨道时，形成的化合物会具有三角形的结构，BF_3 就是这样的分子（表 2.5）。

表 2.5　具有 sp^2 杂化轨道的分子

F—B(—F)—F	[:Ö:—C(—:Ö:)=O]$^{2-}$	O=Ṅ—O	O=Ö—O	O=S̈—O

并非所有的 sp^2 杂化轨道都参与成键。一个轨道可能被一对孤对电子或一个单电子占据。如果不考虑未共享的电子，分子就是弯的而不是直线的。上面与 BF_3 列在一起的三个分子就是这样的分子。在化合物 $H_2C═CH_2$ 中碳原子就利用了 sp^2 杂化轨道。在这个分子中，碳原子中剩下的 p 轨道重叠形成了额外的 π-π 键。其他的离子如 CO_3^{2-} 和 NO_3^- 也可以用相同的方式解释（表 2.6）。

表 2.6　具有 sp^2 杂化轨道的平面形分子

H(H)C=C(H)H	[O—C(—O)=O]$^{2-}$	[O—N(—O)=O]$^-$

（3）sp^3 杂化轨道　混合一个 s 轨道和所有的三个 p 轨道形成四个相同的 sp^3 杂化轨道。四个 sp^3 杂化轨道指向四面体的四个顶点，如图 2.8 所示。

当中心原子利用 sp^3 杂化轨道形成分子时，分子具有四面体的形状。典型的分子是 CH_4，此分子中氢原子的 1s 轨道与一个 sp^3 杂化轨道叠加形成一个 C—H 键。四个氢原子形成四个这样的键，它们都是等同的。CH_4 是四面体形分子中最常被提到的分子。其他具有四面体形

的分子和离子有 SiO_4^{4-}、SO_4^{2-}。

像 sp^2 杂化轨道一样，sp^3 杂化轨道中的一个或两个轨道可能被非键电子占据。水和氨就是这样的分子。CH_4、NH_3、H_2O 中的 C、N 和 O 原子利用了 sp^3 杂化轨道，然而，在氨中是一对孤对电子占据一个 sp^3 杂化轨道，在 H_2O 中是两对孤对电子占据两个 sp^3 杂化轨道。孤对电子可以用价层电子对互斥模型（VSEPR）来考虑，我们可以用 E 代表一对孤对电子，用 E_2 代表两对孤对电子，这样，我们就会分别得到 NH_3E 和 OH_2E_2。

价层电子对的数目等于成键电子对数加上孤对电子数，与键级无关。每一成键的电子对被认为是一个键。对 CH_4、:NH_3、:$\ddot{O}H_2$ 来说，价层电子对都是 4。根据价层电子对互斥理论，孤对电子要占据更大的空间，H—O—H 键角是 105°，小于理想四面体中的 109.5°的键角。

（4）dsp^3 杂化轨道　当一个 3d 轨道、一个 4s 轨道和三个 4p 轨道混合时就会产生 dsp^3 轨道。当一个原子利用 dsp^3 轨道与其他五个原子键合时，分子的几何构型通常是三角双锥形。例如，$PClF_4$ 分子就是这样的结构。在图 2.9 中，Cl 原子位于三角双锥的轴上。在有的结构中 Cl 原子占据图中赤道的位置。排列的改变是通过键角的变化实现的。

图 2.8　sp^3 杂化轨道的示意图

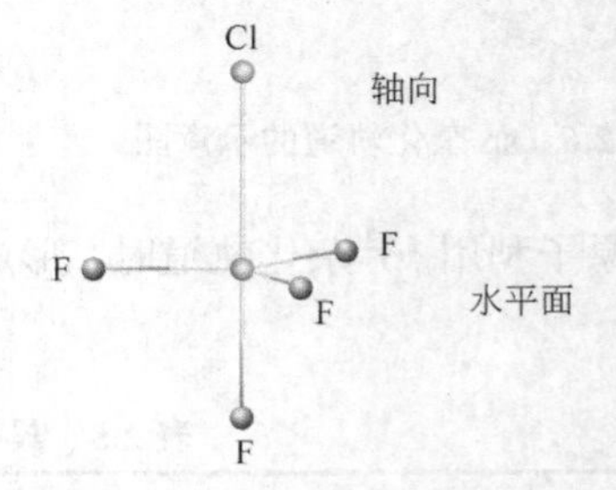

图 2.9　以 $PClF_4$ 分子为例的 dsp^3 杂化轨道的示意图

一些 dsp^3 也可能被电子对占据。这种分子的形状比较有趣。在 $TeCl_4$ 分子中，只有一个 dsp^3 轨道被一对孤对电子占据。这种结构可以用 $TeCl_4E$ 表示，这里 E 代表一对孤对电子。在 BrF_3 中有两对孤对电子占据这样的轨道，可表示为 BrF_3E_2。化合物 SF_4 是 AX_4E 类型的分子，许多混卤素化合物，如 ClF_3 和 IF_3，是 AX_3E_2 型的分子，I_3^- 离子具有 AX_2E_3 型的结构（图 2.10）。

（5）d^2sp^3 杂化轨道　当两个 3d 轨道，一个 4s 轨道和三个 4p 轨道混合时就会产生六个 d^2sp^3 杂化轨道。当一个原子用六个 d^2sp^3 杂化轨道与六个其他的原子键合时形成的分子会具有八面体的形状。气态化合物 SF_6 就是一个典型的这种结构（图 2.11）。

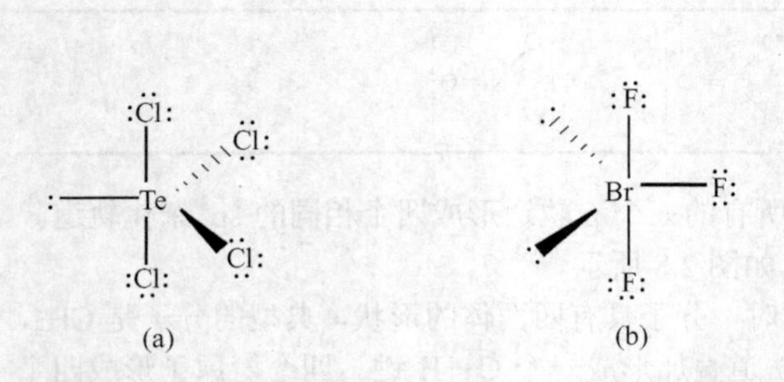

图 2.10　（a）$TeCl_4E$ 分子和（b）BrF_3E_2 分子的结构模型（E 代表孤对电子）

图 2.11　d^2sp^3 杂化轨道的示意图

在 d^2sp^3 杂化轨道中也有轨道被孤对电子占据的情况，这会产生下面几种结构：

AX_6，AX_5E，AX_4E_2，AX_3E_3，AX_2E_4，IOF_5，IF_5E，XeF_4E_2

2.2.2　分子形状

当引入杂化轨道的概念后，价层电子对互斥模型可用于进一步讨论各种分子的形状。在价层电子对互斥理论中，孤对电子和成键电子都作为价层电子。单键、双键和叁键都作为一对价层电子。即便综合考虑杂化轨道和价层电子对互斥理论，有时仅基于价层电子数也不能找到一个系统的方法去合理说明许多分子的形状。

表 2.7 简要说明了杂化轨道、价键理论、价层电子对互斥理论和八隅体规则的概念。表中，用线形、平面三角形、四面体形、三角双锥形、八面体形来描述分子的几何形状。使用的杂化轨道是：sp，sp^2，sp^3，dsp^3 和 d^2sp^3。每一组中，价层电子数都是相等的。我们不用 NH_3E 和 OH_2E_2 而用:NH_3、:$\ddot{O}H_2$ 来强调孤对电子。

表 2.7　杂化轨道、价键理论、VSEPR、共振结构和八面体的总结

线形	平面三角形	四面体形	三角双锥形	八面体形
sp	sp^2	sp^3	dsp^3	d^2sp^3
BeH_2	BH_3	CH_4	PF_5	SF_6
BeF_2	BF_3	CF_4	PCl_5	IOF_5
CO_2	CH_2O	CCl_4	$PFCl_4$	PF_6^-
HCN	(>C=O)	CH_3Cl	:SF_4	SiF_6^{2-}
HC≡CH	>C=C<	NH_4^+	:TeF_4	:BrF_5
	CO_3^{2-}	:NH_3	::ClF_3	:IF_5
	苯	:PF_3	::BrF_3	::XeF_4
	石墨	:SOF_2	:::XeF_2	
	C_{60}	::OH_2	::: I_3^-	
	·NO_2	::SF_2	(:::I I_2^-)	
	N_3^-		::: ICl_2^-	
	:$OO_2(O_3)$	SiO_4^{4-}		
	:SO_2	PO_4^{3-}		
	SO_3	SO_4^{2-}		
		ClO_4^-		

· 一个孤电子　　　　: 一对孤电子对

在线形分子中，只涉及了 Be 和 C。在气态时，BeH_2 和 BeF_2 是能稳定存在的，但是这些分子并不符合八隅体规则。碳原子是 sp 杂化的，在线形分子中它可以形成双键或叁键。在平面三角形和四面体形的碳的化合物中，它使用的是不同的杂化轨道。在这组氮的化合物中，氮有额外的未成对电子或孤对电子。因为氧和硫有更多的电子，所以也有更多的孤对电子。这些五原子阴离子都是四面体形的，可以写出许多它们的共振结构。

三角双锥形和八面体形分子中有 5 对和 6 对价层电子。当中心原子具有多于 5 对或 6 对电子时，额外的电子就会成为孤对电子。用 Lewis 点结构表示出价电子后，很容易得到孤对电子的数目。

在描述这些分子的形状时我们常忽略孤对电子，所以·NO_2、N_3^-、:$OO_2(O_3)$和:SO_2 都是弯曲的分子，而:NH_3、:PF_3 和:SOF_2 都是类似金字塔形的。在:SF_4 中，孤对电子位于赤道位置，这种分子与我们前面提到的:TeF_4 具有相同的结构。如果画出这种分子的模型，可看出它们是蝴蝶状的。由于同样的原因，:$\ddot{C}lF_3$ 和:$\ddot{B}rF_3$ 是 T 形的，:$\ddot{X}eF_2$、:$\ddot{I}I_2^-$ 和:$\ddot{I}Cl_2^-$ 是线形的。同样，:BrF_5 和:IF_5 是四方锥形的，而:$\ddot{X}eF_4$ 是平面正方形的。

一个有趣的问题："什么原子常做中心原子?"

通常，中心原子比终端原子带较多的正电荷。然而，H 和卤原子通常在终端位置，因为它们只能形成一个键。价层电子对互斥理论可扩展到更复杂的分子。用价层电子对互斥理论可推出下面的结果。

- H—C—C 的键角等于 109°；
- H—C═C 的键角等于 120°，围绕碳的几何构型是平面三角形；
- C═C═C 的键角等于 180°，即是直线形的；
- H—N—C 的键角等于 109°，围绕氮的几何构型是四面体形；
- C—O—H 的键角等于 105°或 109°，氧原子周围有两对孤对电子。

2.3 晶体场理论

在解释配合物化学方面，晶体场理论很大程度上取代了价键理论。晶体场理论最早是物理学家 Hans Bethe 于 1929 年提出的。后来 J. H. Van Vleck 于 1935 年加以修改，在相互作用中引入了共价作用。这一修改的理论就是常被提到的配位场理论（Ligand Field Theory）。纯粹的晶体场理论认为金属离子与配体间的相互作用是纯粹的静电作用（离子作用）。配体是作为点电荷处理的。尽管有些不切实际，但考虑对称性，对配位场理论以及分子轨道理论而言都是正确的。最重要的是要正确地画出 d 轨道，才能弄清楚是哪个轨道与配体相互作用（点电荷）。

在气态金属离子中五个 d 轨道是简并的。如果在金属周围放一个带负电的球形场，这些轨道仍然保持简并，但是因为配体和 d 轨道负电荷间的相互排斥作用，会导致它们的能量升高。如果不用球形场而是用离散的点电荷与金属相互作用，简并的 d 轨道就会分开。

分裂的 d 轨道的能量是晶体场理论中的核心。

2.3.1 八面体构型的晶体场

从球形场到八面体场，相对于自由离子，所有 d 轨道的能量都升高。然而，d 轨道与分别位于+x，$-x$，+y，$-y$，+z 和$-z$ 轴上的 6 个点电荷之间的相互作用并不是完全相同的。沿轴分布的轨道（即 x^2-y^2、z^2）不如轴间的轨道（即 xy、xz、yz）稳定。参照 O_h 点群的特征标表，x^2-y^2、z^2 轨道属于 E_g 不可约表示，xy、xz、yz 轨道属于 T_{2g} 不可约表示。这两组轨道的（e_g 和 t_{2g} 用能量较低的轨道）分裂能用Δ_o 或 $10D_q$ 表示（图 2.12）。当质心从球形场向八面体场变化时，t_{2g} 轨道是稳定的，而 e_g 轨道变得不稳定。

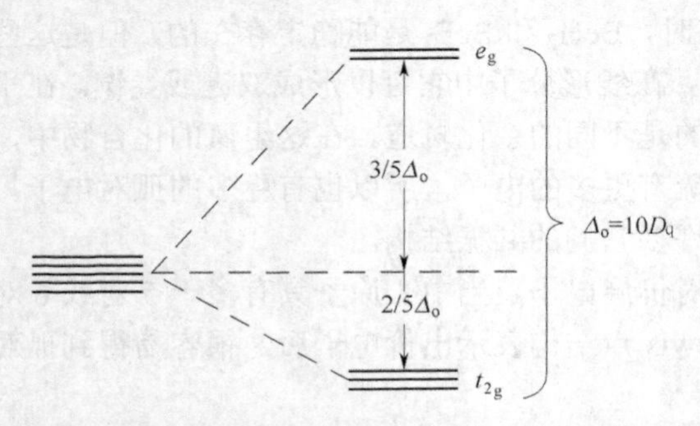

图 2.12 在八面体构型中两组轨道 t_{2g} 和 e_g 的分裂

例如：$[Ti(H_2O)_6]^{3+}$，它的轨道上只有一个 d 电子，电子占据能量最低的轨道，即一个三重简并的 t_{2g} 轨道。吸收紫外线后，t_{2g} 轨道的电子就会跃迁到 e_g 轨道：$t_{2g}^1e_g^0 \rightarrow t_{2g}^0e_g^1$。紫外吸收光谱表明最大吸收波长在 20300cm^{-1}

处，对应的能量是Δ_o=243kJ/mol（图 2.13）。Δ_o 的值与化学键的能量具有相同的数量级（$1000cm^{-1}$=11.96kJ/mol 或 2.86kcal/mol 或 0.124eV）。

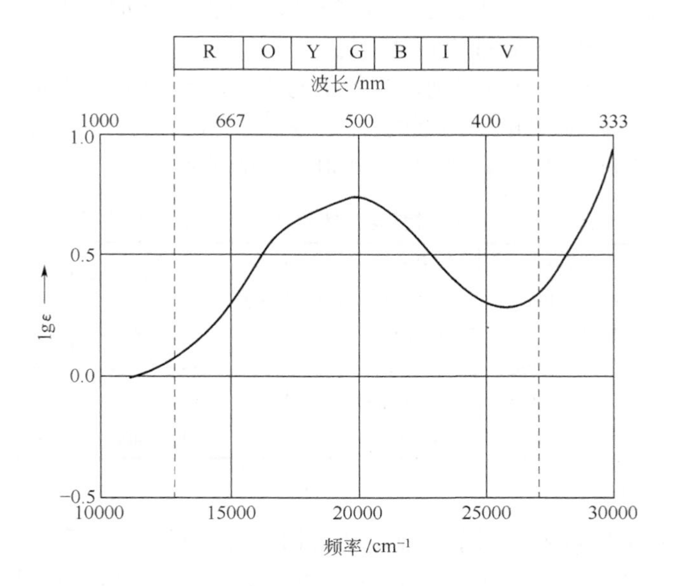

图 2.13　具有 d^1 电子构型的过渡金属的紫外-可见吸收光谱

如果 d 轨道中有多个电子，会怎么样呢?

- 对 d^2-d^9 体系　电子-电子间的相互作用必须要考虑。
- 对 d^1-d^3 体系　根据洪特规则，电子会占据 t_{2g} 轨道且不会配对。
- 对 d^4-d^7 体系　存在两种可能：一是把电子都放在 t_{2g} 轨道上让电子配对，称为低自旋状态或强场位置。一是把电子放到 e_g 轨道，这样能量会高一些，但是电子不配对，称为高自旋状态或弱场位置。因此有两个重要的参数要考虑：电子成对能（P）和晶体场分裂能（即 Δ_o 或 $10D_q$）。

对高自旋和低自旋状态，晶体场稳定化能(CFSE)作为电子数和自旋状态的函数是可以计算出来的（表 2.8）。

表 2.8　作为电子数和自旋状态函数的晶体场稳定化能

	弱场			强场		
d^n	构型	未成对电子数	CFSE	构型	未成对电子数	CFSE
d^1	t_{2g}^1	1	$0.4\Delta_o$	t_{2g}^1	1	$0.4\Delta_o$
d^2	t_{2g}^2	2	$0.8\Delta_o$	t_{2g}^2	2	$0.8\Delta_o$
d^3	t_{2g}^3	3	$1.2\Delta_o$	t_{2g}^3	3	$1.2\Delta_o$
d^4	$t_{2g}^3e_g^1$	4	$0.6\Delta_o$	t_{2g}^4	2	$1.6\Delta_o$
d^5	$t_{2g}^3e_g^2$	5	$0.0\Delta_o$	t_{2g}^5	1	$2.0\Delta_o$
d^6	$t_{2g}^4e_g^2$	4	$0.4\Delta_o$	t_{2g}^6	0	$2.4\Delta_o$
d^7	$t_{2g}^5e_g^2$	3	$0.8\Delta_o$	$t_{2g}^6e_g^1$	1	$1.8\Delta_o$
d^8	$t_{2g}^6e_g^2$	2	$1.2\Delta_o$	$t_{2g}^6e_g^2$	2	$1.2\Delta_o$
d^9	$t_{2g}^6e_g^3$	1	$0.6\Delta_o$	$t_{2g}^6e_g^3$	1	$0.6\Delta_o$
d^{10}	$t_{2g}^6e_g^4$	0	$0.0\Delta_o$	$t_{2g}^6e_g^4$	0	$0.0\Delta_o$

注：该表是相对简化的数据，因为忽略了电子成对能和电子-电子相互作用。

电子成对能包括两部分：

- 相互排斥能　当强制电子去占据相同的轨道时必须要克服这种排斥能。由于轨道弥散性 5d 轨道>4d 轨道>3d 轨道，所以电子成对能随周期数的增加而减小。作为一个规律，具有 4d 和 5d 电子的过渡金属通常都是低自旋的。
- 交换能的损失　根据洪特规则，交换能与自旋平行的电子数成比例。这个数越大，电子越难成对。因此，具有 d^5 电子构型的金属（Fe^{3+}，Mn^{2+}）最容易形成高自旋的配合物。

表 2.9 列出了电子成对能与金属的种类和电子数的关系。

表 2.9　电子成对能与金属的种类和电子数的关系

d^n	离子	$P_{排斥}$	$P_{交换}$	$P_{成对}$	d^n	离子	$P_{排斥}$	$P_{交换}$	$P_{成对}$
d^4	Cr^{2+}	71.2（5950）	173.1（14475）	244.3（20425）	d^6	Mn^+	73.5（6145）	100.6（8418）	174.2（14563）
	Mn^{3+}	87.9（7350）	213.7（17865）	301.6（25215）		Fe^{2+}	89.2（7460）	139.8（11690）	229.1（19150）
d^5	Cr^+	67.3（5625）	144.3（12062）	211.6（17687）		Co^{3+}	113.0（9450）	169.6（14175）	282.6（23625）
	Mn^{2+}	91.0（7610）	194.0（16215）	285.0（23825）	d^7	Fe^+	87.9（7350）	123.6（10330）	211.5（17680）
	Fe^{3+}	120.2（10050）	237.1（19825）	357.4（29875）		Co^{2+}	100（8400）	150（12400）	250（20800）

2.3.2　四面体构型的晶体场

将一个立方体八个顶点中的四个用点电荷占据就会得到一个四面体。在这种情况下，xy，yz，xz 轨道是不稳定的，因为它们指向引入的点电荷。x^2-y^2 和 z^2 轨道是稳定的，所以重心保持不变。这种不可约表示就是 T_d 点群中的 t_2 和 e。四面体场的分裂能小于八面体场的，因为它只有四个配体（而在八面体场中是六个）与过渡金属离子相互作用（图 2.14）。

点电荷模型预测Δ_t=4/9Δ_o。结果，低自旋的情况很少见到。通常如果是强场配体，一般会形成平面正方形。

2.3.3　平面正方形构型的晶体场

可以从八面体的四方对称变形得到平面正方形。从 z 轴移走所有的点电荷，e_g 轨道中的 z^2 轨道会变得稳定。由于剩下的 4 个点电荷的相互作用，x^2-y^2 轨道的能量升高。这里，由于对称性降低，简并度变化，轨道具有不同的不可约表示。

如图 2.15 所示，一个八面体构型的配合物（a）z 轴延长后变为一个变形的四方双锥（b），最后变成了一个平面正方形（c），在平面正方形的配合物中，d_{z^2} 轨道可能在 d_{xz} 和 d_{zy} 轨道下面。

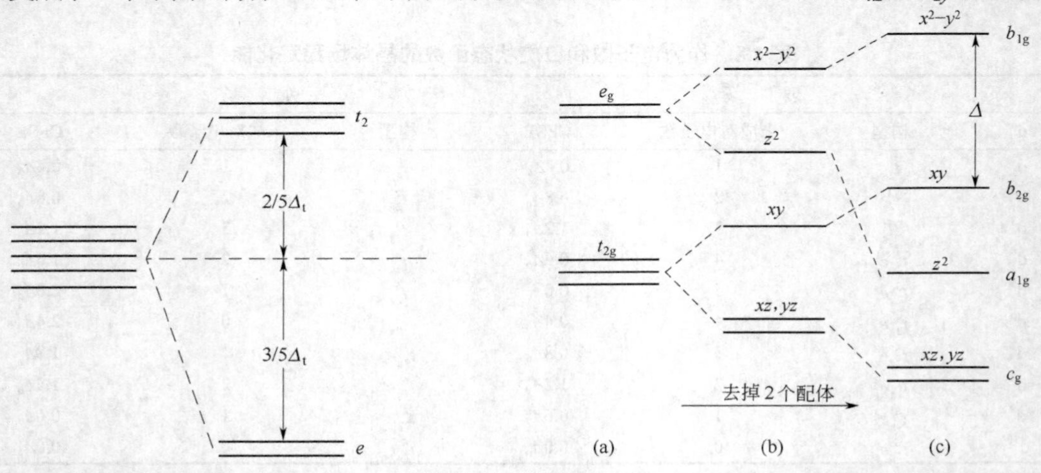

图 2.14　四面体构型中两组轨道 t_2 和 e 的分裂　　**图 2.15**　平面四方形构型中轨道的分裂

2.3.4　影响晶体场分裂能（Δ）大小的因素

● 金属的氧化态：金属价态越高，Δ 越大。有一个规律，氧化态每增加一个单位，Δ 大约增加 50%。

● 金属离子的属性：Δ 的值以 3d、4d、5d 的顺序逐渐增大（从 Co 到 Rh 大约增加 50%，从 Rh 到 Ir 大约增加 25%）。

● 配体的数量和配位几何构型：Δ_o 比 Δ_t 大约大 50%。

● 配体的属性：以各种配合物的参数为基础，可将各种配体按照场强的顺序排列，这一顺序叫光谱化学序列。即

$I^- < Br^- < S^{2-} < SCN^- < Cl^- < N_3^-$，$F^- <$ urea，$OH^- <$ ox，$O^{2-} < H_2O < NCS^- <$ py，$NH_3 <$ en $<$ bpy，phen $< NO_2^- < CH_3^-$，$C_6H_5^- < CN^- < CO$

同样也可以像光谱化学系列那样排列金属离子，大概顺序如下：

$Mn^{2+} < Ni^{2+} < Co^{2+} < Fe^{2+} < V^{2+} < Fe^{3+} < Co^{3+} < Mn^{3+} < Mo^{3+} < Rh^{3+} < Ru^{3+} < Pd^{4+} < Ir^{3+} < Pt^{4+}$

Jørgensen 提供了一个估算 Δ 值的公式：

$$\Delta = fg$$

式中，f 是配体的函数；g 是金属的函数。

表 2.10 列出了根据 Jørgensen 公式所估算的 Δ 值。

表 2.10　配体和金属的晶体场分裂能的估算值

配合物	金属的氧化态	对称性	Δ/cm^{-1}	配合物	金属的氧化态	对称性	Δ/cm^{-1}
$[VCl_6]^{2-}$	4	O_h	15400	$[Ru(ox)_3]^{3-}$	3	O_h	28700
VCl_4	4	T_d	7900	$[Ru(H_2O)_3]^{2+}$	2	O_h	19800
$[CrF_6]^{2-}$	4	O_h	22000	$[Ru(CN)_6]^{4-}$	2	O_h	33800
$[CrF_6]^{3-}$	3	O_h	15060	$[CoF_6]^{2-}$	4	O_h	20300
$[Cr(H_2O)_6]^{3+}$	3	O_h	17400	$[CoF_6]^{3-}$	3	O_h	13100
$[Cr(en)_3]^{3+}$	3	O_h	22300	$[Co(H_2O)_6]^{3+}$	3	O_h	20760
$[Cr(CN)_6]^{3-}$	3	O_h	26600	$[Co(NH_3)_6]^{3+}$	3	O_h	22870
$[Mo(N_2O)_6]^{3+}$	3	O_h	26000	$[Co(en)_3]^{3+}$	3	O_h	23160
$[MnF_6]^{2-}$	4	O_h	21800	$[Co(H_2O)_6]^{2+}$	2	O_h	9200
$[TcF_6]^{2-}$	4	O_h	28400	$[Co(NH_3)_6]^{2+}$	2	O_h	10200
$[ReF_6]^{2-}$	4	O_h	32800	$[Co(NH_3)_4]^{2+}$	2	T_d	5900
$[Fe(H_2O)_6]^{3+}$	3	O_h	14000	$[RhF_6]^{2-}$	4	O_h	20500
$[Fe(ox)_3]^{3-}$	3	O_h	14140	$[Rh(H_2O)_6]^{3+}$	3	O_h	27200
$[Fe(CN)_6]^{3-}$	3	O_h	35000	$[Rh(NH_3)_6]^{3+}$	3	O_h	34100
$[Fe(CN)_6]^{4-}$	2	O_h	32200	$[IrF_6]^{2-}$	4	O_h	27000
$[Ru(H_2O)_6]^{3+}$	3	O_h	28600	$[Ir(NH_3)_6]^{3+}$	3	O_h	41200

配体	f 因子	金属离子	g 因子	配体	f 因子	金属离子	g 因子
Br^-	0.72	Mn(Ⅱ)	8.0	NCS^-	1.02	Ru(Ⅱ)	20
SCN^-	0.73	Ni(Ⅱ)	8.7	$gly^-=NH_3CH_3CO_2^-$	1.18	Mn(Ⅳ)	23
Cl^-	0.78	Co(Ⅱ)	9	py=C_5H_5N	1.23	Mo(Ⅲ)	24.6
N_3^-	0.83	V(Ⅱ)	12.0	NH_3	1.25	Rh(Ⅲ)	27.0
F^-	0.9	Fe(Ⅲ)	14.0	en=$NH_2CH_2CH_2NH_2$	1.28	Te(Ⅳ)	30
ox=$C_2O_4^{2-}$	0.99	Cr(Ⅲ)	17.4	bpy=2,2′-联吡啶	1.33	Ir(Ⅲ)	32
H_2O	1.00	Co(Ⅲ)	18.2	CN^-	1.7	Pt(Ⅳ)	36

2.3.5 晶体场理论的应用

（1）离子半径　对给定的氧化态，周期表中的过渡金属系列，离子半径从左到右依次减小。填入反键轨道（即填充在八面体中的 e_g 能级）会导致离子半径增大。因此，离子半径依赖于金属的自旋状态（即高自旋态或低自旋态）（图 2.16）。

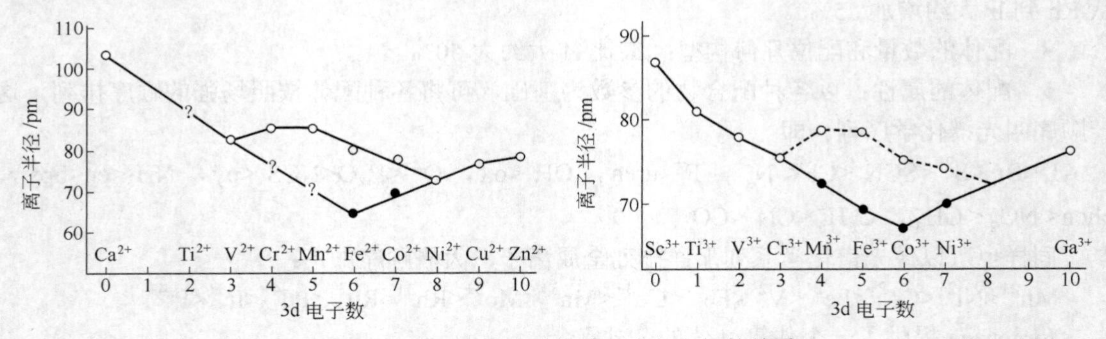

图 2.16　离子自旋状态与离子半径之间的关系

（2）水合焓　下面来看 M^{2+}水合焓的变化。由于水是弱场配体，所以配合物是高自旋态的。

$$M^{2+}(g) + 6H_2O(l) \Longrightarrow [M(OH_2)_6]^{2+}(aq)$$

图 2.17 列出了第一过渡系元素的水合焓曲线。

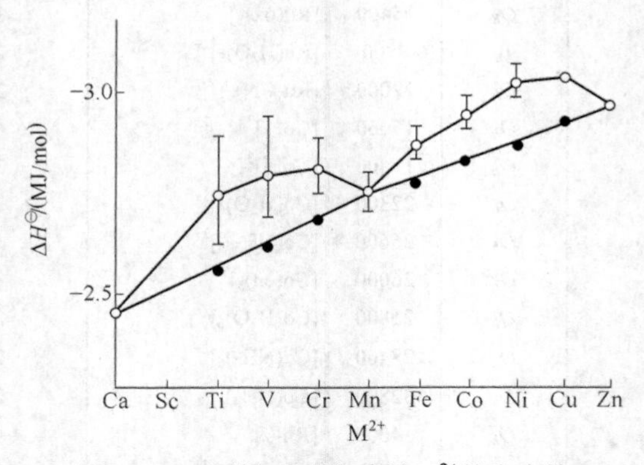

图 2.17　d 区第一过渡系列 M^{2+}的水合焓

图中直线显示了从观察值减去晶体场稳定化能时的变化趋势。在周期表中，从左到右通常的趋势是水合焓（大多放热）逐渐增大。

在第一过渡系列对过渡金属的性质作图时，常常会出现这种双驼峰形曲线。直线（黑圆点）是对于给定电子数，减去晶体场稳定化能所得到的。在过渡系中，稳定性线性增加，主要是由于酸性的逐渐增加而引起的（主要由于金属阳离子的尺寸和静电效应的降低）。这是 Irving-Williams 系列的基础。

Irving-Williams 系列说明了对一个给定的配体，配合物的稳定性从 Ba 到 Cu 依次增加（到 Zn 时是降低的）（图 2.18）。然而，应该注意的是，向右变化时，后部的（靠右边的）过渡金属更倾向于结合软配体。

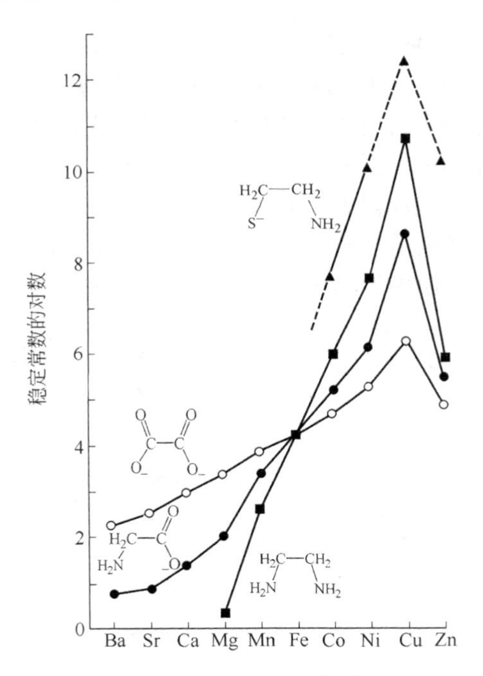

图 2.18　Irving-Williams 效应：从 Ba 到 Cu 稳定性增加，到 Zn 下降

2.4　分子轨道理论

一个好的理论应该能预测分子的物理和化学性质。如：形状、键能、键长和键角等。当然，一个模型不可能描述分子中化学键的所有性质。每种模型可能在说明某些方面的性质时比其他模型更好。对任何一个理论，最终的检验都是实验数据。因为原子轨道的讨论集中在原子的价电子间化学键的形成，所以这种理论称为价键理论。但价键理论不能充分说明一些分子具有介于单键和双键间的两个相同的键的事实。对此，它的最好解释是分子是混合物或是杂化的，对这些分子可写出两种路易斯结构。

这个问题，还有许多其他问题，都可用一种更加成熟的理论——分子轨道理论来说明。分子轨道理论比价键理论更能说明问题，因为它所运用的轨道反映了分子的结构。

分子轨道理论在预测电子光谱和解释顺磁性方面做得很好，而价键理论做不到。分子轨道理论不需要共振结构来描述分子，却也可以预测键长和键角。主要的缺点是只适用于双原子分子（只有两个原子成键），或者用于别的分子时理论会变得很复杂。分子轨道理论将化学键看作是核间共享电子，不像价键理论那样将电子看作是定域在两原子间的球形电子云。分子轨道理论认为电子是非定域的，即它们是分散在整个分子间的。

定性的分子轨道理论是运用一些基本规则来构建分子轨道简图的一个方法。这一部分所讨论的概念可以让我们做一些粗略的预测并对复杂的计算结果进行粗略的解释。

2.4.1　分子轨道

我们从原子轨道的结合可形成分子轨道这一基本原理入手。在构建化合物时，原子轨道间的重叠是由它们的能量、取向和大小来决定的。在氢分子这一最简单的例子中，两个 1s 原子轨道（每个氢原子一个）组合成了两个分子轨道。

氢原子由一个原子核（质子）和一个电子组成。精确确定电子的位置是不可能的，但是可以计算电子在核周围某个位置出现的概率。对氢原子，电子可能的区域是绕核呈球形分布，因此可以画一个球形界面，在这一区域内，电子出现的概率是95%。电子有固定的能量和固定的空间分布，我们称之为一个轨道。在氦原子中，围绕核有两个电子。电子具有相同的空间分布和能量（即它们占据相同的轨道），但是它们的自旋方式不同（鲍林不相容原理）。通常，原子核周围的电子占据具有固定能量和空间分布的轨道，每个轨道最多容纳两个自旋方向相反的电子。在物理学中，周期现象是与波函数相关的，在原子理论中，相关的方程称为薛定谔方程。对一个限定在有限一维势阱中的离子，波动方程的解是分立的，如图2.19所示。

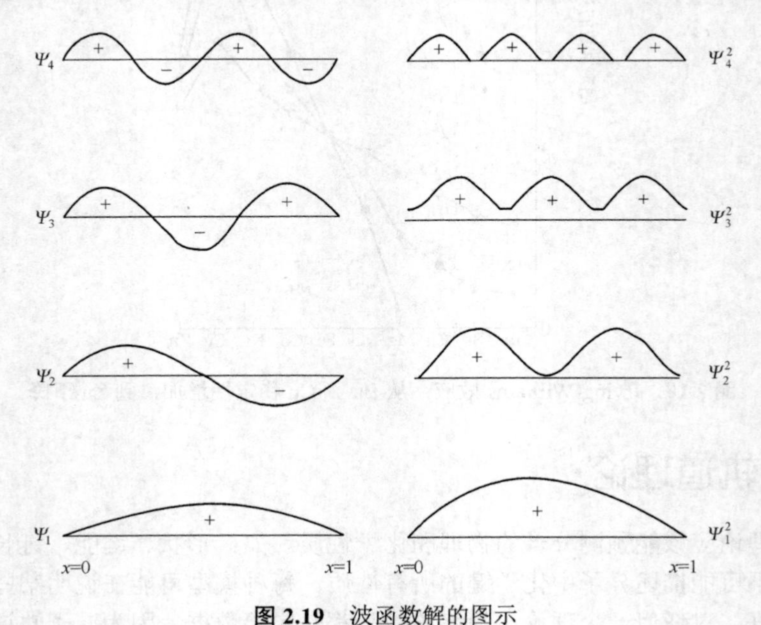

图 2.19 波函数解的图示

Ψ_1~Ψ_4 代表的四个解的能量依次增加。在三维情况下，方程确定了每个电子的能量和空间分布。三维中波动方程的解可以计算出每个轨道的形状。与一个质子相关的一个电子的波函数的前五个解如图2.20所示。

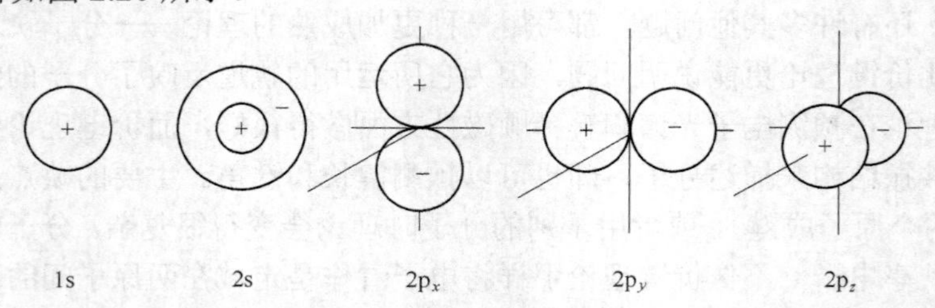

图 2.20 质子中电子的波函数的前五个解的示意图

在氢原子中，1s轨道的能量最低，而剩下的轨道（2s，$2p_x$，$2p_y$和$2p_z$）具有相同的能量（即是简并的）。但对其他所有的原子，2s原子轨道比简并的$2p_x$，$2p_y$和$2p_z$轨道的能量要低。在原子中，电子占据着原子轨道，而在分子中，它们占据着在分子中相似的分子轨道。

最简单的分子是氢分子，它是由两个质子和两个电子构成的。每个氢分子具有两个分子轨道，能量较低的轨道在核间具有较高的电子密度。这样的轨道称为成键分子轨道，它比氢原子的两个1s原子轨道能量低且使分子轨道比两个分离的原子轨道更稳定。另一个分子轨道在电子的波函数中有一个节点，在两个带正电荷的原子核之间的电子密度较低，能量比 1s 原子轨道的高，这样的轨道称为反键轨道。

通常，氢分子中的两个电子占据着成键分子轨道且自旋方向相反。如果用紫外线照射，氢分子会吸收能量，一个电子会激发到反键轨道（σ^*），原子会分开。氢分子中的能级可用图示表示出来——表示出两个原子轨道是怎样结合形成分子轨道的，一个是成键轨道（σ），一个是反键轨道（σ^*）。图 2.21 是两个氢原子结合形成氢分子的示意图。

在这里，每个氢原子的 1s 轨道的能量、方向和大小是相同的。在基态氢分子中，两个电子占据在成键分子轨道中，所以氢分子是一个很稳定的分子。

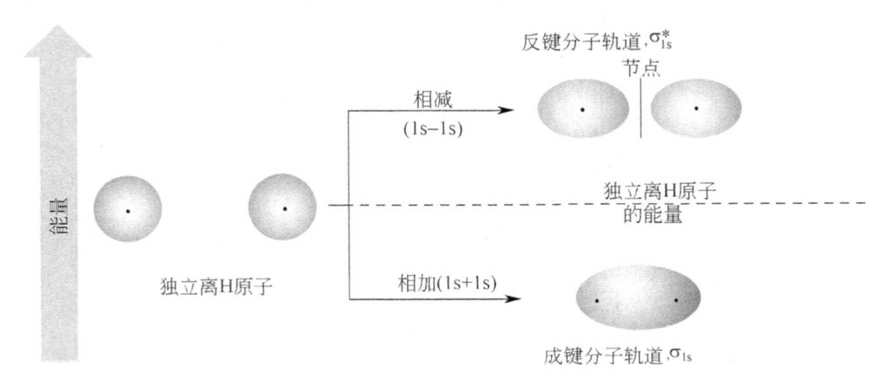

图 2.21　两个氢原子结合形成氢分子的分子轨道图

2.4.2　分子轨道理论的基本原则

两个原子结合时，两个原子轨道互相作用形成了两个可能的分子轨道，其中能量较低的轨道称为成键轨道，它可使分子稳定，因为在这个轨道中电子出现在核间的概率很大。我们称这个轨道为 σ 轨道，因为沿着 H—H 键轴的方向看，它看起来像 s 轨道。另一个轨道称为反键轨道，因为在这一轨道中电子出现在核间以外的概率很大，所以它比原来的原子轨道能量高且使分子不稳定，因此称这一轨道为反键轨道或 σ^*轨道。

分子轨道理论有五个基本原则：

- 分子轨道的数目等于结合的原子轨道的数目。
- 两个分子轨道中，一个是成键轨道（能量较低），一个是反键轨道（能量较高）。
- 电子优先进入能量较低的轨道。
- 一个轨道中最多能容纳两个电子（泡利不相容原理）。
- 电子尽可能分占不同的轨道（洪特规则）。

图 2.22 是氢分子的分子轨道能级图。注意：两个原子轨道结合形成了两个分子轨道——成键轨道和反键轨道。同样也遵循了上述五个原则，电子首先填充在能量最低的轨道（第三条）且成对（第四条），在每个轨道中只有两个电子（第五条）。

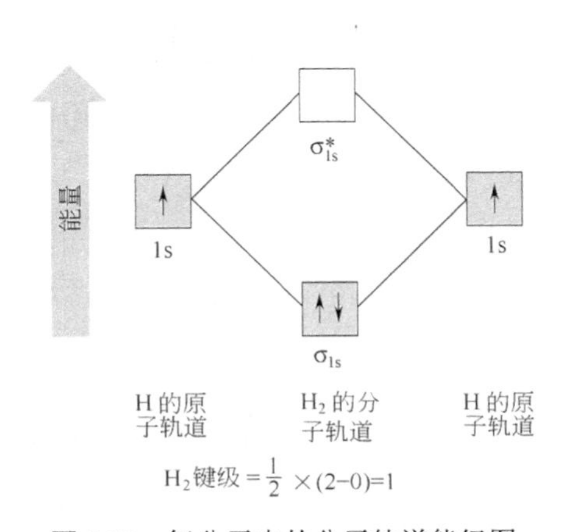

图 2.22　氢分子中的分子轨道能级图

注意图 2.22 的底部，我们提到了“键级”。如果一个分子是稳定的，它的键级一定大于零。键级=1/2×（成键电子数−反键电子数）。如果键级为零，分子不稳定并且不会形成。如果键级为 1，会形成单键；如果键级为 2 或 3，会分别形成双键和叁键。当第二周期的原子相互作用时，一个 2s 轨道和三个 2p 轨道都会参与成键。在这种情况下，会出现两倍的分子轨道。图 2.23 给出了一些分子轨道的能级图。

最后，我们可以将分子轨道用于实践。根据图 2.24，你预测 Li_2 和 Be_2 能形成吗？

答案是 Li_2 可以形成，因为它的键级为 1，所以它是稳定的。Be_2 的键级为 0，是不稳定的，所以不会形成。结合一对具有 $1s^2$ 构型的氦原子会得到一个在成键和反键轨道上都有一对电子的分子，形成 He_2 分子必需的总能量等于一对孤立的氦原子的能量之和，因此，没有额外的能量使两个氦原子结合在一起形成一个分子。

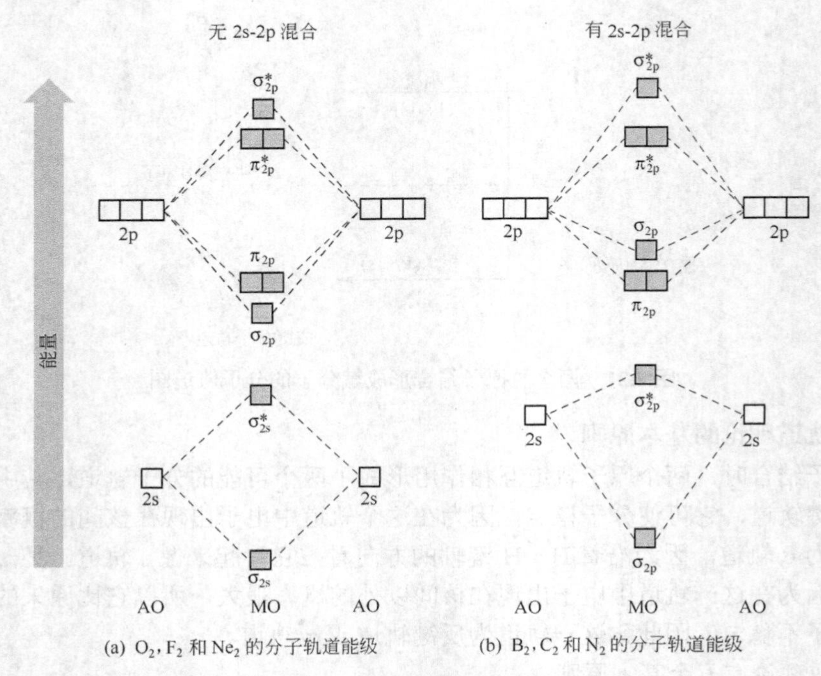

(a) O_2, F_2 和 Ne_2 的分子轨道能级　(b) B_2, C_2 和 N_2 的分子轨道能级

图 2.23　一些分子轨道的能级图

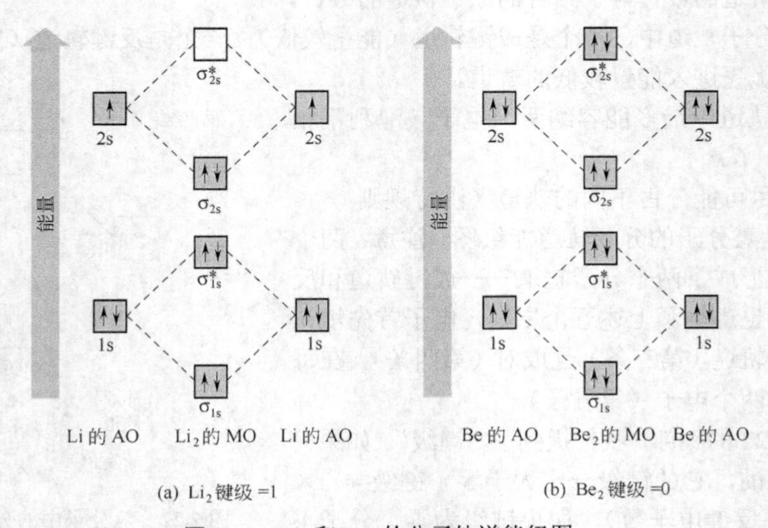

(a) Li_2 键级 =1　(b) Be_2 键级 =0

图 2.24　Li_2 和 Be_2 的分子轨道能级图

He_2 分子与一对孤立的氦原子同样稳定的事实说明了一个重要的原理：原子的内部轨道对分子的稳定性没有贡献。因此，讨论分子轨道时，最重要的是通过价层原子轨道结合而形成分子轨道。因此，氧分子的分子轨道图就忽略了 1s 电子，而关注 2s 和 2p 轨道的相互作用。

2.5　分子间相互作用

分子间相互作用力比共价键弱得多（1～2 个数量级），被称为非共价相互作用，这个概念是在 20 世纪由范德华首先提出的。

共价键和非共价相互作用的本质完全不同。当相互作用的原子的占有轨道部分重叠时，就形成了共价键并构成一对共用电子对。共价键相互作用的距离短，一般小于 2Å（1Å=0.1nm，下同）。非共价相互作用的距离为几埃甚至几十埃，且没有轨道重叠。非共价相互作用来源于固有偶极之间，固有偶极与诱导偶极之间以及瞬时偶极与诱导偶极之间的相互作用。

在最近的二十年里，非共价相互作用在自然界中的作用才得到充分认识。它们在化学和物理学中发挥重要的作用，而且在生物交叉学科中也是至关重要的。液体结构、溶剂化现象、分子晶体、物理吸附、生物大分子诸如 DNA 和蛋白质以及分子识别只是一小部分由非共价相互作用决定的现象。非共价相互作用在超分子化学中起着至关重要的作用，它曾被 Lehn 定义为“超越分子的化学”。

常见的非共价相互作用有：静电相互作用，氢键，堆积作用，范德华力。

2.5.1　静电相互作用

常见的静电相互作用如下。

（1）离子-离子相互作用

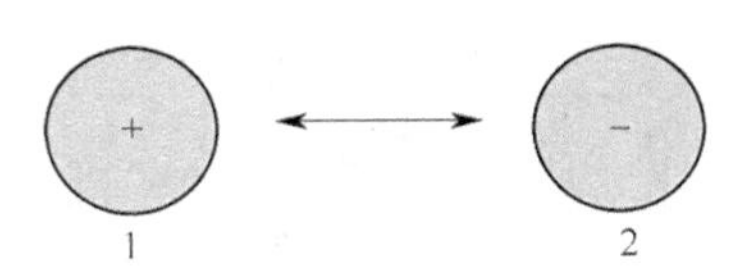

这种相互作用非常强烈，在某些情况下甚至比共价键还强。它是一种长距离（$1/r$）内的吸引或排斥，并且没有方向性。离子-离子相互作用的大小强烈依赖于介质的介电常数。

$$能量=(kz_1z_2e^2)/(\varepsilon r_{12})$$

$$k=1/4\pi\varepsilon_0=库仑常数=q\times10^9(\mathrm{N\cdot m^2/C^2})$$

e 为基本电荷；ε 为介电常数；r_{12} 为电荷间的距离。

离子-离子相互作用的能量与 $1/r$ 成正比，因此是长程作用力。

当设计一种主/客体复合物时，两种带相反电荷的物种在水中距离为 3nm 时的相互作用能为−0.59kJ/mol，距离为 1nm 时相互作用能为−1.77kJ/mol。如果在氯仿中距离为 1nm，这种相互作用能为−28.8kJ/mol。

（2）离子-偶极相互作用　离子-偶极相互作用的特点是：非定向力，可能是吸引力或排斥力，中程相互作用（$1/r^2$），比离子-离子相互作用明显减弱。这种相互作用的能量大小用以下公式来描述：

$$能量=-kQu\cos q/(er^2)$$

当 q=0°或 180°时，有最大值；当 q=90°时，值为 0。

$u=ql$，u=偶极距，l=偶极长度，q=偶极局部电荷，r=电荷到偶极中心的距离，Q=离子电荷（图 2.25）。

（3）偶极-偶极相互作用　偶极-偶极相互作用的特点如下：为非定向力，可能为吸引力或排斥力，短程（$1/r^3$），明显弱于离子-偶极相互作用，发生在具有永久净极性的分子（极性分子）之间。例如，偶极-偶极相互作用发生在 SCl_2 分子间、PCl_3 分子间和 $(CH_3)_2CO$ 分子间。

$$能量=-[ku_1u_2/(er^3)]^2\cos\theta_1\cos\theta_2-\sin\theta_1\sin\theta_2\cos\Phi$$

当 $\theta=0°$时，直线构型有最大值。

简化为：$-2(ku_1u_2/er^3)$。

例如：在氯仿中，两个丙酮分子头尾相连相距 0.5nm，能量为−1.68kJ/mol（图 2.26）。

图 2.25　冠醚和碱金属离子形成的配合物

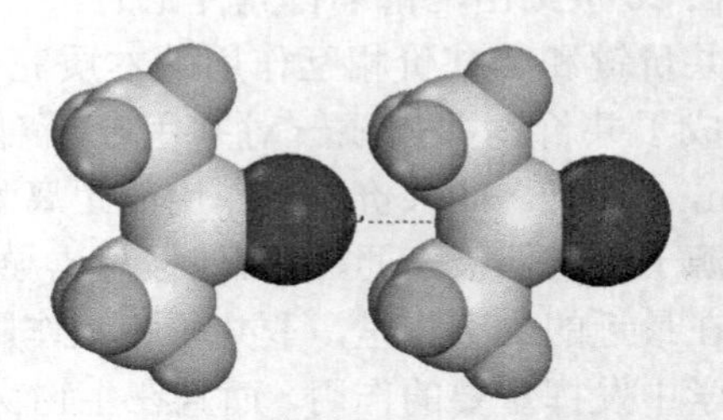

图 2.26　两个丙酮分子之间的偶极-偶极相互作用

2.5.2　氢键

氢键是偶极-偶极相互作用的特例，它有一定的共价键的特征和方向性。它明显强于典型的偶极-偶极相互作用（短距离，2.5～3.5Å），几乎是最重要的分子间相互作用。当氢原子位于两个具有电负性的原子如 O、N、F 之间时就形成氢键 D—H…A（图 2.27）。

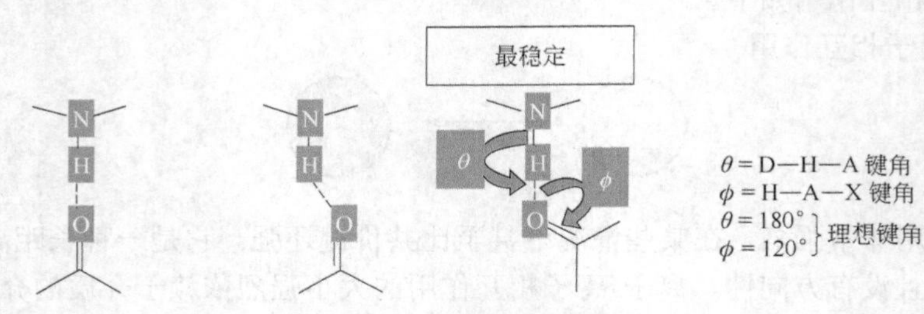

图 2.27　氢键的几何构型

氢原子和氧原子的范德华半径分别为 1.1Å 和 1.5Å，因此，它们的接触距离应为 2.6Å。实际间距（1.76Å）大约少了 1Å，处于范德华半径和 0.96Å 的典型 O—H 共价键长之间（表 2.11）。

表 2.11　氢键中典型的键长与键角（能量 20~40kJ/mol）

NH…O 1.80~2.00Å
OH…O 1.60~1.80Å
θ（D—H—A）150°~160°
ϕ（H—A—X）120°~130°

氢键有经典氢键和非经典氢键两种。

经典氢键：氢键复合物是迄今最重要和最多的非共价键复合物。大多数氢键含有如 X、Y 那样具有一个或两个孤对电子的电负性原子或具有离域电子云的基团（例如，芳香体系的 π 电子）。X，Y 为氟、氧、氮的氢键最常见，例如：O—H…O 和 N—H…O(N)等氢键（图 2.28）。

非经典氢键：氢键的概念已经延伸到 C—H…Y 型键，C—H…Y（Y 为电负性原子）和

C—H⋯π 型氢键也已经被发现。如果 CH 原子团中的氢原子显酸性，则可形成很强的氢键。否则，C—H⋯Y 型氢键将比 O—H⋯Y 或 N—H⋯Y 型氢键弱得多。然而，C—H⋯Y 型氢键因其大量存在而在分子生物结构中起着非常重要的作用。但通常情况下，氢键的判据(C—H 键的伸长和 C—H 伸缩振动频率的红移)并不严密，C—H 仅是与 Y 原子接触而并没有形成，这应称为 C—H⋯Y 接触。

非经典氢键的体系已经扩展到了 C—H⋯O 型非正常氢键和存在于氟仿-环氧乙烷的 C—H⋯X^-（X=卤素）型氢键以及 X^-⋯H_3CY（X，Y=卤素）复合物中的氢键。

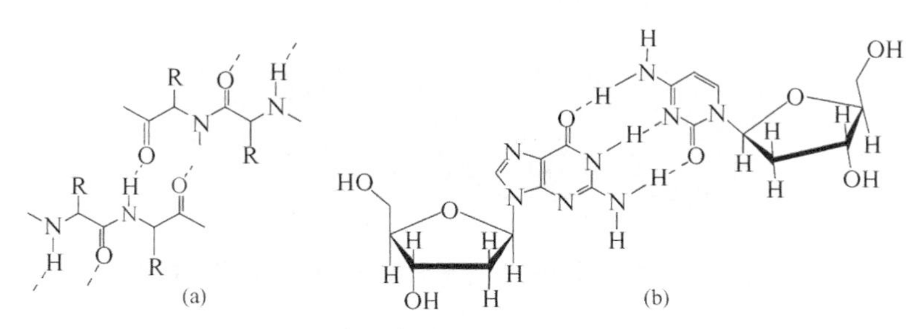

图 2.28　（a）在蛋白质和（b）DNA 中的经典氢键

氢键的几何和光谱形式的动力是什么呢?自然键轨道分析表明：这种动力是从孤对电子或作为电子供体（质子受体）的 π 分子轨道到电子受体（质子供体）的 X—H 反键轨道之间的电子转移。反键轨道上电子密度的增加导致 X—H 键的伸长，也会导致 X—H 伸缩振动的红移。

2.5.3　π-π 堆积

π-π 堆积（0~50kJ/mol）是芳香环之间微弱的静电作用（图 2.29）。有两种基本类型：面对面和边对面。面对面 π 堆积相互作用会使石墨有光滑感。类似的 π 堆积相互作用有助于稳定 DNA 的双螺旋结构。

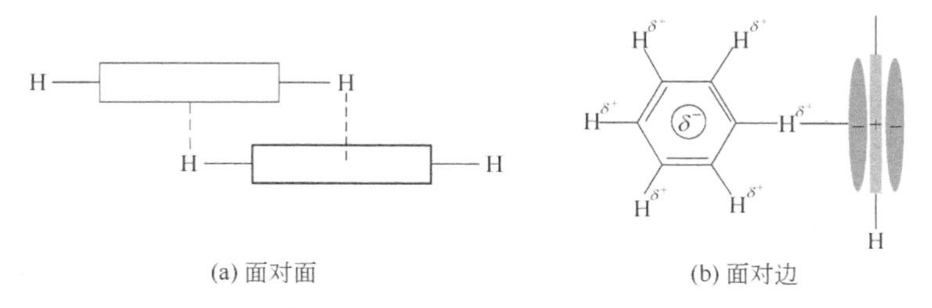

图 2.29　两种 π-π 堆积的类型

2.5.4　范德华相互作用

范德华相互作用几乎存在于所有的原子和分子间，由原子和分子偶极产生。这种波动诱导偶极之间的相互作用是由于瞬时或短暂的振动扭曲产生的，键能非常弱（0.1～3kJ/mol）。

由于电子运动，这种瞬时极化作用还将继续波动并且与诱导极化是同时发生的，因此极性之间的吸引作用一直持续到分子彼此靠近（图 2.30）。

相互作用的强度本质上是接触面积的函数和电子云的极化能力。表面积越大，相互作用越强。在复合物中，不管有没有其他作用力，范德华力总是存在的。正是这种力使分子间消除空间或真空，并使其难以设计成多孔或中空结构，从而产生了“大自然厌恶真空”这一短语。

通常，有三种类型的范德华作用：离子-诱导偶极、偶极-诱导偶极及色散力（图 2.31）。

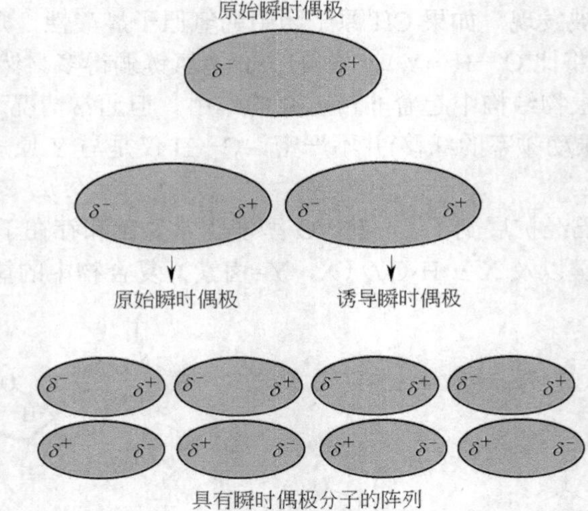

图 2.30 原始瞬时偶极诱导瞬时偶极和瞬时偶极的排列

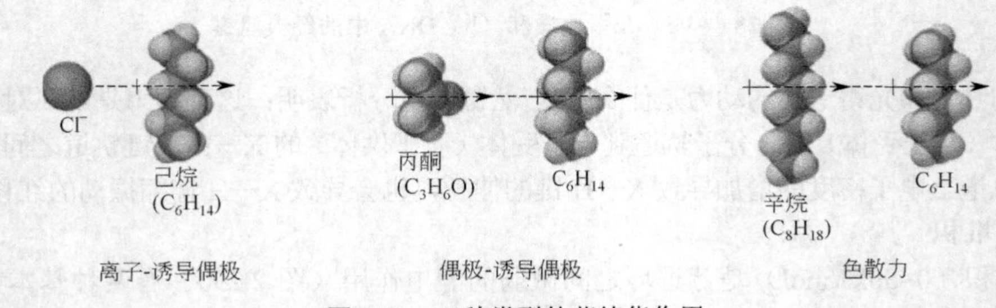

图 2.31 三种类型的范德华作用

第 3 章　配合物的现代分析表征方法

当一束强度为 I_0 的电磁波通过一种物质时，该电磁波或被吸收或被散射，这取决于电磁波的频率和分子的结构。电磁波具有能量，当物体吸收了电磁波，它就获得了能量，从而可从一种能量状态（基态）跃迁到另一种能量状态（激发态）。所吸收的光频率与能量的关系符合普朗克定律（Planck's Law）。

$$E_{终态}-E_{始态}=E=h\nu=hc/\lambda$$

这样，当入射光的频率符合上式时，光波被吸收；相反，不满足上式，则发生散射。入射光的频率对物质（样品）吸收百分率所做的曲线，叫做物质的吸收光谱。

吸收光谱的类型取决于散射的性质和相应的入射光的频率范围。如果伴随转动能级的跃迁，光波频率对应于微波区，该技术叫做微波光谱；如果是振动能级的跃迁，则对应红外区，该技术叫做红外光谱；如果跃迁引起分子的价电子构型发生改变，则对应紫外-可见光区，该技术叫做紫外-可见吸收光谱（或电子吸收光谱）。

3.1　紫外-可见吸收光谱（UV-Vis）

许多分子吸收紫外或可见光，光束衰减越多，溶液的吸收越强。吸光度与光程 b 和吸收的物质浓度 c 成正比例，比尔（Beer's Law）定律反映出它们之间的关系：

$$A=\varepsilon bc$$

这里 ε 为比例常数，叫吸光系数。

不同的分子吸收不同波长的辐射。吸收光谱曲线可以显示分子中不同结构官能团对应的吸收带。例如，丙酮中羰基在紫外区的吸收与戊酮中羰基的吸收峰位置一致。

3.1.1　电子跃迁

分子吸收紫外-可见光引起外层电子的激发，有三种类型的跃迁：

- 含 n、σ、π 电子的跃迁。
- 电荷转移跃迁。
- d 电子和 f 电子的跃迁（本章暂不讨论 f 电子的跃迁）。

当原子或分子吸收能量，电子会从基态跃迁到激发态。在分子中，原子彼此之间可发生相应的转动和振动。这些转动、振动能级同样也是能量不连续的，可认为是在每个电子能级上叠加的（图 3.1）。

3.1.2　含 n、σ、π 电子的物质的吸收

有机分子中能吸收紫外可见光的基团只局限于某些官能团（又叫生色团），生色团含有较低的激发能的价电子。含有这些生色团的有机分子的吸收光谱是复杂的。这是因为电子跃迁时，伴随有转动、振动能级的跃迁，会造成谱线的重叠，从而表现出连续的吸收带（图 3.2）。

（1）$\sigma \to \sigma^*$跃迁　当电子从成键 σ 轨道跃迁到相应的反键 σ^*轨道上时，所需要的能量很大。例如，甲烷（只有 C—H 键，只能发生 $\sigma \to \sigma^*$跃迁）在 125nm 处有最大吸收峰。$\sigma \to \sigma^*$跃迁的最大吸收峰在典型的 UV-Vis 中难以观察到，这是因为 UV-Vis 光谱的波长范围为 200～700nm。

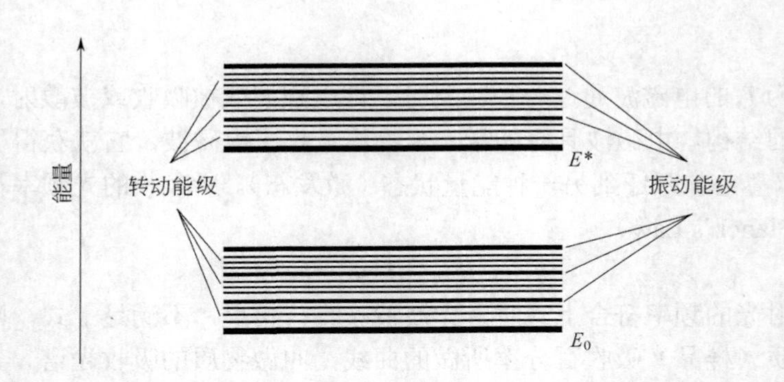

图 3.1　分子中振动和旋转的分立能级示意图

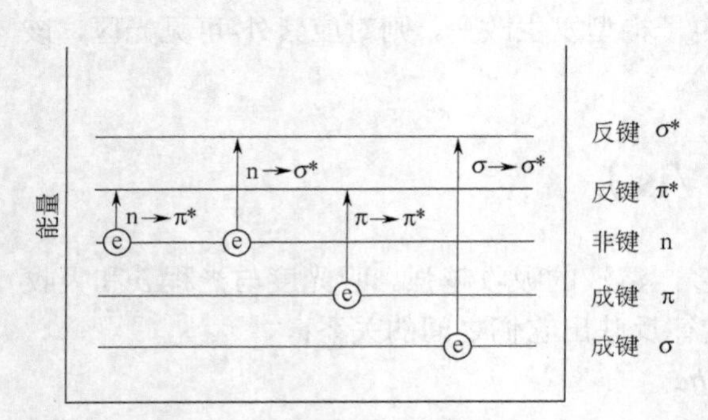

图 3.2　π，σ 和 n 电子可能发生的电子跃迁

（2）$n \to \sigma^*$跃迁　含有孤对电子（非键电子）的饱和化合物可以发生 $n \to \sigma^*$跃迁，这种跃迁的能量低于 $\sigma \to \sigma^*$跃迁，波长 150~250nm 的光即可激发。在紫外区，具有 $n \to \sigma^*$跃迁吸收峰的有机官能团很少。

（3）$n \to \pi^*$和 $\pi \to \pi^*$跃迁　大多数有机化合物的吸收光谱是基于$n \to \pi^*$或$\pi \to \pi^*$跃迁。因为，这些吸收峰恰好位于一般实验的检测范围（200～700nm）内。这种跃迁需要由分子中的不饱和官能团提供 π 电子。$n \to \pi^*$跃迁的摩尔吸光系数通常相对较低，介于 10~100L/（mol·cm）。$\pi \to \pi^*$跃迁的摩尔吸光系数一般介于 1000~10000L/（mol·cm）。溶剂对吸收光谱也会产生影响。溶剂极性增加，$n \to \pi^*$跃迁的最大吸收峰向短波方向移动，称为蓝移（blue shift）。$\pi \to \pi^*$跃迁恰好相反，即红移（red shift）。这种红移对 $n \to \pi^*$跃迁也有影响，但相对于由孤对电子的溶剂化作用引起蓝移（blue shift）而言，可以忽略。一些常见的发色团的吸收和有机发色团列于表 3.1 和表 3.2 中。

表 3.1　**一些常见发色团的吸收**

发色团	实 例	激 态	λ_{max}/nm	ε	溶 剂
C═C	乙烯	$\pi \to \pi^*$	171	15000	环己烷
C≡C	1-己炔	$\pi \to \pi^*$	180	10000	环己烷
C═O	乙醛	$n \to \pi^*$ $\pi \to \pi^*$	290 180	15 10000	环己烷 环己烷
N═O	硝基甲烷	$n \to \pi^*$ $\pi \to \pi^*$	275 200	17 5000	乙醇 乙醇
C—X　X=Br X=I	甲基溴 甲基碘	$n \to \sigma^*$ $n \to \sigma^*$	205 255	200 360	环己烷 环己烷

表 3.2　一些常见的有机发色团

发色团	跃迁	λ_{max}/nm	lgε	发色团	跃迁	λ_{max}/nm	lgε
腈（—C≡N）	$n\to\pi^*$	160	<1.0	醛［—C(H)═O］	$n\to\pi^*$	290	1.0
炔（—C≡C—）	$\pi\to\pi^*$	170	3.0	胺（$—NR_2$）	$n\to\sigma^*$	190	3.5
烯（—C═C—）	$\pi\to\pi^*$	175	3.0	酸（—COOH）	$n\to\pi^*$	205	1.5
醇（ROH）	$n\to\sigma^*$	180	2.5	酯（—COOR）	$n\to\pi^*$	205	1.5
醚（ROR）	$n\to\sigma^*$	180	3.5	酰胺［$—C(═O)NH_2$］	$n\to\pi^*$	210	1.5
酮［—C(R)═O］	$\pi\to\pi^*$	180	3.0	硫醇（—SH）	$n\to\sigma^*$	210	3.0
	$n\to\pi^*$	280	1.5	硝基（$—NO_2$）	$n\to\pi^*$	271	<1.0
醛［—C(H)═O］	$\pi\to\pi^*$	190	2.0	偶氮（—N═N—）	$n\to\pi^*$	340	<1.0

（4）电荷转移吸收　许多无机物都具有电荷转移吸收，这类物质称为电荷转移配合物。对于具有电荷转移行为的配合物，一般来讲，其组分之一应该是电子供体，另一组分应该是电子受体。吸收能量，电子从供体的轨道激发到受体的轨道。电荷转移吸收的摩尔吸光系数通常很大［超过 10000L/(mol·cm)］。

3.1.3　配合物的电子吸收光谱

在固态和溶液中的配合物包含有三种类型的电子跃迁，它们是：基于金属离子的 d-d 跃迁；金属到配体的电荷跃迁（MLCT）［更多的是配体到金属的电荷跃迁（LMCT）］；基于配体的电荷跃迁（LC）。

（1）金属离子的 d-d 跃迁　正如第 2 章中所讨论的，在八面体和四面体配合物中，配位场将导致 d 轨道简并裂分。因此，似乎可以预期所有过渡金属配合物都会表现出基于金属离子的 d→d 跃迁吸收曲线。然而，事实上并非如此简单。一些配合物会表现出强的吸收，另一些配合物仅表现出弱的吸收。例如，T_d 型配合物表现出强的 d→d 跃迁，O_h 型配合物则并非如此。这可以用选择定律来解释。

为了说明选择定律的概念，我们可以用简单的体系——氢原子来讨论。氢原子的能级可用 Grotrian 图表示（图 3.3）。

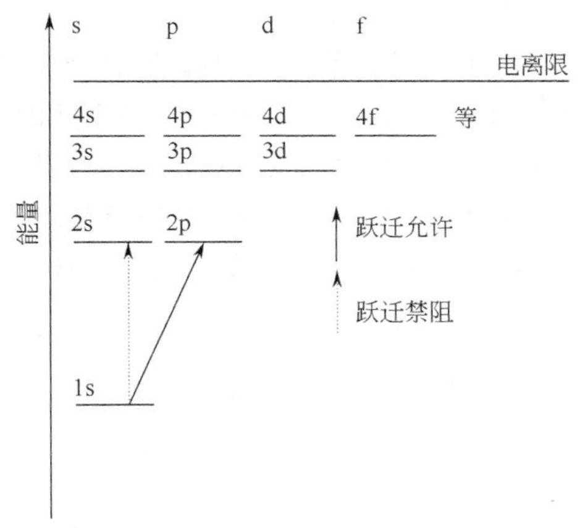

图 3.3　氢原子的能级 Grotrian 图

显然，从 1s→2p 的跃迁是允许的，而 1s→2s 的跃迁是禁阻的。原因是氢原子具有反演中心。这可以通过 SALC（对称性匹配的线性组合）与具有 O_h 和 T_d 对称性的配合物的 d 轨道的结合来考察。如果具有反演中心，线性组合规则要求基态与终态具有不同的宇称。这就意味着 d→d 跃迁是对称性禁阻的——也就是说，跃迁偶极矩为零。但是，考察点群 T_d 的对称性操作，就会发现它并不具有反演中心。因而 T_d 配合物 d→d 跃迁是允许的。而对 O_h 配合物，有一反演中心，因而是对称性禁阻的。这就是 T_d 对称性的配合物比 O_h 对称性的配合物吸收强得多的原因。

但是，这并不能解释 O_h 配合物为什么仍然具有 d→d 跃迁吸收。这一定发生了某一过程，破坏了对称性选择规则。最明显的方式（原因）就是打破了 O_h 对称性。当 O_h 配合物发生振动时，会出现这种可能。一些振动模式在一定程度移动了配体，使得化合物不再具有反演中心。譬如沿着配体-金属轴的振动模式，会导致配体-金属距离改变——某种程度上是随机的。这种振动比电子跃迁要慢得多，因此，实际上，在跃迁过程中，可以认为分子是固定不动的。像这样的非协调的运动方式说明：在大量分子中，有些分子在它们的配体-金属距离上出现了不对称性差别。这种差别以一定的分布状态分布。当存在不对称性时，上述的公式不再适用——分子具有暂时的不对称排列，因而吸收成为可能，从而产生弱的吸收。这同样也可以说明吸收峰变宽，因为不同环境有一个分布，因此不同分子中的偶极和势场也有一个分布。Laporte 选律中的弱“弛豫”现象就是振动偶合，这种振动偶合源于振动模式与电子跃迁模式的相互作用。

更有意思的是，这种弱的吸收峰落在可见光区，因而可以解释配位化合物的颜色。

（2）金属-配体电荷转移跃迁（MLCT 或 LMCT 跃迁） 这些具有最低激发态能量、发生在金属中心的 dπ 基态与配体 π^*态之间的跃迁可以在可见区观察到。选择合适的配体可以调控 MLCT 吸收。确实，π 受体的配体具有低的 π^*轨道，同时通过反馈配位作用来稳定金属中心的 dπ 轨道。这也可以调控配合物的氧化-还原性质，因为还原过程涉及配体的最低 π^*轨道，氧化过程涉及金属中心的 dπ 轨道。

为理解这一点，有必要了解光吸收过程中涉及的有关轨道。以八面体金属配合物为例（图 3.4），最高占据轨道（HOMO）是金属的 dπ 轨道，最低空轨道（LUMO）是配体的 π^*轨道。HOMO 与 LUMO 轨道的本性和能量对于理解吸收光的分子的激发态非常重要，它们均为最低能量激发态。

发生在具有不饱和配体的 d^6 金属配合物的电子跃迁有：配体的 $n\to\pi^*$与 $\pi\to\pi^*$跃迁；金属-配体的电荷转移（MLCT）跃迁和配体场跃迁（图 3.5）。跃迁强度由 Laporte 选律确定。允许跃迁必须同时符合 Laporte 选律与自旋允许。配体的 $\pi\to\pi^*$跃迁和 MLCT 跃迁都符合 Laporte 选律和自旋允许，摩尔吸光系数通常介于 10^3～10^5L/(mol·cm)之间，配体场跃迁自旋允许但 Laporte 禁阻，吸光系数在 100～1000L/(mol·cm)。

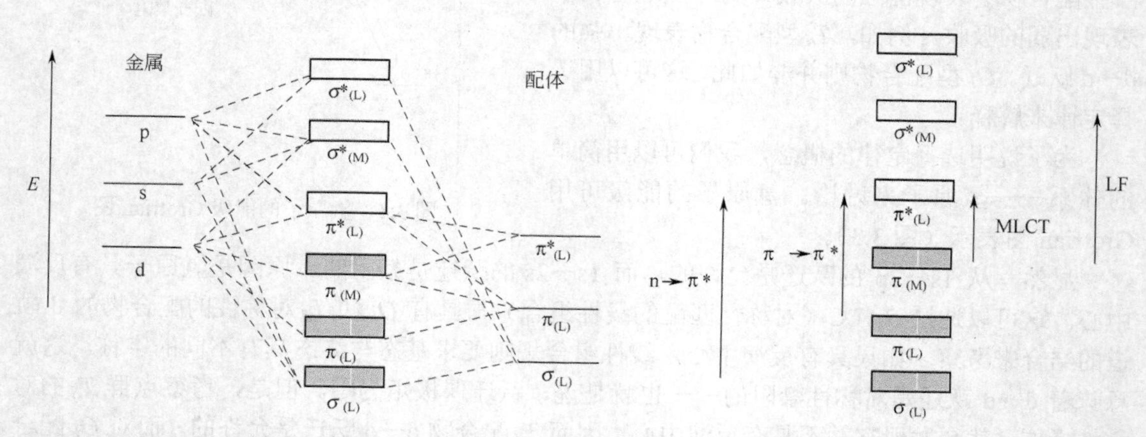

图 3.4 d^6 八面体金属配合物中的分子轨道方块示意图　　**图 3.5** d^6 八面体金属配合物中可能的电子跃迁

随着$[Ru(bpy)_3]^{2+}$光物理性质的发现，对该生色团光引发的能量与电子转移进行了大量的研究工作。此配合物对光稳定，室温下，发射寿命长（τ=640ns，乙腈溶剂中），同时具有相当高的发射量子产率（ϕ=0.062，乙腈溶剂中）。乙腈溶剂中$[Ru(bpy)_3](PF_6)_2$ 的电子吸收光谱如图 3.6 所示，在紫外区有 $n\to\pi^*$与 $\pi\to\pi^*$跃迁，可见光区主要是 MLCT 跃迁。在 450nm 处的

MLCT 跃迁峰实际上包含了几个能量稍有差异的电子跃迁。

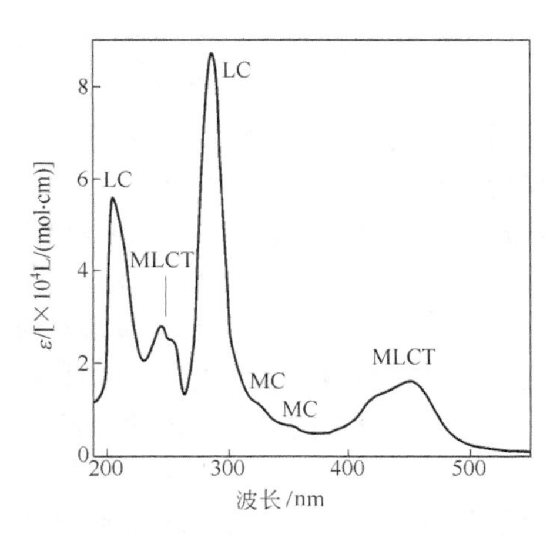

图 3.6 $[Ru(bpy)_3](PF_6)_2$ 在乙腈溶剂中的电子吸收光谱

3.2 红外光谱与拉曼光谱

3.2.1 分子的运动类型

（1）分子的平动 分子作为一个整体可沿着任意方向，以一定的速率运动，这种运动方式叫平动。相应的动能为 $mv^2/2$（v 为质心的运动速度）。运动速度又可沿笛卡尔坐标系的三个轴分解为三个分量。可得：

$$mv^2/2=mv_x^2/2+mv_y^2/2+mv_z^2/2$$

v_x 为 x 轴方向的速度，等等；m 为分子的质量。

上式告诉我们，总平动能 KE 可分解为三个部分。每一部分可看作沿三个轴方向的动能，总动能为三个轴方向动能的和，分别源自 x 轴、y 轴、z 轴，这种情况下可以说分子有三个平动自由度，分别对应笛卡尔坐标系的三个轴。

（2）分子的转动 分子也可绕轴转动，同样也可分解为 x 轴、y 轴、z 轴三个方向，分子转动可视为三个互相垂直方向的矢量和。转动能可写作：

$$KE_{rot}=I_x\omega_x^2/2+I_y\omega_y^2/2+I_z\omega_z^2/2$$

I_x，I_y，I_z 分别为 x，y，z 轴的转动惯量；ω_x，ω_y，ω_z 分别为 x，y，z 轴的角速度。

同样，转动也有三个转动自由度，分别对应于笛卡尔坐标的三个轴。但线形分子例外，因为对于它而言，三个轴方向中一个通常被定为分子轴。如图 3.7 所示，对于双原子分子，其键轴为坐标系中的 z 轴，这种情况下，z 轴方向没有转动能。因为该方向上的转动惯量为 0。所以对线形分子，只有两个转动自由度，而不是三个。

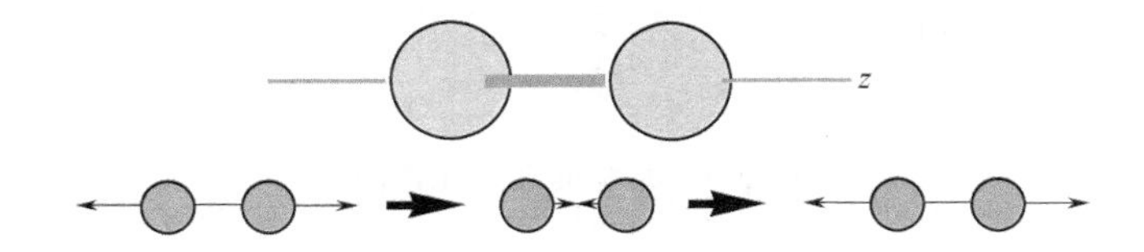

图 3.7 坐标系中的 z 轴做键轴的双原子分子，仅有两个转动自由度

（3）分子的振动 分子也可发生振动。例如，H_2O 分子可沿键轴发生重复的伸缩振动，如图 3.8 所示。

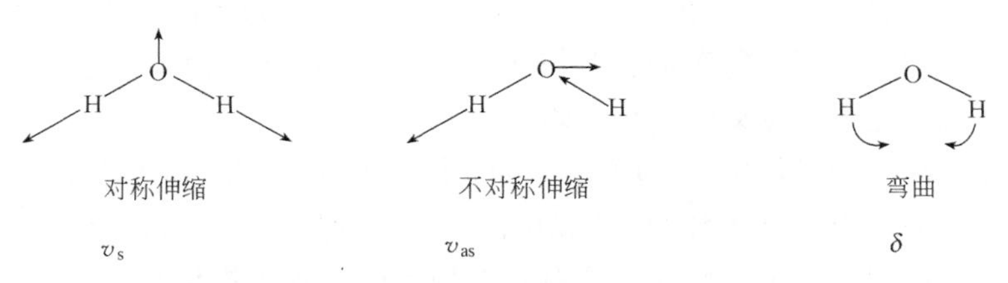

图 3.8 非线形多原子分子中三种独立的振动模式

这类分子，除了三个平动自由度、三个转动自由度外，还有一个振动自由度。一般来讲，对多原子分子，总自由度减去平动自由度和转动自由度即为振动自由度（有时叫振动模式）。总自由度为 $3N$（每个原子可沿三个方向自由运动，有三个自由度，则分子有 $3N$ 个自由度），N 为原子数目。这就有了一个一般的规则：对于非线性多原子，振动自由度为 $3N-6$，而对于线形双原子分子则为 $3N-5$。根据这个原则，对非线形分子水分子，$N=3$，有三个振动自由度即三种振动模式，如图 3.8 所示。

分子中的每一种振动类型都对应于一定的频率，该频率成为分子或某些特定振动的特征。振动能与振幅成比例，振动能越大，振幅就越大。

对应于分子的每一种振动模式，有一系列的振动能级。当分子吸收一定量子化的电磁辐射，如 $E_{终态}-E_{始态}=h\nu$，可从一个能级（基态）跃迁到另一较高能级（激发态）。经历跃迁后，分子获得了振动能，这可从振幅的增加得以证明。

大多数分子的振动频率在 10^{14}Hz，其相应波长 $\lambda=c/\nu=(3\times10^{10}\text{cm/s})/10^{14}\text{s}^{-1}=3\times10^{-4}\text{cm}=3\mu\text{m}$，即引发跃迁的光波在 3μm 处，这一波长恰好位于所谓的红外光谱区。红外光谱研究分子的振动跃迁，因此也称振动光谱。

红外光谱图常以红外线的能量（以 μm 或波数为单位）对透光率作图，如图 3.9 所示。

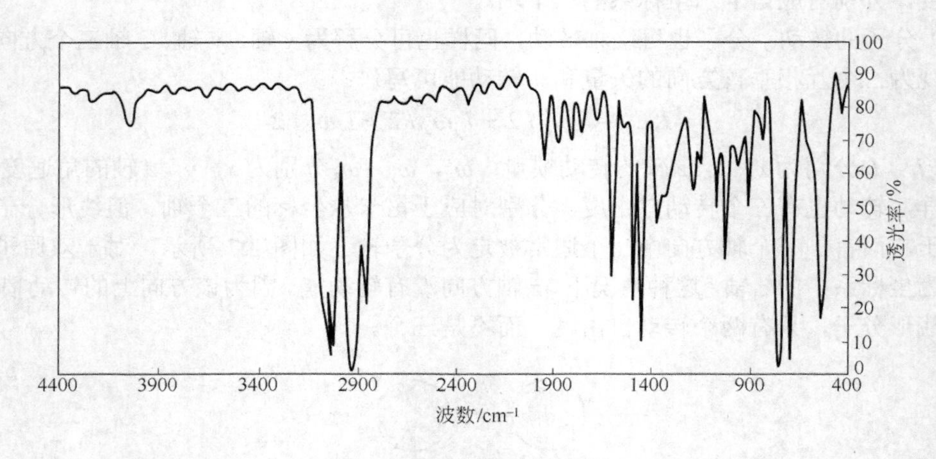

图 3.9 典型的红外光谱图

3.2.2 配合物的红外光谱

有机官能团之间的区别在于所含的原子及键的强度不同。例如，3200～3400cm^{-1} 之间强而宽的峰表明分子中存在 O—H（羟基）官能团。1700cm^{-1} 的强峰表明存在 C═O（羰基）官能团。

对有机分子，其红外光谱可分为三个区。1300～4000cm^{-1} 为特征官能团区，1300～909cm^{-1} 为指纹区（主要来源于分子中复杂的相互作用），909～650cm^{-1} 为苯环区。

一些重要有机官能团的振动频率列于表 3.3 中。

现讨论红外光谱在无机化学中的应用。实际上，许多所谓的无机化合物在很大程度上是有机的。因此，对待这类无机化合物，就如同有机化合物一样可在红外光谱中找到它们的官能团。但是，这里我们只讨论简单的、含有几个原子的纯粹无机化合物或含原子团的简单无机盐类。

表 3.3　有机官能团的振动频率

键　型	特殊环境	振动频率 ν/cm^{-1}
C—H	C(sp^3)—H	2800～3000
	C(sp^2)—H	3000～3100
	C(sp)—H	3300
C—C	C—C	1150～1250
	C═C	1600～1670
	C≡C	2100～2260
C—N	C—N	1030～1230
	C═N	1640～1690
	C≡N	2210～2260
C—O	C—O	1020～1275
	C═O	1650～1800
C—X	C—F	1000～1350
	C—Cl	800～850
	C—Br	500～680
	C—I	200～500
N—H	RNH_2,R_2NH	3400～3500(两个)
	RNH_3^+,$R_2NH_2^+$,R_3NH^+	2250～3000
	$RCONH_2$,RCONHR′	3400～3500
O—H	ROH	3610～3640(游离) 3200～3400(氢键)
	RCO_2H	2500～3000
N—O	RNO_2	1350,1560
	$RONO_2$	1620～1640, 1270～1285
	RN═O	1500～1600
	RO—N═O	1610～1680(两个), 750～815
	C═N—OH	930～960
	$R_3N—O^+$	950～970
S—O	R_2SO	1040～1060
	R_2S(═O)O	1310～1350 1120～1160
	R—S(═O)$_2$—OR′	1330～1420, 1145～1200
累积体系	C═C═C	1950
	C═C═O	2150
	R_2C═N═N	2090～3100
	RN═C═O	2250～2275
	RN═N═N	2120～2160
平面外的弯曲振动		
炔烃	C≡C—H	600～700
烯烃	RCH═CH_2	910,990
	R_2C═CH_2	890
	反-RCH═CHR	970
	顺-RCH═CHR	725,675
	R_2C═CHR	790～840
芳香类	单-	730～770,690～710 (两个)
	邻-	735～770
	间-	750～810,690～710 (两个)
	对-	810～840
	1,2,3-	760～780,705～745 (两个)
	1,3,5-	810～865,675～730 (两个)
	1,2,4-	805 ～ 825, 870, 885 (两个)
	1,2,3,4-	800～810
	1,2,4,5-	855～870
	1,2,3,5	840～850
	五-	870
羰基伸缩振动频率		
醛类	RCHO	1725
	C═CCHO	1685
	ArCHO	1700
酮类	R_2C═O	1715
	C═C—C═O	1675
	Ar—C═O	1690
	四元环	1780
	五元环	1745
	六元环	1715
羧酸类	RCOOH	1760(单体) 1710(二聚体)
	C═C—COOH	1720(单体) 1690(二聚体)
	RCO_2^-	1550～1610,1400(两个)
酯类	RCOOR	1735
	C═C—COOR	1720
	ArCOOR	720
	γ-内酯	1770
	δ-内酯	1735
胺类	$RCONH_2$	1690(游离) 1650(缔合)
	RCONHR′	1680(游离) 1655(缔合)
	$RCONR'_2$	1650
	β-内酰胺	1745
	γ-内酰胺	1700
	δ-内酰胺	1640
酸酐	RCOOCOR′	1820,1760(两个)
酰基卤化物	RCOX	1800

以简单无机盐 KNO_2 为例。从它的经验式（化学式）我们很容易想到它有 6（3×4−6=6）种简正振动类型。然而，我们已经知道亚硝酸钾晶体中不含单独的 KNO_2 分子，所以最初的设想是错误的。事实上，亚硝酸钾是由 K^+ 和 NO_2^- 组成的无限规整的阵列。晶体中的基本单元是 K^+ 和 NO_2^-。这样我们可认为 K^+ 和 NO_2^- 各自具有独立的振动类型。起码在一级近似，情况确实如此。这是因为 K^+ 是单原子（离子），没有振动模式（3×1−3=0）。所以，只需考虑 NO_2^-。根据价层电子对互斥理论，NO_2^- 结构如图 3.10 所示。

NO_2^- 像水分子一样是弯折形的。我们因此可以预测 NO_2^- 有三种红外活性的简正振动，与前面提到的 H_2O 分子的模式相对应。

不出所料，在 KNO_2 红外光谱中发现了三个吸收峰：1335cm^{-1} 的对称伸缩振动；1250cm^{-1} 的不对称伸缩振动；830cm^{-1} 的弯曲振动（相比于伸缩振动，弯曲振动一般在低频处）。

而且，不论与哪种阳离子结合，这三个吸收峰都是相同的，再次证明晶体中阴、阳离子是完全独立振动的，这种独立只是一种近似，我们现在不必考虑一些复杂因素。这样就可以通过红外光谱来判断 NO_2^- 是否存在。这种方法极为有用，但只局限于相对简单的物质。

现在，再来讨论一个稍微复杂一点的例子，如 $NaNO_3$。它有 6 种简正振动类型。然而，其红外光谱中仅发现三个峰：831cm^{-1}、1405cm^{-1}、692cm^{-1}。存在 6 种简正振动类型是毫无疑问的。计算公式仍然有效。但对于 NO_3^-（或对任何其他类似的平面三角形离子或分子），对称伸缩振动是非红外活性的（图 3.11）。

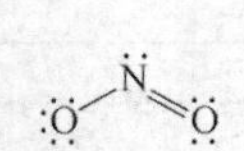

图 3.10 亚硝酸阴离子的结构示意图

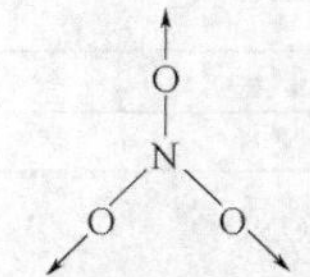

图 3.11 硝酸阴离子的对称伸缩振动模式

这是因为这种伸缩振动并没有引起离子偶极矩的改变，不能发生红外吸收，因此，预期的 6 个吸收峰要减少一个。

在余下的 5 个峰中，有两组是简并的，也即有两种振动发生在相同的频率，因此尽管有 5 种红外吸收，却只能给出三种特征吸收峰，这种对称伸缩振动可以在（下面讲到的）拉曼光谱中观察到。然而，如同 NO_2^-，这些吸收峰也可用于检测 NO_3^- 的存在。

同样的道理，其他相对简单的阴离子（或阳离子）可通过红外光谱检测判断。一些常见无机原子团的红外光谱数据列于表 3.4 中。

我们可以看出未被干扰的（离子化的）NO_2^- 有三个红外吸收带，分别为 1335cm^{-1}，1250cm^{-1}，830cm^{-1}。任一无机亚硝酸盐如 KNO_2、$NaNO_2$ 中 NO_2^- 的红外光谱均是如此。

然而，NO_2^- 也可作为配体，它与金属离子可有两种配位方式，如图 3.12 所示。

能利用红外光谱来检测和判断 NO_2^- 的配位情况吗?答案是肯定的。当 NO_2^- 以 O 原子配位时，一个 N—O 键几乎近于双键。另一情况下，两个 N—O 键则介于单双键间。键强增加，键的振动频率增加。可以预测，NO_2^- 中 N—O 键的频率以下列顺序递增：

$$\text{N—O（O 成键）} < \text{N—O（N 成键）} < \text{N=O（O 成键）}$$

与此相符，已经发现 NO_2^- 通过 O 配位成键的配合物中，两个 N—O 键频率分别为 1500～1400cm^{-1}（N=O）和 1100～1000cm^{-1}(N—O)。

表 3.4 一些常见无机原子团的红外光谱数据（主要的常见多原子离子）

盐	谱峰的位置/cm^{-1}	相对强度
NaSCN	758	w
	940	vw
	1620	m
	2020	s
	3330	m
KSCN	746	m
	945	vw
	1630	m
	2020	s
	3400	m
$NaNO_2$	831	m
	1358	vs
	1790	vw
	2428	vw
$NaNO_3$	836	m
	1358	vs
	1790	vw
	2428	vw
Na_2SO_4	645	w
	1110	vs
K_2SO_4	1110	vs
$NaClO_3$	935	s
	965 990	vs
$KClO_3$	938	w
	962	vs
$NaClO_4$	1100	vs
	1630(H_2O)	s
	2030	vw
$KClO_4$	627	w
	940	vw
	1075	s
	1140	s
	1990	vw
$NaBrO_3$	807	vs
$KBrO_3$	790	vs
$NaIO_3$	767 775	vs
	800	m

注：vs，很强；s，强；m，中等；w，弱；vw，很弱。

NO_2^- 通过 N 配位成键的配合物中，吸收频率介于二者之间，即 1340～1300cm^{-1} 和 1430～1360cm^{-1}。这样，就可相对容易地依据红外光谱判断 NO_2^- 是通过 N 原子还是通过 O 原子配位。

再举一例，如 NO_3^-，作为配体，可有三种配位方式（图 3.13）。

氧原子配位　　氮原子配位

图 3.12 亚硝酸阴离子的两种配位模式

硝酸根离子——所有的氧原子是等价的

图 3.13 硝酸根离子的三种配位模式

当上述三种配位方式发生时，原本在自由的 NO_3^- 中不可区分的三个氧原子，在配位后变得不再相同。在所有这三种情况中，三个氧原子中的两个是等同的，另一个不同。这意味着配位降低了 NO_3^- 的对称性。总的结果是将 AB_3 型变成了 AB_2C 型。由于 C 不同于 B 使得分子（离子）产生了偶极矩。更重要的是，现在的对称伸缩振动很可能引起偶极矩的改变，从而成为红外活性的。类似地，在 AB_3 型中不对称伸缩是二重简并的，AB_2C 型中将有两个不同的能级（如图 3.14 所示）。

自由 NO_3^-：单一谱带。

配位 NO_3^-（任意模式）：两个谱带。

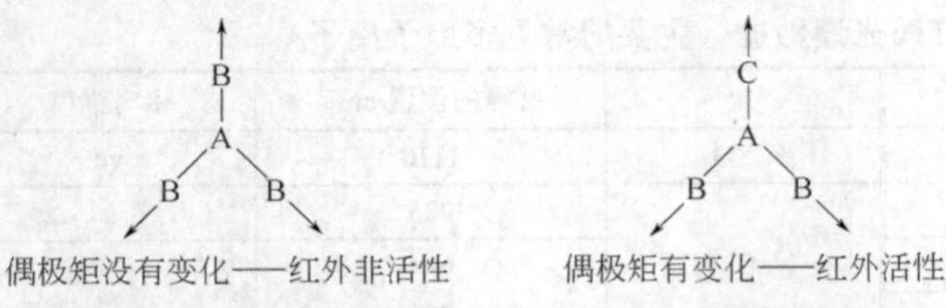

图 3.14 配位的硝酸根阴离子的对称伸缩振动模式

因此，通过红外光谱很容易将自由 NO_3^- 与配位 NO_3^- 区别开来。分子的对称性越高，在红外光谱中出现的峰的数目就越少。

3.2.3 拉曼效应与拉曼散射

当光线被分子散射时，大部分光子发生弹性散射。散射的光子与入射的光子具有相同的能量（即相同的频率），也即相同的波长。但是，极少部分散射光子（约 $1/10^7$ 个光子）的频率不同于（通常是低于）入射光的频率。这种非弹性散射过程称作拉曼效应。分子振动能、转动能或电子能改变时会发生拉曼散射。化学家主要关心振动拉曼效应，本章中所说的"拉曼效应"专指振动的拉曼效应。

入射光与拉曼散射光的能量差等于散射分子的振动能。散射光强度对能量差作图就是拉曼光谱。

（1）散射过程　拉曼效应起因于光子入射到分子并与分子的电偶极矩作用。它是一种电子光谱（更准确地说是电子振动光谱），尽管谱图包含分子振动频率。用经典术语说，这种相互作用可看作对分子电场的扰动。从量子力学观点看，散射可描绘为一种向虚拟状态的激发，该虚拟状态的能量稍低于实际的电子跃迁能量，几乎同步伴随着褪激和振动能量的改变。散射发生在 10^{-14}s 或更短。散射过程的虚拟态描述如图 3.15 所示。

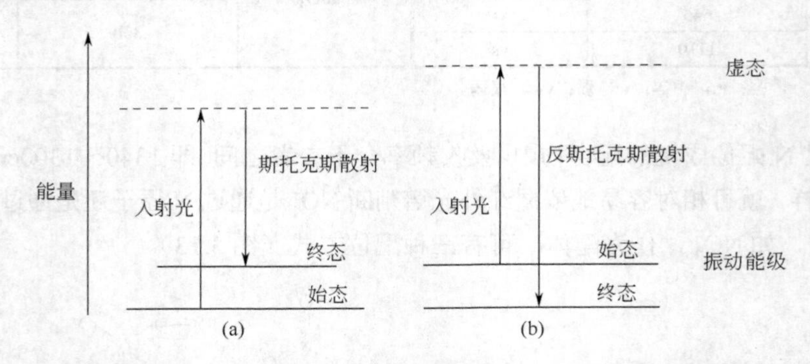

图 3.15 拉曼散射能级图：（a）斯托克斯拉曼散射；（b）反斯托克斯拉曼散射

室温时，振动激发态的热力学布居数很低，但并不为零。因此，始态为基态，散射光子比激发态光子的能量低（更长的波长）。这种斯托克斯位移散射（Stokes Shifted Scatter）是常常在拉曼光谱中看到的［如图 3.15（a）所示］。

小部分分子处于振动激发态，从振动激发态发生的拉曼散射使分子回到基态，散射光子以高能量出现［如图 3.15（b）所示］。反斯托克斯位移拉曼光谱（Anti-Stokes-Shifted Raman Spectrum）常常比斯托克斯位移光谱更弱；但是，对频率低于 $1500cm^{-1}$ 的振动，在室温下，它已足够强，会很有用。

斯托克斯与反斯托克斯拉曼散射含有相同的频率信息，在任何振动频率，反斯托克斯与斯托克斯谱线强度的比例与温度有关。反斯托克斯拉曼散射可用于非接触式测热仪。当检测器响应差而使斯托克斯光谱不能直接观测到时，可用反斯托克斯光谱代替。

（2）振动能量　振动能量与分子结构及分子所处环境有关。原子质量、键级、分子取代基、分子几何构型和氢键都会影响振动力常数，振动力常数反过来可反映振动能量。例如，磷-磷键伸缩频率对单键、双键、叁键分别为 460cm^{-1}、610cm^{-1}、775cm^{-1}。

人们已经做过许多努力来估计或测量键力常数。对小分子，甚至对一些延伸结构如肽，通过商品化专业软件可合理、准确地计算出其振动频率。

振动拉曼光谱不仅仅局限于分子内振动。晶体点阵振动和其他扩展的固体的振动模式，也是拉曼活性的。它们的光谱在聚合物、半导体领域很重要。在气相，转动结构也可通过振动跃迁来解析。振动/转动光谱广泛用于研究燃烧和气相反应。在这种意义上的振动拉曼光谱对物理学、生物化学、材料科学等广大领域来说，是一个非常有用的手段。

3.2.4　拉曼选律与强度

拉曼选律与大家很熟悉的红外活性振动选律类似。红外活性振动选律指出振动过程中永久偶极矩必定发生净变化。根据群论很容易理解具有对称中心的分子，拉曼活性的振动在红外中是非活性的，反之亦然。

之所以发生拉曼散射是因为分子振动会改变极化率。如振动对极化率的改变不大，拉曼带的强度将会很低。高极性部分的振动，如 O—H 键，常常较弱。外加电场不能使偶极矩发生大的改变，键伸缩或弯曲也不会改变偶极矩。

典型的强拉曼散射对应的是带有分布电子云的部分，如 C═C 双键。双键的 π 电子云在外加电场中很容易变形。键的弯曲或伸缩也将改变电子云的密度分布，从而导致诱导偶极矩的极大改变。

化学家通常将量子力学方法引入拉曼散射理论。该理论将散射频率和强度与分子的振动和电子能态联系在一起。标准的微扰理论处理方法认为入射光的频率比第一电子激发态的频率稍低，并用分子所有可能的振动激发态来描绘基态波函数较小的改变。

3.2.5　极化效应

拉曼散射是部分极化的，甚至对气态或液态分子也是如此，在气态或液态中，单分子的方向是随机的。通过平面极化的激发源很容易观察到这种效应（影响）。在各向同性介质中极化发生是因为诱导电偶极矩具有随分子的配位状况不同而使空间结构改变的成分。完全对称的拉曼振动散射，会强烈极化并与入射光的极化面平行取向。非完全对称的振动散射强度只有完全对称振动强度的 3/4。

晶体材料中情况更复杂。在这种情况下，在光学系统中晶体的取向是固定的，极化组分依赖于晶体轴相对于入射光的极化平面的取向，以及入射光和起偏镜的相对极化强度。

3.3　X 射线衍射分析

X 射线是波长很短的电磁波，波长范围在 0.001～10nm。这一波长与原子的大小和晶体内原子间的最小距离接近。1912 年，Max von Laue 证明了由于晶体点阵存在周期性，可以用 X 射线衍射法十分精确地测定晶体内的原子位置。

3.3.1　晶体中的对称性

在 2.1 节，我们知道分子具有对称性。字典中把对称定义为源于物体各部分互相平衡的一种形式美。如果物体中某条线、某个面或某个点对侧的不同部分的大小、形状、相对位置都一一对应，我们说此物体具有对称性。虽然对称对我们而言是一种直观感觉，但是对称在化学及其他学科中却有多种应用。在自然界的各种不同尺度上，从天体运行的轨道，到原子中电子的行为，都会观察到对称性。对称的概念也有广泛的应用，它把一些看起来非常不同、毫无关联的现象联系在了一起，如雪花的形成、音乐的节律以及蜜蜂飞舞的图案。对我们来说，更为重要的是物质的物理性质与其内部原子排列之间有密切联系，一旦知道了分子或晶体的对称性，就可以预测其物理性质。在晶体结构分析中，则经常应用对称性计算点阵常数及晶体中原子的位置。

晶体是对称性物体，具有规则的几何外形。这意味着可以用群论的方法对晶体进行分类并预测其物理性质。而且，晶体是由离子、原子或分子按特定的方式排列而成的，这种排列方式决定了晶体规则的几何外形。因此也有可能用群论的方法来研究晶体中分子排列的局部对称性。

由于晶体内部结构的限制，晶体中有限对称元素的数目只有 8 个：1, 2, 3, 4, 6, −1, −2 和 −4，其中有些是两个简单对称元素的组合。对称元素和对称操作只能在特定的角度上互相作用，产生新的对称元素和对称操作。对称元素的数目是有限的，其互相作用的方式也是有限的，因此对称群的数目也是有限的。实际上，三维晶体学点群共有 32 个，这 32 个晶体学点群可以分为 7 个晶系。晶体对称元素的组合方式及其取向使不同晶系的晶轴长度和其间的夹角都有独有的特征（表 3.5）。

表 3.5 晶体学点群和晶系

晶系	点阵常数特征	点群	点群的阶
三斜	$a \neq b \neq c$ $\alpha \neq \beta \neq \gamma$	C_1 C_i	1 2
单斜	$a \neq b \neq c$ $\alpha = \beta = \pi/2 \neq \gamma$	$C_s\ (C_{1h})$ C_2 C_{2h}	2 2 4
正交	$a \neq b \neq c$ $\alpha = \beta = \gamma = \pi/2$	C_{2v} D_2 D_{2h}	4 4 8
四方	$a = b \neq c$ $\alpha = \beta = \gamma = \pi/2$	C_4 S_4 C_{4h} D_{2d} C_{4v} D_4 D_{4h}	4 4 8 8 8 8 16
三方	$a = b = c$ $\alpha = \beta = \gamma < 2\pi/3 \neq \pi/2$	C_3 S_6 C_{3v} D_3 D_{3d}	3 6 6 6 12
六方	$a = b \neq c$ $\alpha = \beta = \pi/2,\ \gamma = 2\pi/3$	C_{3h} C_6 C_{6h} D_{3h} C_{6v} D_6 D_{6h}	6 6 12 12 12 12 24
立方	$a = b = c$ $\alpha = \beta = \gamma = \pi/2$	T T_h T_d O O_h	12 24 24 24 48

离子、原子或分子组成某种最小单元，并在三维空间按某种方式周期性重复排列，即得到单晶。这样一个无限的周期性结构可用空间点阵描述，点阵中的点都具有相同的环境。点阵的最小重复单元称为晶胞。三维空间点阵可以用 3 个不共面的矢量 $\boldsymbol{a}$, $\boldsymbol{b}$ 和 $\boldsymbol{c}$ 来表达，这 3 个向量通常与晶轴一致。晶轴的长度和轴间夹角分别为 a, b, c, α, β 和 γ（图 3.16）。

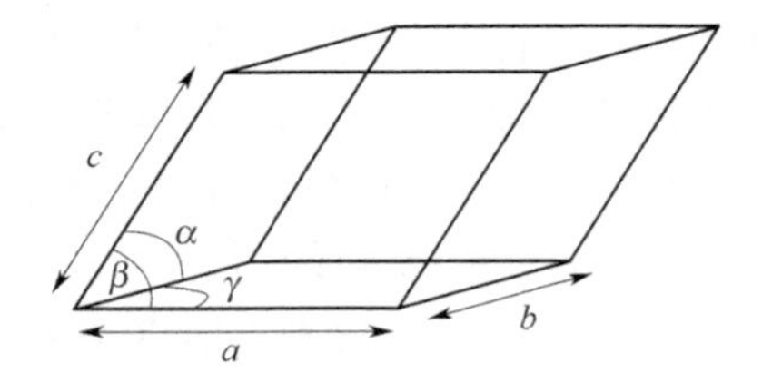

图 3.16　晶胞及点阵常数

系统描述和枚举三维空间点阵的工作是由 Bravais 于 1850 年完成的，他发现可以把所有的空间点阵划分为 14 个不同类型，这就是 14 种 Bravais 点阵（图 3.17）。

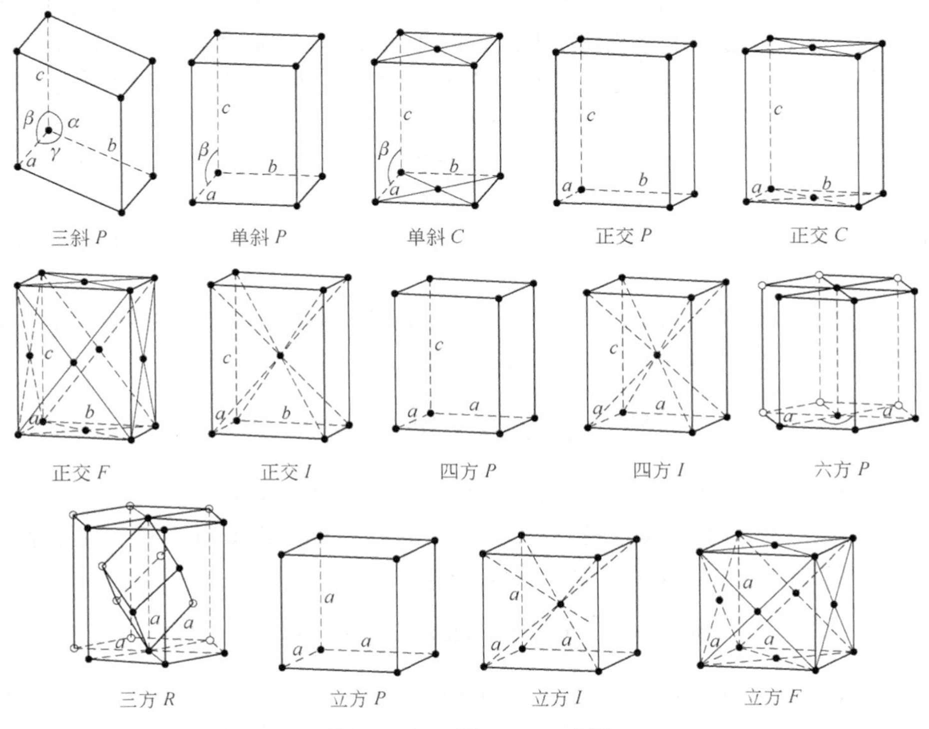

图 3.17　14 种 Bravais 点阵

我们记得，点群描述的是有限图形的对称性，点阵描述的是物体在三维空间的周期性重复。因此有一个问题是，把一个有限图形按 Bravais 点阵进行重复，会有怎样的结果？实际上，这就是晶体中原子排列的真实情况。可以通过点对称、平移对称或二者的组合使一个无限周期性结构与其自身重合。其中，点对称和平移对称的组合可以是螺旋旋转或滑移反映。无限对称元素之间也可以相互作用并产生对称元素，而且这些对称元素不必通过一个公共点。不同的有限和无限对称元素的组合形成空间群。空间群的对称元素分布在晶体内的无限空间内，因此连续物体中的每个点都可以通过空间群的对称操作，在晶体空间内周期性平移。因为旋转和平移的方式都是有限的，所以三维晶体空间群数目也是有限的，共有 230 个。每个空间群的详细信息可以在 International Tables for Crystallography, Volume A 中查到。

3.3.2　X 射线单晶衍射

单晶对 X 射线的衍射效应是 Max von Laue 于 1912 年发现的，Bragg 父子马上把这种方法用来分析晶体结构。现在，这种方法已经成为在原子水平上分析晶体结构最为有效的手段。晶体结构可以给化学家和物理学家提供大量信息，如原子间的连接方式、分子构型、精确的键长键角、结构的对称性及原子的三维堆积方式、立体化学以及晶体密度等。

3.3.2.1　晶体的 X 射线衍射

原子中的电子受到入射 X 射线电磁场的作用，做同频率的受迫振动，因此发射同频率的 X 射线，此即原子对 X 射线的散射。原子不同部分散射 X 射线的强度随着与入射线的夹角逐渐增大而降低，这种现象可用原子散射因子描述。

晶体中的每个原子都散射 X 射线，都对最终的散射图像有贡献。不同原子散射的 X 射线会互相干涉，可以加强，也可以减弱，加强或减弱依赖于衍射线的方向和原子的位置。描述一条衍射线，至少需要方向和振幅两个量，前者由 Laue 方程规定，后者由结构因子给出。

衍射线的方向，或者说衍射线和入射线方向间的关系，最先由 Laue 给出。其形式为一个由 3 个方程组成的方程组，此方程组给出了衍射发生的条件，即 Laue 方程［式（3.1）］。

$$\begin{aligned} \boldsymbol{a} \cdot \boldsymbol{s} &= h \\ \boldsymbol{b} \cdot \boldsymbol{s} &= k \\ \boldsymbol{c} \cdot \boldsymbol{s} &= l \end{aligned} \tag{3.1}$$

其中 $\boldsymbol{a}, \boldsymbol{b}, \boldsymbol{c}$ 是点阵常数，$\boldsymbol{s}$ 是散射矢量，h, k, l 是三个整数。此方程组在 X 射线晶体学中十分重要。

衍射强度受很多因素的影响，其中有一个只和晶体结构有关，称为结构因子。结构因子可以从晶胞中原子的分布情况计算出来［式（3.2）］：

$$F(hkl) = \sum_{j=1}^{N} f_j \exp[2\pi i(hx_j + ky_j + lz_j)] \tag{3.2}$$

式中，f_j、x_j、y_j、z_j 是原子 j 的原子散射因子和分数坐标；h, k, l 是 Laue 方程中的三个整数；N 是晶胞中原子的数目。可以证明，$F(hkl)$是电子密度的 Fourier 变换，且晶胞中任何一点（xyz）处的电子密度可以如下计算［式（3.3）］：

$$\rho(xyz) = \frac{1}{V}\sum_{hkl} F(hkl)\exp[-2\pi i(hx + ky + lz)] \tag{3.3}$$

式中，求和范围是所有 $F(hkl)$；V 是晶胞体积。一旦知道了晶胞中每一点上的电子密度，就可以找到电子密度峰值的位置，这就是晶体中原子的位置（图 3.18）。

晶体结构分析的一般步骤如图 3.19 所示。

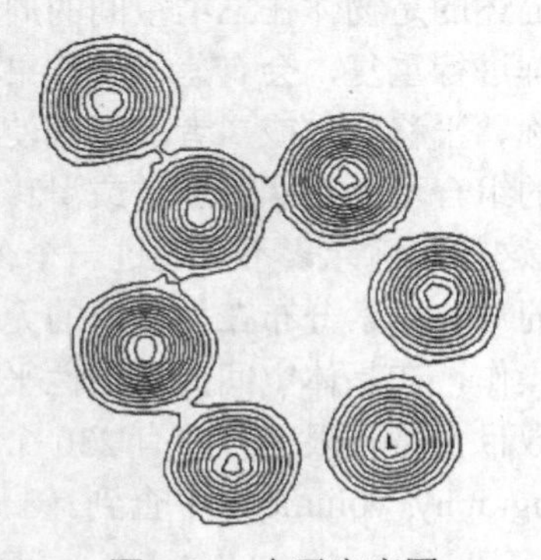

图 3.18　电子密度图

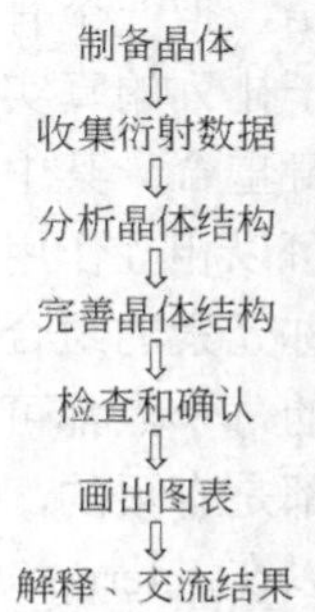

制备晶体
⇩
收集衍射数据
⇩
分析晶体结构
⇩
完善晶体结构
⇩
检查和确认
⇩
画出图表
⇩
解释、交流结果

图 3.19　晶体结构分析的一般步骤

3.3.2.2　晶体培养

衍射实验所用晶体的质量决定了最终结构分析结果的好坏。衍射实验过程中，培养晶体的目的是得到几颗大小合适，通常为0.1～0.4mm，质量足够好的单晶。听起来不难，但是晶体培养不是常规的工作，它更像一门艺术。培养晶体所用的方法很大程度上依赖于化合物本身的性质。一些常用的技术如下。

- 缓慢蒸发　这是培养晶体最简单的方法，适用于温和条件下不敏感的化合物。
- 缓慢冷却　适用于溶质的溶解度低且溶液沸点低于100℃的溶质-溶剂体系。
- 缓慢蒸发-缓慢冷却的变式　若上述两种方法不能从单溶剂体系中得到合适的晶体，可以把这两种方法拓展到双溶剂或三溶剂体系（一定要记下溶液组成）。
- 蒸发扩散　将物质用溶剂S_1制成溶液，并置于试管T中。在烧杯B中放入溶剂S_2，然后把试管T放入烧杯B中，并密封烧杯。S_2慢慢向试管内扩散，S_1则向外扩散，二者相互作用，形成晶体。
- 溶剂扩散法（分层法）　该方法适用于在温和条件下对空气和湿度等敏感的物质。将溶质溶于S_1中，放在试管内。缓慢向试管中滴加S_2，使S_1和S_2形成严格的分层，如CH_2Cl_2/Et_2O。
- 反应物扩散　类似其他扩散法，但是反应物溶液可以互相扩散。如果产物不溶，就会在反应物接触的地方形成晶体。
- 升华　对易挥发且空气敏感的晶体，可以把样品密封于真空玻璃管中，然后将其置于加热炉内几天或几周，就有可能长出晶体。
- 对流　该方法的原理是溶液在容器的热端区域饱和，迁移到冷端时就会成核形成晶体。对流速度与容器的热梯度成正比。
- 共晶法　不适用于所有化合物。若其他方法都不能得到合适的晶体，可以尝试让化合物与结晶助剂共同结晶。例如，三苯基膦氧化物（TPPO）曾被用作质子性有机分子的结晶助剂。此法的本质与缓慢蒸发相同，但是要在溶液中加入等物质的量的TPPO。
- 抗衡离子　采用不同的抗衡离子通常会改善晶体的质量。体积相似的离子会堆积得更紧密，形成更好的晶体，而且还可能会有更好的物理性质。比如，目标化合物的Na盐有吸湿性，而四甲基铵盐就没有或更弱。需要注意的是，不管用什么抗衡离子，确保它不与你的目标离子反应。
- 中性化合物离子化　如果化合物是中性的，且包含质子供体或受体基团，可以先将其质子化或去质子化，以长出更好的晶体。离子化后，化合物可以更好地利用H键形成更好的晶体。这会改变复合物的电学性质，但是如果你只想通过结构分析确定合成实验是否成功，这就不是大问题。

3.3.2.3　单晶X射线衍射实验

培养出晶体后，就可以开始单晶衍射实验。第一步通常是在显微镜下挑选几颗大小合适的晶体，并检查其质量。挑出最好的一颗，将其粘于细玻璃丝的顶端，安置于测角仪上，拍摄衍射照片，进一步确认晶体质量。如果晶体足够好，设置好波长、温度、数据收集方法等实验条件，然后可以收集衍射图像。收集完成后积分提取衍射强度，给定hkl的衍射强度正比于结构因子的平方。为计算结构因子的振幅，进行结构分析，必须对衍射强度进行温度因子、LP因子及吸收因子等校正。校正完成后，得到一套衍射数据，可用于结构分析。

精确测量衍射强度的典型设备是四圆衍射仪。直到20世纪90年代中期，几乎所有的化学结晶学衍射数据都是用四圆衍射仪收集的。四圆衍射仪共有 4 个圆，2θ、ω、χ和ϕ（如图3.20）。ω、χ和ϕ联动，可以把晶体转动到任何可能发生衍射的方向。2θ则可以把探测器转动到合适的位置，去接收衍射强度。

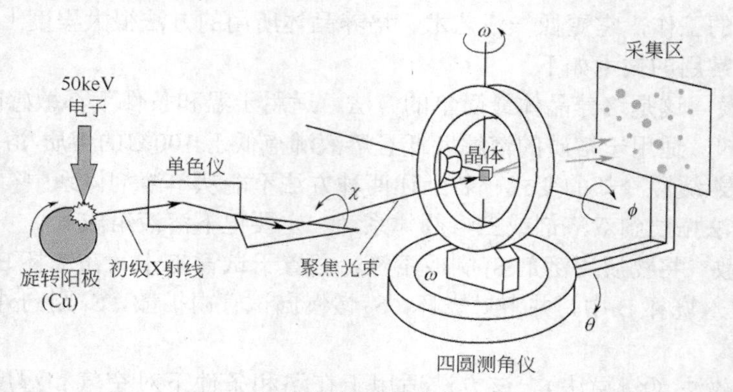

图 3.20　四圆单晶衍射仪示意图

图 3.21　四圆衍射仪

现在的衍射仪通常配备二维探测器如 CCD、CMOS 或者 IP。有些设备保留了 4 个圆，也有一些把χ固定在一个特殊的角度，只允许 2θ、ω和ϕ三个圆转动。图 3.21 是实验室里典型的衍射仪。

在配置 CCD 的单晶衍射仪上，典型的衍射实验通常有如下几步。

① 晶体安置和对心，记录晶体的颜色、形状和尺寸。

② 拍摄静态或有限回摆图，检查晶体质量。

③ 收集部分衍射图像，进行指标化，精修取向矩阵，确定点阵类型，检查晶胞是否合理。

④ 确定数据收集方法，收集衍射图像。

⑤ 积分衍射点，还原原始数据。

3.3.2.4　晶体结构分析

在方程（3.2）中，我们看到电子密度是结构因子 $F(hkl)$的傅里叶变换。但是测量的衍射强度中只能得到结构因子的振幅$|F(hkl)|$，却丢失了相位。因此傅里叶变换无法进行，必须首先通过 Patterson 函数法、直接法或电荷翻转法等找到衍射的相位。这就是晶体结构分析中的相位问题。

几乎所有的相位分析技术都只能给出部分近似相位，因此只能得到不完整的结构模型。确定了近似相位后，可以把这些相位与结构因子的模量$|F(hkl)|$组合，计算新的电子密度图或差值电子密度图。从这些电子密度图上，可以找到新的原子位置。此过程可以多次重复，直到找到所有原子。

此时，结构模型虽然完整，但是原子位置还存在很大误差。这会导致观测的结构因子与计算的结构因子不一致。因此，结构分析的下一步就是优化结构模型，使计算结构因子和观测结构因子的差值尽可能小。这就是晶体结构修正的过程，修正过程最常使用的是最小二乘法。

完成结构修正后，通常会得到一个比较好的结构模型，可以用来计算键长、键角、扭角及最佳平面等结构参数。计算之前，往往先检查并解决可能的错误。结构参数可以用数字、列表等形式表达，也可以画成结构图形。最后，在发表结果时需要提供结构信息文件。图3.22是单晶结构分析得到的一个分子结构。

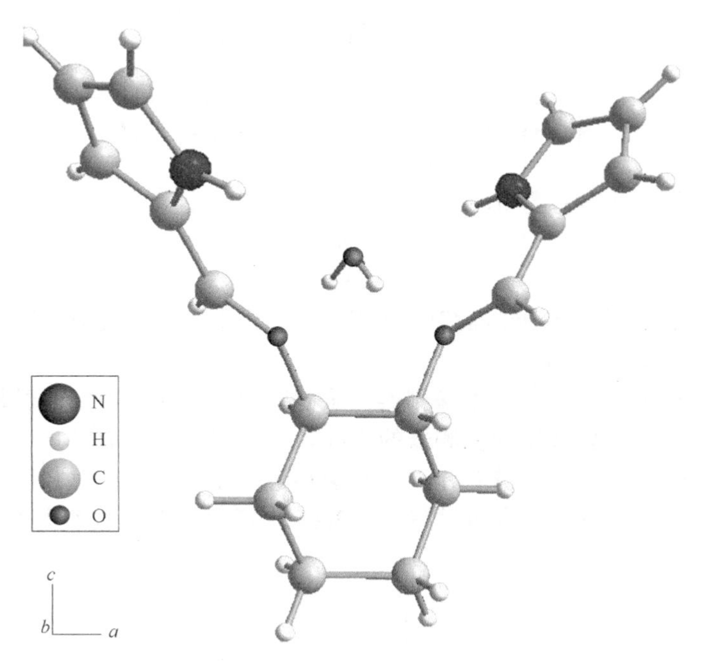

图3.22　单晶X射线衍射得到的一个分子结构

3.3.3　X射线粉末衍射

毫无疑问，X射线单晶衍射是获得高质量结构数据的首选方法。但是很多情况下，得到大小合适的高质量晶体并不容易。一些重要材料如高温超导体、磁性材料、多铁性材料及离子导电聚合物等都不容易制备合适的单晶，其重要结构信息都是从粉末衍射数据得到的。另外，粉末衍射还可以给单晶衍射数据提供补充数据，如块体样品组成等。而且，还有很多情况是需要研究晶体在高温、高压、强磁场等非温和条件下的结构和性质。这时，粉末衍射实验比单晶衍射实验更容易操作。粉末衍射法是Debye、Sherrer（1916）和Hull（1919）分别在欧洲和美国独立实现的。

3.3.3.1　粉末晶体对X射线的衍射

在单晶衍射实验中，用一束X射线照射到单晶上，用探测器记录其衍射线的位置和强度。在粉末衍射实验中，则用X射线照射随机取向的多晶颗粒。在满足特定的几何条件时，散射的X射线互相干涉加强，形成衍射峰（图3.23）。在1912年，W. L. Bragg找到了以下几个实验因素间的关系。

- 晶体中原子平面间的距离d，单位Å（$1Å=10^{-10}m$）。
- 衍射角θ，单位为度。由于实验上的考虑，衍射仪测量的角度一般是2θ。
- 入射X射线的波长λ。在使用Cu做阳极时，为1.54Å。

这几个因子由Bragg公式联系起来：

$$n\lambda=2d\sin\theta \tag{3.4}$$

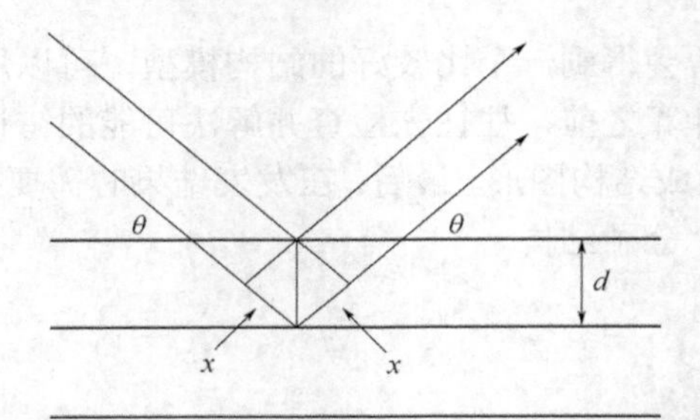

图 3.23 晶体中两个原子平面对 X 射线的反射

把公式进行改写，可以计算 d：

$$d=\frac{n\lambda}{2\sin\theta} \tag{3.5}$$

单晶衍射中的Laue方程和粉末衍射中的Bragg方程都可以确定衍射线的方向。可以证明，虽然二者形式不同，但是互相等价。

如果也像单晶衍射一样，用二维探测器记录衍射强度，就会得到衍射强度的圆环，称为Debye 环。Debye 环是衍射圆锥和探测器的交线。一般情况下，衍射强度通过点探测器或一维探测器扫描 Debye 环的一窄条来记录。结果会得到一条曲线，衍射强度是纵坐标，衍射角 2θ是横坐标，这就是最常见的粉末衍射图（图 3.24）。任何粉末衍射图都有多个 Bragg 峰，其位置、强度、峰形都不同。

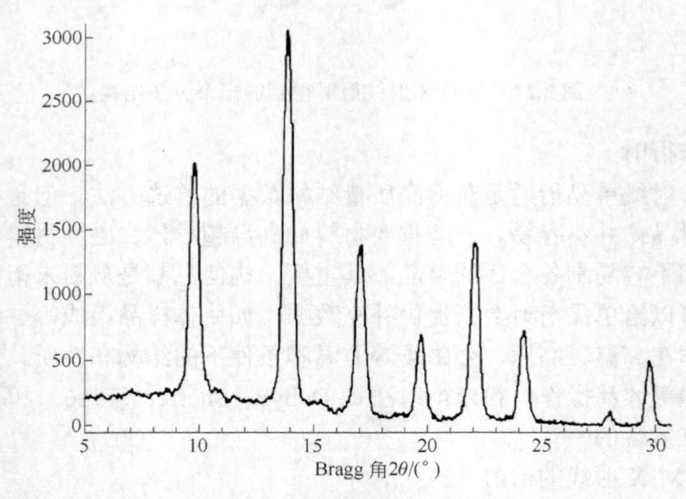

图 3.24 典型的粉末衍射图

衍射峰的位置是 d、λ和 hkl 的函数，但是此函数是不连续的。当 hkl 相同时，点阵常数和辐射波长是决定衍射峰位置的两个主要因素。图 3.25 是 NaCl 和 KCl 粉末衍射图的对比。KCl 衍射图中的衍射峰标出了 Miller 指数 hkl，表示产生此衍射峰的原子平面。与 NaCl 相比，KCl 的衍射峰左移，这是因为 KCl 的点阵常数略大。

衍射强度是结构因子的函数，也与样品和仪器有关。样品因素包括但不限于择优取向、晶粒大小及分布等；仪器因素则包括波长、聚焦几何、狭缝系统等，这些因素都会引起强度偏差。

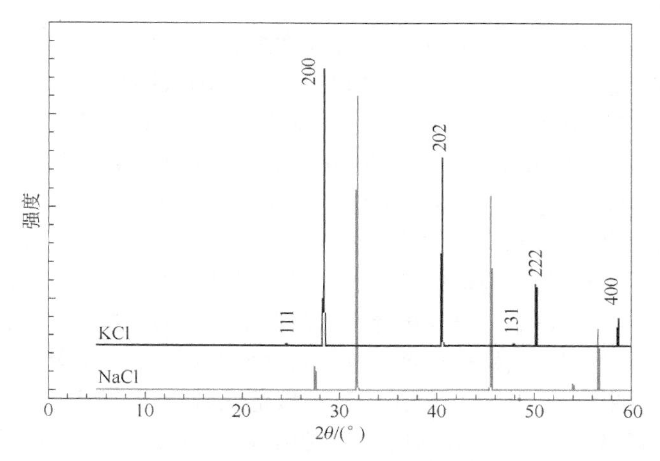

图 3.25　NaCl 和 KCl 粉末衍射图的对比

3.3.3.2　X 射线粉末衍射实验

早期的粉末衍射图使用不同的粉末相机来记录，在 20 世纪 70 年代粉末相机被自动粉末衍射仪取代。粉末衍射仪的种类很多，但是它们有些共同点。以常规的发散 X 射线作为光源的衍射仪，其最普遍构型采用赝聚焦的 Bragg-Brentano 几何（图 3.26）。X 射线管 S 产生发散的 X 射线，经过单色器 M 单色化后聚焦于 F。焦点 F、平板样品和接收狭缝位于聚焦圆上。聚焦圆的半径随θ变化。样品散射的相干 X 射线收敛于探测器 D 前的接收狭缝。探测器和样品都绕测角仪中心轴转动，前者转速是后者转速的 2 倍，此即θ-2θ扫描模式。也有的测角仪设计成射线源和探测器绕测角仪轴同步同速转动，样品保持不动，这是现在最常用的θ-θ扫描。图 3.27 是一台立式测角仪，采用 Bragg-Brentano 赝聚焦几何，扫描方式为θ-θ扫描。

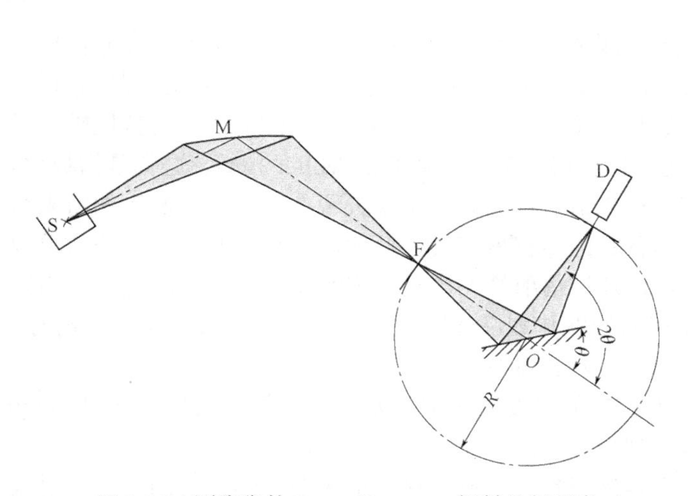

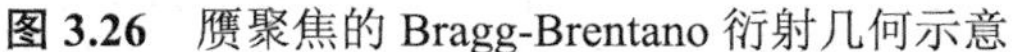

图 3.26　赝聚焦的 Bragg-Brentano 衍射几何示意

图 3.27　一台典型的粉末衍射仪

粉末衍射数据受很多因素影响，如样品状态、辐射波长、扫描速度等。理想的粉末样品应该由数量巨大的细小晶体颗粒组成，尺寸在 50μm 或更小，取向随机。样品可以是颗粒细小的沉淀，或者是粗粒晶体经研磨后的粉末。很多情况下，研磨或球磨会产生形状各向异性的颗粒，因此在把这些颗粒装入样品架时就需要格外小心，以避免择优取向，尤其是对片状或针状样品。平板样品的长度应足够大，确保在数据收集过程中，X射线在样品上的投影在任何Bragg角都不超过样品长度。样品厚度也应足够，确保样品对入射线完全不透明。样品在任何时候都需要恰当安放在测角仪中心。入射线的波长通常根据样品的特点及实验的目的进行选择。最典型的辐射是 CuK_α，波长 0.154nm。长波长多用于精确测定点阵常数，短波长多用于检测较大的倒易空间。另一个重要的考虑是样品是否会产生X射线荧光。例如，所有的含Co材料在用 CuK_α射线照射时都会产生严重的X射线荧光，这显然不利于得到高质量的数据。要恰当选择入射线的狭缝，使其与衍射几何、样品尺寸相匹配。衍射线的狭缝也要恰当选择，确保数据有足够的强度和分辨率。恰当选择所有的仪器参数后，接下来要确定扫描方式、扫描范围、扫描步长和计数时间，以根据仪器、样品及需要得到的信息收集最佳的衍射数据。

典型的粉末衍射图中，强度是衍射角的离散数值函数。很多应用，比如物相分析、点阵常数测定、结构解析等需要的是积分强度、衍射指标和Bragg角。为得到这些信息，需要进行初步的数据处理，包括寻峰、去背景、平滑和 K_{α_2} 剥离。

3.3.3.3　粉末衍射数据分析

粉末衍射数据中包含大量关于样品的信息。峰的位置由晶胞的大小、形状决定，峰的强度则取决于晶胞中原子的种类、数目和位置。峰形是仪器参数和样品微观结构如晶畴的大小、应变的卷积。粉末衍射数据也可以用来测定晶体结构，但是实验的常规目的一般是物相分析、点阵常数精确测定、晶粒的大小和应变以及Rietveld修正。

X射线粉末衍射可以对材料的相组成进行定性和定量分析。材料中每一个结晶相都有其特征的衍射峰，就像人的指纹一样。根据这些特征峰，可以判断材料中是否包含某一物相。峰的强度大约正比于其含量。把测量的粉末衍射图谱与数据库中已知物相的衍射图谱对比，就可以确定材料的相组成。另外，粉末衍射在小分子化学结晶学中也有重要应用。其中一个例子是确定体相材料的结构是否与单晶结构分析结果一致。对于多相样品，可以通过分析粉末衍射数据，得到相含量。

我们说测定了一个晶体结构，一般指测定了其晶胞的大小、形状、对称性及晶胞内原子的排列方式。因此测定点阵常数 a、b、c、α、β 和 γ 是不可避免的。在没有合适单晶的时候，粉末衍射也可以提供这些信息。有了这些信息后，可以直接将其与已知的点阵常数比较，做简单快速的物相鉴定。而且，还可以通过计算得到每个衍射峰的Miller指数，这就是指标化。

如果晶体颗粒足够大且没有应变，粉末衍射峰应该是十分尖锐的。但实验测量的粉末衍射峰却从来不是这样，因为仪器和样品参数的卷积效应会使衍射峰宽化。其中一个非常重要的样品因素就是晶体颗粒的尺寸，这可以通过衍射峰的半高宽来衡量。晶体颗粒越小，衍射峰的角宽度越大。Sherrer首先解决了衍射峰宽化和晶体颗粒大小之间的关系，其关系式为

$$L=K\lambda/(\beta\cos\theta)$$

式中，L 为晶体颗粒在垂直衍射平面方向的厚度；β 和 θ 分别为此衍射峰的半高宽和Bragg角；λ为波长；K 为接近1的常数，通常取0.89或0.94。

很多难以形成单晶的化合物，都很容易得到高质量的粉末衍射图。因此可用粉末衍射数据修正晶体结构，这就是Rietveld方法。这一方法对于无法形成单晶的晶体结构都可以进行修正。Rietveld方法最初是由H. G. Rietveld在20世纪70年代发展起来的，用于根据中子粉末衍射数据修正晶体结构。后来，这一方法很快扩展到X射线数据。此方法中，可以根据结构模型、峰形函数及背景函数等在每个实验数据点 2θ上计算理论衍射强度 y_{calc}，然后进行最小二乘修正，调整结构参数和仪器参数，使计算强度 y_{calc} 和观测强度 y_{obs} 之间的差值最小。当符合程

度足够好时，结束修正，就得到了结构参数和样品参数。图 3.28 是 Rietveld 修正的一个例子。

Rietveld 方法是一个有力的结构优化和修正工具，可以得到样品的结构细节。此方法需要一个结构初始模型，但是它本身却无法建立此模型。为了建立修正的初始模型，需要根据粉末数据，用 Patterson 法、直接法、电荷翻转法和模拟退火法等进行结构分析。其基本步骤如下：

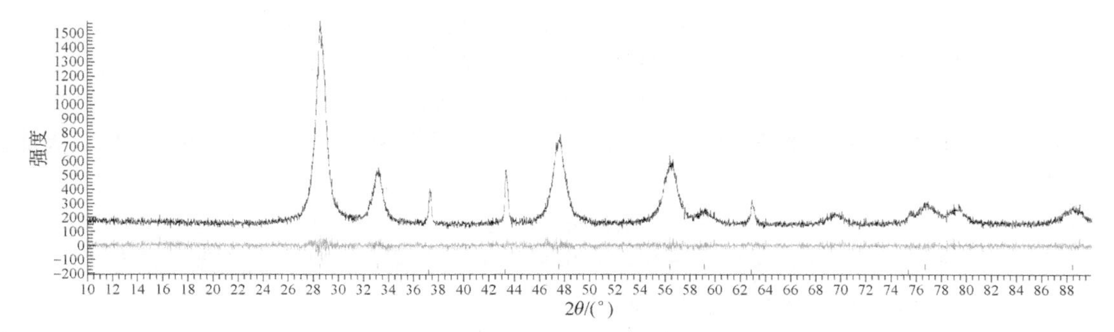

图 3.28　Rietveld 拟合实例

- 测定点阵常数；
- 从粉末衍射图提取积分强度（可选）；
- 根据系统消光确定空间群（可选）；
- 用不同方法建立初始模型；
- 用 Rietveld 法修正结构模型。

从粉末衍射数据测定晶体结构，开辟了晶体结构研究的新道路。但是此方法要比单晶结构分析困难很多。主要的问题是单晶衍射中三维的晶体衍射数据信息被压缩成一维的粉末衍射图，造成了在点阵常数测定及衍射强度提取方面的多解性。但是随着仪器的更新，算法的发展，也有可能仅从粉末衍射数据测定一些晶体结构。

3.4　光电子能谱

光电子能谱利用光电离和发射光电子的能量色散的分析来研究样品表面的组成和电子状态。通常，依据所用激发光源的不同，光电子能谱技术可分为（以下两种）。

- X-射线光电子能谱（XPS）：用软 X 射线（1200～2000eV）研究原子内层电子状态。
- 紫外光电子能谱（UPS）：用紫外线（10～45eV）研究价层电子状态。

同步辐射源技术的发展，可获得更宽、更完整的能量范围的光源射线（5～5000eV），同时获得更高的分辨率。但是，由于造价高昂、技术复杂，在光电子研究中，这种技术的应用受到限制。

光电子能谱基于光子进入和电子发射过程。多数观点认为此过程比俄歇过程更为简单。光子能量可由爱因斯坦公式得出

$$E=h\nu$$

式中，h 为普朗克常数，6.62×10^{-34}J·s；ν 为辐射频率，Hz。

光电子能谱使用单色光源（也即固定能量的光子）。

在 XPS 中，固体或分子中的原子吸收光子，导致内层电子发生电离和发射。与此相应，UPS 中，光子与固体或分子中的价层相互作用，导致失去一个价层电子而发生电离。通过合适的电子能量分析器能测定所发射的电子的动能分布（例如发射的电子数作为动能的一个参

数），记录下来就得到光电子能谱。

光电离过程可认为分几步进行，总过程可简单表示如下：

$$A + h\nu \longrightarrow A^{+} + e^{-}$$

能量守恒要求：

$$E(A) + h\nu = E(A^{+}) + E(e^{-})$$

由于此时电子的能量仅为动能（KE），光电子的 KE 可表示为：

$$KE = h\nu - [E(A^{+}) - E(A)]$$

括号项表明电离了的离子与中性原子的能量差值，通常称为电子的结合能（BE），因此，又可得下式：

$$KE = h\nu - BE$$

这里结合能（BE）可看成把电子从最初的原始能级移到真空能级所需要的能量。

光子发射（XPS 或 UPS）的基本要求是：

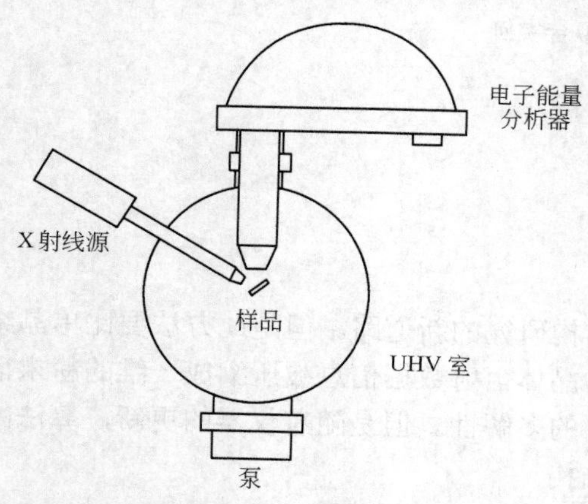

图 3.29　光电子能谱实验系统的装置示意图

- 特定能量的辐射源（X 射线源用于 XPS；氦灯光源用于 UPS）；
- 电子能量分析器（能将发射电子根据它们的动能进行能量色散，从而可以测量到具有一定能量的发射电子流）；
- 真空系统（能保证发射电子在不受到气相碰撞的干扰下进行能量分析）。

这样的装置如图 3.29 所示。

电子能量分析器有许多种类型，但对于光发射实验来说，首选的是浓度型半球分析器(CHA)，它是在两个半球体之间施加一个电场，将电子根据动能进行分布。

3.4.1　X 射线光电子能谱（XPS）

对所有元素，每一个原子都有与内部原子轨道相关的特征结合能。即每一元素在光电子能谱中都将产生一组特征峰。这些特征峰由光子能量与相应结合能所决定。

在特定能量处峰的存在表明所研究样品中含有某一特定元素，因此，峰的强度与样品中元素的浓度成正比例。这样，通过这项技术，可对表面组成进行定量分析。这项技术有时由它的首字母缩略词 ESCA（电子光谱化学分析）来命名。

通常使用的 X 射线源有：Mg K_α（$h\nu$=1253.6eV）和 Al K_α（$h\nu$=1486.6eV）。相应地，发射的电子动能介于 0～1250eV 或 0～1480eV。由于这些电子在固态中有很短的 IMFPs（非弹性散射平均自由程），所以这项技术对表面测定非常灵敏。

例：金属钯的 XPS 光谱

图 3.30 是用 Mg K_α 射线源测定金属钯所得到 XPS 光谱。主要的峰

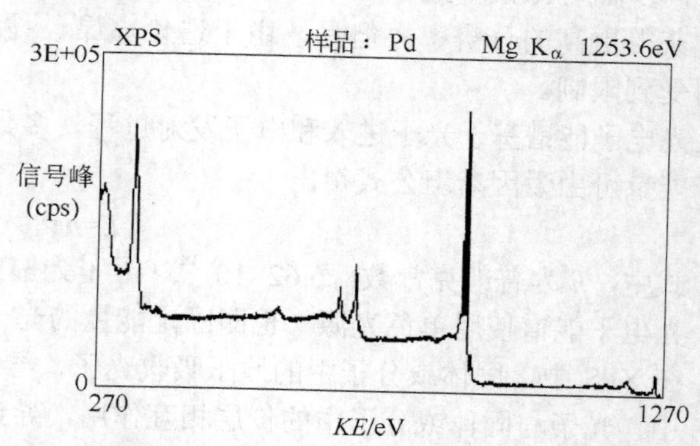

图 3.30　利用 Mg K_α 辐射源得到的金属钯（Pd）的 XPS 谱图

有：330eV；690eV；720eV；910eV 和 920eV。

由于射线的能量是已知的，因此，很容易把电子动能转换为结合能，如图 3.31 所示。

结合能在 335eV 处出现的峰的强度最大，从最高能级往下：

- 价带发射（4d，5s）常在结合能 0～8eV 范围内发生；
- 4p 和 4s 发射分别在 54eV、88eV 处出现很弱的峰；
- 在 335eV 处出现的最强峰是由于 Pd 的 3d 电子发射，3p、3s 分别在 534/561eV 和 673eV 处出现峰；
- 其余的峰根本不是 XPS 峰！是 X 射线诱导俄歇发射形成的俄歇峰。

这些峰的归属如图 3.32 所示。

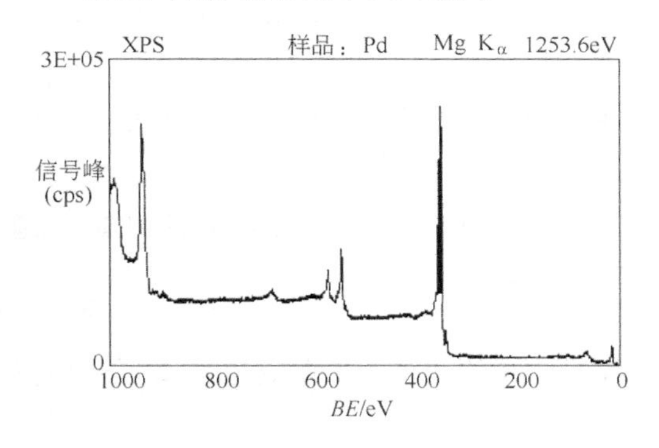

图 3.31　以结合能（*BE*）为横坐标所得到的金属钯（Pd）的 XPS 谱图

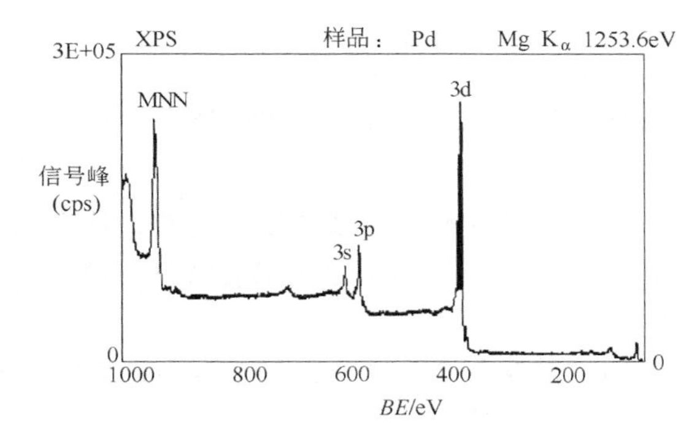

图 3.32　金属钯（Pd）XPS 谱图中的谱峰的归属

需进一步引起注意：不同光子发射峰的自然宽度明显不同；峰强度不仅仅只与轨道中电子占有情况相关。

3.4.1.1　自旋-轨道裂分

对光谱的进一步研究发现，一些能级的发射（3p、3d 最明显）并不给出单一峰，而是很近的双重峰（图 3.33）。

为看得更清楚，我们将 3d 部分的光谱展开。可以发现：3d 峰实际上裂分为二重峰，其中一个峰在 334.9eV 处，另一个峰在 340.2eV 处。两者的强度之比为 3∶2。这是由于终态的自旋轨道偶合作用而引起的。

在初始状态时，金属 Pd 的内层电子排布为：

$$(1s)^2(2s)^2(2p)^6(3s)^2(3p)^6(3d)^{10}\cdots，所有的亚层全满$$

通过光电离，从 3d 亚层去掉一个电子，得到终态的（3d）^{9} 构型。由于 d 轨道（l=2）具有非零的轨道角动量，因此，会发生未成对电子自旋与轨道角动量间的偶合。自旋-轨道偶合通常用两种模型的一种来处理，这两种模型有限制条件，满足了两限制条件，偶合可发生：一是 LS（或 Russell-Saunders）偶合近似，另一是 *j-j* 偶合近似。

如果我们用 LS（或 Russell-Saunders）偶合近似模型来考虑金属 Pd 的电离终态时，其（3d）^{9} 构型将产生两种能态（忽略任何价带间的偶合）。它们在能量和简并度方面有细微的不同。

$^2D_{5/2}$　　$g_J=2\times(5/2)+1=6$

$^2D_{3/2}$　　$g_J=2\times(3/2)-1=4$

这两种状态归因于 L=2 和 S=1/2 间的偶合，分别给出允许 J 值 3/2 和 5/2（图 3.34）。具有最大 J 值的能级（因层上为半满），即 J=5/2，是最低能量终态，并对应于较低键合能峰。

两个峰的相对强度反映出两个终态的简并度（$g_J=2J+1$）。反过来，简并度决定了在光电离过程中发生在这样的能态间的跃迁概率。

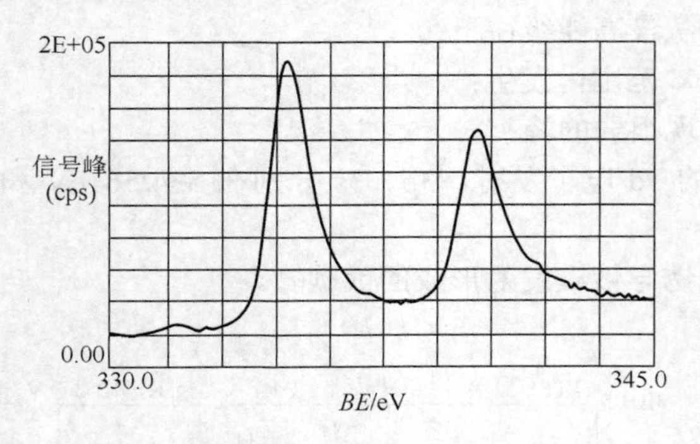

图 3.33 3d 电子光发射的 XPS 谱峰的分裂

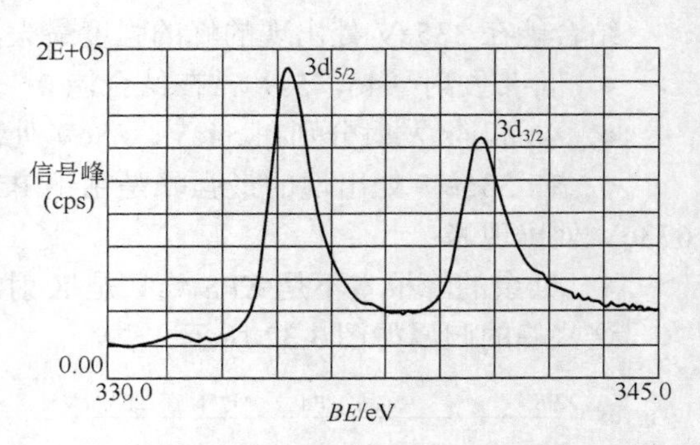

图 3.34 3d 电子光发射的 XPS 谱中两分裂峰的相对强度

注意，这些峰通常如图所示注解，注意用小写字母。自然，s 轨道（$L=0$）的自旋-轨道偶合并不明显，p、d 和 f 则都有特征的自旋-轨道偶合双重峰。

3.4.1.2 化学位移

电子的精确结合能不仅与光发射的能级有关，而且与①原子的表观氧化态和②原子所处的物理、化学环境有关。

条件①或②的改变将会使谱图中峰的位置发生小的位移，这种位移称为化学位移。

这样的位移在 XPS 很容易被观察到（而在俄歇谱中却看不到），也可得到解释。因为该技术具有内在的高分辨率（原子内层的能级是分立的并通常具有确定的能量）并且是一个电子过程。

由于光发射电子与离子壳之间有很大的库仑作用力，元素的氧化态越高，键合能也就越高。这种鉴别不同氧化态及化学环境的能力正是 XPS 技术的主要优点之一。

事实上，分辨具有细微不同的化学位移的原子的能力会受到峰宽的限制，而峰宽又受到一系列因素的影响，尤其是：

- 始态能级的内在宽度和终态寿命；
- 入射光的线宽——传统的 X 射线源只可通过单色器来改进；
- 电子能级分析器的分辨能力。

大多数情况下，第二个因素起主要作用。

例如，钛（Ti）的氧化态分析。不同氧化态的钛（Ti）之间存在很大的化学位移。图 3.35 是金属钛 Ti（0）与 TiO_2 中 Ti（+4）的 2p 谱图。

注意：

- 两个自旋态表现出相同的化学位移（约为 4.6eV）；
- 单质金属常有不对称峰的特征，在较高键合能处有拖尾现象，而在 TiO_2 中却表现出对称的峰；
- 下部谱图中在 450.7eV 附近的弱峰是由能量稍高于 X 射线源（Mg K_α）的发射 X 射线引起的，这种峰常称为 $2p_{3/2}$ 主要峰的“鬼峰”，它是由于附加的 X 射线所引起的电离而产生的。

3.4.1.3 角度分析

基于电子的分析技术如 XPS，其表面灵敏度可通过相对于表面的不同发射角发射的光电子体现出来。这种方法可用于各种表面的无损探测。

例如，图 3.36 所示的硅晶片表面氧化层的角度分析。图中显示了表面发射角从 10°至 90°一系列 Si 的 2p 光谱图。注意：Si 的 2p 氧化峰（*BE* 约 103eV 处）的强度随着发射角的减小而迅速增加，而单质 Si（*BE* 约 99eV）峰的强度几乎不变。

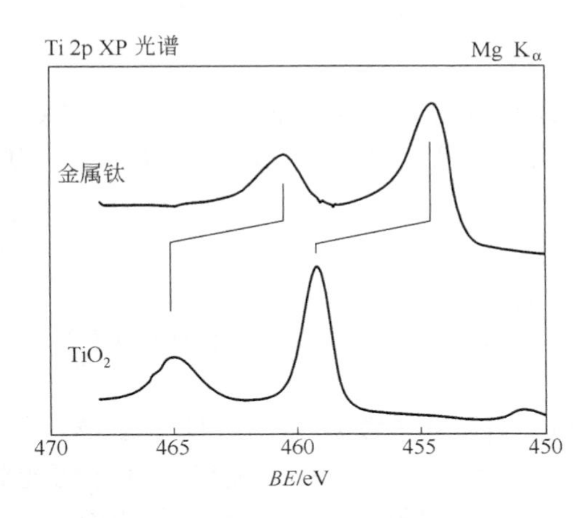

图 3.35　金属钛 Ti（0）与 TiO_2（Ti: +4）XPS 光谱图中的化学位移

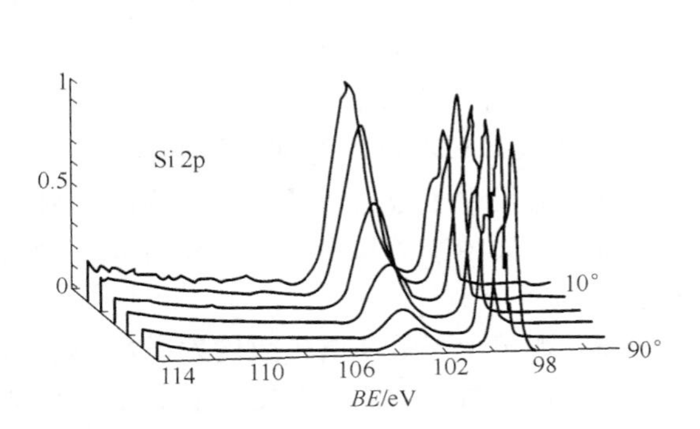

图 3.36　硅晶片表面自然氧化层的角度分析

3.4.2　紫外光电子能谱（UPS）

UPS 通常使用惰性气体作辐射源。经常使用 He-辐射灯，发射出能量为 21.2eV 的 He-I 射线。

这类射线仅能将原子的最外层电子（即价层电子）电离。使用这类紫外线的优点是射线的线宽极窄，并且从简单的辐射源可得到高密度的光电子流。

使用 UPS 主要研究：

- 固体的电子结构——详细的角度分辨研究能勾画出 k 空间完整的能带结构；
- 金属中相对简单分子的吸附——通过比较吸附物种与独立分子的分子轨道。

XPS 光电子键合能对原子序数作图如图 3.37 所示。

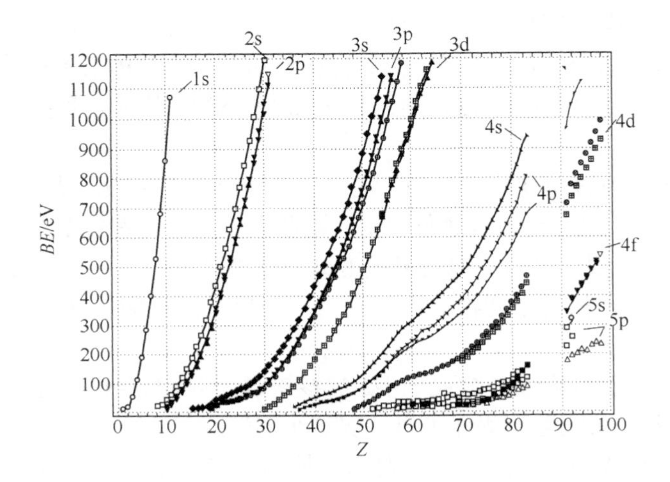

图 3.37　XPS 光电子键合能对原子序数作图

3.5 核磁共振波谱

核磁共振波谱(NMR）是一种功能强大、理论复杂的分析工具。NMR 是对原子核而不是电子进行实验，从某一原子核的信息可推导出该原子核所处的化学环境。

3.5.1 NMR 的基本原理

亚原子微粒（电子、质子、中子等）可想象为绕自身的轴旋转。许多原子中（如 ^{12}C），这些自转与另一个相反的自转配对，使得原子核净自旋为零。但是有些原子（如 ^{13}C 和 ^{1}H）原子核净自旋不为零。

判断原子核净自旋的规则如下。

- 如果中子数与质子数均为偶数，则该原子核净自旋为零。
- 如果中子数与质子数总和为奇数，则该原子核的自旋量子数 I 为半整数（即 1/2，3/2，5/2，…）。
- 如果中子数与质子数均为奇数，则该原子核的自旋量子数 I 为整数（即 1，2，3，…）。
- 自旋量子数 I 很重要，量子力学告诉我们自旋量子数为 I 的核自旋有 $2I+1$ 个可能取向。若 $I=1/2$，应有 2 个可能取向。没有外加磁场，这些取向的能量相等。施加外磁场，能级发生裂分。每一能级对应一个磁量子数 m（图 3.38）。

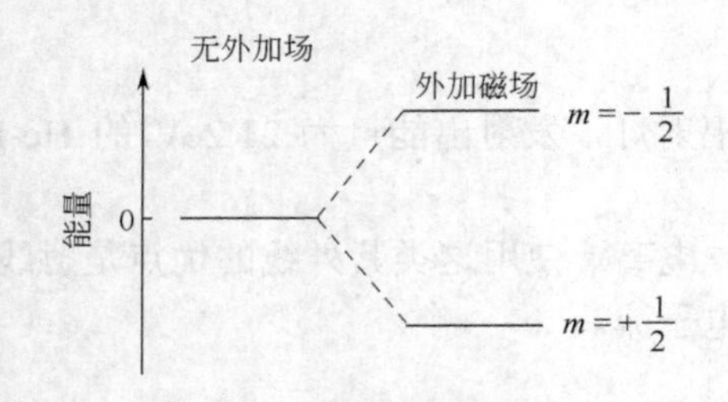

图 3.38 在外加磁场中自旋量子数为 1/2 的核的能级

不同能级之间的能量差（即跃迁能 ΔE）可从下式得出：

$$\Delta E = \frac{\gamma hB}{2\pi}$$

这也就是说，磁场增强（B），能量（ΔE）也增加；原子核磁旋比（γ）越大，相应能量（ΔE）也越大。

除了 ^{1}H 和 ^{13}C 核以外，还有很多的核都可以用 NMR 技术进行研究。非零净自旋原子核并不一定可以得到 NMR 谱图，至少还有四方面的因素需要考虑。

- 同位素丰度　有一些自旋活性的核，如 ^{19}F，其丰度为 100%（^{1}H 为 99.985%）。而另一些，如 ^{17}O 的丰度却很低（0.037%），像这样的核给出的 NMR 信号很弱，除非将它们进行同位素富集。^{13}C 的丰度为 1.1%，这也是为什么做 ^{13}C NMR 谱时，需要大量的样品和进行多次扫描的原因。
- 灵敏度　^{1}H 的灵敏度被定为 1.0。仅有一个同位素的灵敏度大于 1.0（很不寻常），其他同位素的灵敏度小于 1.0。灵敏度越低，就越需要大量的时间和样品量才可以得到信号。有些核，如 ^{103}Rh，其丰度为 100%，但其灵敏度却只有 0.000031。一般来说，该核几乎不可能得到 NMR 谱。但鉴于该核的丰度很高，当它与其他自旋活性的核（^{1}H，^{13}C）进行偶合时，仍然可以给出有用的信息。
- 核四极矩　对于自旋量子数大于 1/2 的核来说，其核的四极矩通常也较大。这样 NMR 谱线的宽度变得较大。在低温下测样可以解决这个问题。
- 弛豫时间　有两个因素决定了激发态的核的自旋弛豫，一是自旋-晶格弛豫，T_1，为核的弛豫时间，二是自旋-自旋弛豫，T_2，为一系列有一定取向的核失去它们的相位匹配的时间。通常 T_1 远大于 T_2，所以，我们常常只考虑 T_1。

这四个因素决定一个核能否给出有用的 NMR 谱图信息。一些常用在 NMR 中的核（I=1/2）列在表 3.6 中。

表 3.6　在 NMR 中较常见的 I=1/2 核

核	自然丰度/%	相对灵敏度	核	自然丰度/%	相对灵敏度
^{1}H	99.985	1.0	^{19}F	100	0.83
^{13}C	1.108	0.016	^{31}P	100	0.07

3.5.2　磁场中原子核对辐射的吸收

假设 I=1/2 的原子核处在磁场中，该原子核处在低能级（即磁矩没有与外加磁场相反）。原子核绕自己的轴旋转，在磁场中，该自旋轴又将绕磁场进动（图 3.39）。

进动能量为：

$$E=-\mu B\cos\theta$$

θ 为外加磁场方向与核旋转轴之间的夹角。

如果原子核吸收能量，进动夹角 θ 将会改变。对 I=1/2 的原子核，吸收辐射能后，磁矩就会“翻”向与外加磁场（高能态）相反的方向（图 3.40）。

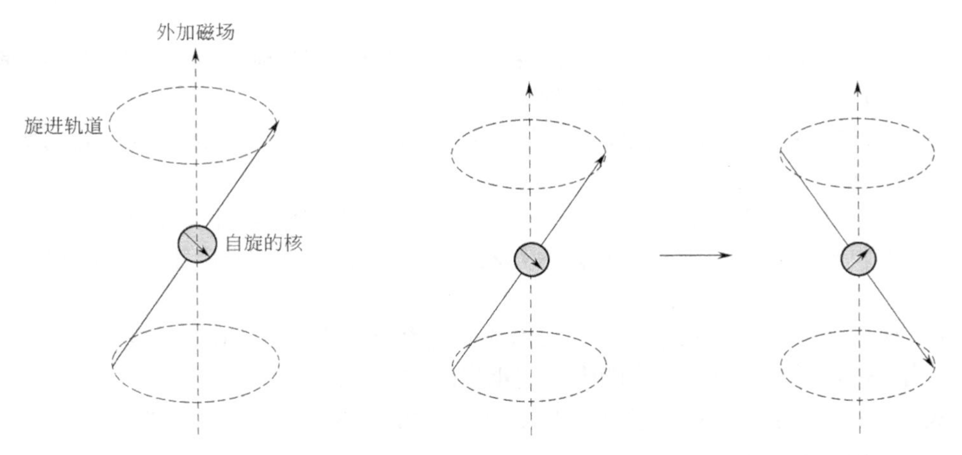

图 3.39　在外加磁场中旋转进动轴的示意图　　**图 3.40**　磁矩在外加磁场中的翻转

3.5.3　化学位移

原子核所受到的磁场并不等于外加磁场的大小，核外电子会屏蔽部分外加磁场。外加磁场与原子核所受磁场的差别，叫做核屏蔽。

以分子中 s 电子为例，s 电子云呈球形对称，在外加磁场中做圆环形运动，产生的磁场与外加磁场的方向相反（图 3.41），这意味着，为使核在其跃迁频率吸收，必须提高外加场的强度，这种向高场的移动叫做抗磁位移。

化学位移定义为核屏蔽与外加磁场之比。化学位移是原子核与其所处环境的函数。通常是通过相对参照物来测量的。对 ^{1}H NMR，该参照物常是 $Si(CH_3)_4$。

3.5.4　自旋-自旋偶合

多核 NMR 的最重要方面是所有自旋活性的核能相互偶合，这种偶合的多重度由 $2nI$+1 来决定（n 为参与偶合的等价核的数目）。例如，一个质子与两个等价的质子相邻，则共振峰将为三重峰，因为 $2nI+1=2\times2\times\left(\frac{1}{2}\right)+1=3$。若两个相邻的核是氟原子而不是氢原子，共振峰仍然是三重峰，因为氟的 I 为 $\frac{1}{2}$，并且 ^{19}F 的丰度为 100%，唯一不同的是偶合常数 J_{HH} 与 J_{HF} 不同。

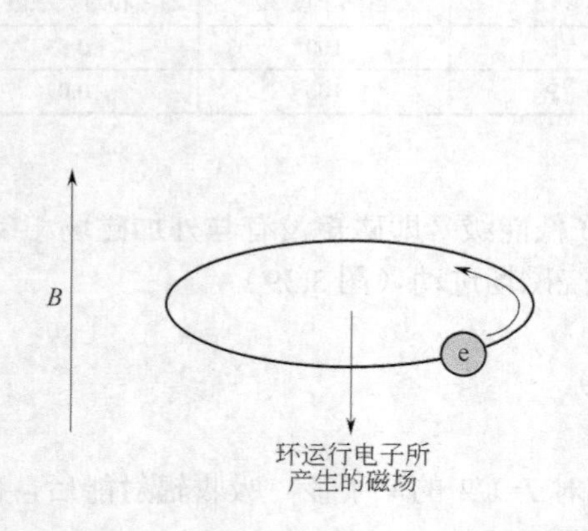

图 3.41 由 s 电子产生的磁场与外加磁场的方向相反

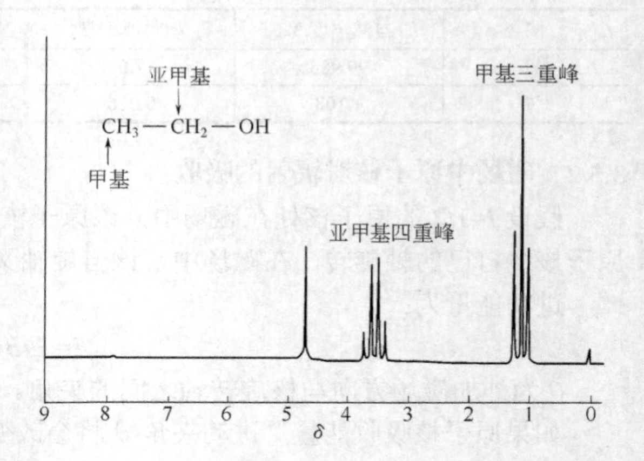

图 3.42 乙醇的 ^{1}H NMR 谱

乙醇的 ^{1}H NMR 谱（图 3.42）表明：甲基峰裂分为三重峰、亚甲基裂分为四重峰。发生裂分是因为两官能团中的质子间有弱相互作用（偶合）。

在一级谱图中（相互作用的官能团间的化学位移远大于它们之间的偶合常数），关于裂分模式的解释是非常清楚的。

- 多重峰的多重度等于邻近原子核的等价质子的数加 1，即 $n+1$ 规则。
- 等价原子核之间没有相互作用。乙醇中甲基的三个质子使邻近亚甲基的两个质子发生裂分，而三个甲基质子之间不会发生裂分。
- 偶合常数与外加磁场无关。从相邻的化学位移峰很容易确定多重峰。

3.5.5 ^{1}H NMR 和 ^{13}C NMR 谱图中一些重要的化学位移

表 3.7 和表 3.8 分别列出了 ^{1}H NMR 和 ^{13}C NMR 谱图中一些重要的化学位移。

表 3.7 质子的化学位移

质子的类型	化学位移 δ	质子的类型	化学位移 δ
烷基，RCH_3	0.8～1.0	酮，$RCOCH_3$	2.1～2.6
烷基，RCH_2CH_3	1.2～1.4	醛，RCOH	9.5～9.6
烷基，R_3CH	1.4～1.7	乙烯基，$R_2C{=}CH_2$	4.6～5.0
烯丙基，$R_2C{=}CRCH_3$	1.6～1.9	乙烯基，$R_2C{=}CRH$	5.2～5.7
苄基，$ArCH_3$	2.2～2.5	芳香烃，ArH	6.0～9.5
烷基氯化物，RCH_2Cl	3.6～3.8	炔烃，$RC{\equiv}CH$	2.5～3.1
烷基溴化物，RCH_2Br	3.4～3.6	醇羟基，ROH	0.5～6.0①
烷基碘化物，RCH_2I	3.1～3.3	羧酸类，RCOOH	10～13①
酯，$ROCH_2R$	3.3～3.9	酚羟基，ArOH	4.5～7.7①
醇，$HOCH_2R$	3.3～4.0	胺基，RNH_2	1.0～5.0①

① 这些质子的化学位移在不同的溶剂中随着温度和浓度的不同而不同。

表 3.8　^{13}C NMR 谱中一些重要的化学位移

碳原子类型	化学位移 δ	碳原子类型	化学位移 δ
烷基，RCH_3	0～40	烯烃，R_2C═	100～170
烷基，RCH_2R	10～50	苄基碳	100～170
烷基，$RCHR_2$	15～50	腈类，—C≡N	120～130
烷基卤化物或胺类，$(CH_3)_3C$—X（X=Cl，Br，NR_2）	10～65	酰胺类，—$CONR_2$	150～180
醇或醚，R_3COR	50～90	羧酸类，酯类，—COOH	160～185
炔烃，—C≡	60～90	酮类或醛类，—C═O	182～215

3.6　电子顺磁共振（EPR）

EPR 是检测具有未成对电子的化学物种的一种光谱技术。

具有正磁化率的物质称为顺磁性的。它们能被磁场吸引。许多物质含有顺磁性成分：未充满导带中的电子、辐射损伤位置的捕获电子、自由基、多种过渡态离子、三线态、半导体中杂质。

样品中未成对电子的自旋磁矩对样品内的本身磁场非常敏感。这些磁场常常是由体介质中各种原子核的核磁矩引起的。EPR 提供了一种能详细研究内部结构的独特手段。

EPR 现已广泛应用于各个学科，如：生物学、物理学、地质学、化学、医学、材料学等，固、液、气均可使用 EPR。通过与 EPR 联用一些特殊技术（如自旋捕获、自旋标记、ESEEM、ENDOR），可满足研究者各种科学研究的需要，如化学动力学、电子交换、电化学过程、晶体结构、基础量子理论、催化及聚合反应。

对顺磁性物质施加强磁场 B，未成对电子的磁矩取向或者与外加磁场一致或反向平行，这使得未成对电子具有确定的能级，从而使得净吸收电磁波（微波）成为可能。当磁场与微波频率相符合时，产生共振（图 3.43）。

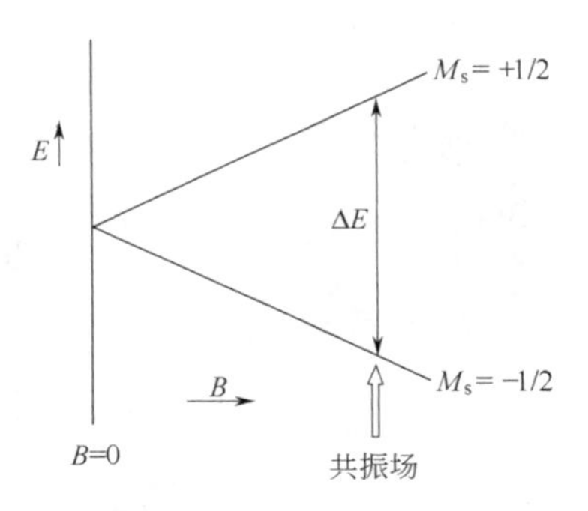

图 3.43　作为外加磁场 B 函数的两种自旋状态的能级图

这表示最简单的 EPR 跃迁（如自由电子）。

当$|\Delta M_s|=1$（M_s 为自旋态的磁自旋量子数）时，跃迁允许。描述两个自旋态之间的能量吸收或发射的方程为

$$\Delta E=h\nu=g\beta B$$

式中，ΔE 为两自旋态之间的能级差；h 为普朗克常数；ν 为微波频率；g 为塞曼因子；β 为玻尔磁子；B 为外加磁场。

3.7　圆二色谱（CD）

圆二色谱是一种测量物质对左旋、右旋圆偏振光的吸收差异与波长关系的波谱方法。

许多生物分子，包括蛋白质、核酸是手性的，并在紫外吸收带表现圆二色谱，从而可以测量其二级结构。金属中心与这些分子结合后，即使原来无手性，由于基于配体的或配体-

金属的电荷转移在紫外吸收带也会表现出圆二色谱（CD），CD 常与吸收光谱和磁圆二色谱（MCD）一起来研究电子的跃迁。

圆二色谱技术特别适合下列研究：

- 判断蛋白质是否折叠。如果是，归属其二级、三级结构及所属结构群。
- 比较来源不同的蛋白质结构，或相同蛋白质的不同突变变种。
- 说明加工或处理后溶液结构的相似性。
- 研究外力下蛋白质结构的稳定性——热稳定性、pH 稳定性、变性剂稳定性及这种稳定性随缓冲剂组成、稳定剂、赋形剂的改变而变化的情况。
- CD 极适合于发现增加熔点温度或热去折叠可逆性的溶剂条件及延长贮存期限的条件。
- 研究蛋白质-蛋白质之间相互作用能否改变蛋白质的结构。如果构造发生改变，波谱将不同于各组分波谱的加和。例如，不同受体-配体化合物生成后，将有小的构象改变。

cis-[CuNa(GMP)(HGMP)]（**1**）是一种手性核苷酸 GMP 为配体的过渡金属配合物，其中一个为 GMP^{2-}配体，另一个为质子化的 $HGMP^{-}$配体，这两种配体都是用配体碱基中的 N7 原子与过渡金属 Cu（Ⅱ）配位。Cu（Ⅱ）的配位几何构型是配位数为五的四方锥形，其中，两个 N 原子来自 GMP，另三个氧原子来自三个配位的水分子。两个 GMP 配体采用了面对面的方式，一个配体中的糖基环在四方锥底面的上方，而另一个在底面下方［图 3.44（a）］。嘌呤环-Cu^{2+}-嘌呤环的二面角为 56.6°。因为手性的 GMP 配体与 Cu（Ⅱ）的配位是非对称的，因此，配体的手性在配合物中得以保留，可通过溶液圆二色谱检测［图 3.44（b）］。

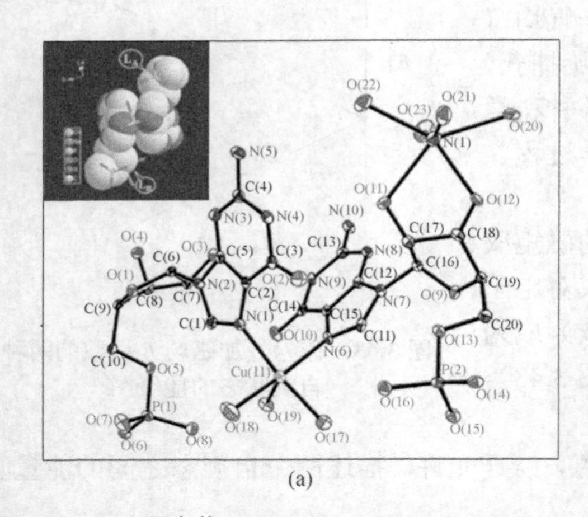

(a)

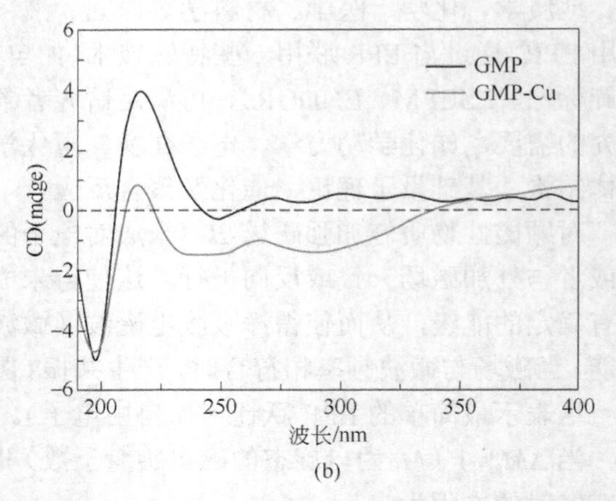

(b)

图 3.44 （a）配合物$[CuNa(GMP)(HGMP)(H_2O)_7]\cdot 6(H_2O)\cdot CH_3OH$（50%可能性）（**1**）的分子结构图，为了清晰，忽略了溶剂分子和氢原子，内插图为配合物分子的空间填充图，其中 Na、P、O 和 Cu 等为球棍模式；（b）GMP 配体和 Cu-GMP 配合物的溶液 CD 谱（0.1mg/mL）

cis-[CuNa(GMP)(HGMP)]（**1**）的单晶结构表明：单核的配合物通过配位的水分子与核苷酸配体功能基团之间的氢键形成了一维的超分子螺旋链［图 3.45（a）和（b）］。GMP 配体和 Cu-GMP 配合物的固态 CD 谱不同于溶液 CD 谱。氢键产生的一维 M-螺旋链的超分子手性可以通过单晶结构分析与固态 CD 谱的测量而进行研究［图 3.45（c）］。可见，超分子 M-螺旋链的手性信号在固态 CD 谱中被成功地捕捉到。所以，对溶液 CD 谱和固态 CD 谱的详细分析是理解配合物手性的一种有效途径。

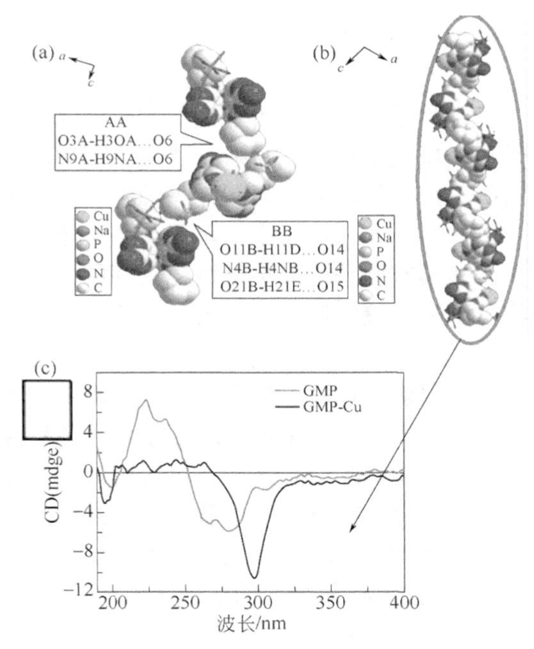

图 3.45　(a) 配合物 **1** 中构成超分子螺旋的分子间氢键的空间填充图；(b) 配合物 **1** 的一维螺旋链的空间填充图；(c) GMP 配体和 Cu-GMP 配合物的固态 CD 谱（样品：KCl=1：40），其中位于 297nm 处强的负 Cotton 效应（CE）是由 M-螺旋链中芳香生色团的 π-π*跃迁的激发偶合产生的

非手性物质仍然能表现出磁圆二色散性（MCD）（Faraday 效应）。因为磁场导致电子轨道与自旋状态简并度的增加，即变成电子的混合态。MCD 常与吸收光谱和圆二色散（CD）联用（图 3.46）。

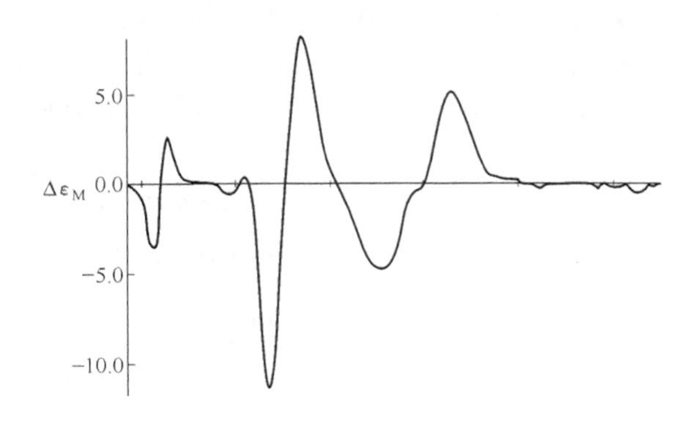

图 3.46　$Pt[P(t\text{-}Bu)_3]_2$ 在乙腈中的 MCD 谱

MCD 谱来自三方面贡献：

- 基态与激发简并态的自旋塞曼裂分；
- 磁致混合态；
- 顺磁基态塞曼亚能级分子分布数的改变。

其中最后一条仅对基态顺磁物质起作用，且低温下强度增大。它随磁场、温度的改变可分析基态的磁参数，如自旋、g 因子、零场分裂。变温 MCD 对于判断与确定源自顺磁生色团的电子跃迁非常有效。

3.8　纳米配合物的先进成像技术

为对包括纳米配合物在内的纳米材料有全面深入的理解，人们做了大量的努力来发展相应的表征技术，建立标准评估方法。现在，有很多先进技术用于材料的表征且各有优点。本节主要介绍几种先进的成像技术，包括扫描电镜、透射电镜和冷冻电镜 3 种电子显微镜以及扫描探针显微镜（扫描隧道显微镜和原子力显微镜）。

3.8.1 电子显微镜

电子显微镜作为一种电子光学系统，广泛用于把非常小的物体转换成放大的图像。为得到可观测的图像，普通光学显微镜通常检测光照射到物体表面后产生的反射、透射、散射及吸收光。电子显微镜则利用电子束照射样品，得到样品的表面形貌和结构信息。可见光是一种电磁波，其波长范围为 400～700nm。电子束与可见光的显著区别是其波长仅为 0.001～0.01nm。这一特征使电子显微镜具有比普通光学显微镜更高的分辨率，因此可以作为一种更为有效的分析系统。在介绍电子显微镜之前，有必要先详细了解一下高能电子和样品中原子的相互作用。

一般情况下，当入射电子束照射到样品上时会产生弹性散射和非弹性散射，同时样品发射电子或电磁波。弹性散射情况下，入射电子在接近原子核的位置穿过原子，只发生方向的改变。发生弹性散射的电子可以用于探测原子序数及衍射成像。在非弹性散射情况下，入射电子会发生方向改变，同时损失一定的能量。弹性散射和非弹性散射现象在成像和分析中都非常有用。图 3.47 给出了这些作用产生的不同信号，其中有代表性的有如下几种：

- 二次电子（SEs）：从非常接近样品表面的区域逃逸出来，能量小于 50eV。因此其数量足够多，可以用作扫描电镜的成像信号。
- 背散射电子（BSEs）：有样品发射，能量很低，可用于扫描电镜中的成像、衍射及分析。
- 透射电子：用于透射电镜。
- X 射线：由外层电子跃迁到内层空位时产生，是原子的特征射线。
- 俄歇电子：是由入射电子和样品中原子核附近的内层电子相互作用产生的一类特殊电子，能量可达 1～2keV，可用于表面元素分析。

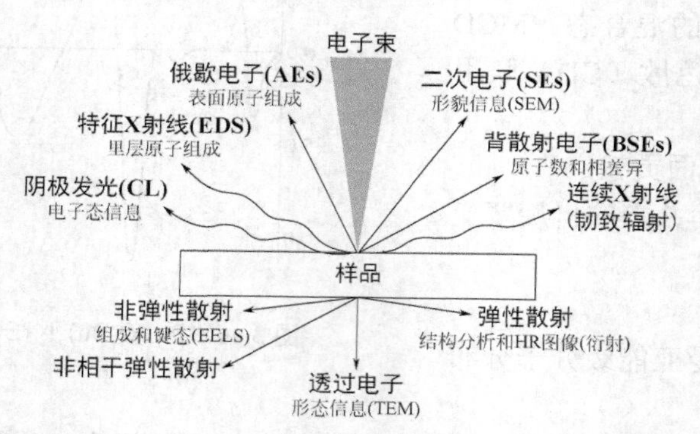

图 3.47 电子显微镜中入射电子束与样品相互作用产生的信号

3.8.1.1 扫描电子显微镜

扫描电子显微镜（SEM）用一束聚焦的高能电子束扫描样品。电子与样品中的原子相互作用，产生反射电子及其他信号，形成图像。图 3.48 是简单的 SEM 系统示意，其主要部分如下。

- 电子枪：由钨热电子灯丝组成，也可以是 La 或 Ce 的六硼化物或 Schottky 场发射电子枪。电子枪产生电子束，电子电流 250μA。电子束的能量在 1～30keV（或 40keV）之间可调。

- 聚光镜系统：通常置于电子枪之下，用于汇聚电子束。电子束通过 1～2 个透镜后，会缩小至原来的 $1/10^5$～$1/10^4$。
- 物镜（也称为终聚光镜）：用于形成一级中间像。它把缩小的电子交叉像在样品表面投影为一点。
- 电子收集系统：用于探测样品表面经入射电子束照射后发射的 SE、BSE 及其他类型的电子。

SEM 成像时，入射电子先由电子枪的热阴极发射，然后加速到 1～30keV。电子通过几个（一般是 2 个）聚光镜后，电子束被聚集，直径仅为 2～10nm。汇聚的电子束轰击样品，产生携带样品形貌信息、成分信息及其他性质的信号。电子束的扫描方向和位置由扫描线圈控制，并通过计算机实现可视化。经过探测和分析来自样品表面的信号形成图像，显示在计算机屏幕上（见图 3.49）。

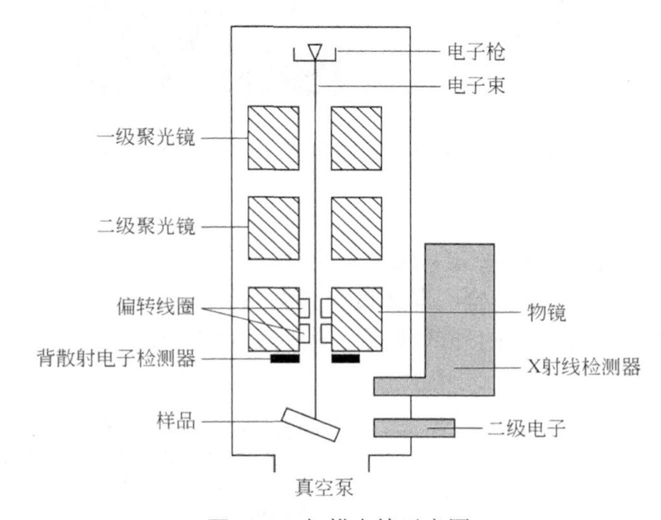

图 3.48　扫描电镜示意图

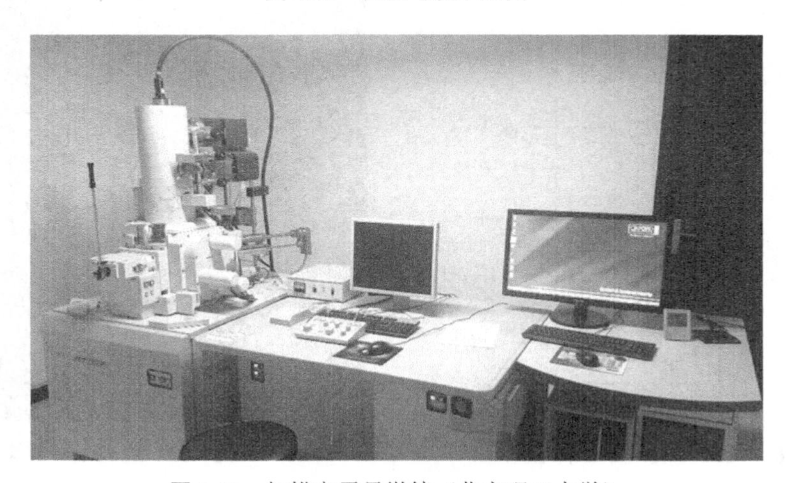

图 3.49　扫描电子显微镜（北京理工大学）

SEM 使用及成像的基本要求如下：

- 镜筒需抽真空至 10^{-7}atm，以减少空气对电子的散射。

● 样品必须导电、干燥。非导电样品需用导电且稳定的材料（如 C、Au、Ag 等）镀膜。磁性样品必须消磁。

通常，SEM 用于研究块体材料的（亚）表面结构。最近这一表征技术开始用于研究纳米尺度配位聚合物的结构。图 3.50 是 MOF-74 的扫描电镜图像，孔径 3.5nm。这一图像比其他间接分析的结果更为直观，易于解释。

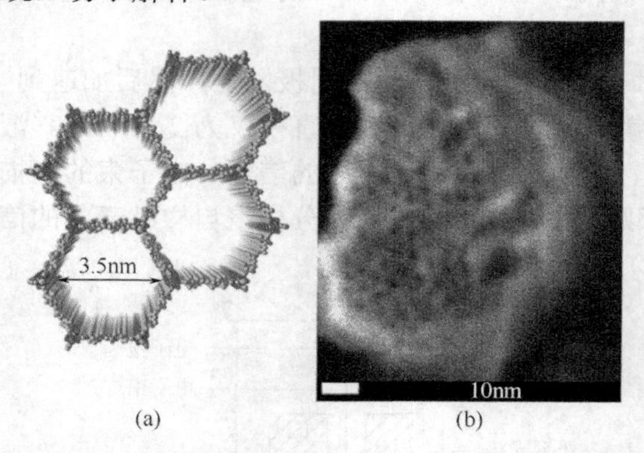

图 3.50 MOF-74 结构的示意图（a）和金属的 SEM 图像（b）

3.8.1.2 透射电子显微镜

与 SEM 类似，透射电子显微镜（TEM）也使用一束电子照射样品。这意味着 TEM 和 SEM 仪器结构类似，也包括电子枪、聚光镜和真空系统。但是，TEM 主要用于研究薄片样品的内部结构，而不是块体样品的外部特征。图 3.51 为 TEM 的主要组成部分示意，图 3.52 是 TEM 的图片。

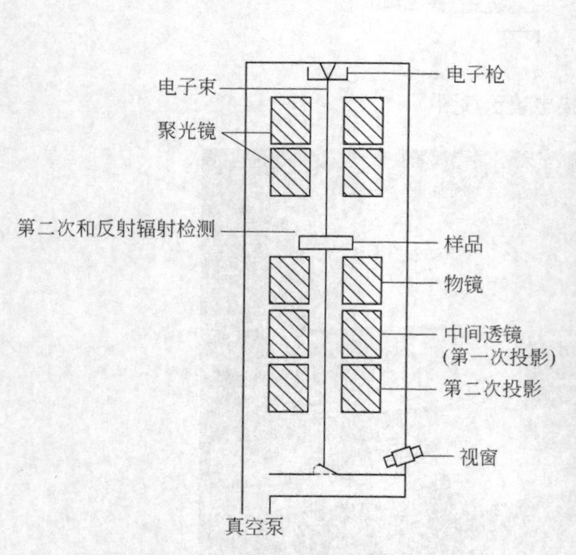

图 3.51 TEM 主要组成部件示意

图 3.52 透射电子显微镜

TEM 使用及成像的基本要求如下：

● 电子束照射样品必须在高真空中进行（10^{-4}mbar 或更低）。

● 样品必须干燥，且足够薄以使足够多的电子穿透样品而形成图像。在高真空中的电子束轰击下，样品必须稳定。样品大小合适，可以安置于样品架上。

TEM 可以进行电子衍射及高分辨成像，是观测纳米材料局部结构，如组分的空间分布、纳米晶的晶体结构、晶体的表面晶面及缺陷等的有力工具。图 3.53 是晶体材料 UiO-66 的高分辨率图像。

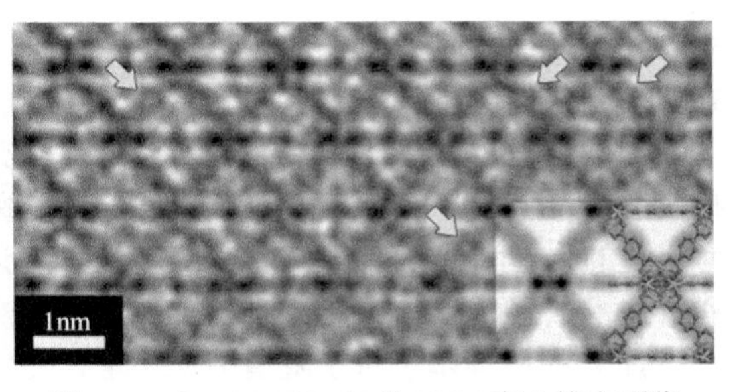

图 3.53　f MOF UiO-66 的 TEM 高分辨率图像

3.8.1.3　冷冻电镜

由于电子会被空气强烈散射，因此常规 SEM 和 TEM 都要求样品干燥且真空度低于 10^{-6}torr。但是，大多数生物分子，尤其是水中的生物体，水一旦被去除，其原始结构就不能保留下来。这对现有的表征技术提出了挑战。配备低温样品台的电子显微镜被认为是检测新鲜生物材料的最佳技术。

冷冻电镜是一种测试深度冷却样品的电子显微镜。样品一般低至液氮温度，且位于玻璃态的水中。样品的冷冻操作非常重要。若冷却太慢或冷冻材料温度超过−80°C，就会形成大量冰晶，破坏样品结构或提供虚假结构信息。因此，冷冻样品的最佳过程需遵循一些最基本的原则和要求。

- 使用防冻剂（如甘油、甲醇、丙酮等）抑制结晶。
- 提高冷冻速率（快速冷却技术），确保水在低温时处于玻璃态，包括加入液氮冷却的液态丙烷避免形成冰晶，向样品喷洒雾状液态丙烷，及使用高压冷冻技术（Moor，1987）。

冷冻电镜尤其适合以下情况：

- 直接观测生物分子、细胞、蛋白质、高分子等。
- 对冷冻的含水离子进行 3D 成像，或在保证结构畸变最小的条件下解析其分子结构。
- 对在真空中具有挥发性或对电子束敏感的样品进行成像。

Jacques Dubochet、Joachim Frank 和 Richard Henderson 因发展了冷冻电镜技术，简化并改进了生物分子成像，获得了 2017 年诺贝尔化学奖。这一技术大大推进了生物化学的发展，使其进入新时代。图 3.54 是病毒 Semliki Forest Virus（SFV）悬浮液在 18mm 网格上的图像。

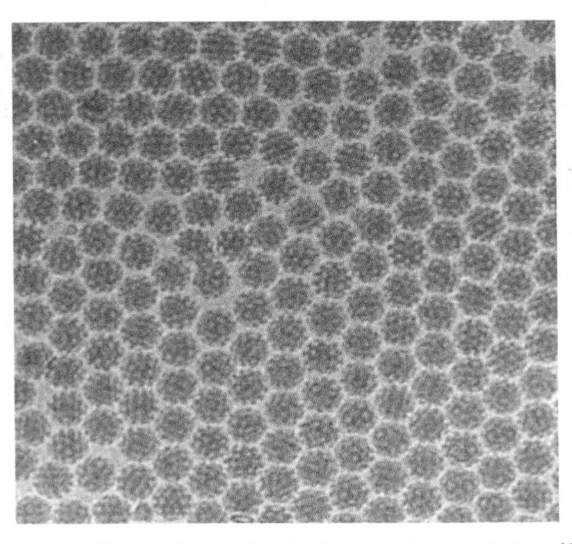

图 3.54　悬浮在新鲜的液体介质中病毒 Semliki Forest Virus（SFV）的冷冻电镜图像

3.8.2 扫描探针显微镜

3.8.2.1 扫描隧道显微镜

扫描隧道显微镜（STM）被认为是20世纪80年代最伟大的科学和技术成就。从G. Binning和H. Rohrer发明STM开始，人类第一次观察到样品表面上单个原子的排列及与表面电子行为相关的物理和化学性质。STM应用量子理论的隧道效应测量样品表面结构，并衍生了一系列新技术。因此，STM的发明者获得了1987年诺贝尔物理学奖。

STM的原理基于量子隧道效应。量子力学认为，即使离子能量低于阈值，当许多离子冲向能垒时，部分离子反弹回去，另外一些则穿过能垒，看起来像有一条隧道，称为量子隧道。一般认为固体导体中的电子分布于其表面，且轻微扩展到自由空间。当两个固体导体靠得很近，且在二者间加一个小的偏压时，它们的电子分布就会在一定距离内互相重叠，形成隧道电流。隧道电流的强弱依赖于导体间的距离，这个距离是STM一个非常重要的参数。

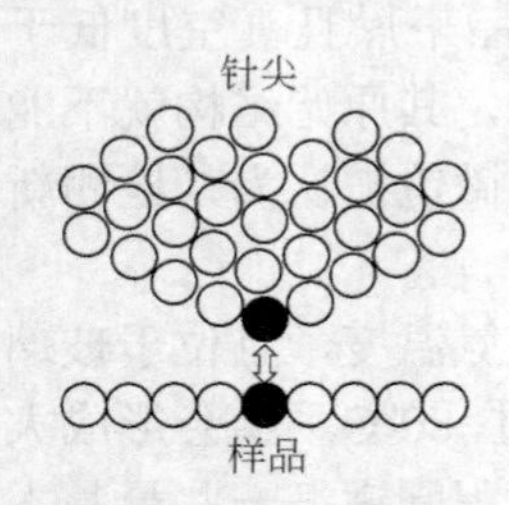

图3.55 STM原子间作用的示意图

在进行STM测试时（图3.55），用一个非常尖的针尖（钨丝或铂铱合金丝）扫描样品表面。针尖和样品表面的间隙距离等效于一个能垒，电子可以通过这一间隙形成隧道电流，电流大小随表面高度的变化而变化。因此探测电流强度就可以提供样品表面信息。

STM测试的基本要求如下：

- 为避免灰尘及其他空气污染，得到清晰的原子图像，STM测试通常在高真空环境中进行。
- 用压电晶体控制针尖在*xyz*方向的移动。*x*、*y*控制扫描，*z*控制针尖在竖直方向的移动。
- 主要测试导电材料。

STM让科学家能在实际应用中观察并定位单个原子，这比类似的原子力显微镜分辨率要高。而且，它不仅是一个重要的测试工具，还是纳米技术中一个重要的处理工具，因为其具有精确操控原子的能力。现在，STM已经和其他技术如SEM结合，在实验室得到广泛应用。

3.8.2.2 原子力显微镜

原子力显微镜（AFM）是一种高分辨扫描探针显微镜，通过感应和放大悬臂探针和样品间的作用力进行探测。从1986年发明AFM开始，这一有力的纳米平台已经吸引了大量注意，并得到了广泛应用。

图3.56是AFM的示意图。使用时，在悬臂上放一个精细针尖。当针尖接近样品表面时，针尖上的原子和样品表面的原子间距通过原子间作用力而保持恒定。悬臂则随样品表面的形貌而上下移动，这是由原子间是引力还是斥力决定的。用来自悬臂和探测器的反射激光束来探测和记录这一运动轨迹。由于AFM对竖直偏差的灵敏度可以小于0.1nm，因此样品表面一个原子的高度都可以很容易探测到。

AFM一般有三种成像模式，即接触模式、非接触模式和敲击模式，以适应不同样品。AFM对样品无特殊要求。柔软样品如生物分子或有机薄膜也可以测量，且分辨率较高。实际上，大多数AFM可以在空气、水或其他液体中测试。AFM可以测导体，也可以测非导体，弥补了STM的缺陷。但是AFM成像需要计算机处理，不像SEM那么直接。尽管如此，多年来它仍然是一种强大而有用的技术。图3.57是一个生物界面的AFM图像。

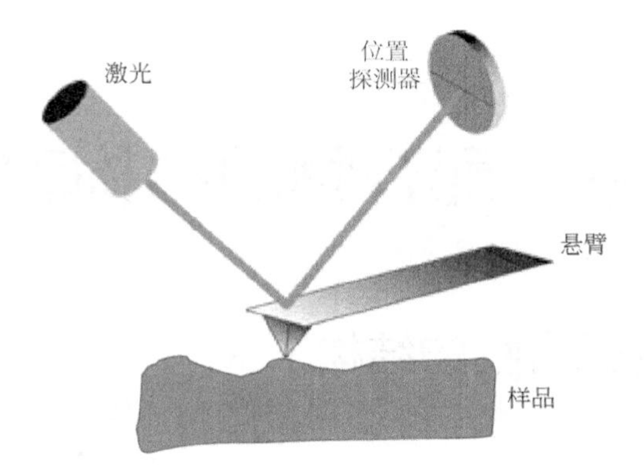

图 3.56　AFM 示意图

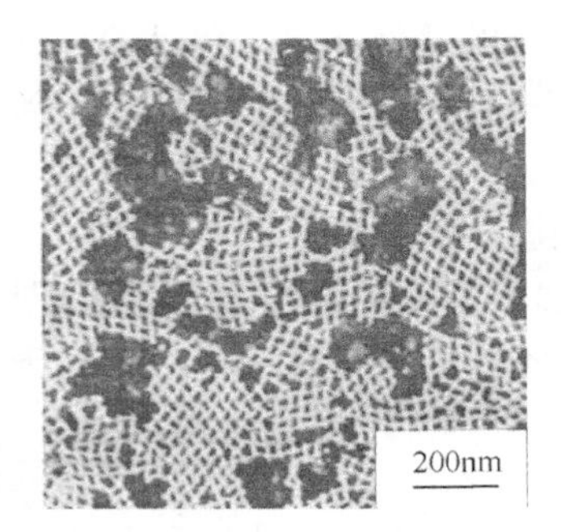

图 3.57　自组装 RNA 条带的 AFM 图像

第 4 章　配合物的结构和性质

4.1　几种类型的配合物的结构

4.1.1　金属有机化合物

金属有机化合物的部分特征是含有一个或多个金属-碳键，其中碳可以认为是有机化合物的一部分，金属则指电负性比碳小的任何元素（包括 B、Si 和 As，但不包括 P 和卤素）。

它们通常具有三种类型：R—M，R—M—R，R—M—X（其中 R 是有机基团；M 为金属离子）。

金属有机化合物中化学键如图 4.1 中说明。

金属-碳键是一种由共价和离子结构的共振杂化所表现出来的极性键。金属-碳键经常用它们具有多少“离子”特性来分类，即通过使用一种相对于共价结构，离子共振贡献重要性的指数来分类（图 4.2）。

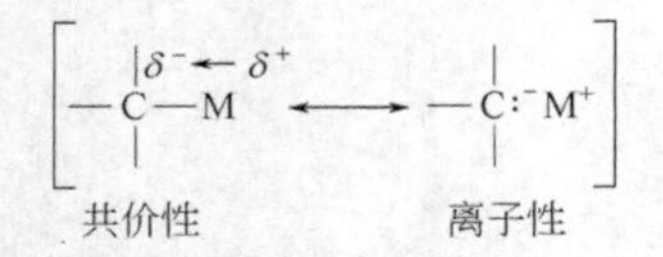

图 4.1　金属有机化合物中化学键的本质

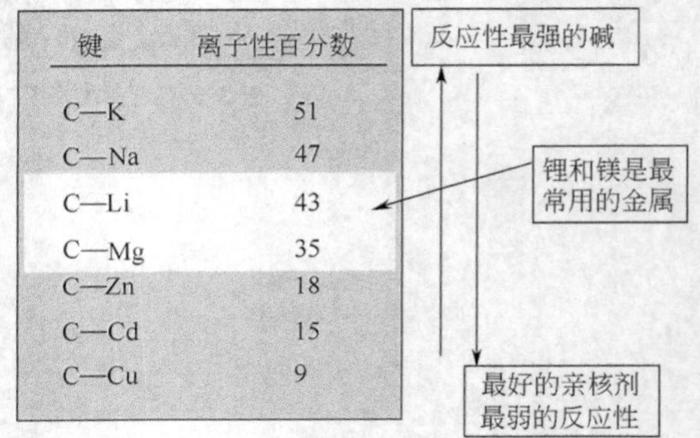

图 4.2　金属-碳键中共价性和离子性特征指数

（1）18 电子规则　大多数过渡金属 π 配合物都遵循一个经验规则，即“18 电子规则”，此规则认为金属原子从配体获得足够的电子而使其能达到紧邻的下一个稀有气体的电子组态。这就意味着金属原子的价层含有 18 个电子。因此，金属 d 电子总数加上由配体提供的电子数之和为 18。这使得金属原子具有恒定的电子排布。

以铁配合物为例，金属有机配合物中电子数的计算如下。

- 计算自由金属原子价层中的电子数，$Fe^0 \rightarrow 8e^-$（$4s^23d^6$ 或计算元素周期表中的族数）。
- 加上或减去金属配合物的总电荷数，$[FeL_x] \rightarrow 8e^-$，$[FeL_y]^+ \rightarrow 7e^-$，$[FeL_z] \rightarrow 9e^-$。
- 加上与金属配位的配体的电子数。末端配位（T）$CO \rightarrow 2e^-$，桥联（B）$CO \rightarrow 1e^-$，金属-金属键 $\rightarrow 1e^-$，有机配体 → 电子数取决于族数。
- 2、3 项中的电子数相加。稳定的配合物的价层具有 18 电子。

计算电子有两种不同的方法：①中性或共价键法；②有效原子或离子数法。

金属有机化学家们认为过渡金属价电子均为 d 电子。确实也有 $4s^23d^x$ 的情况发生，但在一级近似中可以忽略。对于零价金属，我们认为电子数仅简单地对应于它在周期表中所在的列数。Ti^0：d^4，而非 d^2；Fe^0：d^8，而非 d^6；Re^{3+}是 d^4（Re 在第七列，然后再加上三个正电荷或减去三个负电荷）。

在下面的例子中分子因为两个铁原子而被分为两部分，这两个铁原子均想获得紧邻的下一个稀有气体的电子组态（表 4.1）。

表 4.1　$[Fe_2(CO)_9]$分子中的电子数的总和

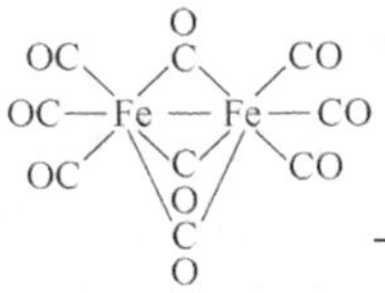

Fe^0：	$8e^-$
$3\times CO_T$：	$6e^-$
$3\times CO_B$：	$3e^-$
Fe—Fe 键：	$1e^-$
总和：	$18e^-$

（2）羰基配合物　羰基配合物的特点如下。

- 羰基由于 π 键而是一种强场配体。
- 金属在羰基配合物中一般以低氧化态存在（有时是负的）。
- 这些配合物几乎都遵循 18 电子规则。

$$M + C\equiv O \longrightarrow M^- {-} C\equiv O^+ \longleftrightarrow M{=}C{=}O$$

稳定的 3d 元素的羰基配合物有：$[V(CO)_6]/[V(CO)_6]^-$，$[Cr(CO)_6]$，$[Fe(CO)_5]$，$[Ni(CO)_4]$，$[Mn_2(CO)_{10}]$，$[Fe_2(CO)_9]$，$[Co_2(CO)_8]$，$[Fe_3(CO)_{12}]$，$[Co_4(CO)_{12}]$，$[Co_6(CO)_{16}]$（图 4.3）。

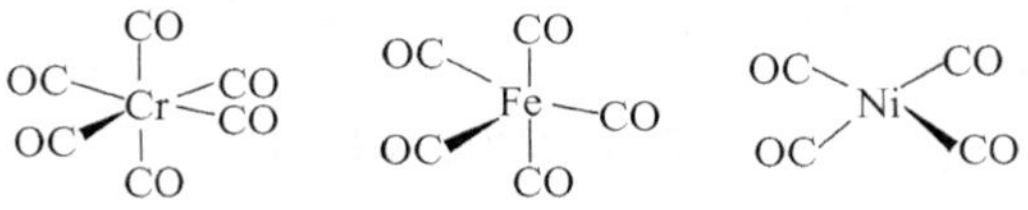

图 4.3　几种 3d 元素的羰基配合物的分子结构示意图

$[V(CO)_6]$：17 电子配合物，在 70℃时分解，还原得到 18 电子配合物$[V(CO)_6]^-$［式(4.1)］。

$$Na + [V(CO)_6] \longrightarrow Na^+ + [V(CO)_6]^- \tag{4.1}$$

重过渡金属具有形成多核配合物的强烈倾向，$Os_3(CO)_{12}$ 比 $Os(CO)_5$ 稳定，而对于 Fe 的配合物则正好相反（可能是尺寸和金属-金属键的原因）。

可以利用 IR 光谱来判断 C—O 键的强度。同样，IR 光谱还可以给出有关结构和对称性方面的信息。$[Fe(CO)_5]$（对称性为 D_{3h} 点群），有两个不同的 C—O 伸缩吸收峰（图 4.4）。

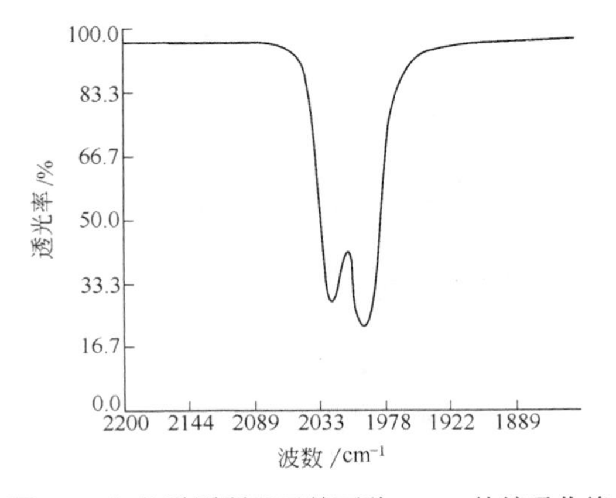

图 4.4　红外谱图所显示的两种 C—O 伸缩吸收峰

（3）多核羰基配合物

具有不止一个金属原子或离子的羰基配合物：$[Mn_2(CO)_{10}]$，$[Fe_2(CO)_9]$，$[Co_2(CO)_8]$，$[Fe_3(CO)_{12}]$，$[Co_4(CO)_{12}]$，$[Os_3(CO)_{12}]$等为多核羰基配合物。多核羰基配合物可以含有桥联的CO配体，如$[Mn_2(CO)_8]$；也可以不含有桥联的CO配体，如$[Co_2(CO)_{10}]$，若不含有，则要有金属-金属键。

桥联的和端配的CO配体：IR光谱表明桥联的C—O键的键级降低，对自由CO，典型的C═O伸缩频率在$2143cm^{-1}$处，端配的CO在$2125\sim1850cm^{-1}$，桥联的CO在$1850\sim1700cm^{-1}$，饱和的酮类在$1715cm^{-1}$。这些数据（结合18电子规则）有助于分析可能的结构。注意：18电子规则不能区分桥联的和端配的CO配体，因为两者都贡献两个电子（图4.5）。

$[Mn_2(CO)_{10}]$中就存在了这两种CO配体，桥联的和端配的CO配体之间的能量差别很小。一般的趋势：较重金属的配合物具有较少的桥联CO配体（图4.6）。

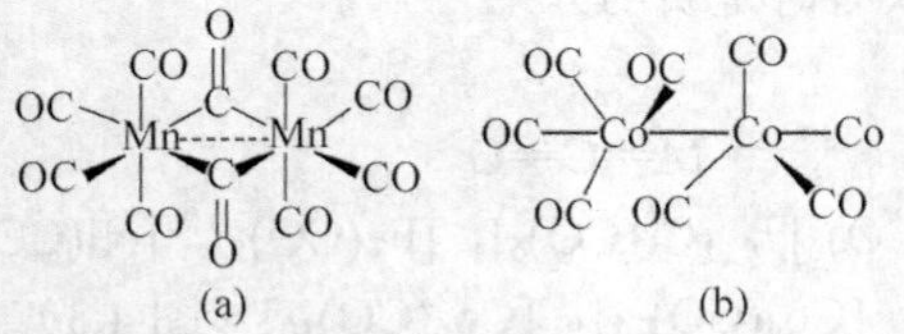

图4.5 $[Mn_2(CO)_{10}]$（a）和$[Co_2(CO)_8]$（b）的分子结构示意图

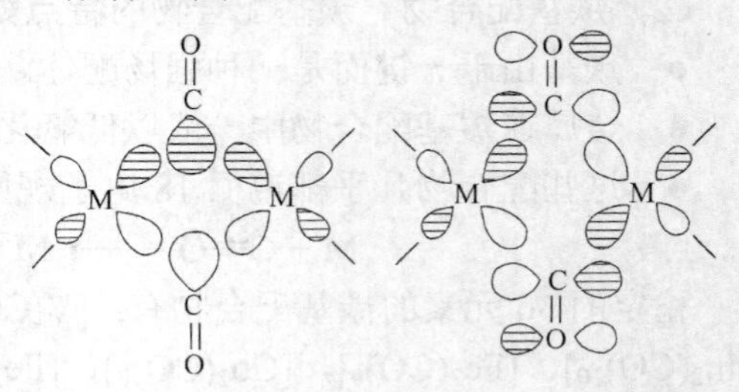

图4.6 羰基（CO）中的σ和π轨道与金属d轨道的示意图

在多核羰基配合物中，由于18电子规则的原因经常需要金属-金属键。

半桥连CO配体（μ_2-CO）：如在$[Os_4(CO)_{14}]$中，CO配体被两个金属中心不等同地分享（图4.7）。

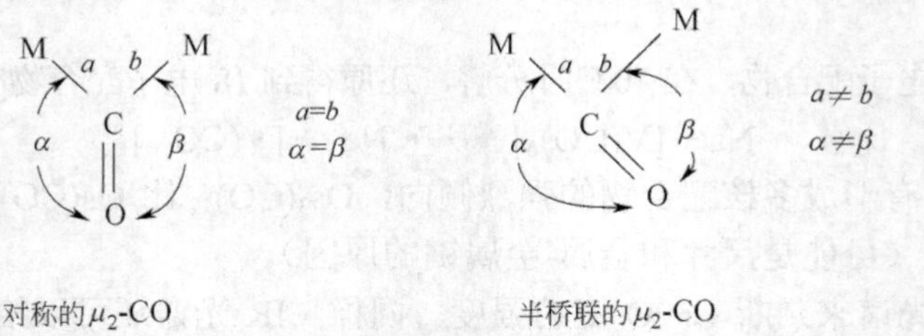

图4.7 对称的和半桥型桥联羰基μ_2-CO的结构示意图

μ_3-CO：见图4.8。

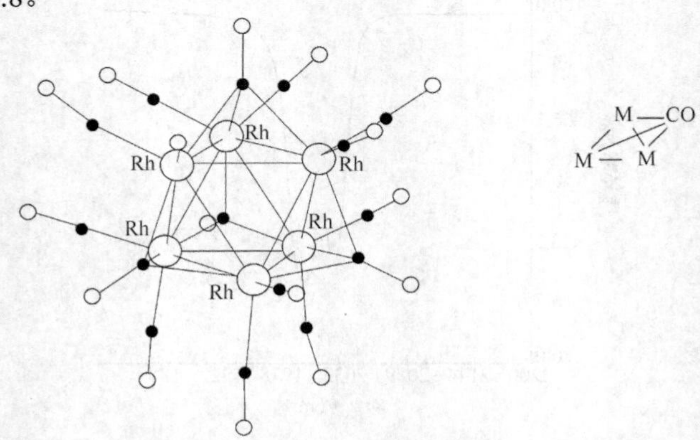

图4.8 $Rh_6(CO)_{16}$分子结构示意图

异核双金属羰基配合物：配合物中具有两个不同的金属中心，例如钌和钼［图 4.9（a）］。

同核双金属羰基配合物：配合物中具有两个相同的金属中心，例如钴［图 4.9（b）］。

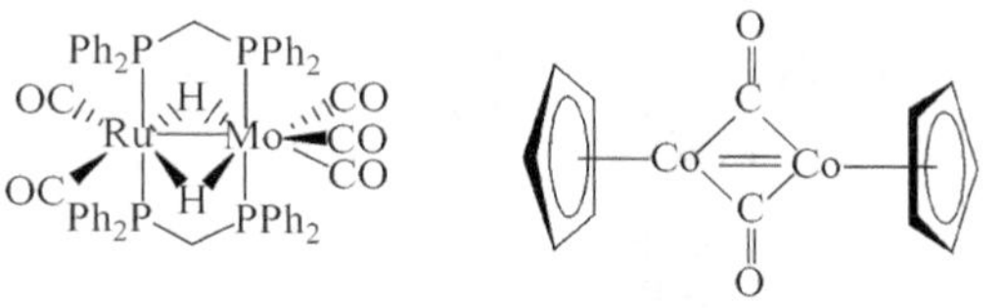

(a) 异核双金属　　(b) 同核双金属

图 4.9　$Ru(CO)_2(\mu\text{-}H)_2[\mu\text{-}CH_2(PPh_2)_2]_2Mo(CO)_3$（a）和 $Co(CO)_2(C_5H_5)_2$（b）的分子结构示意图

这两个金属中心不需要有相同的配体或配位数，但常常以对称的二聚体形式存在。

（4）羰基配合物离子　阴离子型羰基配合物遵守 18 电子规则。在结构和电子上与相应的中性配合物是等电子体和等结构的例子如下：

$[Ti(CO)_6]^{2-}$-$[V(CO)_6]^{-}$-$[Cr(CO)_6]$（IR：1748cm^{-1}-1860cm^{-1}-2000cm^{-1}）

$[V(CO)_5]^{3-}$-$[Cr(CO)_5]^{2-}$-$[Mn(CO)_5]^{-}$-$[Fe(CO)_5]$

$[Cr(CO)_4]^{4-}$-$[Mn(CO)_4]^{3-}$-$[Fe(CO)_4]^{2-}$-$[Co(CO)_4]^{-}$-$[Ni(CO)_4]$

$[Co(CO)_3]^{3-}$

$[Cr_2(CO)_{10}]^{2-}$-$[Mn_2(CO)_{10}]$

$[Fe_2(CO)_8]^{2-}$-$[Co_2(CO)_8]$

（5）二氢配合物　Vaska's 配合物可以与 H_2 发生可逆反应（氧化加成）［式(4.2)］，该过程是通过协同机理进行的［式(4.3)］。

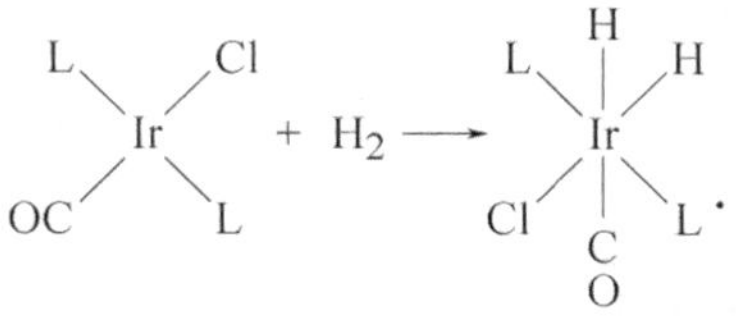

$$trans\text{-}IrCl(CO)L_2 + H_2 \longrightarrow IrH_2Cl(CO)L_2 \tag{4.2}$$

$$M + H_2 \longrightarrow \left[M\cdots(H\cdots H)\right] \longrightarrow M(H)_2 \tag{4.3}$$

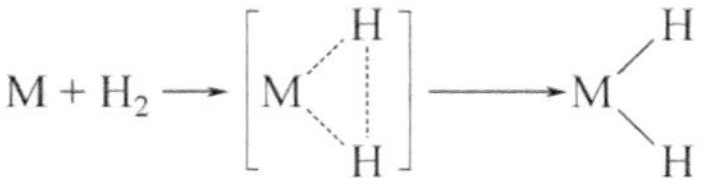

$$Mo(CO)_3(cht) + 2PPr^i_3 \longrightarrow Mo(CO)_3(PPr^i_3)_2 + cht$$

$$Mo(CO)_3(PPr^i_3)_2 + H_2 \longrightarrow Mo(CO)_3(PPr^i_3)_2(H_2)$$

(cht = 环庚烯)

中间体包含有 H—H 键，即在 M（H_2）中的 η_2-H_2 配体[在顺式 *cis*-$M(H)_2$ 中，不是两个分别键合的氢原子或氢离子]，下面的化合物已被分离出来（图 4.10）。

M—H_2 键（类似 M—CO 键）包含有 L—M σ 供体与 L—M π 接受体之间的相互作用，后者将电子引入到 H_2 分子的 σ^* 反键轨道中，从而削弱了 H—H 键，使其断裂。

（6）氢桥配合物　$[(OC)_5Cr—H—Cr(CO)_5]^{-}$类似于乙硼烷 B_2H_6，因两者都有键合电子对对 Lewis 酸的贡献［式(4.4)］。

$$L_nM\text{—}H + ML_n \longrightarrow L_nM\text{—}H\text{—}ML_n \tag{4.4}$$

双氢桥在过渡金属配合物中也是同样存在的（图 4.11）。

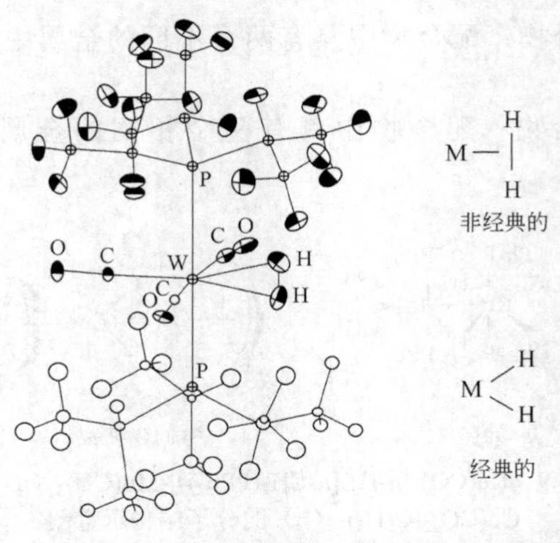

图 4.10 $W[P(i\text{-}Pr)_3]_2(CO)_3(H_2)$的分子结构图

H　H
B　Cr(CO)4
H　H

图 4.11 $Cr(CO)_4(\mu\text{-}H)_2BH_2$ 的分子结构示意图

4.1.2 簇合物

含有金属-金属键的化合物为簇合物，簇合物是多核配合物（图 4.12）。形成金属-金属键的先决条件之一是金属有较低的表观氧化态。较低氧化态的第二个重要因素是 d 轨道的性质，因为 d 轨道的有效重叠对稳定簇合物是必要的。一般来讲，第一过渡系列的金属不利于形成簇合物，因为即使在较低氧化态时，它们 d 轨道的尺寸也比较小。

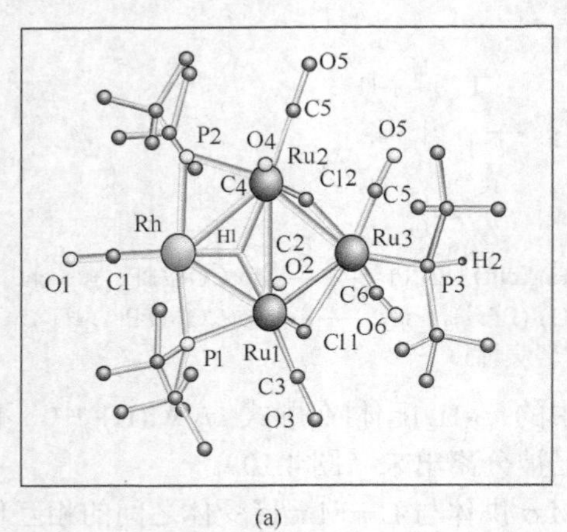

(a)

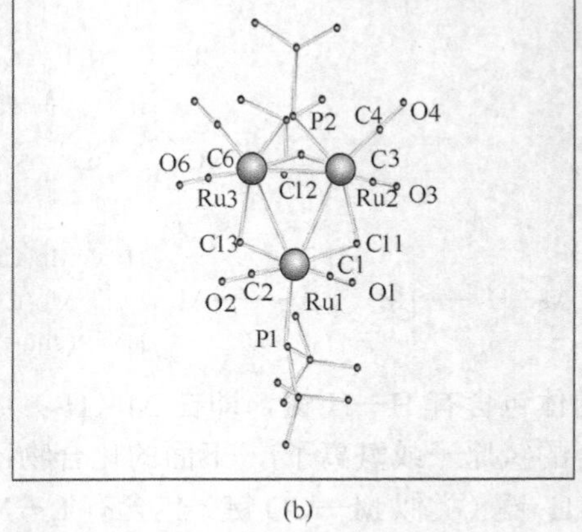

(b)

图 4.12 $[Ru_3Rh(CO)_7(\mu_3\text{-}H)(\mu\text{-}PBu^t_2)_2(Bu^t_2PH)(\mu\text{-}Cl)_2]$（a）和 $[Ru_3(CO)_6(\mu\text{-}Cl)_3(\mu\text{-}PBu^t_2)(Bu^t_2PH)]$（b）的分子结构图

计算多核配合物中 M—M 键的个数的一个方法是计算整个配合物中电子的总数，然后从 $n\times18$ 减去该数，n 为体系中金属的数目。结果就是为达到 18 电子组态而需要的电子数，如果认为这些电子是通过形成金属-金属键而获得的，则该数除以 2 即是键的数目。并不是所有的多核物种都适合定域 18 电子键模型。

簇合物化学的发展受到了超分子化学的影响，例如，异硫金属簇合物（图 4.13）。

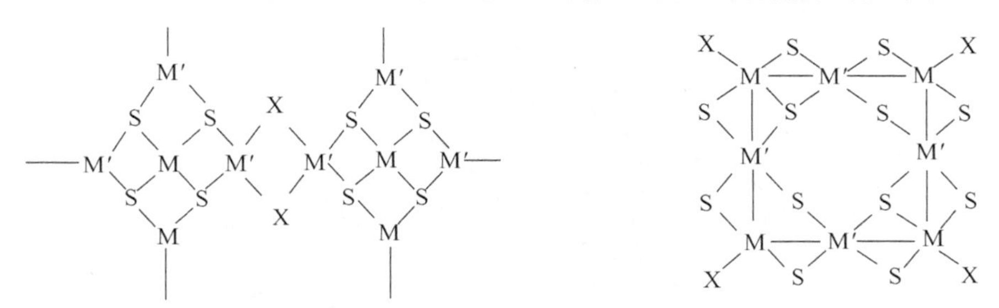

图 4.13　异硫金属簇合物中的分子结构单元示意图

4.1.3　大环配合物

（1）常见的大环配体（图 4.14）　已经为多核配合物特殊设计与合成了一些新型大环配体（图 4.15）。最近有关这种配体的双核混价 Mn（Ⅱ）Mn（Ⅲ）配合物已有报道（图 4.16）。

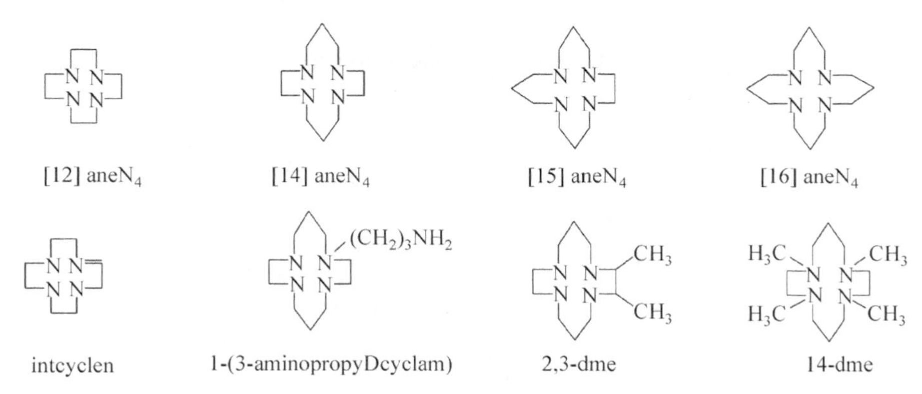

图 4.14　几种四氮杂大环配体

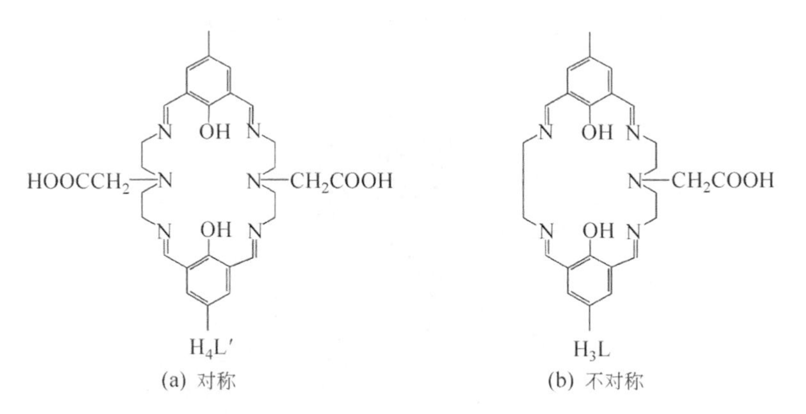

图 4.15　用于形成双核配合物的新型大环配体的结构示意图

另外，还开发出一些双环或三环配体用于酶或蛋白质的金属活性中心的模拟配合物的合成（图 4.17 和图 4.18）。

（2）大环配合物　大环配合物主要归属于生物无机化学（或生物配位化学），这是正在迅速发展着的一个领域。例如，细胞色素活性中心是亚铁血红素基，是一种 Fe^{2+}或 Fe^{3+}的卟啉配合物（图 4.19）。

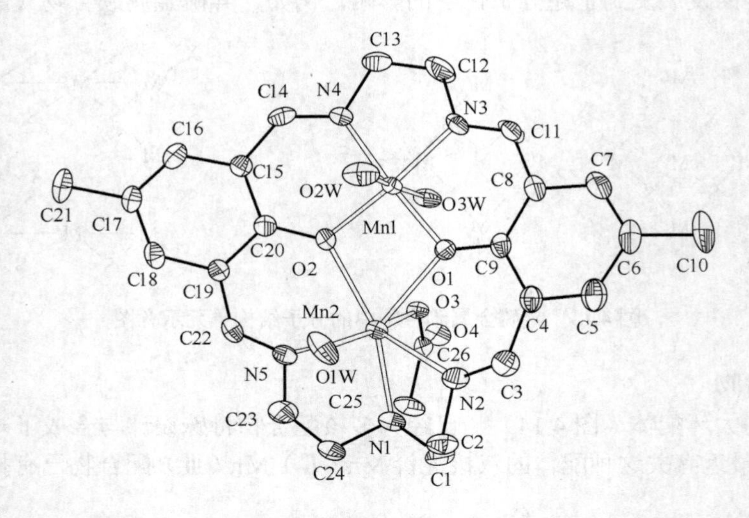

图 4.16　配合阳离子$[Mn(Ⅱ)Mn(Ⅲ)L(H_2O)_3]^{2+}$的晶体结构 ORTEP 图［配体 L 为图 4.15（b）］

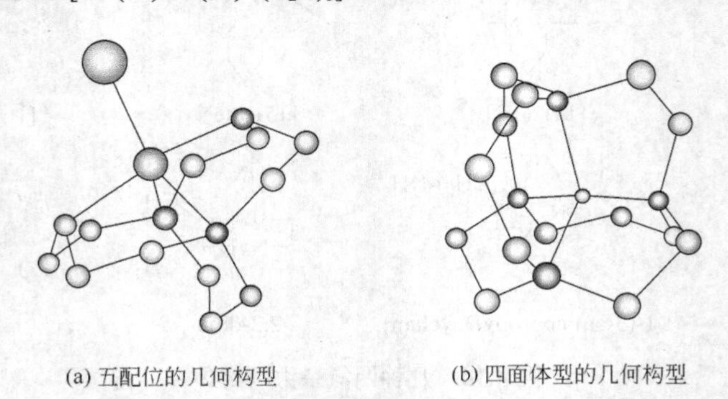

(a) 五配位的几何构型　　(b) 四面体型的几何构型

图 4.17　具有双环和三环配体的配合物的结构示意图

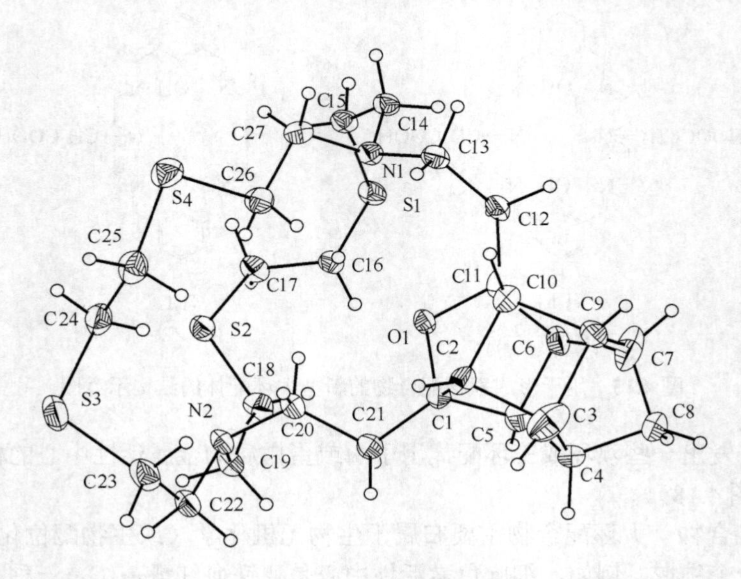

图 4.18　三环配体 $C_{27}H_{41}S_4N_2O$ 的晶体结构 ORTEP 图

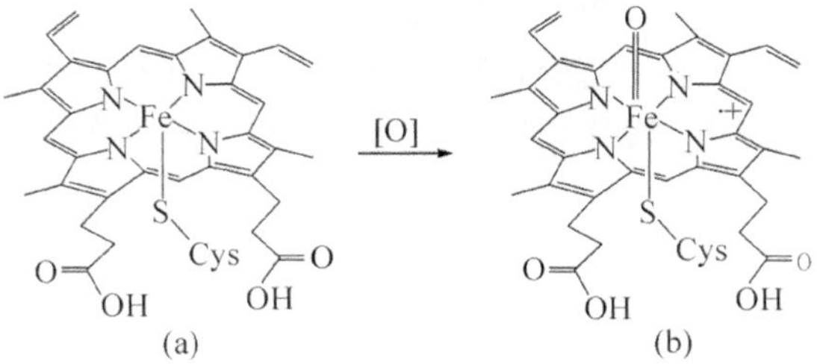

图 4.19　半胱氨酸作为轴向配体的铁（Ⅲ）卟啉配合物的结构示意图（a），它是细胞色素 P450、CPO 和 NOS 酶的代表。这些蛋白质和相关血红素酶中的活性氧成分是氧化的铁（Ⅳ）卟啉阳离子自由基（b）

图 4.20 为由细胞色素 P450 进行的氧的活化与传输的催化循环。

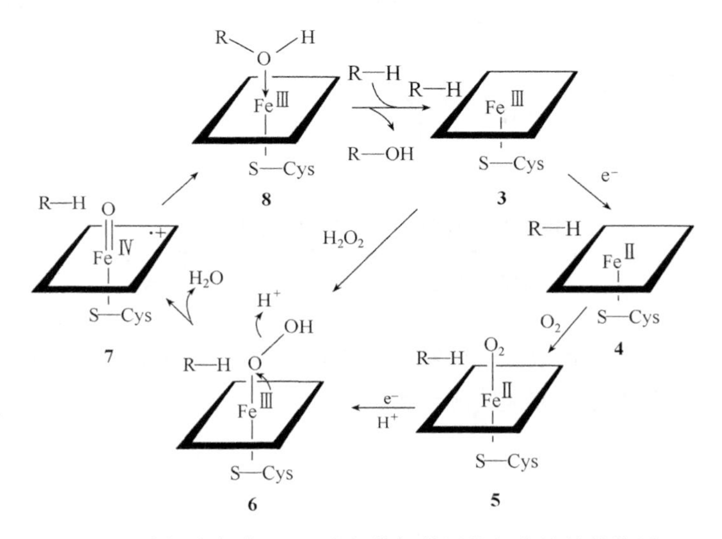

图 4.20　由细胞色素 P450 进行的氧的活化与传输的催化循环

卟啉配合物的稳定性顺序为：$Ni^{2+}>Cu^{2+}>Co^{2+}>Fe^{2+}>Zn^{2+}$，形成动力学遵循以下顺序：$Cu^{2+}>Co^{2+}>Fe^{2+}>Ni^{2+}$。

为什么是铁在细胞色素活性中心?因为其在自然界中丰富的含量和 $Fe^{Ⅱ}/Fe^{Ⅲ}$的氧化还原化学。表 4.2 给出了 $Fe^{Ⅱ}/Fe^{Ⅲ}$的氧化还原电势。

表 4.2　$Fe^{Ⅱ}/Fe^{Ⅲ}$电对的氧化还原性质

化　合　物	E_0/mV	化　合　物	E_0/mV
血红蛋白	170	细胞色素 P450	−400
肌红蛋白	46	$[Fe(H_2O)_6]^{2+/3+}$	771
细胞色素 a_3	400	$[Fe(bpy)_3]^{2+/3+}$	960
细胞色素 c	260	$[Fe(CN)_6]^{4-/3-}$	358
细胞色素 b_5	20	$[Fe(C_2O_4)_3]^{4-/3-}$	20

$$HFeO_4^- \xrightarrow{+2.07} Fe^{3+} \xrightarrow{+0.77} Fe^{2+} \xrightarrow{-0.44} Fe$$

图 4.21 给出了亚铁血红素 A、B、C 的结构示意图。细胞色素 a 有非常高的氧化还原电势，能将氧还原为细胞色素 c 氧化酶配合物中的水。由于细胞色素 c 中的金属原子是 6 配位的，因此细胞色素 c 仅可在电子转移反应（如光合作用和呼吸作用）中发生反应。

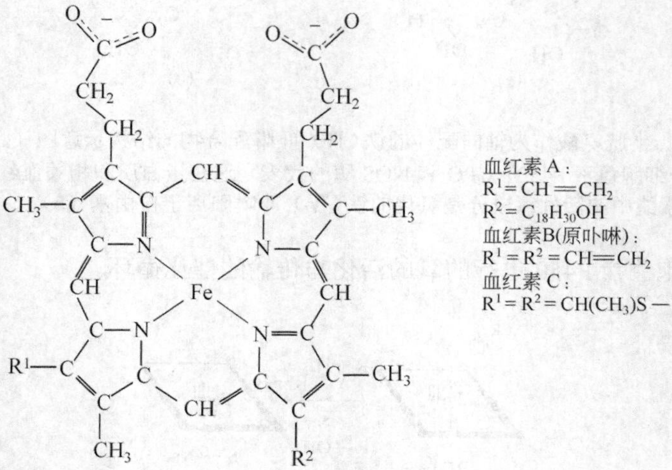

血红素 A：
$R^1 = CH{=}CH_2$
$R^2 = C_{18}H_{30}OH$
血红素B(原卟啉)：
$R^1 = R^2 = CH{=}CH_2$
血红素 C：
$R^1 = R^2 = CH(CH_3)S$—蛋白质

图 4.21 血红素 A、B、C 的结构示意
亚铁血红素 A：细胞色素 a；亚铁血红素 B：血色素，肌球素，过氧化物酶，细胞色素 b；亚铁血红素 C：细胞色素 c

卟啉配体是一种具有共轭双键的四吡咯的大环体系（图 4.22），过渡金属卟啉配合物可能是在生物体系中最重要的一类金属配合物。

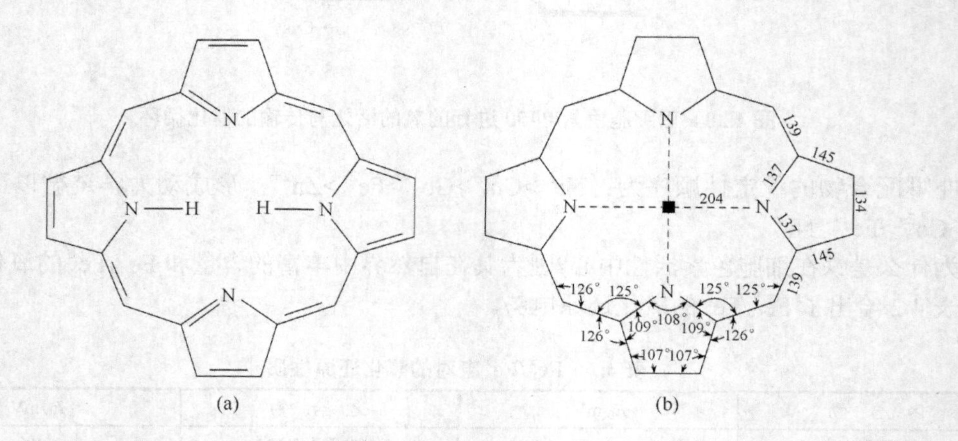

图 4.22 卟啉配体的分子结构的示意图（a）和有关卟啉分子结构的一套合理的数据（b）

叶绿素体系（维生素 B_{12}）是底部带有一个长烷烃链的卟啉大环体系，吡咯环中的一个双键被还原（图 4.23）。

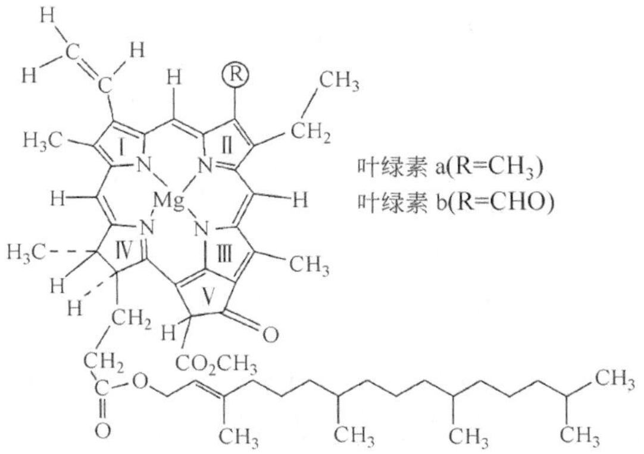

叶绿素 a(R=CH_3)
叶绿素 b(R=CHO)

图 4.23　叶绿素的分子结构示意图

（3）生物无机化学中的氧分子配合物和氮分子配合物　氧气和氮气配体是一类具有弱 π 键的软配体。由于卟啉环体系中的共生作用，使得氧分子可以很好地键合到卟啉环中的二价铁上，由于 CO 键合得更好，它会使血红素（在有 CO 存在下）很难获得氧。

氮分子与一氧化碳是等电子体，但在经历了很长时间以后，才在 1965 年制备出第一个氮分子配合物。氮分子配合物很重要，因为相对于气态分子，这类配合物削弱了 N—N 键级，进而使 N—N 键断裂成为可能（大气固氮）。可以用氮气合成氮配合物，此处亲核的氮取代了弱配体水。桥氮的形成见式(4.5)。

$$[Ru(NH_3)_5Cl]^{2+} \xrightarrow[H_2O]{Zn(Hg)} [Ru(NH_3)_5(H_2O)]^{2+}$$

$$[Ru(NH_3)_5(H_2O)]^{2+} + N_2 \longrightarrow [Ru(NH_3)_5N_2]^{2+} \tag{4.5}$$

氮分子作为配体的可能结构如图 4.24。

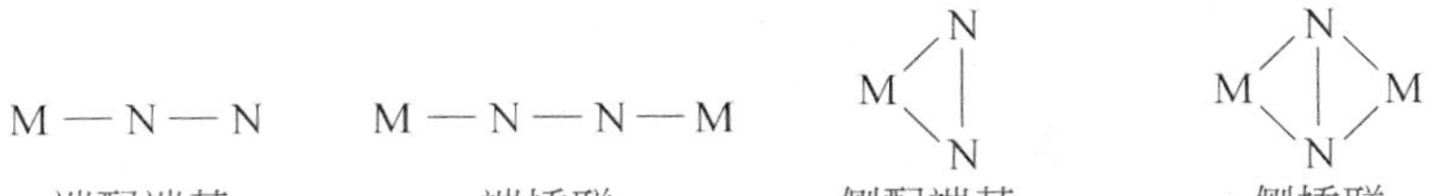

M—N—N	M—N—N—M	M(N≡N 侧配)	M(N≡N)M
端配端基	端桥联	侧配端基	侧桥联

图 4.24　N_2 分子作为配体的可能配位模式

与 CO 相比，键合的 N_2 有 π 反馈键，这可以从较短的 M—N 键和相比于氮气 N—N 键的增长和 N—N 伸缩频率的降低来证明。在$[Ru(NH_3)_5N_2]Cl_2$ 中，N_2 的伸缩频率为 $2105cm^{-1}$，气态氮分子为 $2331cm^{-1}$（仅有拉曼活性）。

然而，CO 既是一个很好的 σ 供体，又是一个很好的 π 受体。由于 CO 较强的 π 接受能力，因此，羰基双氮配合物是罕见的，并且不稳定，如，$[Cr(CO)_5N_2]$和$[Cr(CO)_4(N_2)_2]$仅在低于室温下存在，而$[Cr(CO)_6]$在高达 154℃时才熔化。

图 4.25 示意出羟高铁血红素的形成步骤。在生物体中必须通过立体位阻来抑制这些反应。

4.1.4　含有过渡金属离子的超分子自组装（多核配合物）

超分子化学是在化学、生物化学、物理和技术等交叉领域中迅速发展起来的一门学科。其中心内容是表征、理解和应用各种非共价相互作用。

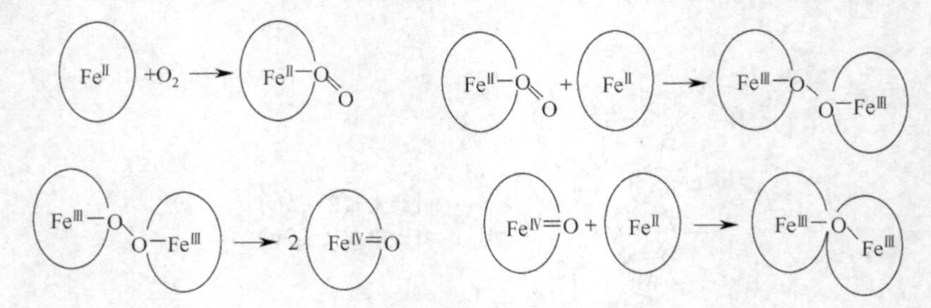

图 4.25 羟高铁血红素的形成步骤

人们在复杂的高对称性结构及这些结构在材料科学、化学技术和医药领域中应用的探索推动了超分子化学的发展。含有过渡金属离子的结构因为其独特的磁学和光学性质（多金属磁中心存在下的超分子自组装）而备受关注。金属离子用于解读包含在具有多个配位点的配体中的信息，这对自组装过程中直接组装精细结构很重要。

超分子化学中最令人感兴趣的结构之一是双螺旋结构。DNA 是由氢键和碱基对相互作用形成的典型生物双螺旋结构。这一领域的研究目标之一就是要创造新的含有过渡金属离子和有潜在纳米应用技术的纳米级结构。构造这种结构的策略之一是基于过渡金属离子和缩氨酸核酸（PNA）的分子识别性质。

纳米结构的空间排列和维数是由过渡金属离子的配位性质和 PNA（DNA 的合成类似物）的结构决定的。金属离子与 PNA 结合是通过取代天然核碱基而获得的。这一步骤可以精确控制 PNA 中金属离子的数目和位置。双聚体或更复杂结构可以通过天然核碱基间的氢键以及 PNA 链中掺杂的配体和桥联这些链的金属离子间的配位键来完成（图 4.26）。

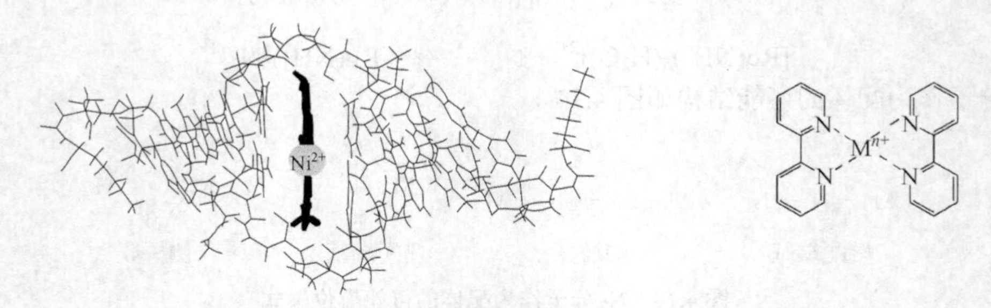

图 4.26 含 Ni^{2+}的 PNA 双螺旋结构示意图

图 4.27 给出了另一种具有一维螺旋结构的超分子组装物。这个结构是由两个长链的有机分子在金属离子的诱导配位下完成的。每个金属离子提供四个配位位置，具有四面体的配位环境。

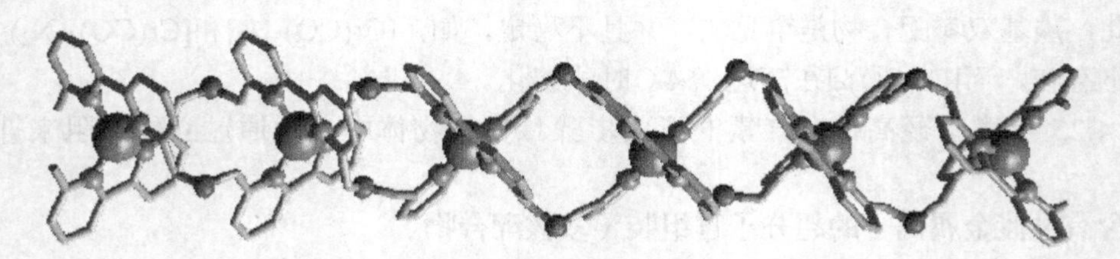

图 4.27 具有螺旋结构的一维超分子组装物的示意图

4.2　配合物的热力学性质

金属离子在溶液中不是孤立的，而是结合了配体（如溶剂分子或简单离子）或螯合剂，形成配离子或配合物。

向含有 $Fe(NO_3)_3$ 的水溶液中加入 SCN^-时会发生什么呢?

在水溶液中，$[Fe(H_2O)_6]^{3+}$（而不是 Fe^{3+}）与 SCN^-发生反应，如式(4.6）所示，SCN^-取代了配位层中的水分子，而形成$[Fe(H_2O)_5(SCN)]^{2+}$配离子。

$$Fe(NO_3)_3(s) \xrightarrow{H_2O} Fe^{3+}(aq) + NO_3^-(aq), \quad Fe^{3+}(aq) \equiv [Fe(H_2O)_6]^{3+} \tag{4.6}$$

$$[Fe(H_2O)_6]^{3+} + SCN^- \longrightarrow [Fe(H_2O)_5(SCN)]^{2+} + H_2O$$

4.2.1　热力学稳定性

“溶液中配合物的稳定性”是指在平衡状态下所涉及的两种物种的缔合程度。定性地讲，缔合程度越大，配合物越稳定。该缔合过程的平衡常数（稳定常数，形成常数）的数值可以定量地说明稳定性。这样，对于反应［式(4.7)]:

$$M + 4L \longrightarrow ML_4 \tag{4.7}$$

平衡常数越大，则溶液中 ML_4 的比例就越高。自由的金属离子在溶液中非常罕见，所以，M 周围通常都有溶剂分子，它们将与配体 L 发生竞争作用而逐步被配体取代掉。为简化起见，我们一般忽略这些溶剂分子，直接写出如式(4.8）所示的四个平衡常数。

$$\begin{aligned} M+L &\longrightarrow ML & K_1&=[ML]/\{[M][L]\} \\ ML+L &\longrightarrow ML_2 & K_2&=[ML_2]/\{[ML][L]\} \\ ML_2+L &\longrightarrow ML_3 & K_3&=[ML_3]/\{[ML_2][L]\} \\ ML_3+L &\longrightarrow ML_4 & K_4&=[ML_4]/\{[ML_3][L]\} \end{aligned} \tag{4.8}$$

K_1，K_2，…为逐级稳定常数。

或者，也可以写出总稳定常数［式(4.9)]:

$$M + 4L \longrightarrow ML_4 \qquad \beta_4=[ML_4]/\{[M][L]^4\} \tag{4.9}$$

逐级稳定常数与总稳定常数之间关系如式(4.10）所示：

$$\beta_4=K_1K_2K_3K_4$$

或一般地

$$\beta_n=K_1K_2K_3K_4\cdots K_n \tag{4.10}$$

以氨合铜离子为例，就有式(4.11）所示的结果：

$$\begin{aligned} Cu^{2+}+NH_3 &\longrightarrow Cu(NH_3)^{2+} & K_1&=[Cu(NH_3)^{2+}]/\{[Cu^{2+}][NH_3]\} \\ Cu(NH_3)^{2+}+NH_3 &\longrightarrow Cu(NH_3)_2^{2+} & K_2&=[Cu(NH_3)_2^{2+}]/\{[Cu(NH_3)^{2+}][NH_3]\} \\ & & \beta_4&=[Cu(NH_3)_4^{2+}]/\{[Cu^{2+}][NH_3]^4\} \end{aligned}$$

$$\lg K_1=4.0, \lg K_2=3.2, \lg K_3=2.7, \lg K_4=2.0 \text{ 或 } \lg\beta_4=11.9 \tag{4.11}$$

因此常用一系列的逐级平衡过程来代表金属的键合过程，其稳定常数的数值差别可以从几百到 10^{35} 甚至更多。由于这个原因，它们常常以对数的形式出现，对数形式的另一有用之处在于它们（$\lg K$）直接与反应的自由能成比例［式(4.12)]:

$$\Delta G^\ominus=-RT\ln b, \Delta G^\ominus=-2.303\,RT\lg 10^\beta, \Delta G^\ominus=\Delta H^\ominus-T\Delta S^\ominus \tag{4.12}$$

例如，醋酸锌的稳定常数在数值上稍大于 10，而葡萄糖酸铜的稳定常数则超过了 10^{18}。

结果，配合物的实际稳定常数的对数便于书写与理解金属-配体的稳定性。显而易见，葡萄糖酸铜是没有生物学意义的。有些物质，如 EDTA 与金属结合时的螯合性、螯合活性和键合能力是非常强的，尤其是锌。

4.2.2 配合物的稳定性

（1）金属本性的影响

- 离子半径越小，配合物越稳定。
- 氧化态越高，配合物越稳定。
- 晶体场影响，稳定性的顺序为：$Ca^{2+}<Sc^{2+}<Ti^{2+}<V^{2+}<Cr^{2+}<Mn^{2+}<Fe^{2+}<Co^{2+}<Ni^{2+}<Cu^{2+}<Zn^{2+}$。
- 硬酸与硬碱、软酸与软碱能形成很稳定的配合物。

（2）配体本性的影响

- 强碱性（对 H^+的亲和性）的配体能形成稳定的配合物。
- 螯合作用：多齿配体与单齿配体一样具有等同的形成配位键的能力。但是，一旦一个配位键形成后，进一步形成配位键的可能性更大。
- 螯合环的尺寸：5 元饱和配体和 6 元不饱和配体能形成稳定的配合物。
- 立体张力：大的配体形成的配合物较不稳定。

（3）螯合作用　螯合作用可以通过螯合剂和金属离子的反应与相应的单齿配体反应的比较而表现出来，例如，2,2′-联吡啶与吡啶或乙二胺与氨。

多年来，这一类的比较表明：螯合剂形成的配合物具有较高的热力学稳定性。

4.2.3 有关物种浓度的计算

形成常数代表了配离子的稳定性，逐级稳定常数 K 代表了单个配体与金属离子的键合［式(4.13)］：

$$Zn^{2+}+NH_3 \rightleftharpoons Zn(NH_3)^{2+} \qquad K_1=\frac{[Zn(NH_3)^{2+}]}{[NH_3][Zn^{2+}]}=3.9\times10^2$$

$$Zn(NH_3)^{2+}+NH_3 \rightleftharpoons Zn(NH_3)_2^{2+} \qquad K_2=\frac{[Zn(NH_3)_2^{2+}]}{[Zn(NH_3)^{2+}][NH_3]}=2.1\times10^2$$

$$Zn^{2+}+2NH_3 \rightleftharpoons Zn(NH_3)_2^{2+} \qquad \beta_2=\frac{[Zn(NH_3)_2^{2+}]}{[NH_3]^2[Zn^{2+}]}=K_1K_2=8.2\times10^4$$

$$Zn^{2+}+3NH_3 \rightleftharpoons Zn(NH_3)_3^{2+} \qquad \beta_3=K_1K_2K_3 \tag{4.13}$$

H^+、金属离子、配体和各种金属离子的沉淀剂之间存在着配位竞争。

金属配合物固体的溶解过程决定于金属螯合物的稳定性、配位剂的浓度、pH 值和不溶金属沉降物的本性。

4.3 分子电子器件——氧化还原活性配合物

4.3.1 分子电子学的概念

分子电子器件为利用单分子执行信号和信息过程的器件或单分子导线。这是一个由全世界范围内的生物学家、化学家、物理学家、数学家、计算机科学工作者和基因工程师共同努力的交叉领域。分子电子器件将为先进技术，如分子导线、分子开关和分子体系，开辟一个崭新的空间。这样的体系也许包括可以植入生物系统（如人体）的生物计算机，它具有在细胞层次上进行修复的功能。

分子电子器件是一种运用分子和分子结构作为建构单元来建立电路系统的概念。分子之间具有导电和传输能量的能力，如果这一过程在某些程度上可以加工和控制，这些分子和分子结构就可以完成信息过程所要求的任务，这将包括一些类似于当今计算机的电路系统中的基本任务。

4.3.2　分子导线

分子电子器件的应用之一是分子导线。分子导线（或纳米线）已被用于描述许多差异很大的物质，包括碳纳米管、其他类型的纳米管、聚炔类化合物、掺杂聚合物以及其他的一些物质种类。这一术语的宽泛性预示着它仅是原子的一种延长排列，其组成和厚度等可能有很大的不同，但都应具有电子传输的通道。

一般来讲，分子导线的定义是：可以沿着分子链的方向传输电荷（电子或空穴）的（预）一维线形分子物质。

无论作何种用途的分子导线都必须具有以下几个功能。

- 必须能沿着其长度方向传输电子。这既可以是通过空穴或电子（电子隧道、共振和非共振）的运动来完成，也可以通过激发光子导线以及在分子水平上其他形式的电子运动来完成。
- 该类导线分子必须容易被氧化或还原。
- 分子导线的第三个特点是：必须有绝缘护套来保护电流不会泄漏到周围环境。将叔丁基连接到卟啉分子的骨架上而形成的卟啉类分子导线正是这样的一个例子。
- 最后，分子导线必须具有合适的和固定的长度。

具有简单电活性终端的一维氧化还原型分子导线的导电机理比光子导线的机理简单（图 4.28）。

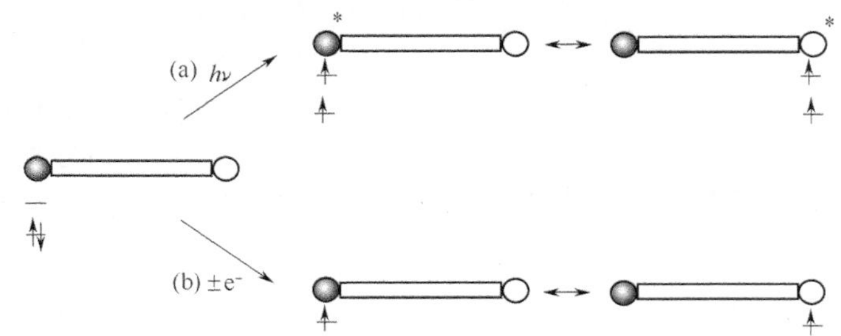

图 4.28　光子型导线（a）和氧化还原型分子导线（b）的导电机理示意图

图 4.29 所示的三组分分子体系就是这样一个例子，它由三部分组成：①供体，氧化还原终端；②导电区域，具有共轭 π 电子的不饱和有机基团（图 4.30）；③受体：氧化还原终端。

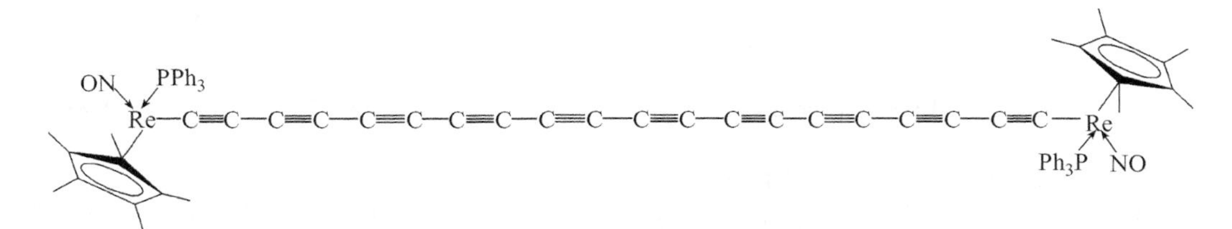

图 4.29　金属配合物组分为发色团、不饱和有机单元为导电空间体的分子导线结构示意图

$M{-}(C{\equiv}C)_n{-}M$　　$M{=}(C)_n{=}M$　　$M{\equiv}C{-}(C{\equiv}C)_n{-}C{\equiv}M$　　$M{=}C{-}(C{\equiv}C)_n{-}M$　　$M{=}C{-}(C{\equiv}C)_n{-}C{=}M$

图 4.30　一些作为导电区域的分子结构组分

4.3.3　分子开关

分子电子器件的另一类装置是开关器件。分子开关是这样一些分子或分子的聚集体，它们在外部刺激下，如光、电、磁或化学环境的影响等，能可逆地改变它们的状态。这些分子应该具有两种不同性质的状态，分别是 ON 状态（允许发生完全的电子转移）和 OFF 状态（电子转移被禁止）。

分子开关的类型有：

- 扭转过程——分子通过单键的扭转来完成在 OFF 状态的去偶合，尽管寻找一种方法在空间上来扭转和锁住分子是不容易的。
- 饱和/非饱和——双键被转变为单键，所以共轭被破坏。这可以通过还原或分子重新组合来完成，ON 和 OFF 状态可以通过价带间的去质子化或质子化来获得。
- 量子界面——基于电子效应和电子波的性质。
- 隧道开关——基于分子激发态的性质。开关可以通过改变能垒的高度或势能井的深度来调整 OFF 状态。

在有电流存在的地方常需要开关，将来在某种程度上它们可能被用作数据记录或记忆储存。

手性光学分子能在两种互为空间镜像的不同状态间转换，这两种镜像状态在电学和光学上具有鲜明的、稳定的开关性质。

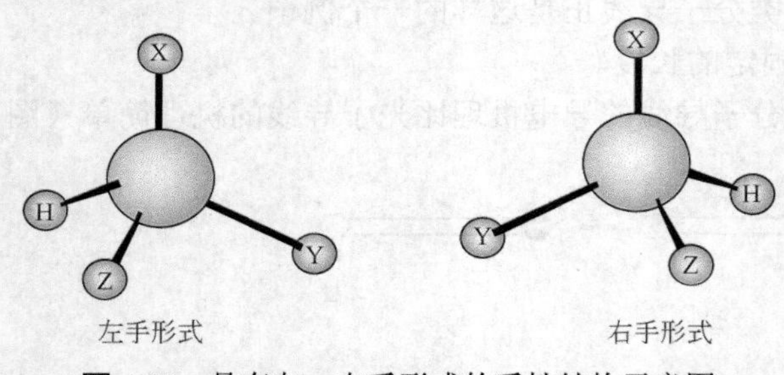

图 4.31　具有左、右手形式的手性结构示意图

手性光学开关是设计用来可以在左手和右手的两种形式间相互转换的一种简单分子，从而能形成基于分子的二进制码，这种分子编码可以在光束通过分子时，由手性效应（又称手性光学性质）读出来（图 4.31）。

这种手性开关过程示意在图 4.32 中。

两个十字箭头分别代表分子的一种状态，组成了开关的二元对（图 4.33）。

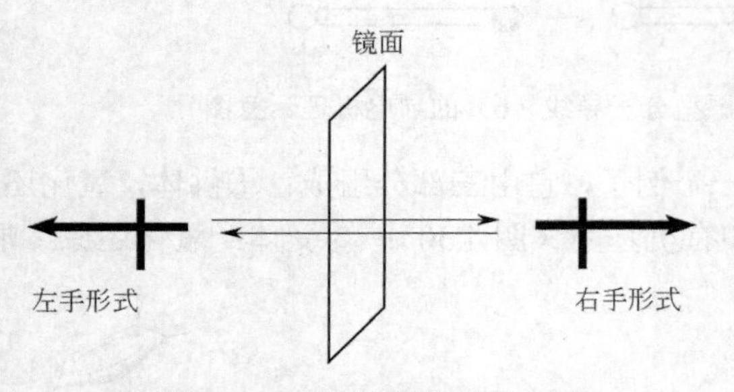

图 4.32　手性转换过程示意图

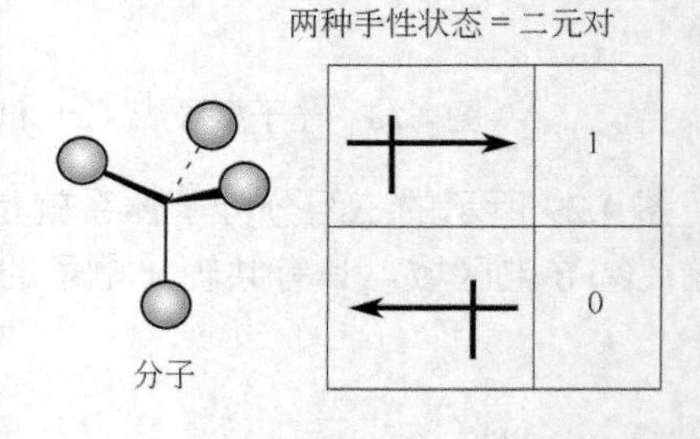

图 4.33　手性光学活性分子的二元对

手性光学分子的开关行为示意在图 4.34 中。

开关过程也可以通过电场翻转偶极来完成，它的状态可以由电场监测器通过监测由偶极诱导的分子电容而读出来。

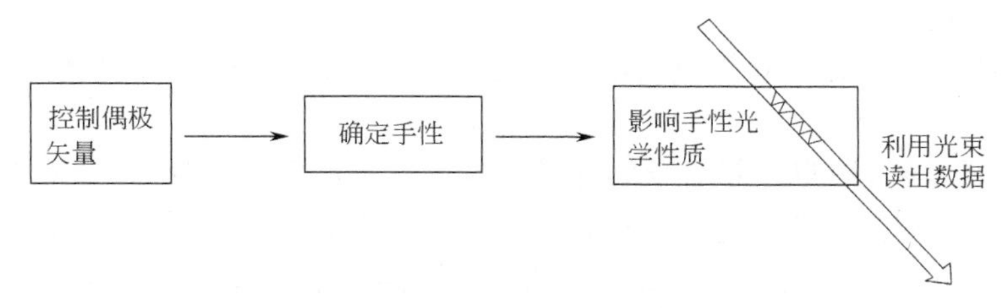

图 4.34　手性光学活性分子的开关过程示意图

4.4　配合物的磁学性质

磁性来自流动的电荷，例如线圈中的电流。在没有电流存在的物质中，仍然有磁相互作用。原子由一直运动的带电粒子（质子和电子）组成，原子中可以产生磁场的过程有如下几个：

- 核自旋　有些核，如氢核有可以产生磁场的净自旋。
- 电子自旋　电子有两种自旋状态（向上和向下）。
- 电子轨道的运动　电子绕核运动可以产生磁场。

这些磁场之间以及与外磁场之间都有相互作用。但是，这些相互作用中有些很强，有些可以忽略不计。

核自旋相互作用的测量在核磁共振谱(NMR）和电子自旋共振谱（ESR）中用来分析化合物。在很多情况下，与核自旋作用产生的影响很小。

电子自旋之间的相互作用对于像锕系这样的重元素是最强的，称为自旋-自旋偶合。对于这些元素，这种相互作用可以改变电子轨道的能级。

电子的自旋与其轨道运动之间的相互作用称为自旋-轨道偶合。自旋-轨道偶合对于许多无机化合物的能级有重要的影响。

宏观的效应，如一块磁铁的吸引，主要源自化合物中未成对的电子数和它们的排布，各种不同的情况为物质的不同磁状态。

4.4.1　物质的磁状态

（1）抗磁性　抗磁性物质中所有的电子都成对，净自旋为零。磁铁对抗磁性物质有弱的排斥作用。抗磁化作用是一种非常弱的磁化作用，仅在有外磁场存在的情况下才表现出来。这是电子轨道运动在外磁场中的变化而引起的。这种诱导磁矩很小并且与外磁场的方向相反。

（2）顺磁性　顺磁性物质中具有未成对的电子，顺磁性物质可以被磁铁吸引。顺磁化作用使原子磁偶极矩倾向于与外磁场的方向相一致，磁偶极矩是由量子自旋和电子轨道角动量引起的。

（3）铁磁性　铁磁性物质中也具有未成对的电子，它们通过铁磁偶合过程而取向一致。铁磁性物质，如铁，能被磁铁强烈地吸引。铁磁化作用是物质能显示出自发磁化作用的一种现象，是一种最强的磁化作用。这种现象的表现是，物质如铁（钴和镍）在磁场中被磁化，磁场被移走以后仍然保持其磁化作用。

（4）亚铁磁性　亚铁磁性物质中同样具有未成对的电子，其中一部分方向向上，一部分方向向下，这就是亚铁磁性偶合。由于亚铁磁性物质中的一些自旋的取向是沿着一个方向的，所以这种物质能被磁铁吸引。

（5）反铁磁性　当自旋方向相反的电子数目相等时，该物质被磁铁强烈地排斥。这被定义为反铁磁体。

（6）超导体　超导体被穿过其自身的磁场排斥，超导体的这一性质，又称 Meissner 效应，用于检测超导状态的存在。关于超导性的起源仍有很多问题需要研究。已经表明：超导体磁性质的机理与这里所讨论的化合物有着重要的差别，因为这一原因，超导体在这里将不再进一步讨论。

4.4.2　与外磁场的相互作用

当物质置于磁场中，物质内的磁场将是外磁场和物质本身产生的磁场的总和。物质内的磁场称为磁感应。用符号“B”表示，有式(4.14)：

$$B=H+4\pi M \tag{4.14}$$

式中，B 为磁感应；H 为外磁场；$\pi=3.14159$；M 为磁化作用（物质的一种性质）。

磁场（H）的单位通常为高斯（G）或特斯拉（T），1T=10000G。

为数学和实验方面的方便起见，该式常写成式(4.15)：

$$B=1+4\pi M=1+4\pi\chi_V \tag{4.15}$$

式中，χ_V 为体积磁化率（$\chi_V=M/H$）。

（1）磁化率　磁化率是物质对磁场响应程度的物理量。质量磁化率可以用符号“χ”来表示（式 4.16）。

$$\chi=M/J \tag{4.16}$$

式中，J 是物质单位质量内的磁偶极矩，A/m。

如果 χ 是正值，该物质为顺磁性，磁场会因该物质的存在而加强。如果 χ 是负值，该物质为抗磁性，磁场会因该物质的存在而削弱（表 4.3）。

表 4.3　一些物质的磁化率

物质	$\chi_V/10^{-5}$	物质	$\chi_V/10^{-5}$
铝	+2.2	氢	−0.00022
氨	−1.06	氧	+0.19
铋	−16.7	硅	−0.37
铜	−0.92	水	−0.90

之所以命名体积磁化率（χ_V）是因为 B、H 和 M 等都是单位体积的物理量。但是，这导致体积磁化率是没有单位的物理量。用磁化率比用磁化度方便，因为抗磁性物质和顺磁性物质的磁化率与外磁场的强度 H 无关。

有多种不同形式的磁化率，主要的两种是质量磁化率（χ_ρ）和摩尔磁化率（χ_m）[式(4.17)]。

$$\chi_\rho=\chi_V/\rho\ (\mathrm{m^3/kg})$$

$$\chi_m=\chi_V M_a/\rho\ (\mathrm{m^3/mol}) \tag{4.17}$$

式中，ρ 是物质的密度，$\mathrm{kg/m^3}$；M_a 是物质的摩尔质量，kg/mol。

表 4.4 列出了几种常见物质的质量磁化率，表 4.5 是根据磁性质进行的物质分类。

（2）有效磁矩　磁相互作用的另一个量度是有效磁矩，μ，即：

$$\mu=2.828(\chi_m T)^{1/2} \tag{4.18}$$

式中，μ 为有效磁矩；χ_m 为摩尔磁化率；T 为温度。

表 4.4　几种常见物质的质量磁化率

物质	χ_ρ/（$10^{-8}m^3/kg$）	物质	χ_ρ/（$10^{-8}m^3/kg$）
铝	+0.82	氢	−2.49
氨	−1.38	氧	+133.6
铋	−1.70	硅	−0.16
铜	−0.107	水	−0.90

表 4.5　基于磁性质的物质分类

分　类	χ 对磁场 B 的依赖	对温度的依赖	磁滞现象	例子	χ
抗磁性	否	否	否	水	-9.0×10^{-6}
顺磁性	否	是	否	氨	2.2×10^{-5}
铁磁性	是	是	是	铁	3000
反铁磁性	是	是	是	铽	9.51×10^{-2}
亚铁磁性	是	是	是	$MnZn(Fe_2O_4)_2$	2500

μ 的单位是波尔磁子（BM）。$1BM=9.274\times10^{-24}J/T$。有效磁矩是一种很方便的物质磁性的测量，因为抗磁性物质和顺磁性物质的有效磁矩与温度和外磁场的强度都无关。

因此，可以检测物质的磁化度、磁化率和有效磁矩与物质结构的关系。

配合物的磁性起源于未满电子层中电子的自旋和轨道角动量。伴随着这种顺磁性，总是存在较小的抗磁性效应。这种抗磁性是由于在外磁场作用下电子的旋进运动而产生的。Pascal 常数提供了一种经验的方法，用于估算过渡金属配合物分子中原子的抗磁化率（表 4.6）。

表 4.6　一些离子和基团的 Pascal 常数

离子	Na^+	K^+	NH_4^+	Hg^{2+}	Fe^{2+}	Fe^{3+}	Cu^{2+}	Br^-	I^-	NO_3^-	ClO_4^-	IO_4^-	CN^-	NCS^-	H_2O	$EDTA^{4-}$
常数	6.8	14.9	13.3	40	12.8	12.8	12.8	34.6	50.6	18.9	32	51.9	13	26.2	13	~150
离子	Co^{2+}	Co^{3+}	Ni^{2+}	VO^{2+}	Mn^{3+}	Cr^{3+}	Cl^-	SO_4^{2-}	OH^-	$C_2O_4^{2-}$	OAc^-	pyr	Me-pyr	$Acac^-$	en	urea
常数	12.8	12.8	12.8	12.5	12.5	12.5	23.4	40.1	12	34	31.5	49.2	60	62.5	46.3	33.4

4.4.3　抗磁性

抗磁性可描述为在外磁场存在下电子形成的电流及绕核的运动。因此，对每一个原子核都可以计算其抗磁性贡献。但是，抗磁性的数值比起顺磁性和其他的作用是非常小的，对于很多物质而言是可以忽略不计的。

4.4.4　顺磁性

决定顺磁性行为的最重要的结构特征是化合物中的未成对的电子数目。顺磁性化合物的磁矩（仅有自旋）如式(4.19）所示：

$$\mu=g\{S(S+1)\}^{1/2} \tag{4.19}$$

式中，μ 为有效磁矩；$g=2.0023$；$S=1/2$（一个电子），1（两个电子），3/2（三个电子）等。

该式中有时 $g=2$，这样不会带来很大的误差，因为简单的“仅有自旋”是一种合理的近似，但常常不精确。

将电子的自旋和轨道运动都考虑进去，则如式(4.20）所示：

$$\mu=\{4S(S+1)+L(L+1)\}^{1/2} \tag{4.20}$$

式中，μ=有效磁矩；S=1/2（一个电子），1（两个电子）等；L=总轨道角动量。

这一公式仅对具有很高对称性的分子才适用，因为在这样的分子中未配对电子的轨道能量是简并的。有关“L”的计算在任何一本量子力学课本中和物理化学教材的量子力学章节中都能找到。

仅有自旋的磁矩公式对所有的顺磁性行为常常是一级粗略近似。若测定磁化率的目的仅仅是决定未成对电子的数目，那么，这一公式是够用的。

4.4.5 铁磁性、反铁磁性和亚铁磁性

运用有效磁矩来描述顺磁性行为的优势在于这种磁行为的测量不依赖于温度和外磁场的强度。建立一套统一的标准同时适用于铁磁性、反铁磁性和亚铁磁性物质是不可能的。

所有这三类物质可以认为是顺磁性行为的特例。顺磁性行为的描述建立在每个分子都相互独立的假设上。这里所讨论的物质有这样一个情形：一个分子所产生的磁场方向受相邻分子所产生的磁场方向的影响，换句话讲，它们的行为是偶合的。如果这种偶合方式在磁场中总是方向一致时，是铁磁性物质，这种偶合称为铁磁性偶合。反铁磁性偶合是指在磁场中两个自旋相反的方向上具有相同的电子数目。亚铁磁性偶合是指在磁场中两个自旋相反的方向中的一个方向上具有比另一个方向上多的电子。

有些例外，磁矩的取向在整个物质中是不一致的。特定的区域，又称为畴，会形成不同的取向。偶合分子中畴的存在将引起在下文中所描述的几种磁行为。

在外磁场存在下，分子倾向于取向一致的趋势，加强了物质的磁化度。这就是为什么铁磁性物质和亚铁磁性物质的磁化率在数值上会高于顺磁性物质几个数量级的原因。这也得出一结论：这些物质的磁化率与外磁场的强度有关，这一点与抗磁性和顺磁性物质是一样的。

对于铁磁性物质，一个已知的磁偶极矩（未配对电子）所感受到的实际场强用“H_t”表示，得到一个与上述磁感应类似的式子［式(4.21)］。

$$H_t=H+N_wM \tag{4.21}$$

式中，H_t为电子所感受到的场强；H为外磁场的场强；N_w为分子场常数，大约为10000；M为磁化强度。

运用这一公式是因为对铁磁性物质的数学处理可以和顺磁性物质类似。该式中的分子场常数 N_w 是为了进行铁磁性偶合计算而定义的一个经验性常数。为了粗略地得到这一常数，需要进行量化计算，这种计算应考虑元素、它们在固体中的排布、电子的动能、核对电子的库仑引力、与其他电子间的排斥力以及自旋相互作用等因素。

随着温度升高，分子振动会微扰磁畴结构。所以，这三种类型物质的磁性在低温时是最强的。在足够高的温度下，所有这三类物质都没有磁畴结构存在，因而在高温时变成顺磁性物质。对于铁磁性和亚铁磁性物质来讲，可以看到顺磁性行为的温度称为居里（Curie）温度，而对反磁性物质来讲，这一温度称为尼耳（Neel）温度。这也是为什么与温度无关的有效磁矩不能用来定义这些物质的原因。

即使在没有外磁场存在的情况下，磁畴中磁矩的取向也可以给物质一净磁矩。这就是永久磁体，如磁铁。对于在放入外磁场之前没有净磁矩的物质来讲，在放入外磁场之后，可能保持有净磁矩。这就是录音带和计算机磁盘储存信息的基础。将磁化强度对磁场强度作图（磁场强度的变化是从一极向另一极，并再循环回来），可以定量地衡量这种记忆效应的大小，强的记忆效应该有一个宽的磁滞回线。

4.4.6　随温度变化的磁行为

随温度变化的磁行为的根源在于原子的热运动对分子磁矩取向的微扰。所以，抗磁性行为不随温度变化就不足为怪了。

（1）顺磁性作用　随温度的升高，顺磁性物质的磁化率下降。

一般的顺磁性化合物，磁化率与温度成反比。它们被称为“正常顺磁体”，其磁性主要是由永久磁矩的存在而引起的。这就是居里定律，其数学形式如式(4.22)。

$$\chi=C/T \tag{4.22}$$

其中

$$C=N_A g^2 b^2/4k$$

式中，χ 为磁化率；C 为居里常数；T 为温度；N_A 为阿伏加德罗常数；g 为电子因子；b 为玻尔磁子；k 为玻尔兹曼常数。

对于大多数的顺磁性化合物，这种反比例的关系都可以观察到。但外推到零度时，就不再遵守居里定律了，而遵守居里-外斯定律［式(4.23)］。

$$\chi=C/(T-\theta) \tag{4.23}$$

式中，χ 为磁化率；C 为物质特殊的居里常数；T 为热力学温度，开尔文；θ 为居里温度，开尔文。

（2）铁磁性和亚铁磁性作用　随着温度的升高，铁磁性和亚铁磁性物质的磁化率同样也是下降。但是，磁化率对温度的曲线与顺磁性物质的不同。顺磁性物质的曲线是正曲率，而铁磁性物质的曲线是负曲率，大致的形状如图 4.35 所示。

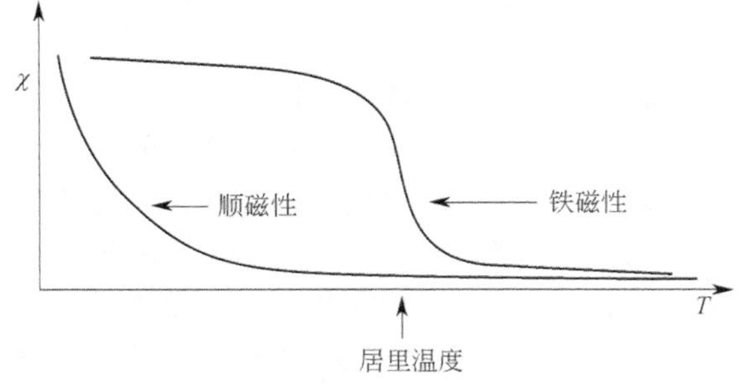

图 4.35　顺磁性和铁磁性物质 χ-T 曲线的粗略示意

当达到了居里温度时，曲线的曲度就会发生变化。在居里温度时，铁磁性和亚铁磁性化合物变成顺磁性的。居里温度的变化范围可以从 Gd 的 16℃到 Co 的 1131℃。

（3）反铁磁性作用　反铁磁性化合物的磁化率是随着温度的升高而升高，一直到其临界温度，也就是尼耳温度。在尼耳温度以上，这些化合物也变成顺磁性的化合物。尼耳温度的变化范围可以从 $MnCl_2 \cdot 4H_2O$ 的 1.66K 到 α-Fe_2O_3 的 953K。

4.5　配合物的光化学性质

光反应是有机化学和无机化学中普遍存在的一种现象。在光的辐射下，化合物的结构和电子性质会发生改变，光诱导电荷分离或光作用下的顺-反式异构化过程都是很好的例子。

结果，就可能发生具有或没有滞后现象的相转变。如果光诱导的新相具有足够长的寿命以及不同的光和（或）磁性质，同时这些变化可以容易地用光或磁的手段监测，则该类物质就具有作为开关或显示器等器件的潜力。

4.5.1　光化学过程的基本性质

分子吸收光子以后，紧跟着就有迅速的振动弛豫过程，这使分子达到一个与其电子激发态相关的平衡的几何构型。在每一个激发态，都存在着几个过程之间的竞争，如物理过程、辐射过程（荧光、磷光）、非辐射过程（内转换、系间穿越）以及失活的化学反应（图 4.36）。

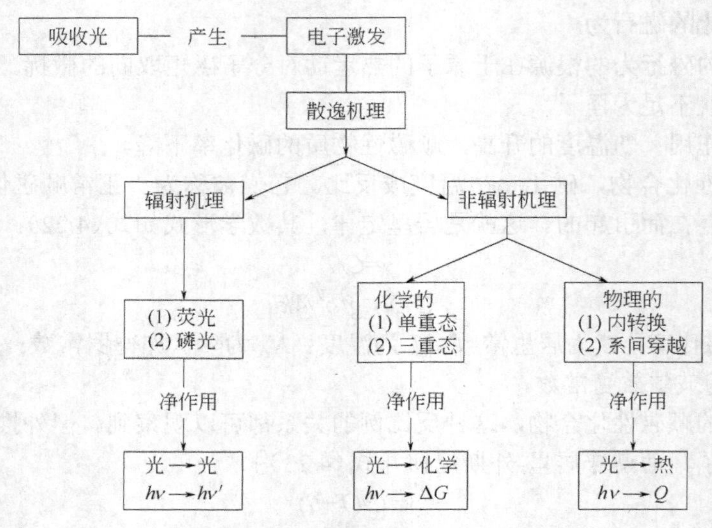

图 4.36 分子光化学过程示意图

利用过渡金属构建超分子体系有以下几个优势：

- 相对于简单的有机分子而言，d 轨道的参与可以提供较多的成键模式和构型的对称性。
- 通过不同的辅助配体，可以在较宽的范围内调整电子性质和立体性质。
- 通过运用不同长度的桥联配体，可以容易地修饰所预期的超分子体系的尺寸。
- 可引入特殊的光谱、磁性、氧化还原性质、光化学和光物理性质。

而且，过渡金属中心所具有的配位键角的多样性以及配体和金属之间成键的高度方向性，也提供了比弱静电作用、范德华力和 π-π 相互作用更多的优点。另一个有趣的方面是：在溶液中，由热力学驱动的各个分子组分能自组装成完好的结构，这与配位化学和生物学中的很多现象相当类似，这使得过渡金属配合物在模拟较复杂的生物体系方面很有价值。

4.5.2 人工光合作用

对廉价和清洁能源日益增长的需求刺激了能有效转化太阳能的新颖化学体系的开发。迄今为止，自然界的光合作用是最清洁和最有效的过程。它包含了收集可见光的发色团（天线）组合、电荷分离系统和氧化还原中心。所有这些元素（天线、电荷分离和反应中心）都含有过渡金属。过渡金属配合物在模拟更复杂的化学体系方面的应用，是因为它们具有丰富和多样的光化学反应过程。

（1）捕获光能的天线 捕获光能的天线是一种有组织的多组分系统，其中发色团分子系列吸收入射光并提供将激发态能量转移给一般接收组分的通道。天线效应仅当超分子排列在空间、时间和能量区域等方面都合理时才能获得。每一分子组分都必须能吸收入射光，由这一途径获得的激发态必须能在进行辐射或非辐射的去活过程前将电子能量转移给邻近的组分。作为能捕获光能的天线的发色团的候选者之一是卟啉和酞菁类，以及绿色植物中天然叶绿素的合成替代物（图 4.37）。

卟啉的特点是摩尔吸收系数高和向系统中其他部分的电子或能量转移快，可以作为大超分子体系中的建构单元。

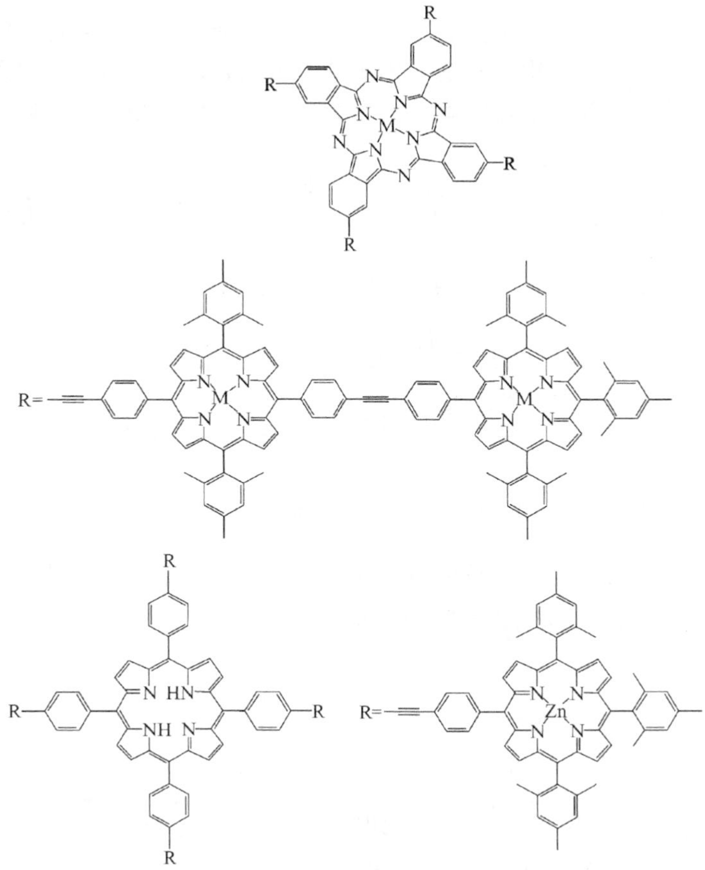

图 4.37　作为光吸收天线的卟啉和酞菁类物质

另一类广泛研究的人工天线是金属树枝状化合物和基于多吡啶配体的多金属类配合物。由于钌(Ⅱ)和锇(Ⅱ)多吡啶配合物高的摩尔吸收系数、光稳定性和激发态性质使其有很好的前景（图 4.38）。

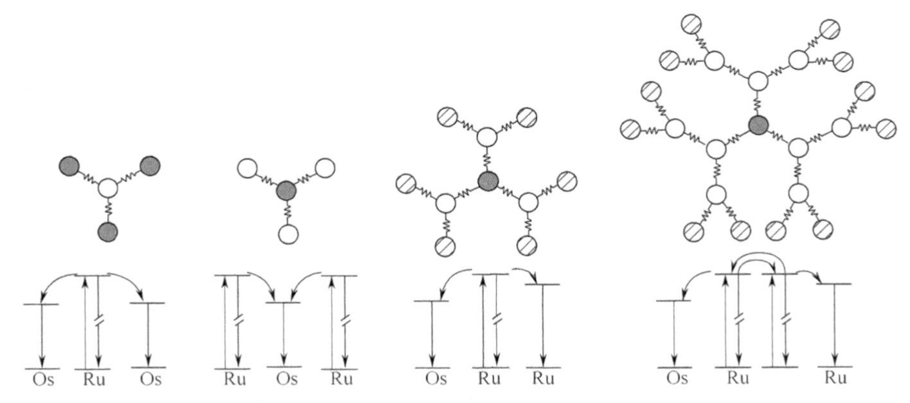

图 4.38　几种 Ru^{II} 和 Os^{II} 的多吡啶树枝类配合物的结构示意以及表示能量转移的能量图

钌中心为 ○ 和 ⊘，而锇中心为 ●，为清楚起见，端配体省略

（2）电荷分离系统　一旦太阳能被捕获并聚集在反应中心，它一定转化成更有用的形式，即化学能。当系统中形成的电荷分离状态具有足够的寿命时，这一目标才能实现。显然，在这样的结构中，电荷再结合是很容易的。为了减慢再结合过程，有必要引入附加的反应步骤将电荷移得更远一些。最简单的途径就是引入第二供体（D′）和第二受体（A′）（图 4.39）。

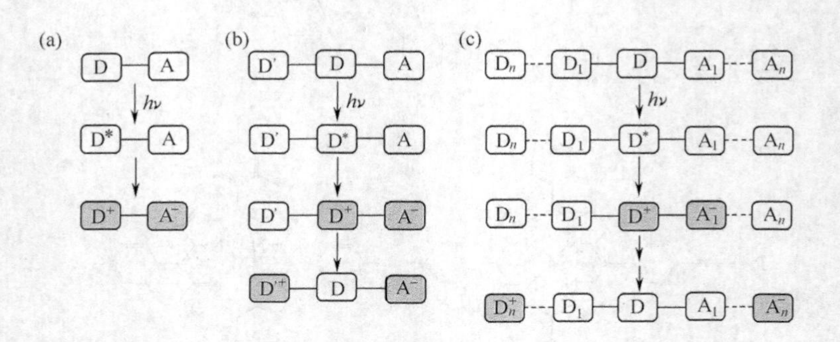

图 4.39　包含第二供体和第二受体的电荷转移途径示意图

二茂铁-卟啉-富勒烯三组分系统能有效地进行光诱导电荷分离，例如，Fe-ZnPH$_2$P-C$_{60}$（图 4.40），其电荷分离状态的寿命长到 380ms。

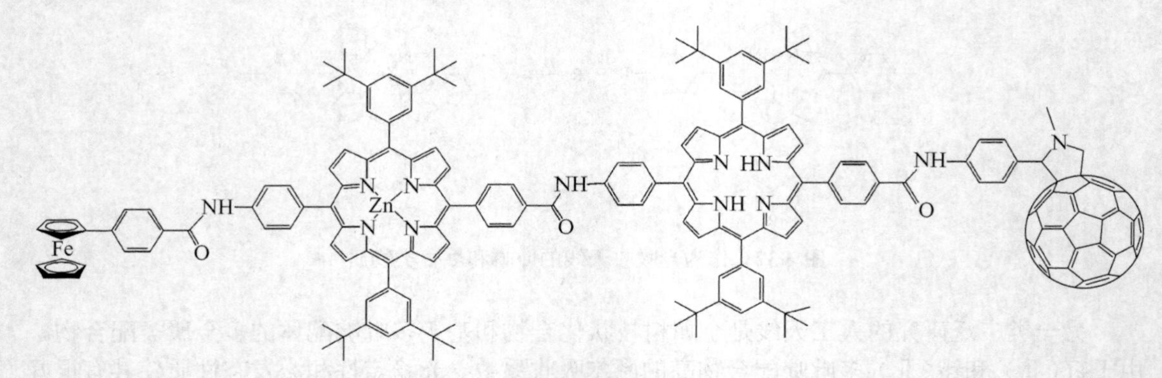

图 4.40　Fe-ZnPH$_2$P-C$_{60}$ 的分子结构示意图

第5章　配位反应的动力学和机理

5.1　简介

（1）配合物的形成　典型的配合物要比其他分子更不稳定或易于发生变化［式(5.1)］。

$$MX + Y \rightleftharpoons MY + X \tag{5.1}$$

X 是离去基团，Y 是进入基团。一个例子是配体的竞争，溶剂分子 L，如水，可作为一个配位单元［式(5.2)］。

$$[Co(H_2O)_6]^{2+} + Cl^- \longrightarrow [Co(H_2O)_5Cl]^+ + H_2O \tag{5.2}$$

（2）形成常数　将配合物的形成看作一系列的平衡反应［式(5.3)］：

$$M + L \rightleftharpoons ML \qquad K_1 = \frac{[ML]}{[M][L]}$$

$$M + L \rightleftharpoons ML_2 \qquad K_2 = \frac{[ML_2]}{[ML][L]} \tag{5.3}$$

总反应[式(5.4)]：

$$M + nL \rightleftharpoons ML_n \qquad \beta_n = \frac{[ML_n]}{[M][L]^n} = K_1K_2K_3\cdots K_n \tag{5.4}$$

通常，K_{n-1} 的值要比 K_n 大得多，只有极少数配合物形成的是 ML_n 而不是 ML_{n-1}。图 5.1 为$[Al(H_2O)_{6-x}F_x]^{(3-x)+}$的分步生成过程中 $\lg K$ 与 x 的关系。

（3）螯合作用　比较 K_1 和 β_2（$M=Cu^{2+}$）［式(5.5)］。

$$[M(H_2O)_2]^{2+} + en \rightleftharpoons [M(en)]^{2+} + 2H_2O$$

$$[M(H_2O)_2]^{2+} + 2NH_3 \rightleftharpoons [M(NH_3)_2]^{2+} + 2H_2O \tag{5.5}$$

$\lg K_1=10.6$，$\lg\beta_2=7.7$。可见，螯合物的稳定性一般要高出 3 个数量级。螯合物比相似的非螯合物稳定性强，这是由于螯合作用用一个成键配体交换了两个限定的溶剂分子，从而引起了熵的改变。

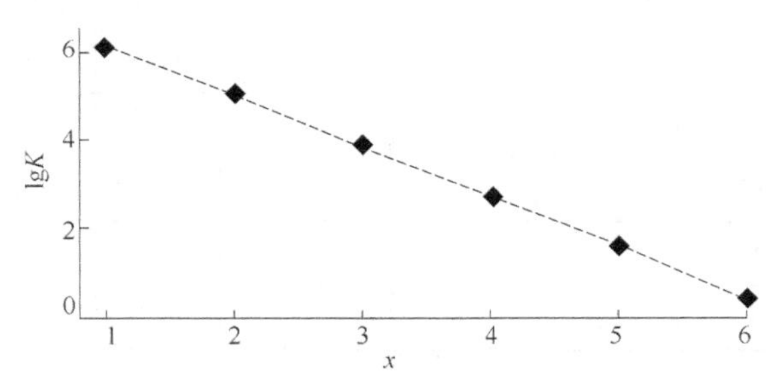

图 5.1　$[Al(OH_2)_{6-x}(F)_x]^{(3-x)+}$ 的分步生成过程中 $\lg K$ 与 x 的关系

（4）环的形成和电子离域　与金属中心形成环使稳定性提高，尤其是五元环或六元环。而且，芳香环的配体还可作为π受体形成具有反馈π键的配合物（图 5.2）。

（5）对给定配体，金属离子对配合物稳定性的影响　Irving-Williams 系列能告诉我们许多信息，这些在第 2 章中已经讨论过。例如，Irving-Williams 系列给出 Cu^{2+}比 Ni^{2+}更稳定。图 5.3 给出了 Ni^{2+}和 Cu^{2+}的水合物中 H_2O 被 NH_3 取代时的一系列 K_f 值。

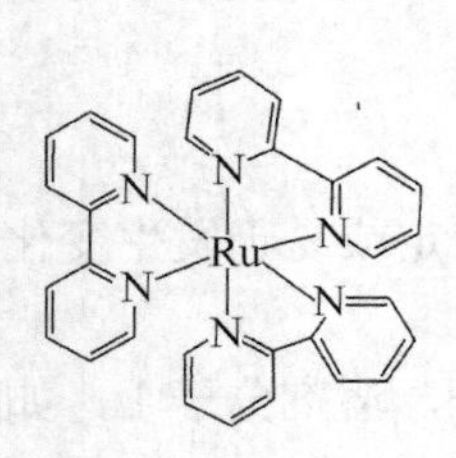

图 5.2　$Ru(bpy)_3^{2+}$ 螯合阳离子的分子结构示意

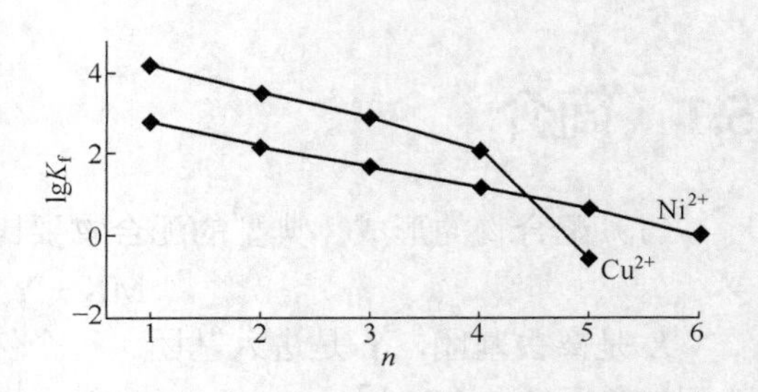

图 5.3　Ni^{2+}和 Cu^{2+}的水合物中 H_2O 被 NH_3 取代时的一系列 K_f 值

5.2　d 区电子金属配合物的反应机理及金属有机反应机理

反应速率是理解配位化学的重要物理量。

惰性配合物：不稳定但能存在几分钟或更长时间的配合物。

不稳定配合物：比惰性配合物反应快的配合物。

一般规则：

- 对正二价离子，d 电子金属不是很稳定，尤其是 d^{10} 的金属（Hg^{2+}，Zn^{2+}）。
- 强场的 d^3 和 d^6 八面体配合物是惰性的，如：Cr（Ⅲ）和 Co（Ⅲ）。
- 配体场稳定化能增加，惰性增加。
- 第二和第三周期的金属通常惰性较强。

配体场稳定化能（LFSE）列于表 5.1，电子构型与稳定性的关系如图 5.4 所示。

表 5.1　配体场稳定化能（LFSE）

d^n	高自旋		低自旋		d^n	高自旋		低自旋	
	构型	LFSE（Δ_o）	构型	LFSE（Δ_o）		构型	LFSE（Δ_o）	构型	LFSE（Δ_o）
d^1	$t_{2g}^1e_g^0$	−0.4			d^6	$t_{2g}^4e_g^2$	−0.4	$t_{2g}^6e_g^0$	−2.4
d^2	$t_{2g}^2e_g^0$	−0.8			d^7	$t_{2g}^5e_g^2$	−0.8	$t_{2g}^6e_g^1$	−1.8
d^3	$t_{2g}^3e_g^0$	−1.2			d^8	$t_{2g}^6e_g^2$	−1.2		
d^4	$t_{2g}^3e_g^1$	−0.6	$t_{2g}^4e_g^0$	−1.6	d^9	$t_{2g}^6e_g^3$	−0.6		
d^5	$t_{2g}^3e_g^2$	0	$t_{2g}^5e_g^0$	−2.0	d^{10}	$t_{2g}^6e_g^4$	0		

5.2.1　缔合反应和离解反应

配体取代反应通常是缔合机理或离解机理。

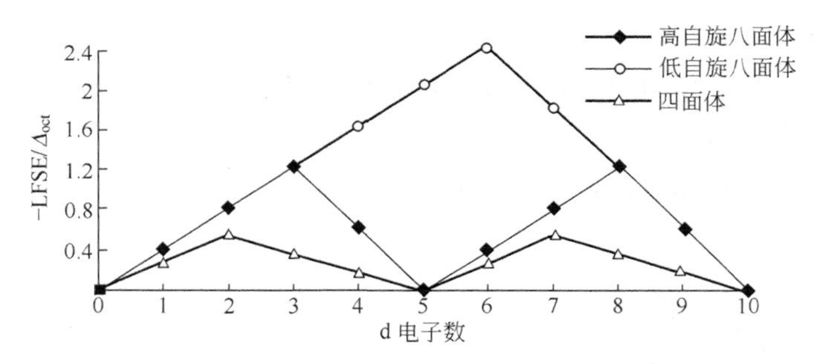

图 5.4　电子构型与稳定性的关系

缔合反应（A 机理，associative mechanism）：反应的中间体比反应物和生成物具有更高的配位数。两个特征是低配位数和反应速率与进入基团有关。

离解反应（D 机理，dissociative mechanism）：反应中间体比反应物和生成物具有更低的配位数。两个特征是八面体型配合物且金属中心小；反应速率与离去基团有关。

5.2.2　反应速率的测量

惰性类： $t_{1/2}$>1min。可用传统的静态技术，如吸光度、pH 值的测量。

不稳定类： $t_{1/2}$ 大约 1ms 到 1min。用停止流量测量法，快速混合、快速光谱测量。

快速反应： 弛豫技术和快速光谱技术。

从反应速率方程来推断反应机理是不可行的，例如，对图 5.5，速率可以为 k[A][B]或 k[A]。

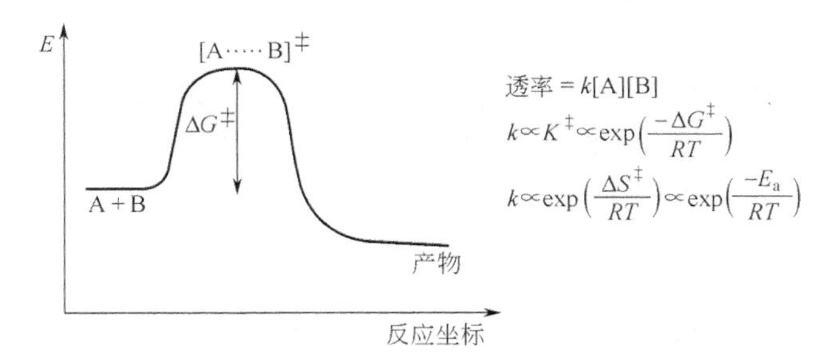

图 5.5　反应 A+B→产物的反应坐标与能量关系图

从图 5.5 中，可以得出以下几点。

- 有溶剂参与（准一级反应）。
- 配合反应仅有一个速率决定步骤。

最后，你只能证明一个机理不正确而不能证明一个机理。速率方程只能很好地与机理保持一致但不能证明它。

区分以下定义：

- “活性”和“惰性”是动力学术语!
- “稳定”和“不稳定”是热力学术语。
- “详细”机理指在分子尺寸上机理的各种细节。

取代反应的速率可以从 1ms 到 10^8s。

由于速率与 $\Delta S^{\ddagger}$和 ΔE_{act} 呈指数关系，用这些非常粗糙的理论来解释机理对速率变化较小的反应意义不大。

5.2.3 典型的反应进程坐标

一级反应的速率与底物的浓度和进入试剂的浓度有关。如果反应速率符合这些特征，说明这种配合物的取代是双分子机理。

这种“详细的”机理的重要几点（图 5.6）如下：

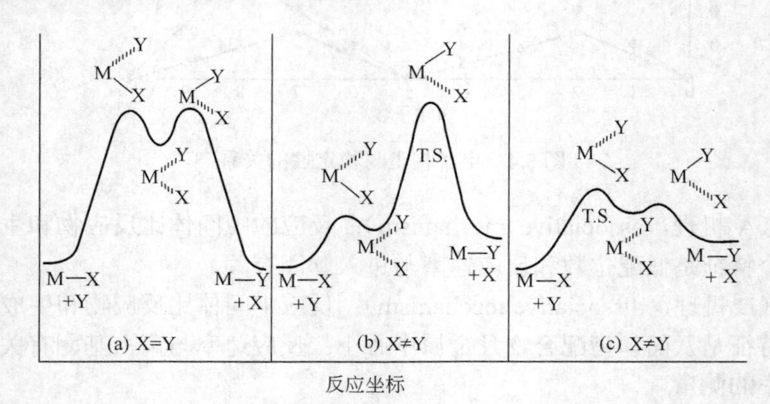

图 5.6 三种紧密反应机理的反应坐标与能量关系曲线

- 当反应物和产物的能量几乎相等时用图（a）。这里过渡态的能量也几乎相等。
- 图（b）适用于键的断裂是速率决定步骤的反应（来自中间体的 d 机理）。
- 图（c）为键的生成是速率决定步骤的情形，这里进入基团 Y 比离去基团有更高的反式效应。

在少数情况下，中间体相当稳定，所以比产物具有更低的能量。例如，这个简图中的反应是在过渡态形成前形成了一个准稳态的中间体。在五配位的中间体中，键的断裂比键结合形成中间体更重要（图 5.7）。

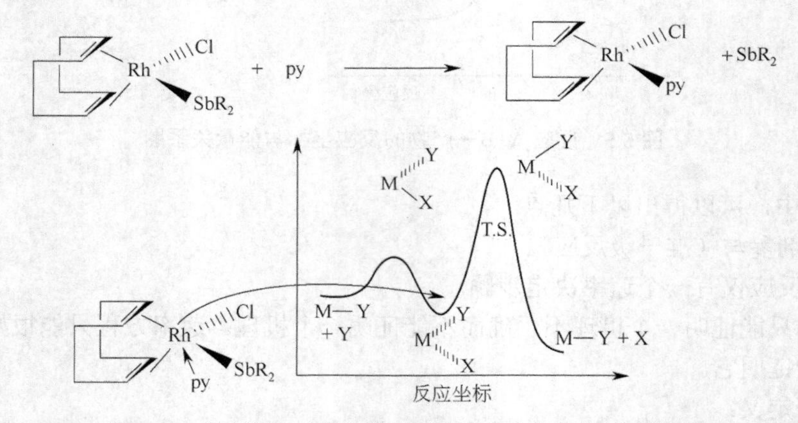

图 5.7 在过渡态形成之前形成准稳态中间体的反应案例

5.2.4 金属有机反应机理

（1）简介

本部分的一些基本概念：Green-Davies-Mingos 规则，配体类型总览和 18 电子规则（在第 4 章中介绍过）。

① Green-Davies-Mingos 规则 在金属有机化学中，对阳离子配合物的亲核加成是最常

见和最重要的反应。烯丙基和乙烯基配体以及不饱和碳氢化合物如乙烯、丁二烯或苯，一般不参与亲核加成反应，它们会受到亲核试剂如 H^-、CN^-或 MeO^-等的进攻，如式(5.6）和式(5.7）所示。

$$\text{M}^+\text{-}(\text{CH})_x \xrightarrow{\text{Nu}^-} \text{Nu-}(\text{CH})_x\text{-M}^+ \qquad (5.6)$$

$$\text{Co} \xrightarrow{\text{Ph}^-} \text{Ph-Co} \qquad (5.7)$$

DGM 规则（1978 年）用三个普遍规则来预测在 18 电子的金属阳离子配合物中动力学控制的亲核进攻的方向。

规则 1　亲核进攻优先发生在偶数配位的多烯中。

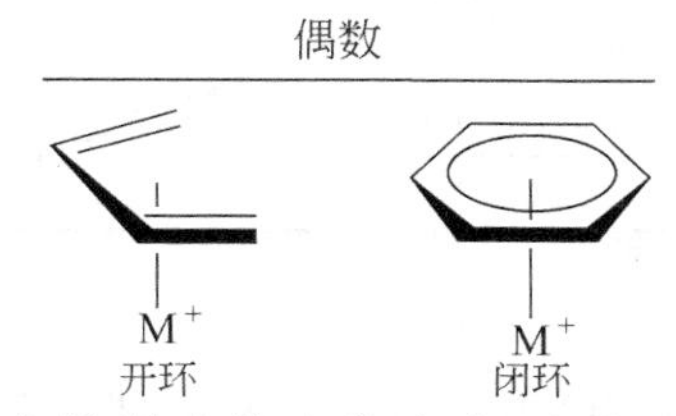

规则 2　对奇数配位的烯，亲核进攻优先发生在开环配体上，而不是闭环配体上。

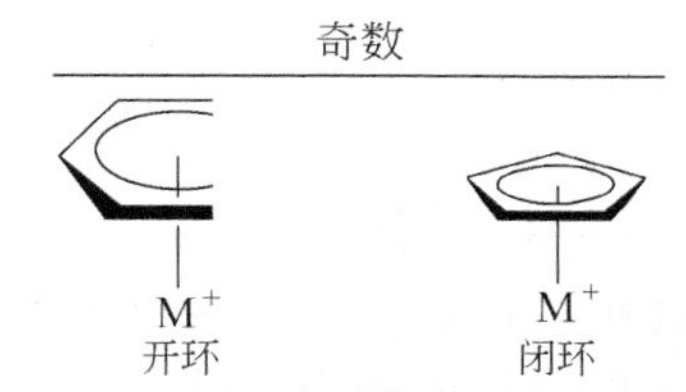

规则 3　对偶数开环的多烯，亲核进攻总发生在端基的碳原子上。对奇数开环的多烯，只有当 L_nM^+是强吸电子基团时，进攻才发生在端基碳原子上。

② 配体类型总览（表 5.2）

表 5.2　金属有机配合物中一些常用的配体

提供的电子数	名称	例　子	提供的电子数	名称	例　子
1	烃基	甲基，乙基，苯基 =	5	二烯基	环戊二烯基 =
2	烯	乙烯 = H₂C=CH₂	6	三烯	苯 = ，环庚三烯 =
3	烯基	烯丙基 = ·CH₂–CH=CH₂ $\xrightarrow{\eta^3\text{-allyl}}$ M	7	三烯基	环庚三烯基 =
4	二烯	丁二烯 = ，环戊二烯 =	8	四烯	环辛四烯 =

离子（电荷）模型 这一方法的基本前提是从金属移走所有的配体，如果必要的话，给每个配体加适当数目的电子使其达到闭壳层态。例如，如果从金属配合物上移走氨，氨有完整的八隅体，是一个中性分子。它可用孤对电子（经典的路易斯酸碱）与金属中心键合，不需要改变金属的氧化态来平衡电荷。我们称氨为中性的两电子供体。

共价（中性）模型 这种方法是从金属上移走所有的配体，但不是使其达到闭壳层态，我们需要做的是使其达到中性。再以氨为例，当我们将其从金属上移走时，它是一个具有孤对电子的中性分子。因此，像离子模型一样，氨是中性的两电子供体。

最关键的一点是氧化态的归属，电子数是形式上的，并不实际反映出电子在分子中的分布，然而，这些形式的东西对我们来讲很有用，两种方法给我们的最终结果都相同（表 5.3）。

表 5.3 两种方法的比较

结构	离子		共价		结构	离子		共价	
Co	Co^{II}	$7e^-$	Co	$9e^-$	Cl, PPh_3, Rh, Ph_3P, PPh_3	Rh^{I}	$8e^-$	Rh	$9e^-$
	$2Cp^-$	$12e^-$	$2Cp\cdot$	$10e^-$		Cl^-	$2e^-$	$Cl\cdot$	$1e^-$
	总	$19e^-$	总	$19e^-$		$3PPh_3$	$6e^-$	$3PPh_3$	$6e^-$
Ti, Cl, Cl	Ti^{IV}	$0e^-$	Ti	$4e^-$		总	$16e^-$	总	$16e^-$
	$2Cl^-$	$4e^-$	$2Cl\cdot$	$2e^-$	CO, OC, CO, W, OC, CO, CO	W^0	$6e^-$	W	$6e^-$
	$2Cp^-$	$12e^-$	$2Cp\cdot$	$10e^-$		6CO	$12e^-$	6CO	$12e^-$
	总	$16e^-$	总	$16e^-$		总	$18e^-$	总	$18e^-$

（2）基本反应类型

① 氧化加成反应　定义［式(5.8)］：（a）外观氧化态增加 2。（b）配位数增加 2。

$$L_nM + X—Y \longrightarrow L_nM(Y)(X) \tag{5.8}$$

$$M^{n+} \longrightarrow M^{(n+2)+} \qquad d^n \longrightarrow d^{n-2}$$

对金属配合物的要求：（a）金属中心必须是 d^2 或更多个 d 电子。（b）金属中心必须配位不饱和。（c）金属必须有合适能量的空轨道和价轨道［式(5.9)］。

空轨道

填充轨道

$$L_nM + X—Y \longrightarrow L_nM(Y)(X) \tag{5.9}$$

② 分子内氧化加成［式(5.10)］

$$(Ph_3P)_3IrCl \xrightarrow{\triangle} (Ph_3P)_2Ir(H)(Cl)(C_6H_4PPh_2) \tag{5.10}$$

这是分子内邻位金属化反应的例子。

③ 双分子反应［式(5.11)］

$$L_nM—ML'_n + X—Y \longrightarrow L_nM—X + L'_nM—Y$$

$$(OC)_5Mn—Mn(CO)_5 \xrightarrow{Br_2} 2\,Mn(CO)_5Br \tag{5.11}$$

④ C—H 键的活化［式(5.12)］

$$(\eta^5\text{-}C_5Me_5)Rh(PMe_3)H_2 \xrightarrow[-H_2]{h\nu} [(\eta^5\text{-}C_5Me_5)Rh(PMe_3)]^{\ne} \tag{5.12}$$

$+C_6H_6 \rightarrow (\eta^5\text{-}C_5Me_5)Rh(PMe_3)(H)(C_6H_5)$；$+CH_4 \rightarrow (\eta^5\text{-}C_5Me_5)Rh(PMe_3)(H)(CH_3)$

注意：C_6H_6 比 CH_4 的反应速率快，但是 Ph—H 键比 CH_3—H 键强。

5.2.5　氧化加成反应的动力学速率公式

（1）d^8 平面正方形配合物　XY 对 $IrCl(CO)(PPh_3)_2$ 的氧化加成（Vaska’s 配合物）：速率=k_2[Ir(I)][XY]。

这是一个简单的二级反应且体系具有配位不饱和的特征［式(5.13)］:

$$IrCl(CO)(PPh_3)_2 \xrightarrow{X—Y} Ir(X)(Y)Cl(CO)(PPh_3)_2 \tag{5.13}$$

（2）d^8 三角双锥配合物　XY 对 $IrH(CO)(PPh_3)_3$ 的氧化加成［式(5.14)］:

$$IrH(CO)(PPh_3)_3 \xrightarrow{X—Y} IrH(X)(Y)(CO)(PPh_3)_2 \tag{5.14}$$

$$反应速率=\frac{k_1k_2[\mathrm{Ir(I)}][\mathrm{XY}]}{k_{-1}[\mathrm{PPh_3}]+k_2[\mathrm{XY}]} \tag{5.15}$$

它具有配位饱和化合物的特征，且符合下面的过程［式(5.16)］:

$$IrH(CO)(PPh_3)_3 \underset{k_{-1}}{\overset{k_1}{\rightleftharpoons}} IrH(CO)(PPh_3)_2 + PPh_3$$

$$IrH(CO)(PPh_3)_2 + XY \overset{k_2}{\rightleftharpoons} IrH(X)(Y)(CO)(PPh_3)_2 \tag{5.16}$$

注意：过量的膦会阻碍反应的进行。

5.2.6　氧化加成反应的机理

对所有的氧化加成反应，没有通用的机理。根据反应物和反应条件，有三个主要的机理。分别为:

- 协同完成;
- （i）亲核进攻（S_N2），（ii）离子机理;

- 自由基反应。

（1）协同机理　它有三个特征：

- 有一个三中心的过渡态，L_nM 从 X—Y 的侧面进攻；
- 分子 X—Y 进攻反应中心的同时或随后电子发生从 X—Y 键到 M—X 键和 M—Y 键的重排［式(5.17)］；

$$L_nM + X—Y \longrightarrow \left[L_nM \begin{matrix} Y \\ \vdots \\ X \end{matrix} \right]^{\ddagger} \longrightarrow L_nM \begin{matrix} Y \\ X \end{matrix} \qquad (5.17)$$

- 在过渡态中有协同的电子流动 X—Y(σ) ⟶ M(σ)，M(π) ⟶ X—Y($σ^*$)。

对式(5.13)，协同反应机理的特征：（a）分子配位不饱和；（b）通常 X—Y 基团键的极性比较低，如 H_2、$R_3Si—H$；（c）立体专一性和顺式加成［式(5.18)］。

$$trans\text{-}Ir(Cl)(CO)(PPh_3)_2 \xrightarrow{H_2} Ir(H)_2(Cl)(CO)(PPh_3)_2 \qquad (5.18)$$

顺式加成可由红外光谱看出，会得到两个 M—H 伸缩峰（对称的和不对称的）。

注意：如果是反式加成，则只有一个不对称的。如果 X—Y 是手性的，反应中构型保持，例如：H_2 的侧面加成活化能低，$ΔG^{\ddagger}$大约为 40kJ/mol［式(5.19)，图 5.8］。

$$(\eta^5\text{-}C_5H_4Me)Mn(CO)_3 + Ph(Me)(Np)Si—H \xrightarrow{h\nu,\ -CO} (\eta^5\text{-}C_5H_4Me)Mn(CO)_2(\eta^2\text{-}H—SiMePhNp) \qquad (5.19)$$

H—H 键的强度（420kJ/mol）意味着具有较小 H—H 键伸缩的“早期过渡态”，因此动力学同位素效应很小［式(5.20)］：

$$Ir(X)(CO)(PPh_3)_2 \xrightarrow{H_2} Ir(H)_2(X)(CO)(PPh_3)_2 \qquad (5.20)$$

$$\frac{K_H}{K_D}=1.09$$

配合物加氢的速率	X：I > Br > Cl
速率比	100　4　0.9

在决定 $ΔG$ 的时候，金属中心的电子云密度很重要。对 H_2（$σ^*$）的贡献是最重要的因素。如果用 CS（好的 π-受体）取代 CO，与 H_2 的反应不再发生。

（2）亲核进攻　当 X—Y 键有极性时，它是有利的，例如：RX，HX，$HgCl_2$，预计是通过偶极过渡态（图 5.9）。

金属中心作为电子对的供体（亲核的），对包含 R—X 键的反应，碳上的立体化学发生翻转，类似于 S_N2 反应［式(5.21)］。

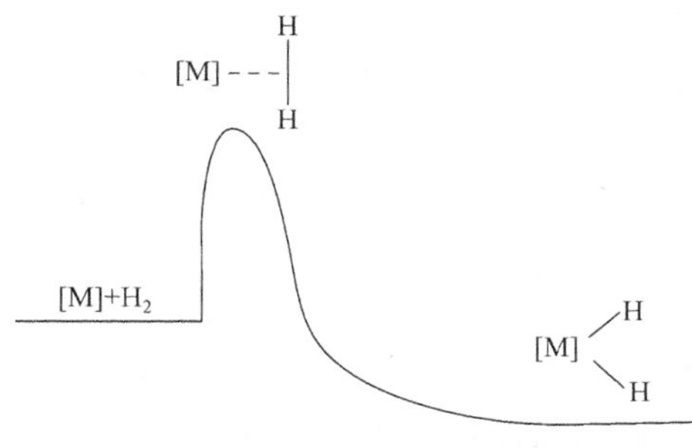

图 5.8　侧向加氢机理的反应坐标和能量图

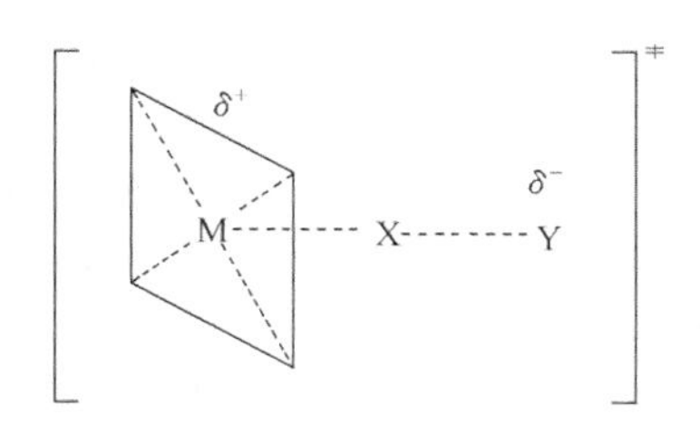

图 5.9　亲核攻击机理中偶极过渡态的示意图

$$L_3Pd: + C(Ph)(H)-X \longrightarrow L_3Pd-C(Ph)(H) + X^- \qquad (5.21)$$

亲核进攻的特征：（a）反式加成；（b）在极性溶剂中速率加快；（c）大的$-\Delta S^{\ddagger}$，因为溶剂围绕在偶极的过渡态周围；（d）反应速率易受其他配体（L）的影响。例如式(5.22)：

$$Ir(Ph_3P)(CO)(Cl)(PPh_3) \xrightarrow[slow]{MeX} [Ir(Me)(Ph_3P)(CO)(Cl)(PPh_3)]^+X^- \xrightarrow{fast} Ir(Me)(Ph_3P)(CO)(Cl)(PPh_3)(X) \qquad (5.22)$$

式(5.22) 具有的一些反应特征：（a）提供给磷化氢的电子越多，加成速率越快；（b）Hammett 图表明增加配体的碱性，会增加 Ir 中心的进攻性；（c）立体影响，配体的尺寸增加，氧化加成的速率降低，这是由于过渡态更加拥挤了。

（3）离子机理　在极性溶剂（如甲醇）中，有时加成是没有立体专一性的。对亲核机理，带正电荷的五配位中间体的寿命增加，所以在式(5.23) 中：

极性溶剂（DMF，MeOH，H_2O，MeCN）⟶ 顺式+反式

非极性溶剂（C_6H_6，$CHCl_3$）⟶ 只有顺式

气相 ⟶ 只有顺式

（4）自由基反应　烷基（不是甲基）、乙烯基、卤化芳基可与 Vaska's 配合物发生自由基链反应。速率取决于：（a）引发剂；（b）抑制剂；（c）自由基的自旋捕获。典型的引发剂包括 O_2 和 Ir（Ⅱ）（图 5.10）。

$$[Ir(Me)(Ph_3P)(CO)(Cl)(PPh_3)]^+ X^- \longrightarrow Ir(Me)(PPh_3)_2(Cl)(CO) + X^- \xrightarrow{A,\ B,\ C} \text{products} \qquad (5.23)$$

B：Ir(Me)(Ph₃P)(X)(Cl)(PPh₃)(CO)；A：Ir(Me)(Ph₃P)(X)(OC)(PPh₃)(Cl)；C：Ir(Me)(Ph₃P)(CO)(Cl)(PPh₃)(X)

$$[Ir^{II}] + O\cdot(\text{痕量自由基}) \longrightarrow [Ir^{II}]-O\cdot$$
$$[Ir^{II}]-O\cdot + R-X \longrightarrow X-[Ir^{III}]-O + R\cdot \quad \text{引发步骤}$$

$$[Ir^{II}] + R\cdot \longrightarrow R-[Ir^{II}]\cdot$$
$$\longrightarrow R-[Ir^{III}]-X + R\cdot \quad \text{增长步骤}$$

$$[Ir^{II}] + R-X \longrightarrow R-[Ir^{III}]-X \quad \text{净反应}$$

图 5.10 包括 O_2 和 Ir（Ⅱ）引发剂的典型自由基反应

链反应的特征：

- R—X 中的 R 完全失去其立体化学特征；
- 17 电子的 ML_5 配合物，如 Co（Ⅱ）- d^7 会发生逐步的自由基加成。

分步反应的特征：

- 动力学 2 级反应；
- 反应物从 X—Y 得到自由基片断；
- 在碳自由基上发生外消旋。

5.2.7 迁移反应（“迁移插入”）

（1）定义　将一个不饱和的配体（Y）插入一个相邻的 M—X 键［式(5.24)］。

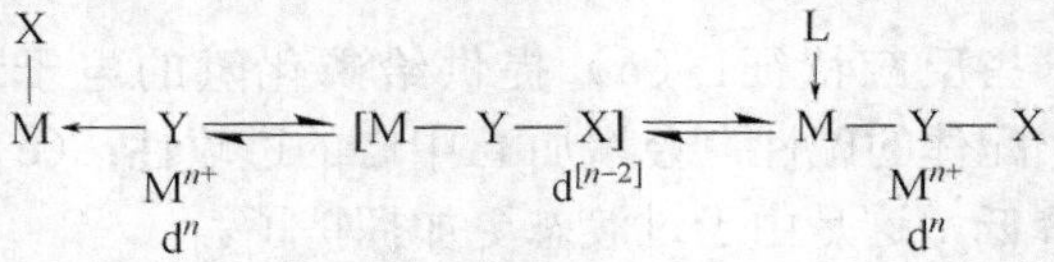

$$\underset{\substack{M^{n+}\\d^n}}{M(X)\leftarrow Y} \rightleftharpoons \underset{d^{[n-2]}}{[M-Y-X]} \rightleftharpoons \underset{\substack{M^{n+}\\d^n}}{(L\rightarrow)M-Y-X} \tag{5.24}$$

Y=CO，C_2H_4，C_2H_2，苯，环戊二烯；X=H，R，Ar，C(O)R；L=路易斯碱，溶剂，PR_3，NR_3，R_2O。

（2）两种主要的类型　1,1-插入［式(5.25)］和 1,2-插入［式(5.26)］。

例如，在式(5.25）中，X=Me，AB=CO；在式(5.47）中，X=H，AB=$H_2C{=}CH_2$。

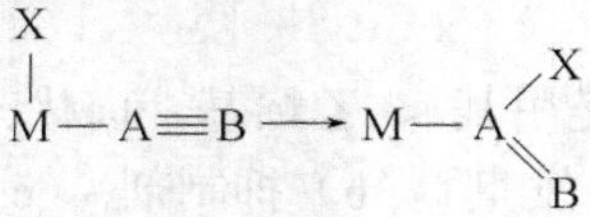

$$M(X)-A\equiv B \longrightarrow M-A(X)=B \tag{5.25}$$

$$M(X)-B(=A) \longrightarrow M-A-B-X \tag{5.26}$$

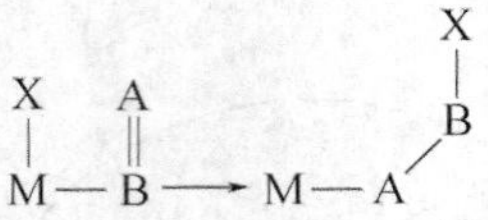

（3）插入反应的一般特征

- 除了插入的配体（Y）是 M=C 或 MC 之外，金属中心的表观氧化态不变。
- 基团必须是相邻的（顺式）。
- 在前期的反应中，必须生成空缺的配位点。
- 当 X 配体有手性时，反应一般保留原构型。
- X-迁移和 Y-插入的例子均已发现。
- 位置的平衡取决于 M—X、M—Y 和 M—（XY）键的强度。
- 一个电子氧化通常加速反应。

（4）迁移反应的例子

氢化物迁移［式(5.27）和式(5.28)］：

Rh　Me_3P　H　⇌　[Rh　Me_3P　CH_2　H　C　H_2]‡　$+PMe_3$ / $-PMe_3$　Rh　Et　Me_3P　PMe_3　(5.27)

[Mo　D　P　P]$^{+}$　$+PMe_3$　→　[H　D　Mo　PMe_3　P　P]$^{+}$　(5.28)

烷基迁移［式(5.29) 和式(5.30)］:

Ru　CH_3　Ph_3P　CH_3　Ph_3C^+ / $-H^-$　→　[Ru$=CH_2$　Ph_3P　CH_3]　→　[Ru　Et　Ph_3P]　↓　[Ru　Ph_3P　H]$^{+}$　(5.29)

Et　Mo　Cl　PR_3　→　Et　PR_3　Mo　Cl　$TlPF_6$　→　[Et　Mo　PR_3]$^{+}$　PF_6　(5.30)

其他例子［式(5.31)］:

Co　OC　Me　ν_{CO}=1780cm^{-1}　k_1 / k_{-1}　[Co　C=O　Me]‡　k_2 / L　→　Co　Ph_3P　C=O　Me　ν_{CO}=1310cm^{-1}　L=PPh_3　(5.31)

k_{-3}　k_3,L

Co　O=C　Me　L　ν_{CO}=1530cm^{-1}　L=PEt_3

（5）氧化加成和烷基迁移反应的结合

烷基 ⟶ 酰基［式(5.32)］

M=Co,Rh

$\xrightarrow[\text{氧化加成}]{\text{MeI}}$ $[\ldots]^{+}I^{-}$ $\xrightarrow{\text{甲基迁移}}$ （5.32）

酰基 ⟶ 烷基［式(5.33)］

$\xrightarrow[\text{氧化加成}]{\text{RC(O)Cl}}$ $\xrightarrow[\text{烷基迁移}]{-PPhMe_2}$ （5.33）

羰基化反应［式(5.34)］

$$(OC)_5MnCH_3 + CO \longrightarrow (OC)_5MnC(O)CH_3 \qquad (5.34)$$

三种可能的机理：（a）直接插入；（b）“敲门”（knock-on）式插入；（c）烷基迁移。式（5.34）的机理可通过用 ^{13}CO 检查产物的立体化学来解释。

红外光谱：乙酰基衍生物具有独立的键，$\nu(CO)=1664cm^{-1}$。^{13}CO 进入/退出的位置可由键的强度和带的位置的改变来推断（式 5.35）。

$$Me—Mn(CO)_5 + {}^{13}CO \longrightarrow \qquad (5.35)$$

红外数据：在 $970cm^{-1}$、$1963cm^{-1}$、$1625cm^{-1}$ 处有强峰。因此，乙酰基与 ^{13}CO 是顺式的；^{13}CO 没有直接被插入；在产物中没有无序现象。

对式(5.36)，红外谱图中 $1976cm^{-1}$（顺式）和 $1949cm^{-1}$（反式）处的峰强度证明了羰基化反应；只有迁移能给出正确的产物分布；通过 CO 的去插入没有得到反式的异构体。

$\xrightarrow{\triangle}$ （5.36）

羰基化反应的速率公式　在通常条件下会得到下面的速率公式［式(5.37)］：

$\underset{k_{-1}}{\overset{k_1}{\rightleftharpoons}}$ $\underset{k_{-2}(-L)}{\overset{k_2(+L)}{\rightleftharpoons}}$ （5.37）

k_1是速率决定步骤，即在烷基-锰中是一级的；在L（进入配体）中是零级的。如果$k_2 \gg k_{-2}$，那么得到式(5.38)。

$$速率=\frac{k_1k_2[Mn(CO)_5R][L]}{k_{-1}+k_2[L]} \tag{5.38}$$

对高浓度的L，在起始原料中是一级反应；低浓度的L是混合级反应。Me—CO—$Mn(CO)_5$中CO的取代和乙酰基化合物中的脱羰基反应，速率常数是相同的［式(5.39)］。

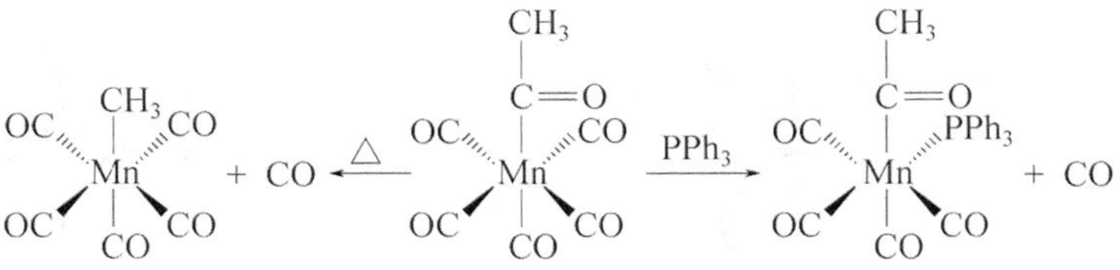

$$CH_3Mn(CO)_5 + CO \xleftarrow{\triangle} CH_3C(=O)Mn(CO)_5 \xrightarrow{PPh_3} CH_3C(=O)Mn(CO)_4(PPh_3) + CO \tag{5.39}$$

路易斯酸催化剂可以加速反应速率，例如：H^+，$AlCl_3$，BF_3。路易斯酸可通过下式使过渡态稳定［式(5.40)］。

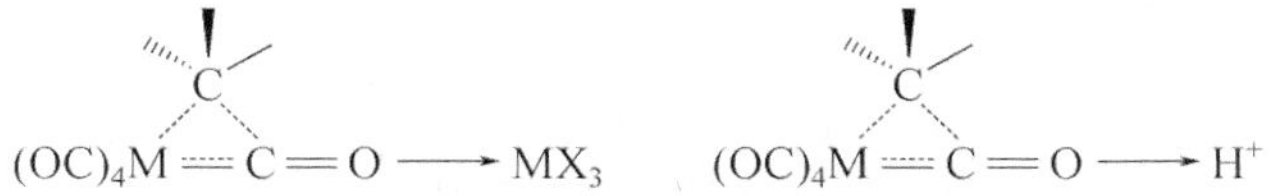

$$(OC)_4M \cdots C=O \longrightarrow MX_3 \qquad (OC)_4M \cdots C=O \longrightarrow H^+ \tag{5.40}$$

几个立体特征　金属中心，碳中心［式(5.41)］；相邻的点［式(5.42)］。

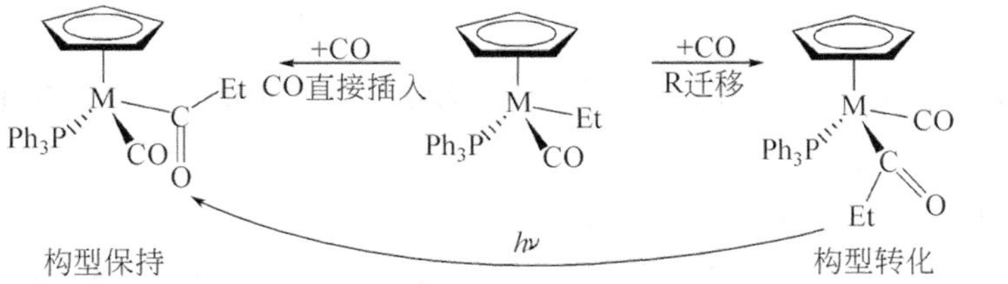

(5.41)

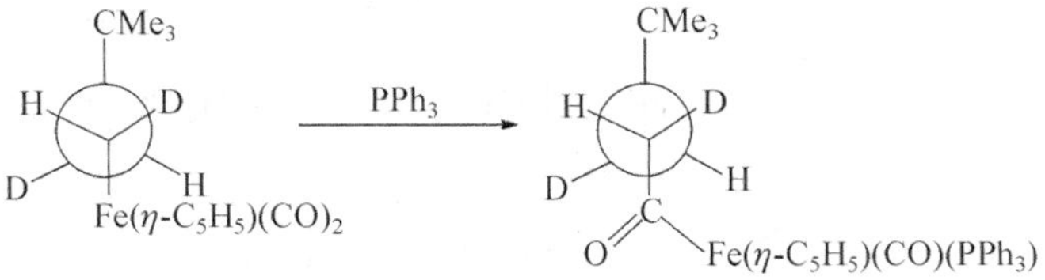

+CO　CO直接插入　+CO　R迁移　hν　构型保持　构型转化

(5.42)

CMe_3　H　D　D　H　$Fe(\eta\text{-}C_5H_5)(CO)_2$　$\xrightarrow{PPh_3}$　CMe_3　H　D　D　H　O　C　$Fe(\eta\text{-}C_5H_5)(CO)(PPh_3)$

5.2.8　消除反应

（1）氢的消除反应

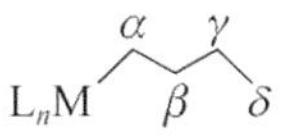

L_nM—α—β—γ—δ

① α-氢消除反应［式(5.43)］

$$L_nM{-}CH_2{-}R \rightleftharpoons L_n\overset{H}{M}{=}CH{-}R \tag{5.43}$$

$$d^n \longrightarrow d^{n-2}\text{；}M^{m+} \longrightarrow M^{(m+2)+}$$

α-消除反应不如β-消除反应具有普遍性，可能是由于金属氢化物有较强的反应活性［式(5.44)］。

$$Cp_2Mo(CH_3)Cl \xrightarrow{PR_3} [Cp_2Mo(CH_3)(PR_3)]^+ \xrightarrow{-PR_3} [Cp_2MoCH_3]^+ \rightleftharpoons [Cp_2Mo(=CH_2)H]^+ \underset{}{\overset{PR_3}{\rightleftharpoons}} Cp_2Mo(H)CH_2-\overset{+}{P}R_3 \tag{5.44}$$

动力学产物

热力学产物

② β-氢消除反应［式(5.45)］

$$L_nM-CH_2-CH_2-R \rightleftharpoons L_nM(H)(CH_2=CHR) \tag{5.45}$$

$$d^n \longrightarrow d^n;\ M^{m+} \longrightarrow M^{m+}$$

通常认为，β-氢的存在与否是决定与过渡金属中心配位的烷基配体稳定性的一个最重要的因素［式(5.46)］。

$$(PEt_3)_2Pt(CH_2CH_3)_2 \rightleftharpoons (PEt_3)Pt(CH_2CH_3)_2 \rightleftharpoons (PEt_3)Pt(CH_2CH_3)(H)(H_2C=CH_2) \tag{5.46}$$

不含 β-氢的配体具有较强的稳定性，例如：—Me、—CH_2Ph、—CH_2Bu-t［式(5.47)］。

$$TaCl_5 \xrightarrow{Zn(CH_3)_2} TaMe_5 \tag{5.47}$$

③ γ-氢消除反应［式(5.48)］

$$L_nM-CH_2-C(CH_3)_3 \rightleftharpoons L_nM(H)(CH_2C(CH_3)_2CH_2) \tag{5.48}$$

$$d^n \longrightarrow d^{n-2};\ M^{m+} \longrightarrow M^{(m+2)+}$$

与 Ta 的 α-消除相比，铂的新戊基的烷基配合物倾向于通过 γ-消除而分解，这可能暗示了在两种情况下有不同的机理。在 Ta 的例子中，α-碳上的一个烷基通过另一个烷基被脱去质子，而 Pt(Ⅱ) 更像是氧化加成的机理［式(5.49)］。

$$\text{L}_2\text{Pt}(\text{CH}_2\text{CMe}_3)_2 \xrightarrow{-\text{CMe}_4} \text{L}_2\text{Pt}(\text{CH}_2\text{CMe}_2\text{CH}_2) \tag{5.49}$$

④ 其他的氢消除反应［式(5.50)］

δ-消除 68%

γ-消除 23%

ε-消除 9%

(5.50)

(2) 还原脱氢反应

$$\text{L}_n\text{M}(\text{X})(\text{Y}) \underset{\text{氧化加成}}{\overset{\text{还原消除}}{\rightleftharpoons}} \text{L}_n\text{M} + \text{X—Y} \tag{5.51}$$

$$\text{d}^n \longrightarrow \text{d}^{n+2};\ \text{M}^{m+} \longrightarrow \text{M}^{(m-2)+}$$

几个特点：

- 与氧化加成相反；
- 在 X 和 Y 基团间形成新键；
- 已有许多应用各种 X 和 Y 配体的例子；
- 协同反应。

例如式(5.52)：

$$\text{L}_n\text{M}(\text{R})(\text{H}) \longrightarrow \text{L}_n\text{M} + \text{R—H} \qquad \text{L}_n\text{M}(\text{C(O)R})(\text{H}) \longrightarrow \text{L}_n\text{M} + \text{RCHO}$$

$$\text{L}_n\text{M}(\text{R})(\text{R}) \longrightarrow \text{L}_n\text{M} + \text{R—R} \qquad \text{L}_n\text{M}(\text{X})(\text{R}) \longrightarrow \text{L}_n\text{M} + \text{R—X} \tag{5.52}$$

为了加强和/或允许分子内的消去反应：

- 需要 X 和 Y 配体在金属中心周围顺式分布［式(5.53)］；
- 金属中心具有高的表观电荷［式(5.54)］。

$$(\text{Ph}_2\text{PCH}_2\text{CH}_2\text{PPh}_2)\text{Pd}(\text{CH}_3)_2 \xrightarrow[\text{DMSO}]{80^\circ\text{C}} \text{CH}_3\text{—CH}_3$$

trans-P—Pd(CH$_3$)$_2$—P $\xrightarrow[\text{DMSO}]{80^\circ\text{C}}$ 不反应

(5.53)

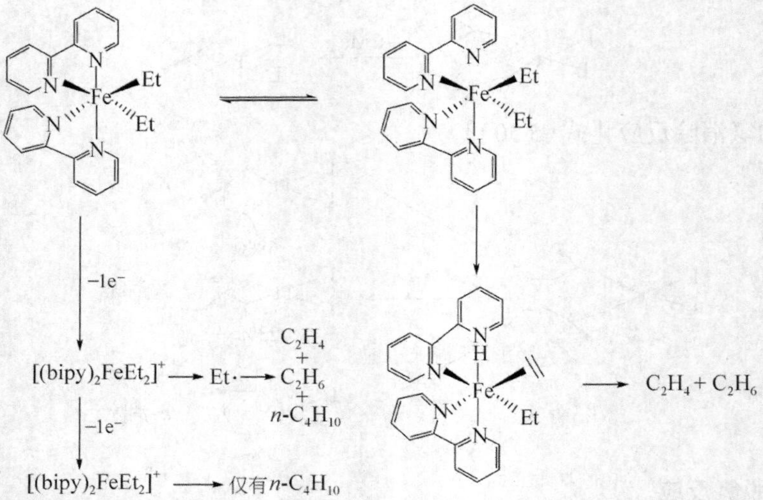

$$[(bipy)_2FeEt_2]^+ \longrightarrow Et\cdot \longrightarrow C_2H_4 + C_2H_6 + n\text{-}C_4H_{10}$$

$$[(bipy)_2FeEt_2]^+ \longrightarrow \text{仅有}\ n\text{-}C_4H_{10}$$

$$\longrightarrow C_2H_4 + C_2H_6 \tag{5.54}$$

在式（5.54）中，中性配合物的热分解给出了乙烯和乙烷，这源自初始的β-氢消除反应。单个阳离子的电化学氧化给出了乙烯、乙烷、丁烷，这是由于 Fe—Et 键均裂产生了乙基自由基。进一步氧化生成的双阳离子则通过还原消除快速生成丁烷。

- 如果金属中心是富电子的，会抑制还原消除反应［式(5.55)］。

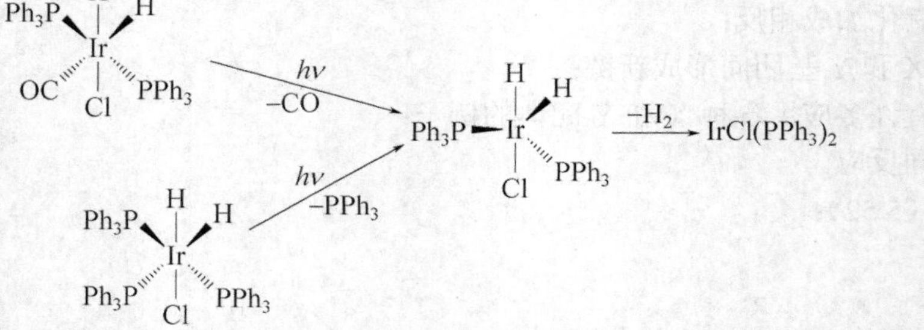

$$\xrightarrow[-CO]{h\nu} \qquad \xrightarrow[-PPh_3]{h\nu} \qquad \xrightarrow{-H_2} IrCl(PPh_3)_2 \tag{5.55}$$

加入 PPh_3 会抑制反应，这意味着有一个包含失去 PPh_3 的预平衡。三个 PPh_3 配体使金属中心对消去来说是富电子的。

标记实验表明它为分子内反应［式(5.56)］并且无交换发生［式(5.57)］。

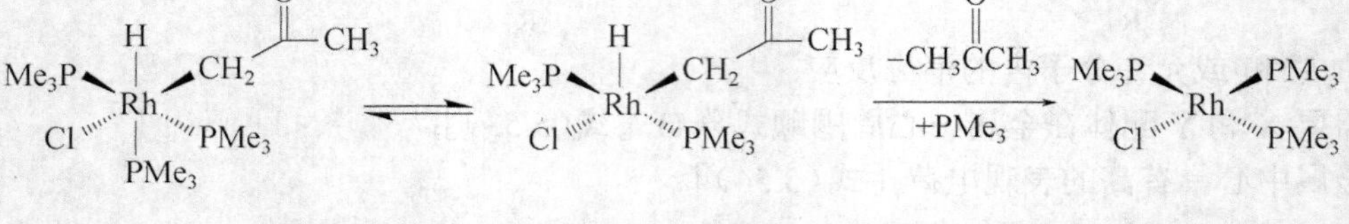

$$\xrightarrow[+PMe_3]{-CH_3CCH_3} \tag{5.56}$$

$$(PPh_3)_2IrClH_2 + (PPh_3)_3IrClD_2 \longrightarrow 2(PPh_3)_2IrCl + H_2 + D_2 \tag{5.57}$$

5.3 配合物的取代反应

亲核取代反应的配体中，有三种类型的配体很重要。

- 进入基团：Y；
- 离去基团：X；
- 旁观配体：既不进入也不离去的配体，当处于反式时尤其重要，以“T”表示。

5.3.1　反应机理的三种模式

缔合机理 A［2 步，式(5.58)］

$$ML_nX + Y \longrightarrow ML_nXY \longrightarrow ML_nY + X \tag{5.58}$$

离解机理 D［2 步，式(5.59)］

$$ML_nX + Y \longrightarrow ML_n + X + Y \longrightarrow ML_nY + X \tag{5.59}$$

交换 I［一个连续的过程，式(5.60)］

$$ML_nX + Y \longrightarrow Y\cdots ML_n\cdots X \longrightarrow ML_nY + X \tag{5.60}$$

5.3.2　平面正方形金属配合物的取代

可能形成平面正方形金属配合物的金属离子列在图 5.11 中。以"■"表示的元素通常形成平面正方形金属配合物。以"□"表示的元素仅和特殊配体形成四配位的平面配合物。

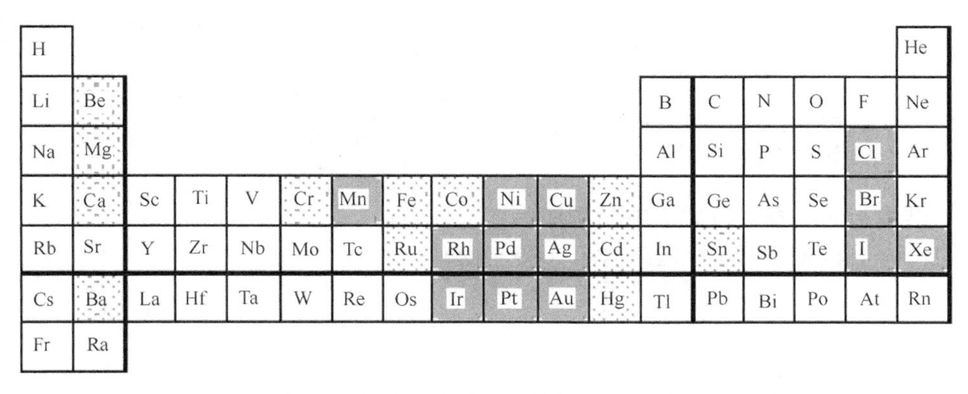

图 5.11　周期表中形成平面正方形金属配合物的可能元素

平面正方形的过渡金属配合物最常见的例子是 Ni(Ⅱ)（主要是 d^8）、Rh(Ⅰ)、Pd(Ⅱ)、Ir(Ⅰ)、Pt(Ⅱ)、Au(Ⅲ)的配合物。

平面正方形配合物的取代几乎都是 A 机理。在进入基团到达时离去基团离去，所以中间产物具有三角双锥的构型。反应速率依赖于进入基团且速率决定步骤是 M—X 键的形成。

速率方程［式(5.61)］：

$$ML_3X + Y \xrightarrow{\text{溶剂}} ML_3Y + X$$

$$\frac{-d[ML_3X]}{dt} = (k_s + K_Y[Y])([ML_3X]) \tag{5.61}$$

影响取代反应速率的因素如下。

- 反式效应：与离去基团（X）处于反位的配体 T 能改变反应速率。
- 进入基团和离去基团的作用。
- 空间效应：大体积的顺式配体降低了 Y 的亲核进攻。
- 立体化学：除非活化态配合物具有长的寿命，顺式/反式在 A 机理中才能保存下来。
- 对 A 机理，$\Delta V^{\ddagger}$和 $\Delta S^{\ddagger}$都是负的。

（1）反式效应　反式效应是一种配体对其反位上配体的取代速率的影响。

平面正方形配合物中与离去基团处于反位的配体 T 影响取代反应的速率。如果 T 是一个强的 σ 供体或 π 受体，取代反应的速率会明显增加。原因是：(a) 如果 T 贡献了一部分电子云密度（T 是一个好的 σ 供体），金属从 X（离去基团）接受电子的能力会降低。(b) 如果

T 是个好的 π 受体，金属上的电子云密度就会减少，就容易受到 Y 的亲核攻击。

对 σ 供体，反式效应的影响按下列顺序增加：

H_2O~OH^-~NH_3~胺~Cl^-<Br^-<SCN^-~I^-<CH_3^-<膦~H^-<烯<CO~CN^-

对 π 受体，反式效应的影响按下列顺序增加：

Br^-<Cl^-<NCS^-<NO_2^-<CN^-<CO

注意“不稳定性效应”是用来强调动力学现象的。由于基态的去稳定性（一个热力学术语）和/或过渡态的稳定化，不稳定性可能出现（图 5.12）。

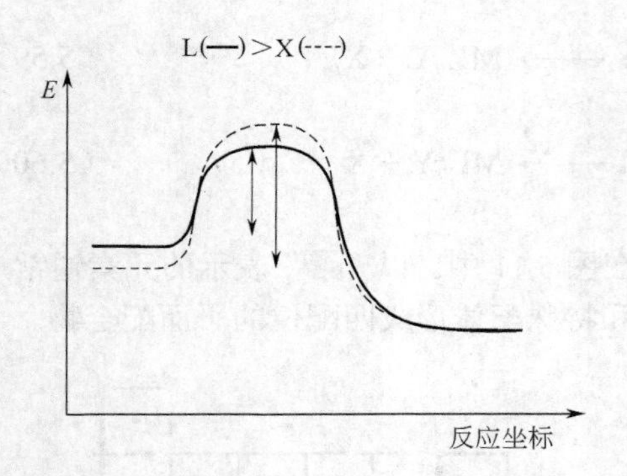

图 5.12 由于基态的去稳定性和/或过渡态的稳定化的不稳定性效应示意图

反式效应纯粹是一种热力学现象，就是说，配体能影响与其处于反位的基团的基态性质。这些性质包括：

- 金属-配体键的键长；
- 振动频率或力常数；
- NMR 偶合常数。

基于结构上的数据，反式影响的顺序为：

R^-~H^->═PR_3>CO~C═C~Cl^-~NH_3（图 5.13）

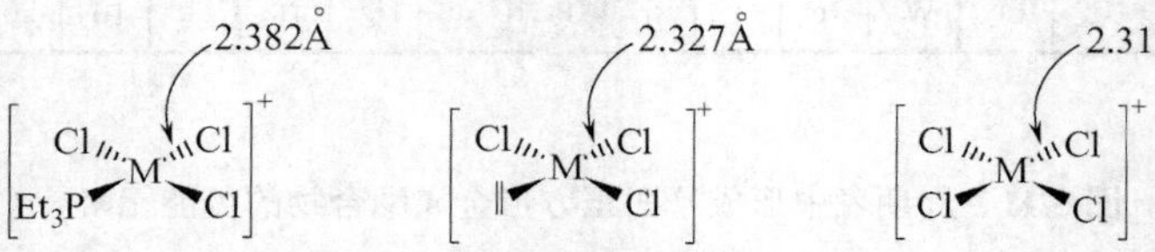

$[M(PEt_3)Cl_3]^+$ (2.382Å)　$[M(C{=}C)Cl_3]^+$ (2.327Å)　$[MCl_4]^+$ (2.317Å)

图 5.13 结构数据显示的反式效应

（2）进入基团和离去基团的作用　这种情况下，一个相对弱的亲核基团作为进入基团进行顺式配位取代［式(5.62)］。

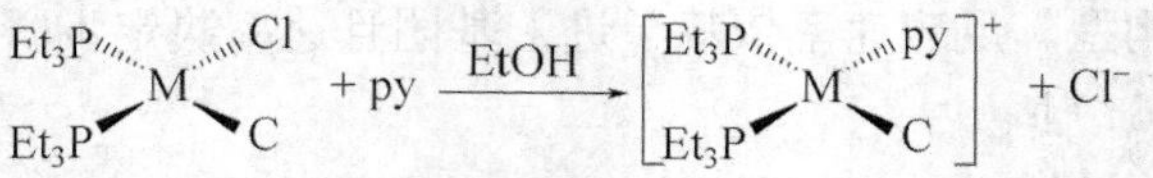

$$M(PEt_3)_2(Cl)(C) + py \xrightarrow{EtOH} [M(PEt_3)_2(py)(C)]^+ + Cl^- \tag{5.62}$$

$$C = CH_3^- > Ph^- > Cl^-$$

$$\frac{k_y(C)}{k_y(Cl^-)} = 3.6 \quad 2.3 \quad 1.0$$

与反式系列相比较［式(5.63)］，顺式配体的作用大致相同。

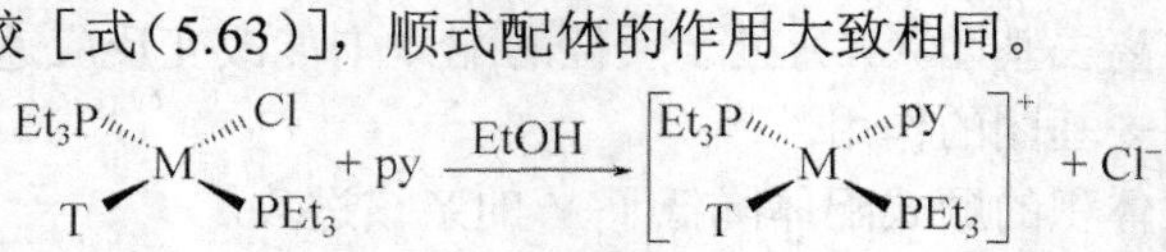

$$M(PEt_3)_2(Cl)(T) + py \xrightarrow{EtOH} [M(PEt_3)_2(py)(T)]^+ + Cl^- \tag{5.63}$$

$$T = H^- > CH_3^- > Ph^- > Cl^-$$

$$\frac{k_y(T)}{k_y(Cl^-)} = 10^4 \quad 1700 \quad 400 \quad 1.0$$

尽管进入和离去配体系列和反式效应系列有很大的相似性，但还是有明显的区别。其中的一些区别是进入和离去基团受配体溶剂化程度的影响，而反式效应系列则不受溶剂化效应影响。

这种缔合方式可用于一个配体取代另一个位于平面正方形反应中心的配体的反应［式(5.64)］。

$$ML_2X_2 + L' \xrightarrow{\text{溶剂}} MLL'X_2 + L \qquad (5.64)$$

图 5.14 说明了溶剂在取代反应中对平面正方形反应中心的重要性。

(3) 空间效应　立体空间阻碍会降低 A 机理的速率而增加 D 机理的速率。金属周围的空间很小意味着高配位数的过渡态有较高的能量，例如：顺式-$[PtXL(PEt_3)_2]$（图 5.15）。

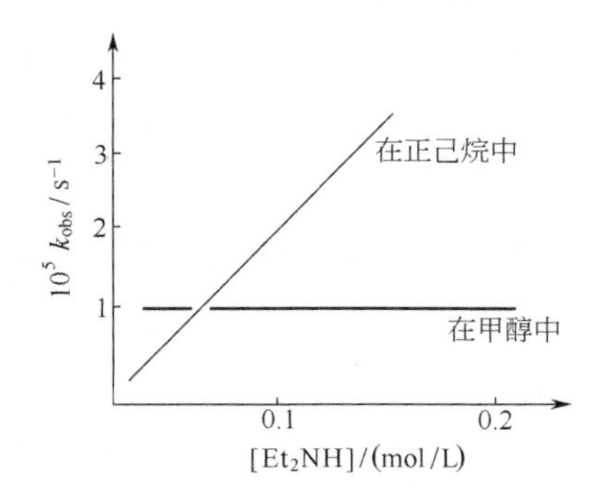

图 5.14　在平面正方形反应中心取代反应中溶剂的重要性

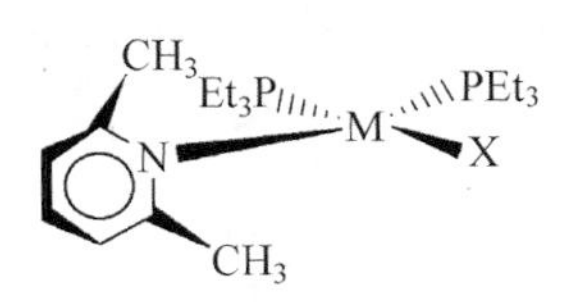

图 5.15　顺式-$[PtX(2,6\text{-二甲基吡啶})(PEt_3)_2]$的分子结构示意图（L=2,6-二甲基吡啶）

速率随 L 的变化：吡啶>2-甲基吡啶>2,6-二甲基吡啶。

平面正方形取代反应的立体化学也很重要。考察最终产物的立体化学可以提供反应机理和中间体寿命的信息（图 5.16）。

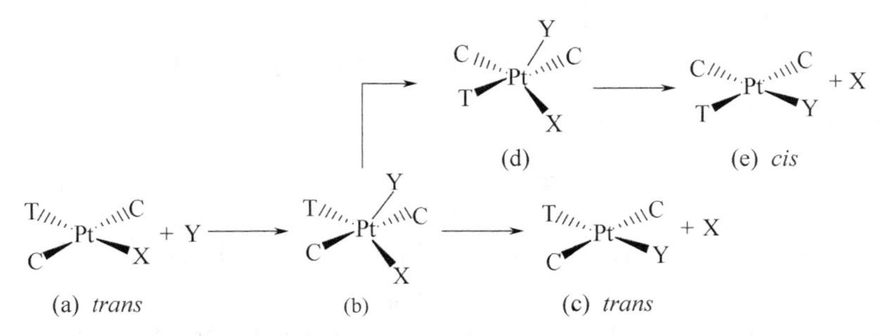

图 5.16　平面正方形取代反应的立体化学示意图

(4) $\Delta V^{\ddagger}$和 $\Delta S^{\ddagger}$

- 沿反应路线体积的改变通常由速率作为压力的函数来决定。$\Delta V^{\ddagger}$为负说明是一个缔合的配合物。
- 从反应物到活化的配合物，熵的改变是 $\Delta S^{\ddagger}$，它可由速率与温度的关系来决定。缔合机理的 $\Delta S^{\ddagger}$为负，这是通过限制进入基团的自由度和未释放离去基团来实现的。

5.3.3　八面体配合物的取代

对 O_h 型配合物的取代来说，交换机理（Ⅰ）是最重要的反应机理，但它可能是交换缔合（I_a）或交换离解（I_d），这依赖于速率决定步骤是 Y—M 的形成还是 M—X 的断裂。缔合反应的速率主要取决于进入基团。离解反应的速率与进入基团无关。

研究八面体型配合物主要限于两类反应。

（1）配位溶剂的取代（如水） 可能这类配合物的取代反应研究的最彻底的是在溶液中由水合金属离子生成一个配合物离子［式(5.65）和式(5.66)]：

$$[Co(NH_3)_5(OH_2)]^{3+} + Br^- \longrightarrow [Co(NH_3)_5Br]^{2+} + H_2O \qquad (5.65)$$

$$[Ni(OH_2)_6]^{2+} + \text{phen} \longrightarrow [Ni(OH_2)_4(\text{phen})]^{2+} + 2H_2O \qquad (5.66)$$

Anation：当进入基团是离子时反应叫 anation。

（2）溶剂解 既然大多数这类反应是在水溶液中发生的，叫水解可能更合适。水解反应可在酸性或碱性条件下发生［式(5.67)]：

$$[Co(NH_3)_5(OSMe_2)]^{3+} + H_2O \longrightarrow [Co(NH_3)_5(OH_2)]^{3+} + OSMe_2 \qquad (5.67)$$

O_h 配合物 I 取代反应的标准机理是 Eigen-Wilkins 机理，这种机理建立在形成一种“相遇配合物”（encounter complexes）的基础上。第一步是预平衡［式(5.68)]：

$$ML_6 + Y \rightleftharpoons \{ML_6, Y\} \quad K_E=\frac{[\{ML_6, Y\}]}{[ML_6][Y]} \qquad (5.68)$$

接着是产物的生成［式(5.69)]：

$$\{ML_6, Y\} \longrightarrow 产物 \quad 速率=k[\{ML_6, Y\}] \qquad (5.69)$$

速率的表达式如下［式(5.70)]：

$$速率=\frac{kK_E[C]_{tot}[Y]}{1+K_E[Y]} \qquad (5.70)$$

这里$[C]_{tot}$是所有配合物的总浓度，如果 $K_E[Y] \ll 1$，速率就变为［式(5.71)]：

$$速率=k_{obs}[C]_{tot}[Y] \qquad (5.71)$$

为什么 Eigen-Wilkins 机理如此重要?因为它对帮助我们预测反应机理十分有用。当$k_{obs}=kK_E$时，我们就得到 k。检查 k 是否随 Y 变化就可确定是 I_a 或 I_d。

O_h 取代的一般规则：

- 大部分 3d 金属属于 I_d 取代，即速率决定步骤与进入基团无关，主要是 M—X 键的断裂。
- 大的金属（4d，5d）倾向于 I_a。
- 离去基团 当 M—X 键的断裂是速率决定步骤时，离去基团 X 的性质十分重要。
- 旁观配体（顺-反效应） O_h 型配合物没有明显的反式效应。一般来说好的旁观 σ 供体在离去基团离去后可使配合物稳定。
- 空间效应 （a）金属中心的立体拥挤可有利于解离活化。（b）解离活化减轻了配合物周围的拥挤。（c）立体效应已被定性和定量的研究——Tolman 锥角（图 5.17）。
- 八面体取代 $\Delta V^‡$ 对机理 I 来说，$\Delta V^‡$不大，但是 I_a 倾向于为负值，I_d 倾向于为正值。d 的数目降低倾向于 I_a 机理。

- O_h 配合物取代的立体化学 它比 T_d 配合物更复杂。例如：顺式-或反式-$[CoAX(en)_2]^{2+}$，顺式配合物倾向于保持顺式，反式配合物受旁观配体和活化配合物的构型影响，会异构化。三角双锥是否发生异构与 Y 进入的位置有关，平面正方形会保持原来的立体化学（图 5.18）。

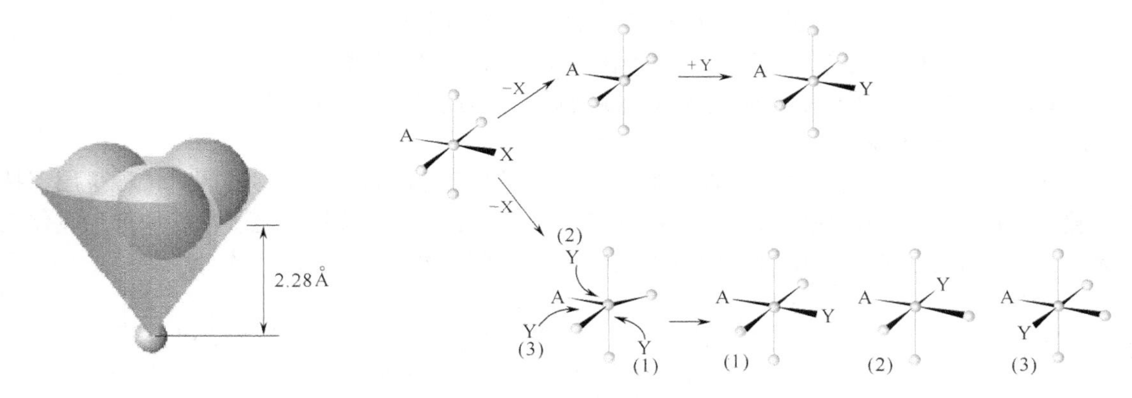

图 5.17 Tolman 锥角——定性和定量地分析立体效应

图 5.18 顺式-和反式-$[CoAX(en)_2]^{2+}$的 O_h 配合物取代反应的立体化学

5.3.4 异构化反应

异构化反应与取代反应类似。在五配位的 TBP 系列中，Berry 假旋转混合了轴向和赤道位置。经历 D 或 I_d 机理的平面正方形配合物都涉及五配位态，所以，异构化是可能的（图 5.19）。

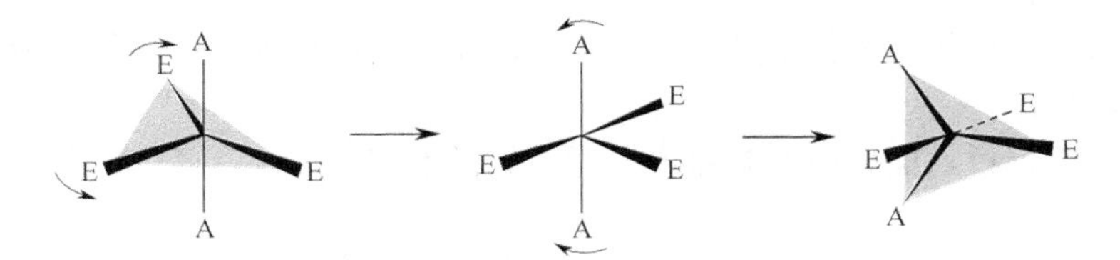

图 5.19 在五配位的 TBP 物种中，假旋转混合了轴向和赤道位置

O_h 配合物借助“扭曲”机理也可以异构化。它不需要失去配体或断键，仅与 Bailar 扭曲（a）和 Ray Dutt 扭曲（b）之间的能垒有关。它们都是通过三棱柱中间体而发生的（图 5.20）。

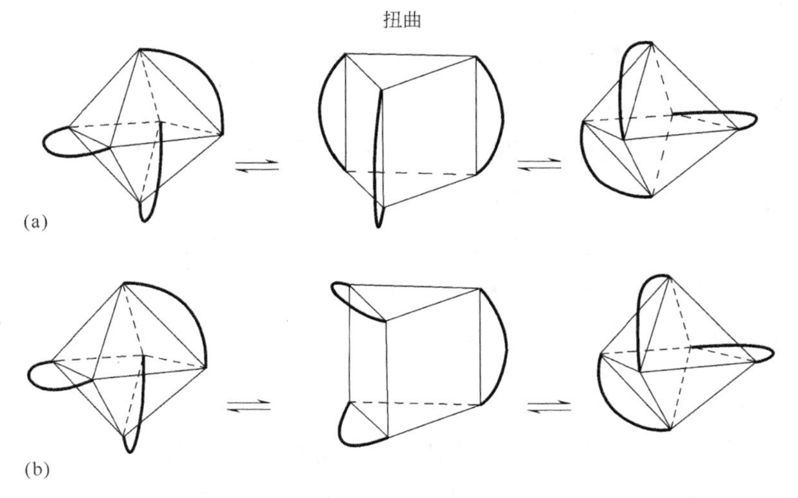

图 5.20 O_h 配合物异构化中的 Bailar 扭曲（a）和 Ray Dutt 扭曲（b）

5.4 配合物的电子转移反应

电子转移反应可以通过下面两个机理中的一个或两个发生。

- 外层机理：没有桥连配体，直接在两个金属中心间转移电子。
- 内层机理：需要形成桥连的双金属物质，同时会导致配体转移。

5.4.1 外层电子转移

在原则上，所有的外层机理都包含了从还原剂到氧化剂的电子转移，同时，配位层保持完整。也就是说一种反应物进入另一种反应物的外部或第二配位层，电子从还原剂向氧化剂流动。在两个惰性取代的配合物间发生快速电子转移时就是这个机理［式(5.72)］。

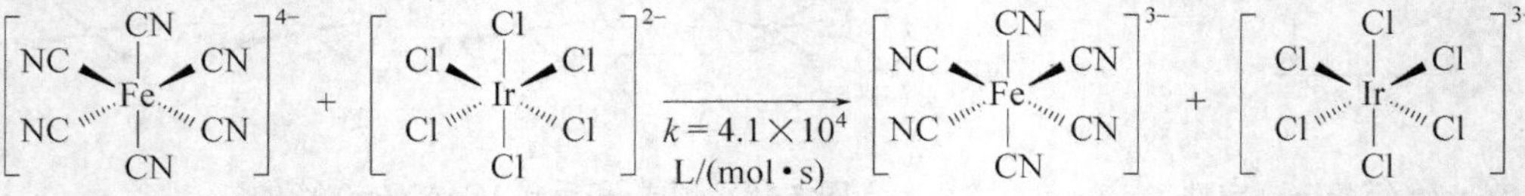

$$[Fe(CN)_6]^{4-} + [IrCl_6]^{2-} \xrightarrow[L/(mol\cdot s)]{k=4.1\times10^4} [Fe(CN)_6]^{3-} + [IrCl_6]^{3-} \tag{5.72}$$

当两个物种间没有配体转移时就可以很快确定这种机理。当配合物对配体取代是惰性时，更容易确认。在这种情况下可以用 Born Oppenheimer 近似，因为电子的运动比核快，配合物的重组可在与电子转移独立的步骤中完成。其他的模型——Marcus 方程对这一机理也是适合的。电子的转移需要振动激发态，势能曲线的形状决定转移速率。

5.4.2 内层电子转移

内层机理认为反应物和氧化剂共享一个配体，在它们内层或主要配位区域，电子通过一个桥连基团被转移。

在内层反应中，桥连配体应有多对电子对，如图 5.21 所示。

$$:\ddot{\underset{..}{Cl}}^-: \quad :S—C\equiv N^-: \quad :N\equiv N: \quad :C\equiv N^-:$$

图 5.21 一些具有多电子对的配体

电子的转移速率依赖于存在的配体。

通常，内层反应分为 3 步。

- 形成桥连配合物［式(5.73)］

$$M^{II}L_6+XM^{III}L'_5 \longrightarrow L_5M^{II}—XM^{III}L'_5+L \tag{5.73}$$

- 电子转移［式(5.74)］

$$L_5M^{II}—XM^{III}L'_5 \longrightarrow L_5M^{III}—XM^{II}L'_5 \tag{5.74}$$

- 分解形成最终产物［式(5.75)］

$$L_5M^{III}—XM^{II}L'_5 \longrightarrow L_5M^{III}—X+M^{II}L'_5 \tag{5.75}$$

与我们期望的一样，速率决定步骤通常是电子转移步骤。然而，桥连配合物的形成或分解也会影响速率。好的共轭能给电子提供一个简单的路径。因此，怎样去构建桥连体系的模型非常重要。

通过外层机理，六水合铬（Ⅱ）还原六氨合钴（Ⅲ）的反应很慢［$k=10^{-3}$L/(mol·s)］。然而，如果六氨合钴（Ⅲ）中的一个氨被 NCS^- 或 Cl^- 取代后，反应速率就会显著加快［$k=6\times10^5$L/(mol·s)］［式(5.76)，式(5.77)，式(5.78)］。

$$[Co(NH_3)_5NCS]^{2+} + [Cr(H_2O)_6]^{2+} \longrightarrow \left[(NH_3)_5Co\text{—}NCS\text{—}Cr(OH_2)_5\right]^{4+} \longrightarrow [Co(H_2O)_6]^{2+} + 5NH_4^+ + [Cr(H_2O)_5SCN]^{2+} \tag{5.76}$$

$$\underset{t_{2g}^6}{[Co(NH_3)_5Cl]^{2+}} + \underset{t_{2g}^3e_g^1}{[Cr(H_2O)_6]^{2+}} \xrightleftharpoons{H^+} \underset{t_{2g}^5e_g^2}{[Co(H_2O)_6]^{2+}} + \underset{t_{2g}^3}{[Cr(H_2O)_6]^{3+}} + 5NH_4^+ \tag{5.77}$$

$$\underset{t_{2g}^6}{[Co(NH_3)_6]^{3+}} + \underset{t_{2g}^3e_g^1}{[Cr(H_2O)_6]^{2+}} \xrightleftharpoons{H^+} \underset{t_{2g}^5e_g^2}{[Co(H_2O)_6]^{2+}} + \underset{t_{2g}^3}{[Cr(H_2O)_6]^{3+}} + 6NH_4^+ \tag{5.78}$$

5.5　均相催化

对金属有机化学研究的兴趣是源于金属有机化合物作为催化剂在化学工业中有潜在应用。在世界范围内催化剂具有上亿美元的财政收入。催化剂是可以改变反应速率而不在产物中出现的物质，它可以加速或减缓反应。对一个可逆反应，催化剂可改变达到平衡的速率而不改变平衡的位置（图 5.22）。

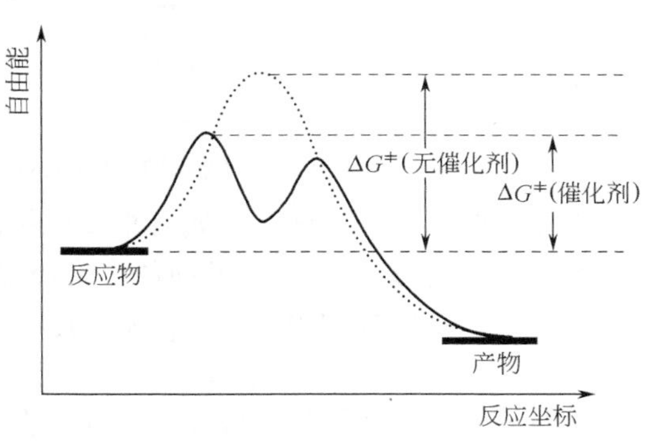

图 5.22　未加催化剂（点线）与加催化剂的反应的过渡态比较

催化过程可粗略地分为两类。

● 均相催化　在这个过程中，催化剂和反应组分处于同一相中。均相催化反应一般比较明确且在溶液中进行。因此，均相催化可以用光谱来研究。但是从混合产物中分离催化剂比较困难。均相催化剂主要是金属有机配合物。一些常见的均相催化的例子列在表 5.4 中。

表 5.4　均相催化剂的例子

均相催化剂	应　用	均相催化剂	应　用
$RhCl(PPh_3)_3$（Wilkinson's 催化剂）	烯烃加氢	$HRh(CO)_4$	羰基化（支链烯烃）
cis-$[Rh(CO)_2I_2]^-$	蒙桑托醋酸合成	$[HFe(CO)_4]^-$	水-气迁移反应
Hco(CO)$_4$	羰基化	Cp_2TiMe_2	烯烃聚合
$Pd(PPh_3)_4$	Heck、Suzuki、Stille、Negishi 偶合反应（很重要的合成）	Cp_2ZrH_2	烯烃/炔烃加氢

● 非均相催化　在这个过程中，催化剂和反应组分在不同的相中。在大多数非均相催化剂体系中，催化剂一般在固相中，而反应物一般是液态或气态。因此，产物和催化剂的分离很容易。

均相催化必须有容易产生开放配位的单元。通常是 ML_4（D_{4h}）或容易分离的配体。均相催化通常具有比较好的选择性和调节能力。

5.5.1 烯烃加氢

到目前为止，最成功的均相金属有机催化剂是 Wilkinson 催化剂 $RhCl(PPh_3)_3$，它可催化烯烃的氢化［式(5.79)］。

$$RCH{=}CH_2 + H_2 \longrightarrow RCH_2CH_3 \tag{5.79}$$

有意思的是，这种催化剂是由 Wilkinson 和 Coffey 于 20 世纪 60 年代在催化烯烃的氢化时独立且几乎同时发现的，现在这种化合物一般称为“Wilkinson 催化剂”。反应的主要机理可以用 Tolman 的催化循环来表示（图 5.23）。

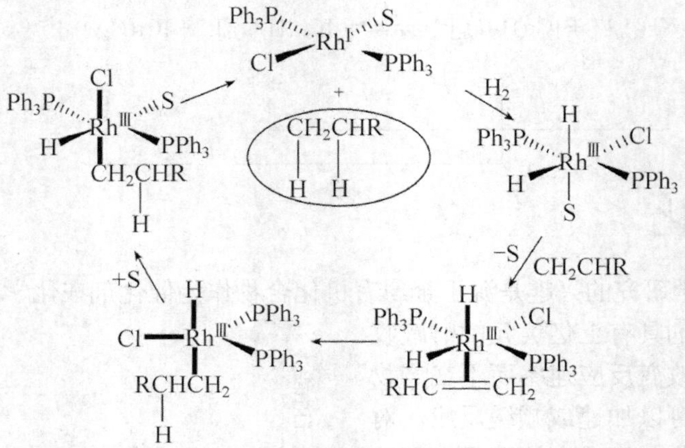

图 5.23 Wilkinson 催化剂——烯烃氢化

在催化循环中，对许多金属有机反应机理来说有几个重要的步骤。一个重要的例子是氢气和溶剂配合物形成顺式-二氢物种［式(5.80)］。

$$\underset{A}{RhCl[P(C_6H_5)_3]_2(S)} + H_2 \longrightarrow \underset{B}{cis\text{-}RhCl[P(C_6H_5)_3]_2(S)(H)_2} \tag{5.80}$$

这个反应是一个氧化加成反应。注意在这个化学转化中，A 只与 4 个配体相连，A 称为 4-配位的“配位不饱和”配合物，而 B 是与 6 个配体相连的，是配位饱和的。Wilkinson 催化剂 $RhCl(PPh_3)_3$ 是一个具有 16 个价电子的平面正方形结构的 Rh(Ⅰ)配合物。因此，可以在金属中心上再添加 2 个电子［式(5.81)］。

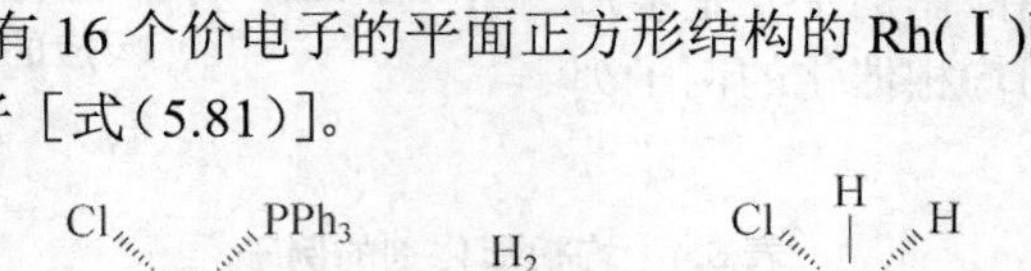

（5.81）

B 物种是具有 18 个价电子的 Rh(Ⅲ)配合物，且没有空的配位位置。因此，在氧化加成反应中，金属的配位数从 4 变到了 6，且金属的氧化态增加了 2。所以在其他反应发生前，配体需要分离［式(5.82)］。

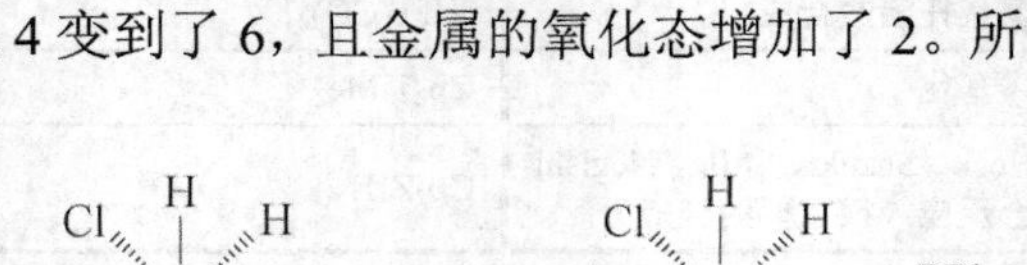

（5.82）

现在，16 电子的$[Rh(PPh_3)_2Cl(H_2)]$配合物能增加一个 2 电子的配体，如 $H_2C{=}CH_2$［式

（5.83）]。

$$RhClH_2(PPh_3)_2 \xrightarrow{H_2C=CH_2} RhClH_2(PPh_3)_2(H_2C=CH_2) \qquad (5.83)$$

18 电子的[Rh(PPh$_3$)$_2$Cl(H$_2$)(η^2-H$_2$C═CH$_2$)]配合物可以发生一些有趣的化学反应：乙烯插入 Rh—H 键，这是金属有机配合物独有的性质。这种迁移插入反应中，形成了一个 C—H 键和一个 Rh—C σ 键。因此，[Rh(PPh$_3$)$_2$Cl(H$_2$)(η^1-H$_2$C—CH$_3$)]是 18 电子的配合物［式(5.84)]。

$$RhClH_2(PPh_3)_2(H_2C=CH_2) \longrightarrow RhClH(CH_2-CH_3)(PPh_3)_2 \qquad (5.84)$$

在[Rh(PPh$_3$)$_2$Cl(H$_2$)(η^1-CH$_2$—CH$_3$)]中，氢和乙基顺式排列，这意味着 CH$_2$—CH$_3$ 和 H 可以离开金属［式(5.85)]。

$$RhClH(CH_2-CH_3)(PPh_3)_2 \longrightarrow RhClH(PPh_3)_2 + CH_3-CH_3 \qquad (5.85)$$

注意：金属的配位数减少了 2，电子数也减少了 2。也就是说，这个反应是第一步中氧化加成的逆反应。因此，氧化加成反应的逆反应也较常见，称为还原消去反应。在这个反应中可以观察到颜色的变化。

为了回到起始点，在催化循环的最后步骤涉及 PPh$_3$ 的加成［式(5.86)]。

$$[Rh(PPh_3)_2Cl]+PPh_3 \longrightarrow [Rh(PPh_3)_3Cl] \qquad (5.86)$$

一旦 Wilkinson 催化剂可以再生，整个循环就可以重复，在催化剂分解之前，循环可以重复大约 1000 次。大多数金属有机催化反应发生在 16-和 18-电子的中间体中。

催化循环的示意：

加氢 ⟶ 膦离解 ⟶ 烯烃加成 ⟶ 氢化物迁移(RDS)
↑———— 烷烃还原消去 ←————┘

可使用手性膦把手性引入到前手性链烯中，William Knowles 和 Ryoji Noyori 分别因为不对称氢化 DIPMAP Rh 配合物和 BINAP Ru 配合物而获得了 2001 年的诺贝尔化学奖。

手性的 Wilkinson-型催化剂见图 5.24。

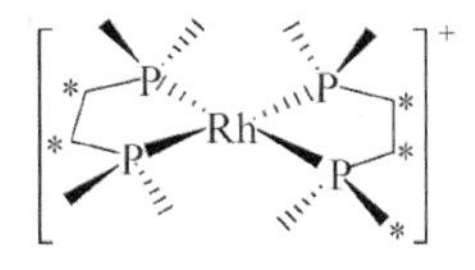

图 5.24　手性的 Wilkinson-型催化剂的分子结构示意图

5.5.2　Monsanto 醋酸合成

醋酸是使醋具有特征味道和酸味的化学物质。醋中醋酸的浓度约为 4%~8%。作为醋的特定组分，自从有历史记录开始，人类就生产和利用了醋酸。实际上，醋酸的名字来源于醋的拉丁文 acetum。醋可通过稀的乙醇溶液，如酒，在氧的存在下与特定的细菌反应得到。这种细菌需要氧气，总的化学变化是乙醇和氧反应生成醋酸和水［式(5.87)]。

$$CH_3CH_2OH + O_2 \longrightarrow CH_3COOH + H_2O \qquad (5.87)$$

醋也可以从其他饮料的发酵得到，比如麦芽或苹果酒。多年来，大批量的商业醋酸都是由乙醇的氧化得到的。

现在，大多数醋酸的工业生产都通过 Monsanto 过程，在这个反应中，CO 与甲醇在

30~60atm、180℃、Rh 配合物作催化剂的条件下反应。每年大约生产 3.85 亿吨的醋酸。大约 60%的醋酸生产都是在 Rh 催化剂下把甲醇转变为醋酸。机理如图 5.25 所示。

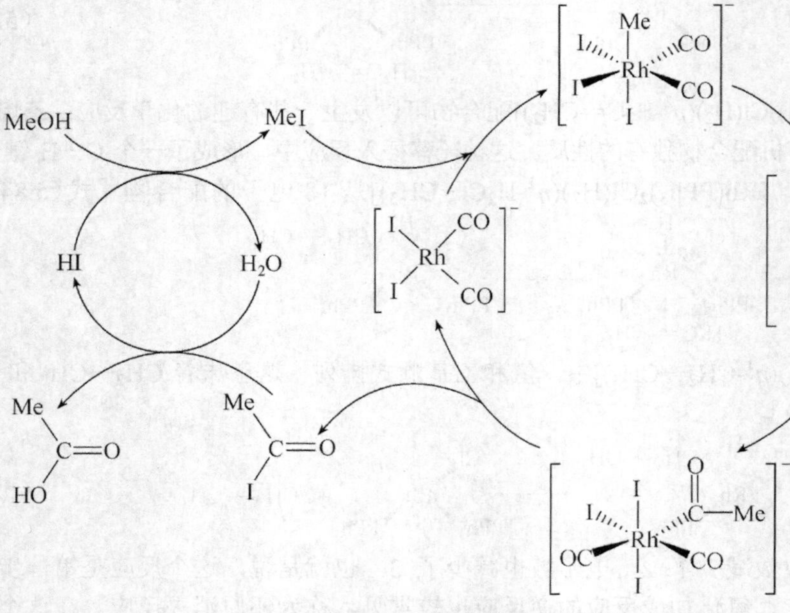

MeOH　MeI　HI　H_2O
$[Rh(CO)_2I_2]^-$：I　CO　Rh　I　CO
Me　I　CO　Rh　I　CO　I
O　I　C—Me　Rh　I　CO　I
CO
I　O　I　C—Me　Rh　OC　CO　I
Me　C=O　I
Me　C=O　HO

图 5.25 Monsanto 醋酸合成的机理示意图

反应的催化剂是顺式-$[Rh(CO)_2I_2]^-$，这是一个 16 电子的配合物，可以参与氧化加成。在此例中，被氧化加成的是 CH_3I。由于 CH_3I 是一个极性分子，氧化加成的产物是反式的而不是顺式的［式(5.88)］。

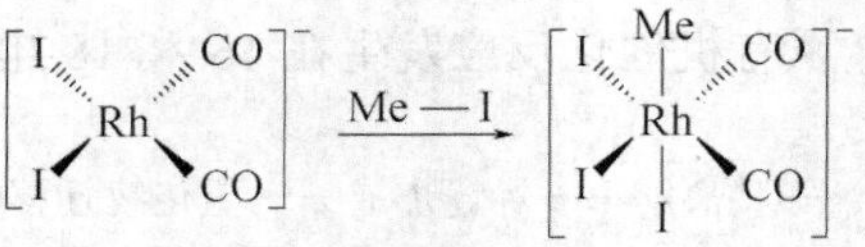

$[Rh(CO)_2I_2]^-$ —— Me—I ——→ $[RhMe(CO)_2I_3]^-$　(5.88)

甲基基团可以迁移到其中的一个 CO 配体上，从而形成 Rh—COMe 基团（酰基配体）。产物$[Rh(COMe)I_3(CO)]^-$是 16 电子的配合物，它可以再加一个 2 电子的配体，如 CO［式(5.89)］。

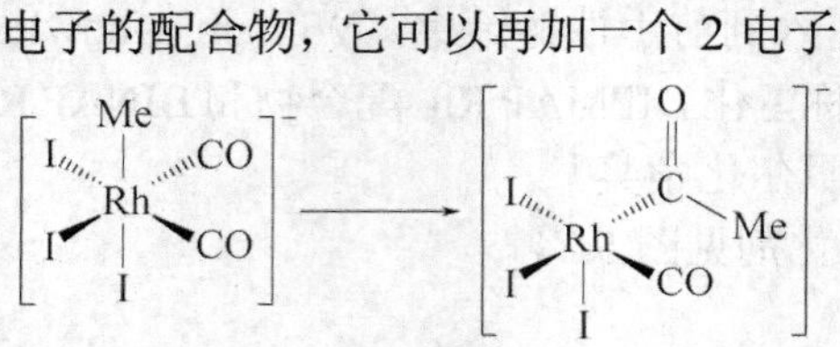

$[RhMe(CO)_2I_3]^-$ ——→ $[Rh(COMe)I_3(CO)]^-$　(5.89)

COMe 和 I 的排列方式适合于 I—COMe 的还原消去［式(5.90)和式(5.91)］。

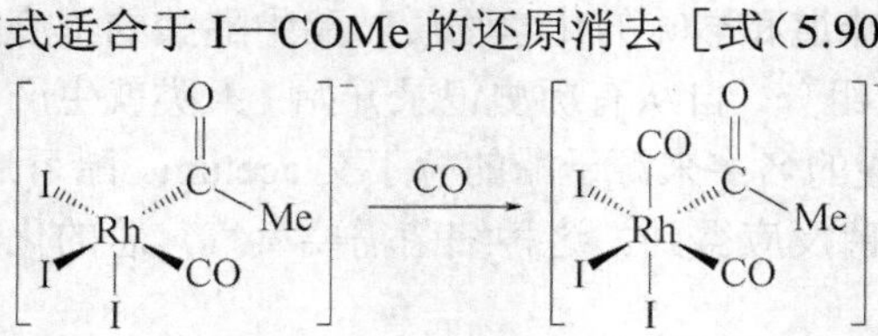

$[Rh(COMe)I_3(CO)]^-$ —— CO ——→ $[Rh(COMe)I_3(CO)_2]^-$　(5.90)

$[Rh(COMe)I_3(CO)_2]^-$ ——→ $[Rh(CO)_2I_2]^-$ + I—C(O)—Me　(5.91)

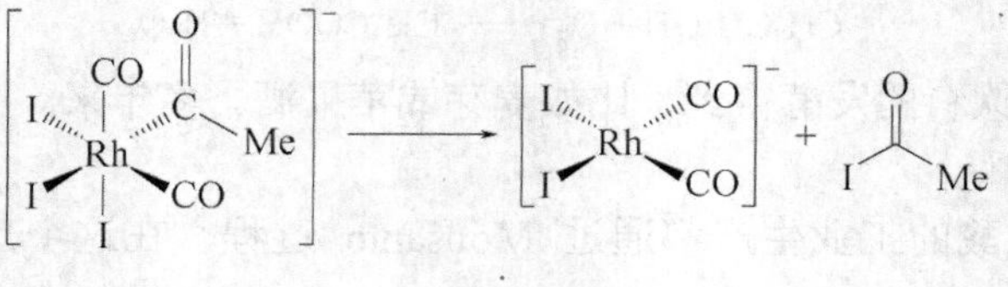

注意：已经再生了催化剂，但是没得到目标产物。醋酸可以从 ICOMe 和水的反应得到［式(5.92)］。

$$\mathrm{ICOMe + H_2O \longrightarrow MeCOOH + HI} \quad (5.92)$$

HI 可用来将甲醇转化为碘甲烷［式(5.93)］。

$$\mathrm{HI + MeOH \longrightarrow MeI + H_2O} \quad (5.93)$$

5.5.3　酰氢化反应

（1）一般反应［式(5.94)］

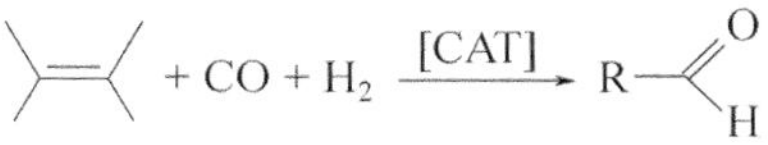

$$\mathrm{C{=}C + CO + H_2 \xrightarrow{[CAT]} R{-}CHO} \quad (5.94)$$

在化学工业中，均相酰氢化催化过程是最大的化学工业过程，在 1997 年，全世界的氧化醛产量是 7.8×10^6t。这种工艺最初是由 Roelen 在 20 世纪 30 年代开发的。最初的催化剂是 $Co_2(CO)_8$，现在仍在广泛使用。酰氢化是形成 C—C 键的一个重要反应，有两种可能的异构产物：直链的和支链的。通常，想要的主要产物是直链。在催化剂和合适的条件下，醛可以在反应中直接还原为醇［式(5.95)］。

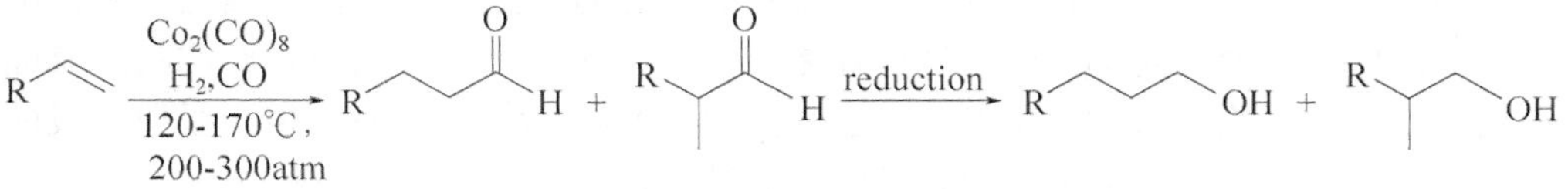

$$\mathrm{RCH{=}CH_2 \xrightarrow[120\text{-}170^\circ C,\ 200\text{-}300atm]{Co_2(CO)_8,\ H_2,CO} RCH_2CH_2CHO + RCH(CH_3)CHO \xrightarrow{reduction} RCH_2CH_2CH_2OH + RCH(CH_3)CH_2OH} \quad (5.95)$$

三种主要的典型均相催化剂：

- H—Co—CO；
- H—Co—Ph_3P；
- H—Rh—Ph_3P—CO。

（2）H—Co—CO　尽管一些步骤的确切过程还未确定，催化循环也可用图 5.26 表示。

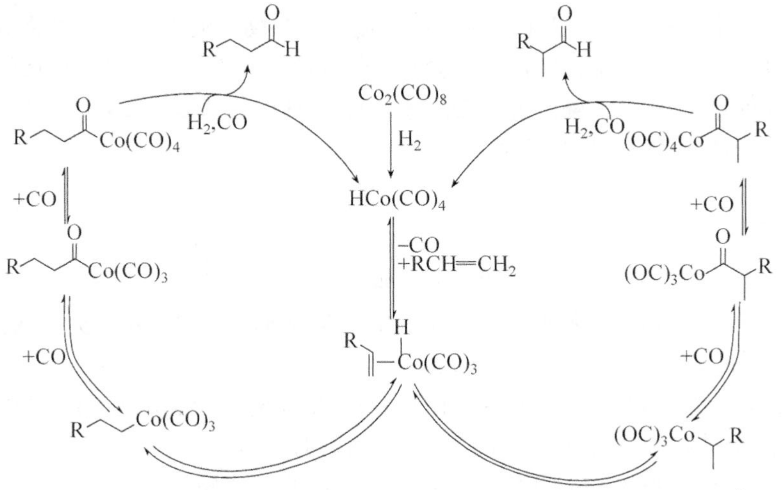

图 5.26　$Co_2(CO)_8$ 氢化催化循环图

$Co_2(CO)_8$ 氢化后得到酸性的氢化物 $HCo(CO)_4$，它也可以通过直接制备用作催化剂。第一步是用 CO 取代烯烃。迁移插入可以得到一个一级或二级的金属烷基配合物。尽管这是确定产物配位化学的步骤，它是一个快速的可逆步骤。这种快速的可逆性可导致烯烃的异构和 H/D 交换。β-H 的消去需要一个开放的配位单元，异构和同位素交换可通过增加 CO 的压力来抑制。

下一步是第二个迁移插入形成不饱和的酰基配合物，它能和其他的 CO 配位形成 18 电子的酰基配合物。在标准的 1-辛烯催化条件下，这是由红外观测到的唯一物种。

一个可能的机理是和氢气通过氧化加成/还原消去反应生成醛和 $HCo(CO)_4$，另一个可能的机理会涉及一个 $HCo(CO)_4$ 的双核反应，这两种机理的证据都已观测到［式(5.96)］。

$$RCH_2CH_2C(O)Co(CO)_4 \underset{}{\overset{-CO}{\rightleftharpoons}} RCH_2CH_2C(O)Co(CO)_3 \begin{cases} \xrightarrow{H_2} RCH_2CH_2CHO + HCo(CO)_3 \\ \xrightarrow{HCo(CO)_4} RCH_2CH_2CHO + Co_2(CO)_7 \end{cases} \tag{5.96}$$

（3）H—Co—膦　通常，直链状的异构体比支链状的更有工业价值。由于催化循环是不可逆的，观察到的区域选择性是动力学产物的比率，而不是热力学的优先性。通常，区域选择性并不是由烯烃插入形成一级和二级 Co—烷基配合物的速率决定的，而是由一级和二级烷基迁移插入 CO 的速率决定的。位阻小的烷基比大的基团迁移更快。

向 Co 催化酰氢化体系中加入膦可以改变区域选择性。用 $HCo(CO)_3PBu_3$ 作催化剂，得到的产物中直链支链比较高。可能是由于最初迁移插入 Co—H 的烯烃有较大程度的区域控制性。这有两个原因：①立体性，用 PBu_3 取代 CO 增加了金属中心的空间体积，这有利于形成一组 Co—烷基；②电子效应，用 PBu_3 取代 CO 明显降低了 Co—H 的酸性。所以，就不容易将 H—Co 加到烯烃上。

（4）H—Rh—膦—CO　由于简单的 Rh 酰基化合物倾向于形成一个稳定的、催化不活泼的 Rh—酰基簇，所以它们对酰氢化反应是不活泼的。在高压的氢气和一氧化碳下，可以得到 $HRh(CO)_4$，它是非常活泼的催化剂，但它会异构化烯烃，且得到的直链支链比较低。

在这些 Rh 的体系中添加膦会明显提高化学选择性和区域选择性，同时还保持了初始 Rh—酰基化合物的高反应速率［式(5.97)］。

$$CH_3CH{=}CH_2 \xrightarrow[\text{熔融 } PPh_3,\ 100^{\circ}C,\ 50atm]{(Ph_3P)_2Rh(CO)H,\ H_2,CO} CH_3CH_2CH_2CHO\ (92\%\ 选择性) \tag{5.97}$$

对 Rh 体系来说，普遍接受的机理是 Geoffrey Wilkinson 在 1968 年提出的，它与 Co 催化剂非常相似（图 5.27）。

Rh 配合物是五配位的 18 电子体系。由于强的顺式效应配体(CO,H)，大体积的膦占据了顺式配位点，两个膦可以都在平伏位或一个占据轴向位置。

这些具有平面正方形结构的催化剂，常有较高的直链支链比。烯烃的迁移插入主要决定区域同分异构的比率，因为在反应条件下，平面正方形的 Rh—烷基迅速被 CO 在一个不可逆的反应中捕获。对一级的 Rh—烷基来说，紧跟着是 CO 迁移插入，然后氢化裂解。二级的 Rh—烷基 CO 插入较慢，所以可观察到烯烃异构化产物。发生在一级 Rh—烷基上的 CO 插入的强动力学优先性非常有用。

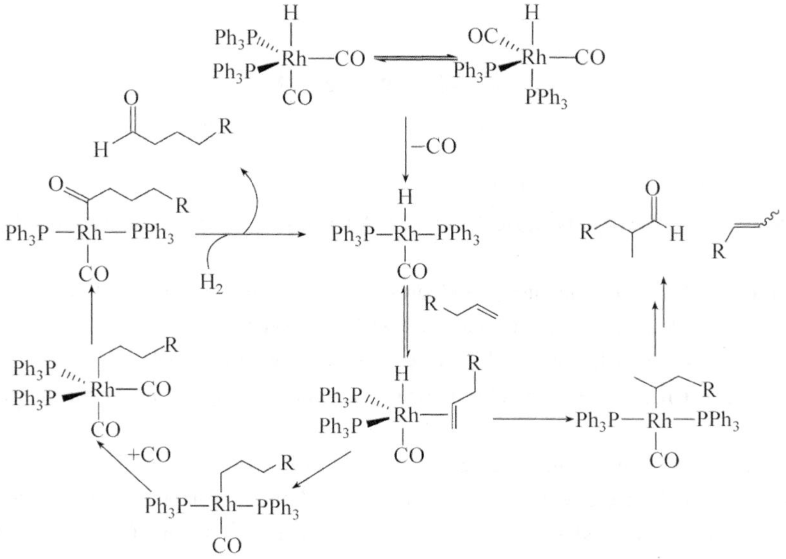

图 5.27　H—Rh—PPh_3—酰基体系的催化循环示意图

最后，在催化体系中确定真正的活性催化剂非常困难。只有实验者通过详细的机理研究才能得到活性催化剂的一些证据。一些科学文献报道的“催化剂”实际上并不是催化剂，通常是杂质或降解产物催化了反应。

NMR 是监测反应体系得到机理和动力学细节的有效手段。这些细节是理解和改善催化剂和反应条件的关键。尽管将 NMR 方法应用于均相和非均相催化剂问题是一种已经建立的技术，它仍然是一个有极大吸引力的领域，因为，NMR 具有独特的传递结构和动力学信息的能力，同时还可以作定性和定量分析。

参 考 文 献

[1] 游效曾. 配位化合物的结构和性质. 北京: 科学出版社, 1992.

[2] Cotton F A, Wilkinson G, Murillo C A, Bochmann M, Advanced Inorganic Chemistry. 6thed. New York: Wiley, 1999.

[3] Kettle S F A. Physical Inorganic Chemistry: A Coordination Chemistry Approach. Oxford: Spektrum, 1996.

[4] Atwood J L, Steed J W. Supramolecular Chemistry. Chichester: Wiley, 2000.

[5] Jolly W L. Modern Inorganic Chemistry. 1sted. New York: McGraw-Hill, 1984.

[6] Huheey J E. Inorganic Chemistry—Principles of Structure and Reactivity. 3rded. Harper International SI Edition, 1983.

[7] Müller U. Inorganic Structural Chemistry. 2nded. New York: Wiley, 1993.

[8] Smart L. Moore E. , Solid State Chemistry—An Introduction. London: Chapman & Hall, 1992.

[9] Constable E C. Metals and Ligand Reactivity—An Introduction to the Organic Chemistry of Metal Complexes. 2nded. Weinheim: VCH, 1996.

[10] Glusker J P, Lewis M, Rossi M. Crystal Structure Analysis for Chemists and Biologists. Weinheim: VCH, 1994.

[11] Tobe M L, Burgess J. Inorganic Reaction Mechanism. Harlow, Essex: Longman, 1999.

[12] Wilkins R G. Kinetics and Mechanism of Reaction of Transition Metal Complexes. Weinheim: VCH, 1991.

[13] Carter R L. Molecular Symmetry and Group Theory. New York: Wiley, 1998.

[14] Nakamoto K. Infrared and Raman Spectra of Inorganic and Coordination Compounds. 5thed. New York: Wiley, 1997.

[15] Figgis B N, Hitchman M A. Ligand Field Theory and its Applications. New York: Wiley-VCH, 2000.

[16] Desiraju G R, Steiner T. The Weak Hydrogen Bond in Structural Chemistry and Biology. New York: Oxford University Press, 1999.

[17] Desiraju G R. The Crystal as a Supramolecular Entity. Chichester: Wiley, 1996.

[18] Simon J, Bassoul P. Design of Molecular Materials—Supramolecular Engineering. London: John Wiley & Sons Ltd, 2000.

[19] Rajesh Chakrabarty, Partha Sarathi Mukherjee, and Peter J. Stang. Supramolecular Coordination: Self-Assembly of Finite Two- and Three-Dimensional Ensembles, Chem. Rev. 2011, 111, 6810-6918.

[20] Yilei Wu, Rufei Shi, Yi-Lin Wu, James M. Holcroft, Zhichang Liu, Marco Frasconi, Michael R. Wasielewski, Hui Li, and J. Fraser Stoddart. Complexation of Polyoxometalates with Cyclodextrins, J. Am. Chem. Soc. 2015, 137, 4111-4118.

[21] Klaus Müller-Dethlefs, and Pavel Hobza. Noncovalent Interactions: A Challenge for Experiment and Theory, Chem. Rev. 2000, 100, 143-167.

[22] Brian Moulton and Michael J. Zaworotko. From Molecules to Crystal Engineering: Supramolecular Isomerism and Polymorphism in Network Solids, Chem. Rev. 2001, 101, 1629-1658.

[23] Anthony K. Cheetham, G. Kieslich, and H. H.-M. Yeung. Thermodynamic and Kinetic Effects in the Crystallization of Metal-Organic Frameworks, Acc. Chem. Res. 2018, 51, 659-667.

[24] Sven M. J. Rogge, Michel Waroquier, and Veronique Van Speybroeck. Reliably Modeling the Mechanical Stability of Rigid and Flexible Metal-Organic Frameworks, Acc. Chem. Res. 2018, 51, 138-148.

[25] Ekaterina A. Dolgopolova, Allison M. Rice, Corey R. Martin and Natalia B. Shustova. Photochemistry and Photophysics of MOFs: Steps towards MOF-based sensing Enhancements, Chem. Soc. Rev., 2018, 47, 4710-4728.

[26] Gui-lei Liu, Yong-jie Qin, Lei Jing, Gui-yuan Wei and Hui Li. Two Novel MOF-74 Analogs Exhibiting Unique Luminescent Selectivity, Chem. Commun., 2013, 49, 1699-1701.
[27] Chaoyuan Chen, Yilei Wu, and Hui Li. Fine-Tuning Aromatic Stacking and Single-Crystal Photoluminescence Through Coordination Chemistry, Eur. J. Org. Chem. 2019, 1778-1783.
[28] Jian-Biao Song, Gui-lei Liu, Liang Hao, Fang Zhang and Hui Li. Crystal Structures and Luminescence Properties of a D-A Type CIEgen and its Zn[Ⅱ] Complexes, Cryst. Eng. Comm., 2019, 21, 3322-3329.
[29] Salma Begum, Zahid Hassan, Stefan Brase, Christof Wöll, and Manuel Tsotsalas. Metal-Organic Framework-Templated Biomaterials: Recent Progress in Synthesis, Functionalization, and Applications, Acc. Chem. Res. 2019, 52, 1598-1610.
[30] Juan-Ding Xiao and Hai-Long Jiang. Metal-Organic Frameworks for Photocatalysis and Photothermal Catalysis, Acc. Chem. Res. 2019, 52, 356-366.
[31] Amarajothi Dhakshinamoorthy, Zhaohui Li and Hermenegildo Garcia. Catalysis and Photocatalysis by Metal-organic Frameworks, Chem. Soc. Rev., 2018, 47, 8134-8172.
[32] Hao Wang, William P. Lustig and Jing Li. Sensing and Capture of Toxic and Hazardous Gases and Vapors by Metal-organic Frameworks, Chem. Soc. Rev., 2018, 47, 4729-4756.
[33] Meiting Zhao, Ying Huang, Yongwu Peng, Zhiqi Huang, Qinglang Ma and Hua Zhang. Two-dimensional Metal-organic Framework Nanosheets: Synthesis and Applications, Chem. Soc. Rev., 2018, 47, 6267-6295.
[34] Meili Ding, Robinson W. Flaig, Hai-Long Jiang and Omar M. Yaghi. Carbon Capture and Conversion Using Metal-organic Frameworks and MOF-based materials, Chem. Soc. Rev., 2019, 48, 2783-2828.
[35] Yuanjing Cui, Jun Zhang, Huajun He and Guodong Qian. Photonic Functional Metal-organic Frameworks, Chem. Soc. Rev., 2018, 47, 5740-5785.
[36] Zhi-Gang Gu, Caihong Zhan, Jian Zhang and Xianhui Bu. Chiral Chemistry of Metal-camphorate frameworks, Chem. Soc. Rev., 2016, 45, 3122-3144.
[37] Pei Zhou, Hui Li. Chirality Delivery from a Chiral Copper(II) Nucleotide Complex Molecule to Its Supra-molecular Architecture, Dalton Trans. 2011, 40, 4834-4837.
[38] Jeanne Crassous. Chiral Transfer in Coordination Complexes: Towards Molecular Materials, Chem. Soc. Rev., 2009, 38, 830-845.
[39] Pei Zhou, Rufei Shi, Jian-feng Yao, Chuan-fang Sheng, Hui Li. Supramolecular Self-assembly of Nucleotide-metal Coordination Complexes: From Simple Molecules to Nanomaterials, Coord. Chem. Rev. 2015, 292, 107-143.
[40] Qi-ming Qiu, Pei Zhou, Leilei Gu, Liang Hao, Minghua Liu, and Hui Li. Cytosine-Cytosine Base-Pair Mismatch and Chirality in Nucleotide Supramolecular Coordination Complexes, Chem. Eur. J. 2017, 23, 7201-7206.
[41] Pei Zhou, Jian-feng Yao, Chuan-fang Sheng and Hui Li. A Continuing Tale of Chirality: Metal Coordination Extended Axial Chirality of 4,4′-bipy to 1D Infinite Chain under Cooperation of a Nucleotide Ligand, Cryst, Eng. Comm., 2013, 15, 8430-8436.
[42] Li-Jun Chen, Hai-Bo Yang and Mitsuhiko Shionoy. Chiral Metallosupramolecular Architectures, Chem. Soc. Rev., 2017, 46, 2555-2576.
[43] Watt, Ian M. The Principles and Practice of Electron Microscopy. New York: Cambridge University Press (Books), 1997.
[44] Peter J. Goodhew, John Humphreys, Richard Beanland. Electron Microscopy and Analysis. New York: Taylor & Francis (Books), 2001.
[45] Daliang Zhang, Yihan Zhu, Lingmei Liu, Xiangrong Ying, Chia-EnHsiung, Rachid Sougrat, Kun Li, Yu Han. Atomic-resolution Transmission Electron Microscopy of Electron Beam-sensitive Crystalline Materials. Science, 59, 675-679 (2018).
[46] Mitsuo Suga, Shunsuke Asahina, Yusuke Sakuda, et al. Progress in Solid State Chemistry 42, 1-21(2014).

[47] David Alsteens, Hermann E. Gaub, Richard Newton, Moritz Pfreundschuh, Christoph Gerber, Daniel J. Müller. Atomic Force Microscopy-based Characterization and Design of Biointerfaces. Nature Reviews Materials, 2, 17008 (2017).

[48] Kyung Joo Lee, Jae Hwa Lee, Sungeun Jeoung, and Hoi Ri Moon. Transformation of Metal-Organic Frameworks/Coordination Polymers into Functional Nanostructured Materials: Experimental Approaches Based on Mechanistic Insights, Acc. Chem. Res. 2017, 50, 2684-2692.

[49] Xiao Xiao, Lianli Zou, Huan Pang and Qiang Xu. Synthesis of Micro/nanoscaled Metal-organic frameworks and Their Direct Electrochemical Applications, Chem. Soc. Rev., 2020, 49, 301-331.

元素周期表

IUPAC 2013

+2 +3 +4 +5 +6	95
Am	镅▲
$5f^7 7s^2$	243.06138(2)✦

氧化态(单质的氧化态为0，未列入；常见的为红色) — +2 +3 +4 +5 +6
95 — 原子序数
Am — 元素符号(红色的为放射性元素)
镅▲ — 元素名称(注▲的为人造元素)
$5f^7 7s^2$ — 价层电子构型
243.06138(2)✦ — 以 $^{12}C=12$ 为基准的原子量(注✦的是半衰期最长同位素的原子量)

s区元素 | p区元素 | d区元素 | ds区元素 | f区元素 | 稀有气体

周期＼族	1 ⅠA	2 ⅡA	3 ⅢB	4 ⅣB	5 ⅤB	6 ⅥB	7 ⅦB	8 ⅧB(Ⅷ)	9 ⅧB(Ⅷ)	10 ⅧB(Ⅷ)	11 ⅠB	12 ⅡB	13 ⅢA	14 ⅣA	15 ⅤA	16 ⅥA	17 ⅦA	18 ⅧA(0)	电子层
1	1 H 氢 $1s^1$ 1.008 −1 +1																	2 He 氦 $1s^2$ 4.002602(2)	K
2	3 Li 锂 $2s^1$ 6.94 +1	4 Be 铍 $2s^2$ 9.0121831(5) +2											5 B 硼 $2s^2 2p^1$ 10.81 +3	6 C 碳 $2s^2 2p^2$ 12.011 −4 +2 +4	7 N 氮 $2s^2 2p^3$ 14.007 −3 −2 −1 +1 +2 +3 +4 +5	8 O 氧 $2s^2 2p^4$ 15.999 −2 −1	9 F 氟 $2s^2 2p^5$ 18.998403163(6) −1	10 Ne 氖 $2s^2 2p^6$ 20.1797(6)	L K
3	11 Na 钠 $3s^1$ 22.98976928(2) −1 +1	12 Mg 镁 $3s^2$ 24.305 +2											13 Al 铝 $3s^2 3p^1$ 26.9815385(7) +3	14 Si 硅 $3s^2 3p^2$ 28.085 −4 +2 +4	15 P 磷 $3s^2 3p^3$ 30.973761998(5) −3 −2 +1 +3 +5	16 S 硫 $3s^2 3p^4$ 32.06 −2 +2 +4 +6	17 Cl 氯 $3s^2 3p^5$ 35.45 −1 +1 +3 +5 +7	18 Ar 氩 $3s^2 3p^6$ 39.948(1)	M L K
4	19 K 钾 $4s^1$ 39.0983(1) −1 +1	20 Ca 钙 $4s^2$ 40.078(4) +2	21 Sc 钪 $3d^1 4s^2$ 44.955908(5) +3	22 Ti 钛 $3d^2 4s^2$ 47.867(1) −1 0 +2 +3 +4	23 V 钒 $3d^3 4s^2$ 50.9415(1) 0 ±1 +2 +3 +4 +5	24 Cr 铬 $3d^5 4s^1$ 51.9961(6) −3 0 ±1 ±2 +3 +4 +5 +6	25 Mn 锰 $3d^5 4s^2$ 54.938044(3) −2 0 ±1 +2 +3 +4 +5 +6 +7	26 Fe 铁 $3d^6 4s^2$ 55.845(2) −2 0 ±1 +2 +3 +4 +5 +6	27 Co 钴 $3d^7 4s^2$ 58.933194(4) 0 ±1 +2 +3 +4 +5	28 Ni 镍 $3d^8 4s^2$ 58.6934(4) 0 ±1 +2 +3 +4	29 Cu 铜 $3d^{10} 4s^1$ 63.546(3) +1 +2 +3 +4	30 Zn 锌 $3d^{10} 4s^2$ 65.38(2) +1 +2	31 Ga 镓 $4s^2 4p^1$ 69.723(1) +1 +3	32 Ge 锗 $4s^2 4p^2$ 72.630(8) +2 +4	33 As 砷 $4s^2 4p^3$ 74.921595(6) −3 +3 +5	34 Se 硒 $4s^2 4p^4$ 78.971(8) −2 +2 +4 +6	35 Br 溴 $4s^2 4p^5$ 79.904 −1 +1 +3 +5 +7	36 Kr 氪 $4s^2 4p^6$ 83.798(2) +2 +4	N M L K
5	37 Rb 铷 $5s^1$ 85.4678(3) −1 +1	38 Sr 锶 $5s^2$ 87.62(1) +2	39 Y 钇 $4d^1 5s^2$ 88.90584(2) +3	40 Zr 锆 $4d^2 5s^2$ 91.224(2) +1 +2 +3 +4	41 Nb 铌 $4d^4 5s^1$ 92.90637(2) 0 ±1 +2 +3 +4 +5	42 Mo 钼 $4d^5 5s^1$ 95.95(1) 0 ±1 ±2 +3 +4 +5 +6	43 Tc 锝▲ $4d^5 5s^2$ 97.90721(3)✦ 0 ±1 +2 +3 +4 +5 +6 +7	44 Ru 钌 $4d^7 5s^1$ 101.07(2) 0 +1 ±2 +3 +4 +5 +6 +7 +8	45 Rh 铑 $4d^8 5s^1$ 102.90550(2) 0 +1 +2 +3 +4 +5 +6	46 Pd 钯 $4d^{10}$ 106.42(1) 0 +1 +2 +3 +4	47 Ag 银 $4d^{10} 5s^1$ 107.8682(2) +1 +2 +3	48 Cd 镉 $4d^{10} 5s^2$ 112.414(4) +1 +2	49 In 铟 $5s^2 5p^1$ 114.818(1) +1 +3	50 Sn 锡 $5s^2 5p^2$ 118.710(7) +2 +4	51 Sb 锑 $5s^2 5p^3$ 121.760(1) −3 +3 +5	52 Te 碲 $5s^2 5p^4$ 127.60(3) −2 +2 +4 +6	53 I 碘 $5s^2 5p^5$ 126.90447(3) −1 +1 +3 +5 +7	54 Xe 氙 $5s^2 5p^6$ 131.293(6) +2 +4 +6 +8	O N M L K
6	55 Cs 铯 $6s^1$ 132.90545196(6) −1 +1	56 Ba 钡 $6s^2$ 137.327(7) +2	57~71 La~Lu 镧系	72 Hf 铪 $5d^2 6s^2$ 178.49(2) +1 +2 +3 +4	73 Ta 钽 $5d^3 6s^2$ 180.94788(2) 0 ±1 +2 +3 +4 +5	74 W 钨 $5d^4 6s^2$ 183.84(1) 0 ±1 ±2 +3 +4 +5 +6	75 Re 铼 $5d^5 6s^2$ 186.207(1) 0 ±1 +2 +3 +4 +5 +6 +7	76 Os 锇 $5d^6 6s^2$ 190.23(3) 0 +1 ±2 +3 +4 +5 +6 +7 +8	77 Ir 铱 $5d^7 6s^2$ 192.217(3) 0 ±1 +2 +3 +4 +5 +6	78 Pt 铂 $5d^9 6s^1$ 195.084(9) 0 +2 +3 +4 +5 +6	79 Au 金 $5d^{10} 6s^1$ 196.966569(5) +1 +2 +3 +5	80 Hg 汞 $5d^{10} 6s^2$ 200.592(3) +1 +2 +3	81 Tl 铊 $6s^2 6p^1$ 204.38 +1 +3	82 Pb 铅 $6s^2 6p^2$ 207.2(1) +2 +4	83 Bi 铋 $6s^2 6p^3$ 208.98040(1) −3 +3 +5	84 Po 钋 $6s^2 6p^4$ 208.98243(2)✦ −2 +2 +4 +6	85 At 砹 $6s^2 6p^5$ 209.98715(5)✦ ±1 +5 +7	86 Rn 氡 $6s^2 6p^6$ 222.01758(2)✦ +2	P O N M L K
7	87 Fr 钫▲ $7s^1$ 223.01974(2)✦ +1	88 Ra 镭 $7s^2$ 226.02541(2)✦ +2	89~103 Ac~Lr 锕系	104 Rf 𬬻▲ $6d^2 7s^2$ 267.122(4)✦	105 Db 𬭊▲ $6d^3 7s^2$ 270.131(4)✦	106 Sg 𬭳▲ $6d^4 7s^2$ 269.129(3)✦	107 Bh 𬭛▲ $6d^5 7s^2$ 270.133(2)✦	108 Hs 𬭶▲ $6d^6 7s^2$ 270.134(2)✦	109 Mt 鿏▲ $6d^7 7s^2$ 278.156(5)✦	110 Ds 𫟼▲ 281.165(4)✦	111 Rg 𬬭▲ 281.166(6)✦	112 Cn 鿔▲ 285.177(4)✦	113 Nh 鿭▲ 286.182(5)✦	114 Fl 𫓧▲ 289.190(4)✦	115 Mc 镆▲ 289.194(6)✦	116 Lv 𫟷▲ 293.204(4)✦	117 Ts 石田▲ 293.208(6)✦	118 Og 气奥▲ 294.214(5)✦	Q P O N M L K

	57	58	59	60	61	62	63	64	65	66	67	68	69	70	71
★ 镧系	57 La★ 镧 $5d^1 6s^2$ 138.90547(7) +3	58 Ce 铈 $4f^1 5d^1 6s^2$ 140.116(1) +2 +3 +4	59 Pr 镨 $4f^3 6s^2$ 140.90766(2) +3 +4	60 Nd 钕 $4f^4 6s^2$ 144.242(3) +2 +3 +4	61 Pm 钷▲ $4f^5 6s^2$ 144.91276(2)✦ +3	62 Sm 钐 $4f^6 6s^2$ 150.36(2) +2 +3	63 Eu 铕 $4f^7 6s^2$ 151.964(1) +2 +3	64 Gd 钆 $4f^7 5d^1 6s^2$ 157.25(3) +3	65 Tb 铽 $4f^9 6s^2$ 158.92535(2) +3 +4	66 Dy 镝 $4f^{10} 6s^2$ 162.500(1) +3 +4	67 Ho 钬 $4f^{11} 6s^2$ 164.93033(2) +3	68 Er 铒 $4f^{12} 6s^2$ 167.259(3) +3	69 Tm 铥 $4f^{13} 6s^2$ 168.93422(2) +2 +3	70 Yb 镱 $4f^{14} 6s^2$ 173.045(10) +2 +3	71 Lu 镥 $4f^{14} 5d^1 6s^2$ 174.9668(1) +3
★ 锕系	89 Ac★ 锕 $6d^1 7s^2$ 227.02775(2)✦ +3	90 Th 钍 $6d^2 7s^2$ 232.0377(4) +3 +4	91 Pa 镤 $5f^2 6d^1 7s^2$ 231.03588(2) +3 +4 +5	92 U 铀 $5f^3 6d^1 7s^2$ 238.02891(3) +3 +4 +5 +6	93 Np 镎 $5f^4 6d^1 7s^2$ 237.04817(2)✦ +3 +4 +5 +6 +7	94 Pu 钚 $5f^6 7s^2$ 244.06421(4)✦ +3 +4 +5 +6 +7	95 Am 镅▲ $5f^7 7s^2$ 243.06138(2)✦ +2 +3 +4 +5 +6	96 Cm 锔▲ $5f^7 6d^1 7s^2$ 247.07035(3)✦ +3 +4	97 Bk 锫▲ $5f^9 7s^2$ 247.07031(4)✦ +3 +4	98 Cf 锎▲ $5f^{10} 7s^2$ 251.07959(3)✦ +2 +3 +4	99 Es 锿▲ $5f^{11} 7s^2$ 252.0830(3)✦ +2 +3	100 Fm 镄▲ $5f^{12} 7s^2$ 257.09511(5)✦ +2 +3	101 Md 钔▲ $5f^{13} 7s^2$ 258.09843(3)✦ +2 +3	102 No 锘▲ $5f^{14} 7s^2$ 259.1010(7)✦ +2 +3	103 Lr 铹▲ $5f^{14} 6d^1 7s^2$ 262.110(2)✦ +3